Tracers in Geomorphology

British Geomorphological Research Group Symposia Series

Tracers in Geomorphology

Edited by

Ian D. L. Foster

Coventry University, UK
with the Coventry University Editorial Panel:
Sue Charlesworth, Alastair Dawson, Stephan Harrison, David Keen, Joan Lees,
Tim Mighall and Adrian Wood

JOHN WILEY & SONS, LTD
Chichester · NewYork · Weinheim · Brisbane · Singapore · Toronto

Other Wiley Editorial Offices

John Wiley & Sons, Inc., 605 Third Avenue, New York, NY 10158–0012, USA

WILEY-VCH Verlag GmbH, Pappelallee 3,
D-69469 Weinheim, Germany

Jacaranda Wiley Ltd, 33 Park Road, Milton,
Queensland 4064, Australia

John Wiley & Sons (Canada) Ltd. 22 Worcester Road,
Rexdale, Ontario M9W 1L1, Canada

John Wiley & Sons (Asia) Pte Ltd, 2 Clementi Loop 02–01,
Jin Xing Distripark, Singapore 129809

Library of Congress Cataloging-in-Publication Data

Foster, Ian (Ian D. L)
Tracers in gemorphology / edited by Ian D.L. Foster.
p. cm.—(British Geomorphological Research Group symposia series)
Includes bibliographical references.
ISBN 0-471-89602-0
1. Geomorphological tracers.   I. British Geomorphological Research Group. General
Meeting. (1998 : Conventry University)   II. Title.   III. Series

GB400.42.T73 F67 2000
551.41'028—dc21   00-032470

British Library Cataloguing in Publication Data

A catalogue record for this book is available from the British Library

ISBN 0 471 89602–0

Typeset in 10/12pt Times by Kolam Information Services Pvt Ltd, Pondicherry, India
Printed and bound in Great Britain by Bookcraft Ltd. Midsomer Norton, Somerset
This book is printed on acid-free paper responsibly manufactured from sustainable forestry, in which at
least two trees are planted for each one used for paper production.

Contents

List of Contributors

G. N. Bailey Department of Archaeology, University of Newcastle, Newcastle-upon-Tyne NE1 7RU, UK

D. N. Barlow Department of Geology and Geophysics, The University of Edinburgh, West Mains Road, Edinburgh EH9 3JW, UK

J. Barron Earth Resources Centre, University of Exeter, North Park Road, Exeter EX4 4QE, UK

S. Black PRIS, University of Reading, Whiteknights, PO Box 227, Reading RG6 9JT, UK

L. J. Bottrill Department of Geography, University of Exeter, Amory Building, Rennes Drive, Exeter EX4 4RJ, UK

M. J. Bray Department of Geography, University of Portsmouth, Buckingham Building, Lion Terrace, Portsmouth PO1 3HE, UK

E. A. Bryant School of Geosciences, University of Wollongong, Wollongong, NSW 2522, Australia

T. P. Burt Hatfield College, University of Durham, North Bailey, Durham DH1 3RQ, UK

D. P. Butcher Department of Land-Based Studies, Nottingham Trent University, Brackenhurst Campus, Southwell, Nottinghamshire, NG 25 0QF, UK

P. A. Carling Department of Geography, University of Lancaster, Bailrigg, Lancaster, Cumbria LA1 4YW, UK

A. D. Carter ADAS Rosemaund, Preston Wynne, Hereford HR1 3PG, UK

A. Chappell Department of Geography, University of Salford, Manchester M5 4WT, UK

S. M. Charlesworth Centre for Environmental Research and Consultancy, NES (Geography), Coventry University, Priory Street, Coventry CV1 5FB, UK

M. B. Collins Present address: AZTI, Strustegui 8, 20008 Donostia (San Sebastian), Gipuzkoa, Spain

R. H. F. Curr Quaternary Research Unit, School of Geography and Development Studies, Bath Spa University College, Newton Park, Newton St Loe, Bath BA2 9BN, UK

A. Dean Earth Resources Centre, University of Exeter, North Park Road, Exeter EX4 4QE, UK

J. A. Dearing   Environmental Magnetism Laboratory, Department of Geography, University of Liverpool, Liverpool L69 3BX, UK

T. Demir   Department of Geography, University of Durham, Science Laboratories, South Road, Durham DH1 3LE, UK

R. M. Dils   Environment Agency, EHS National Centre, Evenlode House, Howbery Park, Wallingford, Oxfordshire OX10 8BD, UK

S. Forsyth   Centre for Quaternary Science, NES (Geography), Coventry University, Priory Street, Coventry CV1 5FB, UK

I. D. L. Foster   Centre for Environmental Research and Consultancy, NES (Geography), Coventry University, Priory Street, Coventry CV1 5FB, UK

S. W. Franks   Department of Civil, Surveying and Environmental Engineering, University of Newcastle, Callaghan 2308, NSW, Australia

V. N. Golosov   Department of Geography, Moscow State University, Vorob'evy Gory, Moscow 119899, Russia

P. Goodwill   Environment Agency (Southern), Worthing BN11 1LD, UK

J. P. Grattan   Institute of Earth Studies, University of Wales, Aberystwyth SY23 3DB, UK

A. M. Gurnell   School of Geography and Environmental Sciences, University of Birmingham, Edgbaston, Birmingham B15 2TT, UK

R. H. B. Hamlin   School of Geography, University of Leeds, Leeds LS2 9JT, UK

I. A. J. Hardy   Environmental Sciences, Aventis CropScience UK Ltd, Fyfield Road, Ongar, Essex CM5 0HW, UK

S. K. Haslett   Quaternary Research Unit, School of Geography and Development Studies, Bath Spa University College, Newton Park, Newton St Loe, Bath BA2 9BN, UK

Q. He   Department of Geography, University of Exeter, Amory Building, Rennes Drive, Exeter EX4 4RJ, UK

A. L. Heathwaite   Department of Geography, University of Sheffield, Sheffield S10 2TN, UK

C. M. Heppell   Department of Geography, Queen Mary and Westfield College, University of London, London E1 4NS, UK

D. L. Higgitt   Department of Geography, University of Durham, Science Laboratories, South Rd, Durham DH1 3LE, UK

K. A. Hudson-Edwards   Department of Geology, Birkbeck College, University of London, Malet Street, London WC1E 7HX, UK

J. Jones   Earth Resources Centre, University of Exeter, North Park Road, Exeter EX4 4QE, UK

C. Kosmas Laboratory of Soils and Agricultural Chemistry, Agricultural University of Athens, Iera Odos 75, Athens 11855, Greece

J. C. Labadz Centre for Water and Environmental Management, Department of Geographical and Environmental Sciences, University of Huddersfield, Queensgate, Huddersfield HD1 3DH, UK

M. W. E. Lee School of Ocean and Earth Science, University of Southampton, Southampton Oceanography Centre, European Way, Southampton SO14 3ZH, UK

P. B. Leeds-Harrison School of Agriculture, Food and the Environment, Cranfield University, Silsoe, Bedford MK45 4DT, UK

G. J. L. Leeks Institute of Hydrology, Crowmarsh Gifford, Wallingford, Oxfordshire OX10 8BB, UK

J. A. Lees Centre for Environmental Research and Consultancy, NES (Geography), Coventry University, Priory Street, Coventry CV1 5FB, UK

X. X. Lu Present address: Department of Geography, University of Western Ontario, London, Ontario, Canada N6A 5CT

M. G. Macklin Institute of Geography and Earth Sciences, University of Wales, Aberystwyth SY33 3DB, UK

J. R. Merefield Present address: Advance Environmental, Wolfson Laboratories, University of Exeter, Higher Hoopern Lane, Exeter EX4 45G, UK

T. M. Mighall Centre for Quaternary Science, NES (Geography), Coventry University, Priory Street, Coventry CV1 5FB, UK

R. J. Oakey Department of Geography, University of Lancaster, Bailrigg, Lancaster, Cumbria LA1 4YW, UK

M. A. Oliver Department of Soil Science, The University of Reading, Whiteknights, PO Box 233, Reading RG6 2DW, UK

D. J. Oostwoud Wijdenes Laboratory for Experimental Geomorphology, K.U. Leuven, Redingenstraat 16, B-3000, Leuven, Belgium

L. M. Ormerod Department of Geography and Environmental Science, University of Newcastle, Callaghan 2308, NSW, Australia

P. N. Owens Department of Geography, University of Exeter, Amory Building, Rennes Drive, Exeter EX4 4RJ, UK

A. V. Panin Department of Geography, Moscow State University, Vorob'evy Gory, Moscow 119899, Russia

J. Poesen Laboratory for Experimental Geomorphology, K.U. Leuven, Redingenstraat 16, B-3000, Leuven, Belgium

D. Pope Department of Civil Engineering, University of Brighton, Cockroft Building, Brighton, Moulsecoomb, Brighton BN2 4GJ, UK

L. J. Pu　Department of Geography, Nanjing University, Hankou Rd, Nanjing 210093, People's Republic of China

J. Roberts　Earth Resources Centre, University of Exeter, North Park Road, Exeter EX4 4QE, UK

J. S. Rowan　Department of Geography, University of Dundee, Dundee DD1 4HN, UK

R. M. Sanders　Centre for Environmental Research and Consultancy, NES (Geography), Coventry University, Priory Street, Coventry CV1 5FB, UK

A. Sawyer　Liverpool John Moores University, I.M. Marsh Campus, Barkhill Road, Liverpool L17 6BD, UK

C. Schell　Present address: Shell International Exploration and Production BV, Technical Applications and Research, Volmerlaan 8, 2288 Rijswijk, The Netherlands

D. A. Sear　Department of Geography, University of Southampton, Highfields, Southampton, Hampshire SO17 1BJ, UK

M. C. Slattery　Department of Geology, Texas Christian University, Fort Worth, Texas 76129, USA

I. M. Stone　Present address: Advance Environmental, Wolfson Laboratories, University of Exeter, Higher Hoopern Lane, Exeter EX4 45G, UK

T. Stott　Liverpool John Moores University, I.M. Marsh Campus, Barkhill Road, Liverpool L17 6BD, UK

R. Thompson　Department of Geology and Geophysics, The University of Edinburgh, West Mains Road, Edinburgh EH9 3JW, UK

S. Timberlake　Early Mines Research Group, 98 Victoria Road, Cambridge CB4 3DU, UK

L. Vandekerckhove　Laboratory for Experimental Geomorphology, K.U. Leuven, Redingenstraat 16, B-3000, Leuven, Belgium

J. Walden　School of Geography and Geology, University of St Andrews, Purdie Building, North Haugh, St Andrews KY16 9ST, UK

D. E. Walling　Department of Geography, University of Exeter, Amory Building, Rennes Drive, Exeter EX4 4RJ, UK

J. Warburton　Department of Geography, University of Durham, Science Laboratories, South Road, Durham DH1 3LE, UK

R. J. Williams　Institute of Hydrology, Crowmarsh Gifford, Wallingford, Oxfordshire OX10 8BB, UK

D. A. Wilson　Environment Agency, Kings Meadow House, Kings Meadow Road, Reading RG1 8DQ, UK

J. C. Woodward   School of Geography, University of Leeds, Leeds LS2 9JT, UK

M. Workman   Department of Oceanography, University of Southampton, Southampton Oceanography Centre, European Way, Southampton SO14 3ZH, UK

Preface

The British Geomorphological Research Group (BGRG) Annual Conference, Tracers in Geomorphology, was held at Coventry University over the weekend of the 18–20 September 1998 and attracted more than 90 delegates from countries as far afield as Australia, Brazil and the USA. Thirty-seven oral presentations and 14 posters were given at the meeting and this volume contains a representative cross-section of the papers presented and subsequently submitted for publication.

The theme chosen for the conference reflects a growing interest amongst the geomorphological community in studies of sediment provenance and in the development of methods for tracing coarse particulate sediments in a range of environments. Surprisingly, whilst several hydrological meetings over the last decade have chosen a 'tracers' theme, no meeting of geomorphologists has attempted to pool the knowledge and expertise of scientists from a range of geomorphological sub-disciplines working with tracers. The meeting proved personally stimulating from many perspectives, not least of which was to learn about the potential of new, and the limitations of existing, methodologies, and to debate the issues and problems in more detail over a pint or two of fine English beer in the spectacular illuminated surroundings of the medieval and modern cathedrals of the City of Coventry.

In organising the conference there are many groups and individuals who deserve special mention and thanks for their help and support. In particular I would like to thank the BGRG Executive Committee for moral and financial support and Coventry University for hosting the meeting. Martin Townsend of the University catering office gave much needed support and guidance and organised accommodation and social events for delegates over the weekend. Gillian West provided many hours of much valued secretarial support, and all technicians, secretaries and research students in the Geography Subject Area of the School of Natural and Environmental Sciences are thanked for contributing so much to the smooth running of the meeting.

All of the chapters published in this volume have been read by at least two referees. A special thanks to all involved in the refereeing process; a time-consuming and largely thankless task, which remains an essential part of the publication process. I am also indebted to members of the editorial panel who helped with the production of this volume and with the organisation of the meeting, and to Sally Wilkinson of John Wiley & Sons Ltd for supporting the proposal to publish the conference proceedings.

The volume has been divided into five sections: background and recent developments, atmospheric and hydrological tracers, tracers for investigating soil erosion and hillslope processes, tracing fluvial sediments, tracers for coastal transport studies and tracers in palaeoenvironmental investigations. The chapters deal with a number of technical problems and provide examples from a wide range of environments and climatic zones. Tracer technologies are still in their infancy and hopefully this volume will stimulate new research to further develop techniques and modelling strategies in order to improve our understanding of geomorphological processes.

Happy New Millennium.

Ian Foster
Coventry
29 December 1999

Section 1

BACKGROUND AND RECENT DEVELOPMENTS

1 Tracers in Geomorphology: Theory and Applications in Tracing Fine Particulate Sediments

I. D. L. FOSTER and J. A. LEES

Centre for Environmental Research and Consultancy, NES (Geography), Coventry University, UK

INTRODUCTION

Efforts to quantify and model the movement of sediment and pathways of water in a wide range of earth systems have increased substantially over the last 40 years. At a basic level, studies using tracer methods can be classified into four major types:

- Studies of whole-system behaviour (e.g. measurement of soil creep and mass movement processes on hillslopes and rates of movement and deformation of glaciers) (cf. Meier 1960; Johnson, 1970; Carson and Kirkby, 1972; Anderson and Finlayson, 1975).
- Studies of coarse sediment transport on hillslopes and in fluvial systems and littoral zones (e.g. displacement of individual fragments, determination of entrainment thresholds, analysis of particle size and shape controls on transport distance, sediment transport estimation and intra-event activity) (see Sear *et al.*, Chapter 2).
- Studies of fine particulate sediment transport and sediment provenance (e.g. fingerprinting airborne particles, estimating hillslope erosion rates, mapping soil redistribution and estimating sediment delivery ratios from field systems, quantifying sediment-associated contaminant movement, and determining the provenance of a range of fluvial, mass movement, limnic and aeolian deposits) (see below).
- Tracing water movement through soil profiles, groundwater systems and caves and in river channels (cf. Church, 1975; Peters *et al.*, 1993; Leibundgut, 1995; Kranjc, 1997; Wilson *et al.*, see Chapter 7).

The collection of papers brought together in this volume demonstrates the breadth of interest in tracing technologies in a wide range of geomorphological settings. These papers also show that fundamental assumptions in tracer methodologies have a

Tracers in Geomorphology. Edited by Ian D. L. Foster. © 2000 John Wiley & Sons, Ltd.

certain degree of commonality. Sear *et al.* (Chapter 2) provide a comprehensive review of coarse sediment tracing technologies and this chapter will focus upon some of the fundamental issues involved in tracing the movement and origins of fine particulate sediments.

Interest in fine sediment stems from one of two requirements. First, a wide range of fundamental research hypotheses can be tested using fingerprinting techniques, including those that address issues relating to the way in which soil and sediment are redistributed and transported from catchments over a wide range of temporal and spatial scales. Secondly, there is often a practical requirement to identify the principal sources of fine particulate sediment, and sediment-associated nutrients and contaminants, for the purposes of improving erosion management and pollution control strategies.

Although not widespread in the geomorphological literature, some research projects have used artificially emplaced fine particulate tracers such as fluorescent microspheres, bacteriophages and thermally enhanced magnetic minerals (cf. Ingle, 1966; Ingle and Gorsline, 1973; Tsoar, 1978; Harvey *et al.*, 1989; Marsh *et al.*, 1991, 1993; Van der Post and Oldfield, 1994; Bricelj and Misic, 1997) as alternatives to more conventional direct survey and monitoring methods (e.g. visual appraisal, ground survey, photogrammetry, remote sensing, erosion pins, erosion plots, direct monitoring) or the use of fingerprinting methods. However, this chapter will focus on two issues relating to the use of fingerprinting methods for determining the provenance of fine particulate sediments in geomorphology: first, the properties that might be used to identify provenance and, secondly, the fundamental assumptions that underpin their application.

CHARACTERISING FINE SEDIMENTS: THE NEW FORENSIC GEOMORPHOLOGY?

There are two fundamental requirements in using the fingerprinting technique for tracing fine particulate sediments. The first is the selection of a property, or suite of properties, which uniquely characterises sediment derived from a particular source. The second involves comparison of these properties with actively transported or deposited sediments (cf. Walling *et al.*, 1993; Collins *et al.*, 1998). A range of tracer properties have been identified as offering potential for discriminating fine particulate sediment sources including colour, mineralogy, geochemistry, mineral magnetic signatures, stable isotopes, radionuclides, particle size and biogenic properties (e.g. C concentrations, C:N ratios, plant and animal microfossils and macrofossils) (Table 1.1). Essentially these signatures reflect one of three basic attributes of the sediment, namely their nuclear properties, molecular identity or orbital electron properties. These three basic properties may be further divided into a number of property subgroups as identified in Figure 1.1. Many of these properties, particularly radionuclides, have applications for dating sedimentary deposits and surfaces as well as for providing opportunities to determine sediment provenance.

The applications of such tracer techniques are widespread in the geomorphological literature which includes case studies demonstrating an ability to discriminate

Table 1.1 Examples of properties used in fine sediment provenance studies

Sediment property	Examples
Colour	Moore (1961), Grimshaw and Lewin (1980), Coard *et al.* (1983), Peart (1993), Udelhoven and Symader (1995).
Mineralogy/heavy minerals	Hathaway (1972), Wall and Wilding (1976), Setty and Raju (1988), Burrus *et al.* (1990), Eisma *et al.* (1993), Merefield (1995).
Geochemistry	Nichol *et al.* (1967), Symader and Thomas (1978), Peart and Walling (1986), Foster and Walling (1994), Foster and Lees (1999a).
Mineral magnetism	Walling *et al.* (1979), Dearing *et al.* (1981), Thompson and Oldfield (1986), Beckwith *et al.* (1986), Foster *et al.* (1988), Yu and Oldfield (1989), Lees and Pethick (1995), Walden *et al.* (1997), de Jong *et al.* (1998), Foster *et al.* (1998), Dearing (see Chapter 3).
Radionuclides	Santschi (1989), Walling (1990), Stanton *et al.* (1992), Murray *et al.* (1993), Walling and He (1994), Walling *et al.* (see Chapter 9).
Stable isotopes	Degens *et al.* (1968).
Biogenic properties	Crowder and Cuddy (1973), Ongley (1982), Peart and Walling (1986), Burrus *et al.* (1990), Oldfield and Clark (1990), Santiago *et al.* (1992), Peart (1993), Foster *et al.* (2000), Haslett *et al.* (see Chapter 23).
Particle size	Peart and Walling (1986), Walling and Moorhead (1989), Stone and Sanderson (1992), Higgitt and Walling (1993).

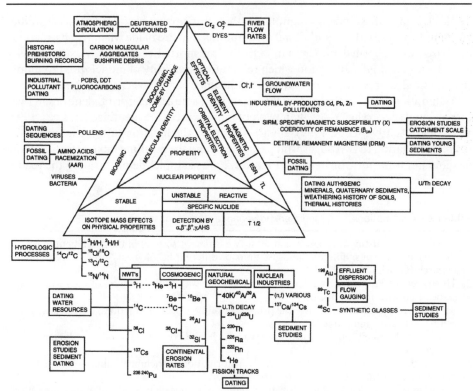

Figure 1.1 Fundamental properties of earth materials which might be used for dating and tracing sediment sources. (Based on an original unpublished diagram by courtesy of Brian Campbell and Bob Loughran)

catchment topsoils and subsoils, fingerprint arable, grassland and woodland sediment sources, quantify patterns of erosion and sedimentation at the plot and field scale, assess the relative importance of individual erosion processes such as sheetwash and gullying, and identify the provenance of deposited material over a range of timescales (loess, floodplain, wetland, lake and marine deposits).

Whilst much recent geomorphological literature has been devoted to the finger-printing approach, insufficient information currently exists to enable geomorphologists to predict which will be the most suitable tracer for different environments. What appears evident, however, is that single tracer properties rarely provide good discrimination (cf. Lees and Pethick, 1995) and that multiple fingerprint signatures probably provide the best opportunity to discriminate sources (Walling *et al.*, 1993; Collins *et al.*, 1998). Despite an increase in the number of tracer studies published, it is essential that assumptions which underpin their application are fully identified and tested. The following section attempts to highlight some of the key issues in the new forensic geomorphology.

SOME ASSUMPTIONS

Table 1.2 identifies those assumptions that are applicable to all tracer studies (Assumptions 1–3), those additional assumptions that are made when reconstructions are based on deposited rather than actively transported sediments (Assumptions 4 and 5), and the final assumption which relates to the validity of models used to unmix the potential sources from the transported sediment or deposit.

Assumption 1 may be limited by the location of source fingerprint identities in two-dimensional space when more than two potential sources are identified (cf. Lees, 1994; Walden *et al.*, 1997; Lees, 1999). Figure 1.2 shows the location of three hypothetical sources in relation to two fingerprint parameters and the resulting mixture(s) following erosion and transport. Included in Figure 1.2 are examples demonstrating the reality

Table 1.2   Assumptions in fine sediment provenance studies (based on Foster, 1998)

1. The tracer can distinguish between at least two different sources in the catchment (e.g. topsoils and subsoils, different land-use units or different lithologies) of relevance to the research problem
2. The tracer is transported and deposited in the same way as the medium of interest (i.e. in association with fine sediment)
3. Selective erosion does not change the properties of the tracer or, if it does, only in a way that can be measured and modelled
4. Differences between source properties have not changed over the period of sediment deposition
5. Once deposited, the tracer undergoes no transformation (enrichment, dilution or depletion) in its new environment
6. The un-mixing models used to reconstruct sediment source changes through time are able to deal with inherent variability in source properties and provide estimates of source contributions within known or predictable tolerances

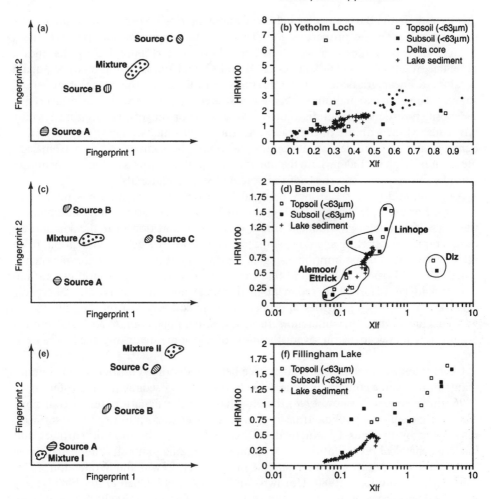

Figure 1.2 Hypothetical mixtures of fine sediment derived from three potential sources and examples from Yetholm Loch and Barnes Loch (Tweed Headwaters) and from Fillingham Lake (Lincolnshire). (See Foster and Lees (1999a,b) for site details and the text for a full explanation)

of the hypothetical problems based on a mineral magnetic analysis of lake sediments and potential source materials (screened to $< 63\mu m$) in three UK lake-catchments. Full descriptions of the sites from which these examples are drawn can be found in Foster and Lees (1999a,b). In Figure 1.2(a), all sources lie along a linear gradient which means that the resulting mixture can be predicted from a range of combinations of sources A, B and C. The model does suggest, however, that the sources have been correctly identified. Figure 1.2(b) demonstrates this linearity problem in an attempt to identify the sources of bottom-sediments in Yetholm Loch. Figure 1.2(c) shows a situation where three sources can be discriminated quantitatively since they are uniquely defined on the basis of the two tracer properties. Figure 1.2(d) exemplifies this scenario in the Barnes Loch catchment where three soil types are discriminated by two mineral magnetic properties and the Loch bottom-sediments appear to be a mixture of two of the three soil types. Scenario three (Figure 1.2(e)) identifies a potential situation in which sources A, B and C, and the resulting mixtures, again lie along a linear gradient, but the mixtures fall outside of the range of measured sources. This scenario may suggest a missing source or, if the mixture was measured in a deposit, post-depositional dissolution (Mixture I) or diagenesis (Mixture II) may have affected the stability of the tracer signature over time (see below). Figure 1.2(f) demonstrates this problem in attempting to model the sources of the bottom-sediments of Fillingham Lake where the mineral magnetic signatures of the sediments are considerably weaker than those of the potential source materials.

Assumption 2 (Table 1.2) may be invalidated for some tracers for a number of reasons. For example, Higgitt (1995) suggested that a key assumption underpinning the ^{137}Cs technique as a sediment tracer was that adsorption near the soil surface occurs rapidly, restricting any downslope movement of ^{137}Cs before adsorption. In an attempt to test this assumption using laboratory rainfall simulation experiments, Dalgleish and Foster (1996) found that during periods of surface runoff generation, ^{137}Cs could be adsorbed to sediment in transport and that significant differences existed between ^{137}Cs activities in eroded soil and soil retained on the hillslope. It was concluded that the differences in ^{137}Cs activity between source and eroded sediments could lead to an overestimation of erosion rates using the ^{137}Cs technique.

Chemical exchanges may also take place between dissolved nutrients and contaminants in the water column and the actively transported sediment (cf. Foster *et al.*, 1996; Environment Agency, 1998) thereby changing the signature during transport. The mechanisms controlling transfers include direct adsorption by clay surfaces, scavenging by hydrous Fe/Mn oxides, direct adsorption by organic particles and bacteria, chemical precipitation, association with organic material (humic and fulvic acids), and incorporation into crystalline matrices (cf. Lewin and Wolfendon, 1978; Förstner and Salomans, 1980; Horowitz and Elrick, 1987; Chen *et al.*, 1989; Horowitz, 1991; Tanizaki *et al.*, 1992). Whilst such transfers are unlikely to affect many fingerprint properties, there remains a paucity of information regarding the significance of such processes in altering tracer signatures during transport (cf. Bubb and Lester, 1991; Zhang and Huang, 1993).

Assumption 3 (Table 1.2) may be invalidated as a result of selective erosion and transport caused by differences, for example, in particle size or in density. Particle size selective transport is well documented in the geomorphological literature and it is well known that mineralogical, geochemical, radionuclide and mineral magnetic signatures are in part controlled by the particle size distribution of eroded and transported sediments (cf. Wall and Wilding, 1976; Komar *et al.*, 1989; Ritchie and McHenry, 1990; Walling and He, 1993; Oldfield and Yu, 1994; Oldfield *et al.*, 1995; Foster *et al.*, 1998). However, such assumptions may be empirically tested in contemporary erosion studies and signatures adjusted to account for the selectivity of sediment transport (cf. Walling and Kane, 1984; Owens *et al.*, see Chapter 15).

Assumptions 4 and 5 remain essentially untestable in most circumstances although in-situ pedological diagenesis, for example, has been identified as a potential cause of change in the signatures of both source materials and deposits (Stallard, 1988; Johnson and Stallard, 1989; Higgitt *et al.*, 1991; Woodward *et al.*, 1992) and it is well known that soil-forming processes can lead to both the creation and destruction of magnetic minerals over Holocene timescales (cf. Dearing *et al.*, 1985; Maher and Taylor, 1988; Maher and Thompson, 1992, 1994; Verusob *et al.*, 1993; White and Walden, 1994; Dearing, see Chapter 3). Atmospherically derived signatures (e.g. heavy metals and ^{137}Cs) may have a complex temporal fallout history and are unlikely to be suitable for sediment provenance studies other than those restricted to very recent timescales. By contrast, point-source contamination within catchments, for example from prehistoric or historical mining episodes, may lead to the input of sediment with diagnostic heavy metal signatures which might be preserved in a range of environments including valley sediments (cf. Lewin *et al.*, 1977; Rang and Schouten, 1989; Bradley, 1995).

A considerable body of literature has also questioned a purely catchment-derived detrital interpretation of some sedimentary records in lake basins. In hyper-eutrophic lakes, for example, magnetite dissolution may take place within strongly reduced organic sediments (Anderson and Rippey, 1988; Foster *et al.*, 1998), whilst the formation of authigenic ferrimagnetic iron sulphides, such as greigite, appears to occur in freshwater muds in contact with sulphur-rich marine sediments (Snowball and Thompson, 1988). There may also be situations in which the magnetic properties of lake sediment cores may include contributions from bacterial magnetite (Kirschvink *et al.*, 1985; Lovely *et al.*, 1987). In addition to potential problems caused by chemical mobilisation and additions from bacterial magnetites, physical disturbance of sediments may be caused, for example, by bioturbation, slumping of marginal sediments or sedimentation from turbidity currents (turbidites). Further detailed discussion of these issues and their relevance to the interpretation of preserved signatures in lake sediments is given by Håkanson and Jansson (1983), Duck (1987), Hilton (1987), Vali *et al.* (1987), Chang *et al.* (1989), Oldfield (1991, 1994), Duck *et al.* (1993), Van der Post *et al.* (1997), Foster *et al.* (1998), Lees *et al.* (1998a,b) and Dearing (see Chapter 3).

Changes in equilibrium concentrations and changes in other important environmental conditions, including temperature, pH, the oxidising–reducing potential (Eh), the aqueous and sediment-associated concentration of the element and the availability of complexing agents will all affect the post-depositional stability of geochemical and other tracers (cf. Salomons and Förstner, 1984; Bourg, 1988; Wolff *et al.*, 1988; Warren and Zimmermann, 1994a,b; Foster *et al.*, 1998). As yet, the significance of these processes in disrupting the stability of the tracer signal have yet to be fully evaluated and tested.

One of the fundamental requirements of forensic geomorphology is to statistically unmix the potential sources from the resultant mixtures. Attempts to model sediment source linkages on the basis of their fingerprint characteristics have simplified the problem by analysing the bulk source materials (cf. Shankar *et al.*, 1994), by functionally separating source materials into particle size bands which approximately replicate those of the transported or deposited materials (cf. Yu and Oldfield, 1989; Foster *et al.*, see Chapter 17), by statistically correcting for particle size effects (cf. Walling and He, 1993; Owens *et al.*, see Chapter 15) or by analysing fractions which are thought to be representative of different hydraulic transport components (cf. Caitcheon, 1993, 1998).

The first stage in the development of reliable unmixing models is to identify a fingerprint, or a number of fingerprints, which provide good discrimination of source materials. Statistical methods, including difference of means tests (parametric or non-parametric), cluster analysis, principal component (or factor) and discriminant analysis have all been used successfully to optimise the number of potential sources (cf. Birks, 1987a,b for a discussion of multivariate statistical methods and Idbeken and Schleyer, 1991; Lees, 1999; Owens et al., see Chapter 15). Such procedures allow the elimination of redundant variables which provide no additional discriminatory power and allow a reduction in the number of variables used in order to generate a parsimonious solution.

Selected fingerprints representing individual sources should have low variability about the mean to minimise errors in model outcomes, and careful consideration should be given to the design of an appropriate spatial sampling strategy in order to minimise errors in the description of source properties. In many studies, a purposive sampling strategy is used to identify likely sediment sources where sources are sampled on the basis of an *a priori* classification (cf. Walling et al., 1979; Foster et al., 1990; Yu and Oldfield, 1993; Foster and Walling, 1994). More recently, attempts have been made to selectively sample effective contributing areas based, for example, on a measured distance from a water course or the existence of one or more intervening field boundaries between a source and a river or stream (cf. Lees, 1994, 1999; Jenns, 1997).

A key issue in designing an effective sampling strategy is to define the variability in source properties (cf. Beckett and Webster, 1971; Oldfield et al., 1989; Sutherland, 1991, 1994; Webster and Oliver, 1992; Foster et al., 1998). Whilst conventional sampling methods have attempted to define the variability in source properties, McBratney and Webster (1981, 1983) suggest that there is a need to account for the spatial dependence of material properties using geostatistical techniques and variograms to optimise sample numbers and/or identify the location of soil boundaries (cf. Oliver, 1987; Williams and Cooper, 1990; Oliver and Webster, 1991; Lees, 1994, 1999; Chappell and Oliver, see Chapter 11).

Having identified the most appropriate signatures and obtained representative samples of source materials, a variety of approaches have been used to unmix the sources from the signatures of the mixtures. These include simple models based, for example, on single parameters, such as frequency-dependent magnetic susceptibility (Dearing and Foster, 1986), bivariate regression models to unmix river tributary sources (cf. Caitcheon, 1993, 1998), multiple regression (cf. Yu and Oldfield, 1989, 1993) or linear programming techniques for the solution of simultaneous equations (Thompson, 1986; Lees, 1994; Walden et al., 1997; He and Owens, 1995; Lees, 1999; Walling et al., 1999; Owens et al., see Chapter 15). Whilst it is normally assumed that mixtures of sources are linearly additive, recent research by Lees (1997) has raised doubts about the linear additivity of some mineral magnetic properties (Figure 1.3).

Despite significant developments in modelling strategies, there remain important questions concerning the statistical uncertainty in predictions derived from linear

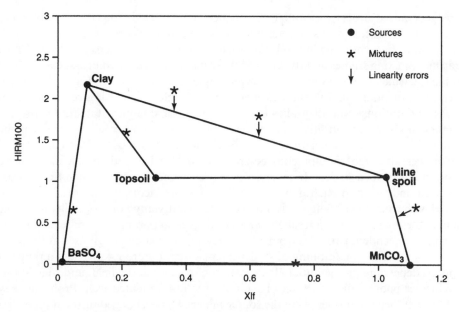

Figure 1.3 Mineral magnetic signatures of laboratory mixed 'sources' demonstrating the deviation of the measured mixture from the linearly additive predictions (cf. Lees, 1997)

programming since a single optimised solution may be only one subset of statistically equivalent solutions, yet each may give different model results in terms of estimated source contributions (cf. Beven, 1996; Rowan *et al.*, see Chapter 14). There is clearly a need to find statistically robust methods that minimise the uncertainty in model outcomes.

DISCUSSION AND CONCLUSIONS

Tracers offer tremendous potential. The forensic geomorphologists' toolbox is one that has been added to a suite of field and laboratory techniques developed in a wide range of disciplines and used by geomorphologists over the last 40 years. The above review, and chapters contained in this volume, provide ample evidence for the widespread use of tracers in geomorphology and the complementary role that these methods can play in association with more traditional techniques in understanding and modelling geomorphological processes at a range of temporal and spatial scales. However, the future success of fingerprinting techniques demands that more attention be paid to the testing of many of the assumptions listed in Table 1.2. Few studies have as yet been established to provide guidelines on the suitability of specific tracers for geomorphological applications in general or to specifically test assumptions in a controlled and rigorous way. Unfortunately, several assumptions may remain untestable, particularly where palaeoenvironmental interpretations are sought from sedimentary deposits whose original source properties are unknown and will remain unknowable.

A generalized 'tracer methodology' has been constructed from the examples and discussion presented in this chapter and from the chapters concerned with fine sediment source fingerprinting in this book. Figure 1.4 outlines the steps involved in most tracing/unmixing studies and the methodology that should be used in association with Table 1.2, which lists assumptions underpinning tracer applications; Figure 1.1, which identifies potentially useful tracer properties; and Table 1.1, which gives examples of tracer applications. Not all studies use, or need to use, complex statistical classification methods or quantitative modelling routines. In some cases, sources can be matched visually or numerically using bivariate relationships and/or curve-fitting techniques although simultaneous equations can also be used to calculate source proportions in such studies. In most cases, however, there are more than two sources involved and this approach does not easily allow the incorporation of errors associated with source variability. It is important to identify any non-normality in source data and eliminate any interrelated variables (e.g. ratio parameters are related to the original independent variables from which they are calculated and, if used at the same time, can cause ill-conditioning of statistical analysis). Figure 1.4 provides a range of options depending on the initial identification of sources in the field, and suggestions for further pilot studies if sources have not been discriminated well. Problems may also occur when the properties of the source material have changed through time for a range of reasons described earlier in this chapter.

Natural variability in source populations will often lead to uncertainty in the proportion of a source contribution calculated from the unmixing model. Uncertainties due to variability in source properties can be estimated by using appropriate measures of variation about the mean value (e.g. mean proportion estimates + estimates at two standard deviations about the mean). These errors, which are often quite large, demonstrate the difficulty in providing absolute estimates of source proportions in mixtures. In some cases, extreme tracer values may be recorded in actively transported or deposited sediments. These may derive from the input of point-source or diffuse pollutants, from occasional inputs derived from clearance or construction work and/or from mobilisation of new sediment sources as a result of extreme weather conditions. Such inputs add further complicating factors in interpreting the modelled outcomes. In sedimentary deposits, dating and other environmental reconstruction methods (e.g. pollen analysis) may offer a means of establishing the cause of these extreme tracer signatures. However, the success or failure of a study having followed a methodology such as that given in Figure 1.4 serves only to extend our knowledge of the suitability of specific tracer techniques, of the assumptions that are made in their use, and of the way in which sources could be identified, discriminated, modelled and interpreted in different environmental settings.

ACKNOWLEDGEMENTS

We wish to express our sincere thanks to Brian Campbell and Bob Loughran for allowing us to use an unpublished diagram as the basis for Figure 1.1. We also wish to thank Erica Milwain of Coventry University who produced the high quality artwork for this chapter from less than perfect drafts.

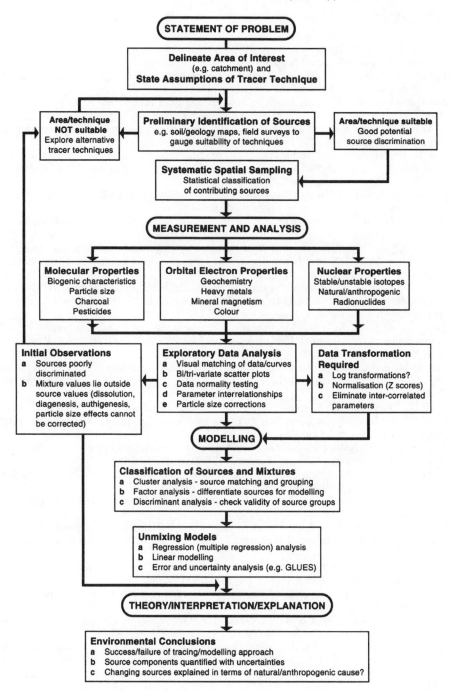

Figure 1.4 A proposed fine-sediment tracer methodology (modified from Lees, 1999). (Refer to Table 1.2 for assumptions, Figure 1.1 for tracer properties and Table 1.1 for examples of applications)

REFERENCES

Anderson, E. W. and Finlayson, B. L. 1975. Instruments for measuring soil creep. *BGRG Technical Bulletin*, **16**, Geo Abstracts, Norwich.

Anderson, N. J. and Rippey, B. 1988. Diagenesis of magnetic minerals in recent sediments of a eutrophic lake. *Limnology and Oceanography*, **33**, 1476–1492.

Beckett, P. H. T. and Webster, R. 1971. Soil variability: a review. *Soils and Fertilizers*, **34**, 1–13.

Beckwith, P. R., Ellis, J. B. and Revitt, D. M. 1986. Heavy metal and magnetic relationships for urban source sediments. *Physics of Earth and Planetary Interiors*, **42**, 67–75.

Beven, K. 1996. Equifinality and uncertainty in geomorphological modelling. In Rhoads, B. L. and Thorn, C. E. (eds) *The Scientific Nature of Geomorphology*. John Wiley, Chichester, 289–313.

Birks, H. J. B. 1987a. Multivariate analysis of stratigraphic data in geology: a review. *Chemometrics and Intelligent Laboratory System*, **2**, 109–126.

Birks, H. J. B. 1987b. Multivariate analysis in geology and geochemistry: an introduction. *Chemometrics and Intelligent Laboratory System*, **2**, 15–28.

Bourg, A. C. M. 1988. Adsorption of trace elements by suspended particle matter in aquatic systems. In West, T. S. and Nurnburg, H. W. (eds) *The Determination of Trace Metals in Natural Waters*. Blackwell, Oxford, 257–284.

Bradley, S. B. 1995. Long-term dispersal of metals in mineralised catchments by fluvial processes. In Foster, I. D. L., Gurnell, A. M. and Webb, B. W. (eds) *Sediment and Water Quality in River Catchments*. Wiley, Chichester, 161–177.

Bricelj, M. and Misic, M. 1997. Movement of bacteriophage and fluorescent tracers through underground river sediments. In Kranjc, A. (ed.) *Tracer Hydrology*. Balkema, Rotterdam, 3–9.

Bubb, J. M. and Lester, J. N. 1991. The impact of heavy metals on lowland rivers and the implications for man and the environment. *Science of the Total Environment*, **100**, 207–233.

Burrus, D., Thomas, R. L., Dominik, B., Vernet, J-P. and Dominik, J. 1990. Characteristics of suspended sediment in the Upper Rhone River, Switzerland, including the particulate forms of phosphorus. *Hydrological Processes*, **4**, 85–98.

Caitcheon, G. 1993. Sediment source tracing using environmental magnetism: a new approach with examples from Australia. *Hydrological Processes*, **7**, 349–358.

Caitcheon, G. 1998. The significance of various sediment magnetic mineral fractions for tracing sediment sources in Killimicat Creek. *Catena*, **32**, 131–142.

Carson, M. A. and Kirkby, M. J. 1972. *Hillslope Form and Process*. Cambridge University Press, London.

Chang, S., Chang, B. R. and Kirschvink, J. L. 1989. Magnetofossils, the magnetization of sediments and the evolution of biomineralization. *Annual Review of Earth and Planetary Science*, **17**, 169–195.

Chen, J., Dong, L. and Deng, B. 1989. A study on heavy metal partitioning in sediments from Poyand lake in China. *Hydrobiologia*, **176/177**, 159–170.

Church, M. 1975. Electrochemical and fluorometric tracer techniques for streamflow measurements. *BGRG Technical Bulletin*, **12**, Geo Abstracts, Norwich.

Coard, M. A., Cousin, S. M., Cuttler, A., Dean, H. J., Dearing, J. A., Eglington, T. I., Greaves, A. M., Lacey, K. P. B., O'Sullivan, P. E., Pickering, D. A., Rhead, M. M., Rodwell, J. K. and Simola, H. 1983. Palaeolimnological studies of annually laminated sediments in Loe Pool, Cornwall, UK. In Merilainen, J., Huttunen, P. and Battarbee, R. W. (eds) *Palaeolimnology (Developments in Hydrobiology, No 15)*. Dr W. Junk, The Hague, 185–191.

Collins, A. L., Walling, D. E. and Leeks, G. J. L. 1998. Use of composite fingerprints to determine the provenance of the contemporary suspended sediment load transported by rivers. *Earth Surface Processes and Landforms*, **23**, 31–52.

Crowder, A. A. and Cuddy, D. G. 1973. Pollen in a small river basin: Wilton Creek, Ontario. In Birks, H. J. B. and West, R. G. (eds) *Quaternary Plant Ecology*. Blackwell Scientific, Oxford, 61–77.

Dalgleish, H. Y. and Foster, I. D. L. 1996. [137]Cs losses from a loamy surface water gleyed soil (Inceptisol): a laboratory simulation experiment. *Catena*, **26**, 227–245.

Dearing, J. A. and Foster, I. D. L. 1986. Limnic sediments used to reconstruct sediment yields and sources in the English Midlands since 1765. In Gardiner, V. (ed.) *International Geomorphology*, John Wiley, Chichester, 853–868.

Dearing, J. A., Elner, J. K. and Happey-Wood, C. M. 1981. Recent sediment flux and erosional processes in a Welsh upland lake catchment based on magnetic susceptibility measurements. *Quaternary Research*, **16**, 356–372.

Dearing, J. A., Maher, B. A. and Oldfield, F. 1985. Geomorphological linkages between soils and sediments: the role of magnetic measurements. In Richards, K., Arnett, R. R. and Ellis, S. (eds) *Geomorphology and Soils*, Allen and Unwin, London, 245–266.

Degens, E. T., Gaillard, R. R. L., Sackett, W. M. and Hellebust, J. A. 1968. Metabolic fractionation of carbon isotopes in marine plankton. I Temperature and respiration experiments. *Deep Sea Research*, **15**, 1–10.

De Jong, E., Nestor, P. A. and Pennock, D. J. 1998. The use of magnetic susceptibility to measure long term soil redistribution. *Catena*, **32**, 23–36.

Duck, R. W. 1987. Aspects of physical processes of sedimentation in Loch Earn, Scotland. In Gardiner, V. (ed.) *International Geomorphology 1986, Part I*. John Wiley, Chichester, 801–821.

Duck, R. W., McManus, J. and Lord, J. 1993. Seismicity and bed sediment morphology in a Perthshire reservoir. In McManus, J. and Duck, R. W. (eds) *Geomorphology and Sedimentology of Lakes and Reservoirs*. John Wiley, Chichester, 225–239.

Eisma, D., Van der Gaast, S. J., Martin, J. M. and Thomas, A. J. 1993. Suspended matter and bottom sediments of the Orinoco delta: turbidity, mineralogy and elementary composition. *Netherlands Journal of Sea Research*, **12**, 224–251.

Environment Agency 1998. *Eutrophication in controlled waters in the Warwickshire Avon Catchment (Final Report)*, Volumes 1 and 2. Environment Agency/Coventry University, Environment Agency, Tewkesbury.

Förstner, U. and Salomans, W. 1980. Trace metal analysis on polluted sediments Part I: assessment of sources and intensities. *Environmental Technology Letters*, **1**, 494–505.

Foster, I. D. L. 1998. *Rivers of mud – geomorphological perspectives on the sediment problem*. Coventry University Professorial Lectures No. 19, 24pp.

Foster, I. D. L. and Lees, J. A. 1999a. Physical and geochemical properties of suspended sediments delivered to the headwaters of LOIS River Basins over the last 100 years: an analysis of lake and reservoir bottom-sediments. *Hydrological Processes*, **13**, 1067–1086.

Foster, I. D. L. and Lees, J. A. 1999b. Changing headwater suspended sediment yields in the LOIS catchments over the last century: a palaeolimnological approach. *Hydrological Processes*, **13**, 1137–1153.

Foster, I. D. L. and Walling, D. E. 1994. Using reservoir deposits to reconstruct changing sediment yields and sources in the catchment of the Old Mill reservoir, South Devon, UK over the past 50 years. *Hydrological Sciences Journal*, **39**, 347–368.

Foster, I. D. L., Dearing, J. A. and Grew, R. 1988. Lake-catchments: an evaluation of their contribution to studies of sediment yield and delivery processes. In Bordas, M. P. and Walling, D. E. (eds) *Sediment Budgets*. IAHS Publication **174**, 413–424.

Foster, I. D. L., Grew, R. and Dearing, J. A. 1990. Magnitude and frequency of sediment transport in agricultural catchments: a paired lake catchment study in Midland England. In Boardman, J., Foster, I. D. L. and Dearing, J. A. (eds) *Soil Erosion on Agricultural Land*. John Wiley, Chichester, 153–171.

Foster, I. D. L., Owens, P. N. and Walling, D. E. 1996. Sediment yields and sediment delivery in the catchments of Slapton Lower Ley, South Devon, UK. *Field Studies*, **8**, 629–661.

Foster, I. D. L., Lees, J. A., Owens, P. N. and Walling, D. E. 1998. Mineral magnetic characterisation of sediment sources from an analysis of lake and floodplain sediments in the catchments of the Old Mill reservoir and Slapton Ley, South Devon, UK. *Earth Surface Processes and Landforms*, **23**, 685–703.

Foster, I. D. L., Mighall, T. M., Wotton, C., Owens, P. N. and Walling, D. E. 2000. Evidence for medieval soil erosion in the South Hams region of Devon, UK. *The Holocene*, **10**(2), 255–265.

Grimshaw, D. L. and Lewin, J. 1980. Source identification for suspended sediment. *Journal of Hydrology*, **47**, 151–162.

Håkanson, L. and Jansson, M. 1983. *Principles of Lake Sedimentology*. Springer-Verlag, Berlin.

Harvey, R. W., Leah, H. G., Smith, R. L. and LeBlanc, D. R. 1989. Transport of microspheres and indigenous bacteria through a sandy acquifer: results of natural- and forced-gradient tracer experiments. *Environmental Science and Technology*, **23**, 51–56.

Hathaway, J. C. 1972. Regional clay mineral facies in estuaries and continental margins of the United States east coast. In Nelson, B. W. (ed.) *Environmental framework of coastal plain estuaries. Memoirs of the Geological Society of America*, **133**, 203–316.

He, Q. and Owens, P. N. 1995. Determination of suspended sediment provenance using caesium-137, unsupported lead-210 and radium-226: a numerical mixing model approach. In Foster, I. D. L., Gurnell, A. M. and Webb B. W. (eds) *Sediment and Water Quality in River Catchments*. John Wiley, Chichester, 207–227.

Higgitt, D. L. 1995. The development and application of caesium-137 measurements in erosion investigations. In Foster, I. D. L., Gurnell, A. M. and Webb, B. W. (eds) *Sediment and Water Quality in River Catchments*. John Wiley, Chichester, 287–305.

Higgitt, D. L. and Walling, D. E. 1993. The value of caesium-137 measurements for estimating soil erosion and sediment delivery in an agricultural catchment, Avon, UK. In Wicherek, S. (ed.) *Farm Land Erosion in Temperate Plains Environment and Hills*. Elsevier, Amsterdam, 301–305.

Higgitt, S. R., Oldfield, F. and Appleby, P. G. 1991. The records of land use change and soil erosion in the late-Holocene sediments in the Petit Lac d'Anncy, eastern France. *The Holocene*, **1**, 14–28.

Hilton, J. 1987. A simple model for the interpretation of magnetic records in lacustrine and ocean sediments. *Quaternary Research*, **27**, 160–166.

Horowitz, A. J. 1991. *A Primer on Trace Element Chemistry*. Lewis, Chelsea Michigan.

Horowitz, A. J. and Elrick, K. 1987. The relationship of stream sediment surface area, grain size and composition to trace element chemistry. *Applied Geochemistry*, **2**, 427–451.

Idbeken, H. and Schleyer, R. 1991. *Source and Sediment: A Case Study of Provenance and Mass Balance at an Active Plate Margin (Calabria, S. Italy)*. Springer Verlag, Berlin.

Ingle, J. C. 1966. The movement of beach sand; an analysis using fluorescent grains. *Developments in Sedimentology*, 5, Elsevier, Amsterdam.

Ingle, J. C. and Gorsline, D. S. 1973. Use of fluorescent tracers in the nearshore environment. In *Tracer Techniques in Sediment Transport*. International Atomic Energy Agency, Vienna, 125–148.

Jenns, N. 1997. Sediment source ascription and quantification using mineral magnetics: a palaeolimnological reconstruction of the sedimentation of Slapton Higher Ley. Unpublished MSc Thesis, Keble College, University of Oxford.

Johnson, A. M. 1970. *Physical Processes in Geology*, Freeman-Cooper, New York.

Johnson, M. J. and Stallard, R. F. 1989. Physiographic controls on the composition of sediment derived from volcanic and sedimentary terrains on Barro Colorado Island, Panama. *Journal of Sedimentary Petrology*, **59**, 768–781.

Kirschvink, J. L., Jones, D. S. and McFadden, B. J. (eds) 1985. *Magnetite Biomineralization and Magnetoreception in Organisms*. Plenum, New York.

Komar, P. D., Clements, K. E., Li, Z. and Shih, S.-M. 1989. The effects of selective sorting on factor analysis of heavy mineral assemblages. *Journal of Sedimentary Petrology*, **59**, 590–596.

Kranjc, A. 1997. *Tracer Hydrology*. Balkema, Rotterdam.

Lees, J. A. 1994. Modelling the magnetic properties of natural and environmental materials. Unpublished PhD Thesis, Coventry University.

Lees, J. A. 1997. Mineral magnetic properties of mixtures of environmental and synthetic materials: linear additivity and interaction effects. *Geophysical Journal International*, **131**, 335–346.

Lees, J. A. 1999. Evaluating magnetic parameters for use in source identification, classification and modelling of natural and environmental materials. In Walden, J., Oldfield, F. and Smith, J. (eds) *Environmental Magnetism: A Practical Guide.* QRA Technical Guide 6, Quaternary Research Association, London, 113–138.

Lees, J. A. and Pethick, J. S. 1995. Problems associated with quantitative magnetic sourcing of sediments of the Scarborough to Mablethorpe coast, northeast England, UK. *Earth Surface Processes and Landforms*, **20**, 795–806.

Lees, J. A., Flower, R. J. and Appleby, P. G. 1998a. Mineral magnetic and physical properties of surficial sediments and onshore samples from the southern basin of Lake Baikal, Siberia. *Journal of Palaeolimnology*, **20**, 175–186.

Lees, J. A., Flower, R. J., Ryves, D., Vologina, E. and Sturm, M. 1998b. Identifying sedimentation patterns in Lake Baikal using whole core and surface scanning magnetic susceptibility. *Journal of Palaeolimnology*, **20**, 187–202.

Leibundgut, Ch. 1995. *Tracer Technologies for Hydrological Systems.* IAHS Publication **229**. IAHS Press, Wallingford.

Lewin, J. and Wolfendon, P. J. 1978. The assessment of sediment sources: a field experiment. *Earth Surface Processes*, **3**, 171–178.

Lewin, J., Davies, B. E. and Wolfendon, P. J. 1977. Interaction between channel change and historic mining sediments. In Gregory, K. J. (ed.) *River Channel Changes.* John Wiley, Chichester, 353–367.

Lovely, D. R., Stolz, J. F., Nord, G. L. and Phillips, E. J. P. 1987. Anaerobic production of magnetite by a dissimilatory iron-reducing micro-organism. *Nature*, **330**, 279–281.

Maher, B. A. and Taylor, R. M. 1988. Formation of ultrafine-grained magnetite in soils. *Nature*, **336**, 368–370.

Maher, B. A. and Thompson, R. 1992. Paleoclimatic significance of the mineral magnetic record of the Chinese loess and palaeosols. *Quaternary Research*, **45**, 143–153.

Maher, B. A. and Thompson, R. 1994. Pedogenesis and palaeoclimate: interpretation of the magnetic susceptibility record of Chinese loess-paleosol sequences: comment on Verusob *et al.*, 1993. *Geology*, **22**, 857–858.

Marsh, J. K., Bale, A. J., Uncles, R. J. and Dyer, K. R. 1991. A tracer technique for the study of suspended sediment dynamics in aquatic environments. In *Proceedings of the International Association for Hydraulic Research, International Symposium on the Transport of Suspended Sediments and its Mathematical Modelling*, Florence, Italy, 2–5 Sept 1991, 665–682.

Marsh, J. K., Bale, A. J., Uncles, R. J. and Dyer, K. R. 1993. Particle tracing experiment in a small, shallow lake: Loe Pool, UK. In McManus, J. and Duck, R. W. (eds) *Geomorphology and Sedimentology of Lakes and Reservoirs.* John Wiley, Chichester, 139–153.

McBratney, A. B. and Webster, R. 1981. Spatial dependence and classification of the soil along a transect in northern Scotland. *Geoderma*, **26**, 63–82.

McBratney, A. B. and Webster, R. 1983. How many observations are needed for regional estimation of soil properties? *Soil Science*, **135**, 177–183.

Meier, M. F. 1960. Mode of flow of Saskatchewan Glacier, Alberta. *US Geological Survey Professional Paper*, No. 351, Washington DC.

Merefield, J. R. 1995. Sediment mineralogy and the environmental impact of mining. In Foster, I. D. L., Gurnell, A. M. and Webb, B. W. (eds) *Sediment and Water Quality in River Catchments.* John Wiley, Chichester, 145–160.

Moore, J. E. 1961. Petrography of north-eastern Lake Michigan bottom sediments. *Journal of Sedimentary Petrology*, **31**, 402–436.

Murray, A. S., Olive, L. J., Olley, J. M., Caitcheon, G. G., Wasson, R. J. and Wallbrink, P. J. 1993. Tracing the source of suspended sediment in the Murrumbidgee River, Australia. In Peters, N. E., Hoehn, E., Liebundgut, Ch., Tase, N. and Walling, D. E. (eds) *Tracers in Hydrology*, IAHS Publication **215**, 293–302.

Nichol, I., Horsnail, R. F. and Webb, J. S. 1967. Geochemical patterns in stream sediment related to precipitation of manganese oxides. *Transactions of the Institute of Mining and Metallurgy*, **76**, 113–115.

Oldfield, F. 1991. Environmental magnetism – a personal perspective. *Quaternary Science Reviews*, **10**, 73–85.

Oldfield, F. 1994. Towards the discrimination of fine grained ferrimagnets by magnetic measurements in lake and near-shore marine sediments. *Journal of Geophysical Research*, **99**, 9045–9050.

Oldfield, F. and Clark, R. L. 1990. Lake sediment based studies of soil erosion. In Boardman, J., Foster, I. D. L. and Dearing, J. A. (eds) *Soil Erosion on Agricultural Land*. John Wiley, Chichester, 201–228.

Oldfield, F. and Yu, L. 1994. The influence of particle size variations on the magnetic properties of sediments from the north-eastern Irish Sea. *Sedimentology*, **41**, 1093–1108.

Oldfield, F., Maher, B. A. and Donoghue, J. 1985. Particle size related mineral magnetic source sediment linkages in the Rhode River catchment, Maryland, USA. *Journal of the Geological Society of London*, **142**, 1035–1046.

Oldfield, F., Maher, B. A. and Appleby, P. G. 1989. Sediment source variations and lead-210 inventories in recent Potomac Estuary sediment cores. *Journal of Quaternary Science*, **4**, 189–200.

Oliver, M. A. 1987. Geostatistics and its application to soil science. *Soil Use and Management*, **3**, 8–19.

Oliver, M. A. and Webster, R. 1991. How geostatistics can help you. *Soil Use and Management*, **7**, 206–217.

Ongley, E. D. 1982. Influence of season, source and distance on physical and chemical properties of suspended sediment. In Walling, D. E. (ed.) *Recent Developments in the Explanation and Prediction of Erosion and Sediment Yield*. IAHS Publication **137**, 371–383.

Peart, M. R. 1993. Using sediment properties as natural tracers for sediment source: two case studies from Hong Kong. In Peters, N. E., Hoehn, E., Liebundgut, Ch., Tase, N. and Walling, D. E. (eds) *Tracers in Hydrology*, IAHS Publication **215**, IAHS Press, Wallingford. 313–318.

Peart, M. R. and Walling, D. E. 1986. Fingerprinting sediment source: the example of a drainage basin in Devon, UK. In *Drainage Basin Sediment Delivery*. IAHS Publication **159**, IAHS Press, Wallingford. 41–55.

Peters, N. E., Hoehn, E., Liebundgut, Ch., Tase, N. and Walling, D. E. 1993. *Tracers in Hydrology*. IAHS Publication **215**. IAHS Press, Wallingford.

Rang, M. C. and Schouten, C. J. 1989. Evidence for historical heavy metal pollution in floodplain soils: the Meuse. In Petts, G. E. (ed.) *Historical Change of Large Alluvial Rivers: Western Europe*. John Wiley, Chichester, 127–142.

Ritchie, J. C. and McHenry, J. R. 1990. Application of fallout caesium-137 for measuring soil erosion and sediment accumulation rates and patterns – a review. *Journal of Environmental Quality*, **19**, 215–233.

Salomans, W. and Förstner, U. 1984. *Metals in the Hydrocycle*. Springer-Verlag. New York.

Santiago, S., Thomas, R. L., McCarthy, L., Loizeau, J. L., Larbaigt, G., Corvi, C., Rossel, D., Tarradellas, J. and Vernet, J-P. 1992. Particle size characteristics of suspended and bed sediments in the Rhone river. *Hydrological Processes*, **6**, 227–240.

Santschi, P. H. 1989. Use of radionuclides in the study of contaminant cycling processes. *Hydrobiologia*, **176/177**, 307–320.

Setty, K. B. and Raju, D. R. 1988. Magnetite content as a basis to estimate other heavy mineral content in the sand deposit along the Nizampatnam Coast, Guntur District, Andrha Pradesh. *Journal of the Geological Society of India*, **31**, 491–494.

Shankar, R., Thompson, R. and Galloway, R. B. 1994. Sediment source modelling: unmixing of artificial magnetisation and natural radioactivity measurements. *Earth and Planetary Science Letters*, **126**, 411–420.

Snowball, I. and Thompson, R. 1998. The occurrence of Greigite in sediments from Loch Lomond. *Journal of Quaternary Science*, **3**, 121–125.

Stallard, R. F. 1988. River chemistry, geology, geomorphology and soils in the Amazon and Orinoco basins. In Drever, J. J. (ed.) *Chemistry of Weathering*. NATO ASI Series, Reidel, Dordrecht.

Stanton, R. K., Murray, A. S. and Olley, J. M. 1992. Tracing the sources of recent sediment using environmental magnetism and radionuclides in the karst of the Jenolan Caves, Australia. In Bogen, J., Walling, T. E. and Day, T. (eds) *Erosion and Sediment Transport Monitoring Programmes in River Basins*. IAHS Publication **210**, IAHS Press, Wallingford. 437–445.

Stone, M. and Sanderson, H. 1992. Particle size characteristics of suspended sediment in southern Ontario rivers tributary to the Great Lakes. *Hydrological Processes*, **6**, 189–198.

Sutherland, R. A. 1991. Examination of caesium-137 areal activities in control (uneroded) locations. *Soil Technology*, **4**, 33–50.

Sutherland, R. A. 1994. Spatial variability of ^{137}Cs and the influence of sampling on estimates of sediment redistribution. *Catena*, **21**, 57–71.

Symader, W. and Thomas, W. 1978. Interpretation of average heavy metal pollution in flowing waters and sediment by means of hierarchical grouping analysis using two different error indicators. *Catena*, **5**, 131–144.

Tanizaki, Y., Shimokawa, T. and Yamazaki, M. 1992. Physico-chemical speciation of trace elements in urban streams by size fractionation. *Water Research*, **26**, 55–63.

Thompson, R. 1986. Modelling magnetisation data using SIMPLEX. *Physics of Earth and Planetary Interiors*, **42**, 113–127.

Thompson, R. and Oldfield, F. 1986. *Environmental Magnetism*. Allen and Unwin, London.

Tsoar, H. 1978. The dynamics of longitudinal dunes. US Army European Research Office Grant DA-ERO 76-G-072. Geography Department, Ben Gurion University of the Negev. (Cited by Goudie, A. 1981. *Geomorphological Techniques*. Allen and Unwin, London, 243.)

Udelhoven, T. and Symader, W. 1995. Particle characteristics and their significance in the identification of suspended sediment sources. In Leibundgut, Ch. (ed.) *Tracer Technologies for Hydrological Systems*. IAHS Publication **229**, IAHS Press, Wallingford. 153–162.

Vali, H., Förster, O., Amarantides, G. and Petersen, N. 1987. Magneto-tactic bacteria and their magnetofossils in sediments. *Earth and Planetary Science Letters*, **86**, 389–400.

Van der Post, K. and Oldfield, F. 1994. Magnetic tracing of beach sand. *Coastal Dynamics*, **94**, 323–330.

Van der Post, K., Oldfield, F., Hawarth, E. Y., Crooks, P. R. J. and Appleby, P. G. 1997. A record of accelerated erosion in the recent sediments of Blelham Tarn in the English Lake District. *Journal of Palaeolimnology*, **18**, 103–120.

Verusob, K. L., Fine, M. J. and TenPas, J. 1993. Pedogenesis and palaeoclimate: interpretation of the magnetic susceptibility of Chinese loess-palaeosol sequences. *Geology*, **21**, 1011–1014.

Walden, J., Slattery, M. C. and Burt, T. P. 1997. Use of mineral magnetic measurements to fingerprint suspended sediment sources: approaches and techniques for data analysis. *Journal of Hydrology*, **202**, 353–372.

Wall, G. J. and Wilding, L. P. 1976. Mineralogy and related parameters of fluvial suspended sediments in Northwestern Ontario. *Journal of Environmental Quality*, **5**, 168–173.

Walling, D. E. 1990. Linking the field to the river. In Boardman, J., Foster, I. D. L. and Dearing, J. A. (eds) *Soil Erosion on Agricultural Land*. John Wiley, Chichester, 129–152.

Walling, D. E. and He, Q. 1993. The use of caesium-137 as a tracer in the study of rates and patterns of floodplain sedimentation. In Peters, N. E., Hoehn, E., Liebundgut, Ch., Tase, N. and Walling, D. E. (eds) *Tracers in Hydrology*, IAHS Publication **215**, IAHS Press, Wallingford. 319–328.

Walling, D. E. and He, Q. 1994. Rates of overbank sedimentation in the floodplains of several British rivers during the past 100 years. In Olive, L. J., Loughran, R. J. and Kesby, J. A. (eds) *Variability in Stream Erosion and Sediment Transport*. IAHS Publication **224**, IAHS Press, Wallingford. 203–210.

Walling, D. E. and Kane, P. 1984. Suspended sediment properties and their geomorphological significance. In Burt, T. P. and Walling, D. E. (eds) *Catchment Experiments in Fluvial Geomorphology*. GeoBooks, Norwich, 311–334.

Walling, D. E. and Moorhead, P. W. 1989. The particle size characteristics of fluvial suspended sediments – an overview. *Hydrobiologia*, **176/177**, 125–149.

Walling, D. E., Peart, M. R., Oldfield, F. and Thompson, R. 1979. Suspended sediment sources identified by magnetic measurements. *Nature*, **281**, 110–113.

Walling, D. E., Woodward, J. C. and Nicholas, A. P. 1993. A multiparameter approach to fingerprinting suspended sediment sources. In Peters, N. E., Hoehn, E., Liebundgut, Ch., Tase, N. and Walling, D. E. (eds) *Tracers in Hydrology*. IAHS Publication **215**, IAHS Press, Wallingford. 329–338.

Walling, D. E., Owens, P. N. and Leeks, G. J. L. 1999. Fingerprinting suspended sediment sources in the catchment of the River Ouse, Yorkshire, UK. *Hydrological Processes*, **13**, 955–975.

Warren, L. A. and Zimmermann, A. P. 1994a. The influence of temperature and NaCl on cadmium, copper and zinc partitioning among suspended particulate and dissolved phases in an urban river. *Water Research*, **28**, 1921–1931.

Warren, L. A. and Zimmermann, A. P. 1994b. Suspended particulate grain size dynamics and their implications for trace metal sorption in the Don river. *Aquatic Science*, **56**, 348–362.

Webster, R. and Oliver, M. A. 1992. Sample adequately to estimate variograms of soil properties. *Journal of Soil Science*, **43**, 177–192.

White, K. and Walden, J. 1994. Mineral magnetic analysis of iron oxides in arid zone soils. In Pye, K. and Millington, A. C. (eds) *The Effects of Environmental Change on Geomorphic Processes and Biota in Arid and Semi-arid Regions*. John Wiley, Chichester, 43–65.

Williams, R. D. and Cooper, J. R. 1990. Locating soil boundaries using magnetic susceptibility. *Soil Science*, **150**, 889–895.

Wolff, C. W., Seager, J., Cooper, V. A. and Orr, J. 1988. *Proposed Environmental Quality Standards for List II Substances in Water: pH*. TR59. Water Research Centre, Medmenham.

Woodward, J. C., Lewin, J. and Macklin, M. G. 1992. Alluvial sediment sequences in a glaciated catchment: the Voidomatis Basin, Northwestern Greece. *Earth Surface Processes and Landforms*, **16**, 207–226.

Yu, L. and Oldfield, F. 1989. A multivariate mixing model for identifying sediment sources from magnetic measurements. *Quaternary Research*, **32**, 168–181.

Yu, L. and Oldfield, F. 1993. Quantitative sediment source ascription using magnetic measurements in a reservoir catchment system near Nijar, S.E. Spain. *Earth Surface Processes and Landforms*, **18**, 441–454.

Zhang, J. and Huang, W. W. 1993. Dissolved trace metals in the Huanghe: the most turbid large river in the world. *Water Research*, **27**, 1–8.

2 Coarse Sediment Tracing Technology in Littoral and Fluvial Environments: A Review

D. A. SEAR,[1] M. W. E. LEE,[2] R. J. OAKEY,[3] P. A. CARLING[1] and M. B. COLLINS[2]

[1]Department of Geography, University of Southampton, UK
[2]School of Ocean and Earth Science, University of Southampton, Southampton Oceanography Centre, UK
[3]Department of Geography, University of Lancaster, UK

INTRODUCTION

The evolution of morphology and physical habitat in river and coastal environments is intrinsically linked to the processes of sediment transport. In many cases these environments are dominated by coarse sediments (here defined as particles having a diameter of 4 mm and over), or at least owe much of their form to the presence of coarse sediment (Leopold, 1992). Understanding coarse sediment transport processes has direct ecological and economic benefits through improvement in the prediction of changes in physical habitat and the location and magnitude of erosion and deposition, the latter accounting for substantial costs in terms of land loss and disruption of infrastructure (Sear *et al.*, 1995; Cooper, 1996).

The study of sediment transport processes has become increasingly sophisticated, with progress in both littoral and fluvial research converging on the prediction of sediment transport rates through improvements in the physically based understanding of entrainment, transport and deposition. Techniques to further this understanding have generic similarities between the two environments although the specific technologies and methods of deployment are different. Recognition of this has led to this review, which seeks to draw together the published literature on a specific research tool, the deployment of coarse tracer particles, and attempts to summarise the characteristics and performance of the different tracing techniques deployed.

COARSE SEDIMENT TRANSPORT IN LITTORAL AND FLUVIAL ENVIRONMENTS

As Hassan and Church (1992) state, 'sediment transport consists fundamentally of movements of individual particles', and it is this basic fact that supports the deploy-

Tracers in Geomorphology. Edited by Ian D. L. Foster. © 2000 John Wiley & Sons Ltd.

ment of tracers as a research tool for coarse sediment transport. Paradoxically, although their statement is fundamentally correct, much of fluvial and coastal geomorphology and engineering is concerned with the bulk transport or accumulation of particles in relation to measures of synoptic or local hydrodynamic forces. At one level the transport of coarse particles in rivers or coasts depends on the application of the same physical principles, but it is the timing and spatial patterns of sedimentology and driving forces that differ (Figure 2.1). In fluvial environments, sedimentology is heterogeneous and flow is predominantly unidirectional, although turbulence and secondary flows complicate this general pattern. In littoral environments, sedimentology is also

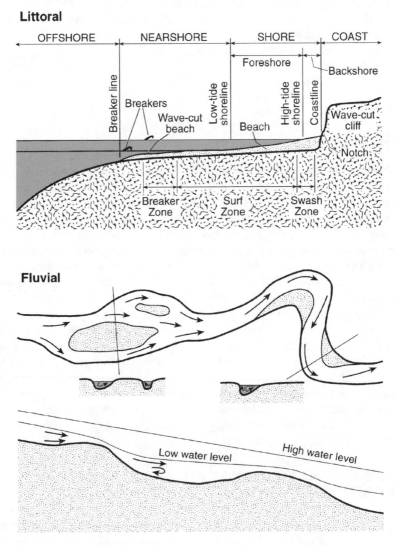

Figure 2.1 Schematic diagram illustrating the contrasting hydrodynamics experienced by coarse sediment tracers in littoral and fluvial environments

Table 2.1 Sediment transport monitoring technologies and limitations

	Trapping	Tracing	Topo-survey	Modelling
Entrainment	Timing and grain sizes	Timing, sizes, location	Very limited	Poor validation
Transport or displacement	Quantity and grain sizes	Distance, 2-D routes, step lengths, velocities	Budget approach	Weak theory with very poor validation
Deposition	Timing	Timing, sizes, 3-D location	Good spatial extent	Poor validation
Cost	High	Low–high	Low–Moderate	High
Application	Small-scale/ local	Extensive	Extensive	Local–extensive depending on sophistication
Continuous	Yes (load only)	Yes	No	No
Intrusive	Yes	No	No	N/A
Spatial cover	Poor	Good	Good	Poor

often heterogeneous, but the flows are oscillatory, with energy transferred via the action of waves (Figure 2.1).

The movement of a particle requires the operation of three basic processes, entrainment, transport and deposition, which together describe the sediment transport process. Techniques for the determination of these processes and for the acquisition of bulk sediment transport rate data are generic to both littoral and fluvial environments and include trapping, tracing, topographic re-surveys and modelling, the latter including both numerical and physical models (Table 2.1). Modelling will not be discussed further since this review is concerned with acquisition of field data; suffice it to say that models require high-quality field data sets for validation.

COARSE SEDIMENT TRACING RESEARCH

The deployment of tracers in the acquisition of data on sediment transport has been undertaken since as early as the late 19th century (Richardson, 1902), although the main period of application in both littoral and fluvial environments has been post-1950 (Figure 2.2). Interest in coarse sediment tracing has grown, largely as a result of increased interest in process studies of fluvial and coastal environments but also due to the development of more robust and widely applicable technologies. For completeness, the main studies in both littoral and fluvial environments are documented in Tables 2.4 and 2.5 at the end of this chapter while their references are included in the References section to facilitate further research.

In littoral research, emphasis has been on the determination of transport rates and, in particular, the direction of movement. Tracer experiments can be classified according to the purpose of their application:

- determination of entrainment thresholds
- descriptions of displacement

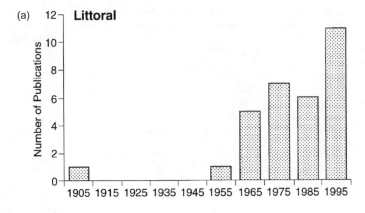

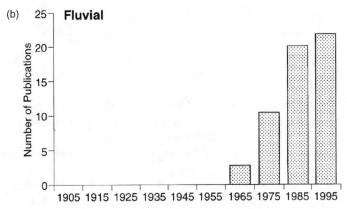

Figure 2.2   Publication records for (a) littoral and (b) fluvial experiments using coarse sediment tracing technology

- particle controls on distance of transport
- morphological controls on transport distance
- sediment transport estimation
- intra-event activity

Use of Coarse Sediment Tracing Technology for Studies of Particle Entrainment

Among the earliest and main applications of coarse sediment tracers has been the determination of flow competence and the hydrodynamic conditions associated with entrainment of particles (Figure 2.3). In their simplest form, these experiments consist of a number of painted particles of varying size, laid out across a section, or deployed in a group. The maximum size moved after a given event is considered to represent the competence of the last flow maximum (Wilcock, 1971; Carling, 1983; Sear, 1996). Some encouraging results have been collected in this way, although there is always the possibility that the tracers were in over-loose positions on the streambed.

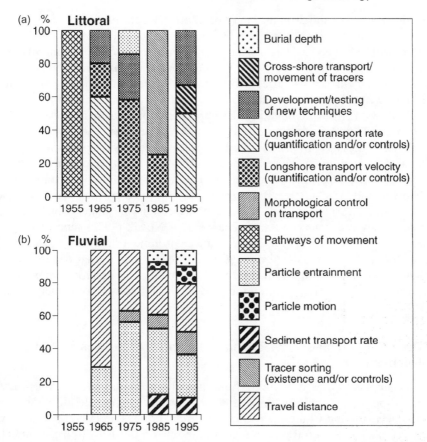

Figure 2.3 The history of coarse sediment tracing applications in (a) littoral and (b) fluvial environments

Brayshaw (1985), Hassan *et al.* (1991) and Sear (1996) have identified bed structure as a control on entrainment using coarse sediment tracers and note that thresholds may be increased above those measured for particles in more exposed or over-loose positions. Tracer-based studies of the hydraulics associated with specific particle entrainment have tended to focus on surface particles. Although the motion of buried particles within a mobile bed have been investigated in terms of the probability of exposure during events (Schick *et al.*, 1987), the timing and hydraulics associated with this motion remain to be established.

Information on particle entrainment has also been derived using magnetically tagged and radio-tagged tracing technologies. An advantage of the latter technology is that individual particles may be identified in motion. In these experiments, tagged particles or naturally magnetic sediments are logged as they pass over a detector placed across the river bed. With tagged particles, these are emplaced a short distance upstream of the detector, whilst with natural sediments critical values of shear stress or discharge are recorded at the onset of transport. The advantage of tagged particles is that the grain size and structure may be controlled. Radio-tagged particles have been tracked, and the discharge at which movement is detected may be used to infer entrainment thresholds. Using relatively small populations (seven maximum) of large particles (grain size represented $\gg D_{50}$), Schmidt and Ergenzinger (1992) confirmed that entrainment thresholds varied substantially between individual particles of the same shape/weight/size and even for the same particles after periods of rest during a single flood event, thus highlighting the importance of local sedimentological and hydrodynamic factors in entraining and transporting bedload.

Although studies have been conducted which have been specifically concerned with the initial movement of coarse beach material (Novak, 1972) these are very rare. Far more common are experiments that are aimed at measuring longshore transport rate (e.g. Nicholls and Wright, 1991; Bray et al., 1996) but these, by their very nature, also provide useful information with respect to the conditions under which sediment movement does and does not occur. The work of Novak (1972) supported the use of the Hjulström (1935) curve for the marine environment landward of the breaker zone for the cobble to boulder size range, while studies such as that of Lee et al. (see Chapter 22) have shown that transport of 34–55 mm (b axis) material can occur even under very low wave energy conditions (when the longshore component of wave power is as low as $5.0\,\mathrm{W\,m^{-1}}$).

The Use of Coarse Sediment Tracing Technology for Studies of Particle Displacement

The main application to date of coarse sediment tracers has been in the study of particle displacements during flood and tidal events (Figure 2.3). In the fluvial literature, Hassan and Church (1992) and Church and Hassan (1992) have reviewed this aspect extensively, utilising a range of data sets on particle displacement. Their study, along with most other displacement studies, found considerable variation in the distances moved by particles even under relatively simple conditions (e.g. exposed particles moving during single events in rivers with limited bedform development), but that the distribution of displacements could be modelled using a gamma function (Figure 2.4(a)). As the field conditions become more complex, the distributions deviate from the gamma model due largely to the presence of multiple or long-duration flow events and bedforms that selectively trap or route particles (Figure 2.4(b)) (Schmidt and Gintz, 1995; Sear, 1996).

Although there has been little attention paid to the distribution of transport distances within the littoral literature, a major focus of tracer studies in the coastal zone has been the identification of factors (sedimentological and hydrodynamic) that influence particle displacement distance (e.g. Richardson, 1902; Jolliffe, 1964; Carr, 1971; Caldwell, 1983; Nicholls and Wright, 1991; Bray et al., 1996; Van Wellen et al., 1999). Studies investigating the effect of particle size on displacement have, in the majority of cases, suggested that coarser tracers move faster longshore than fine tracers (Richardson 1902; Jolliffe, 1964; Caldwell, 1983; Cooper et al., 1996). However, Carr (1971, 1974) found that both positive and negative relationships could exist between particle size and longshore travel distance (Figure 2.5). Nordström and Jackson (1993) found that particle size did not appear to affect cross-shore position,

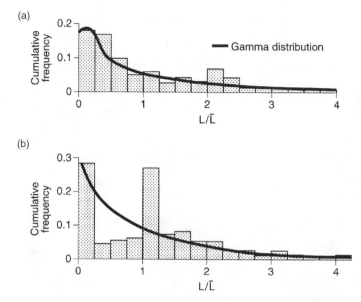

Figure 2.4 (a) Example of a typical gamma function model of short-distance, single-event tracer displacement in fluvial environments. (b) Example of the effects of sediment storage in gravel bars on tracer displacement and the concomitant deviation from the gamma model (after The movement of individual grains on the streambed by M.A. Hassan and M. Church. © Copyright John Wiley & Sons Limited. Reproduced with permission.)

but in contrast Matthews (1980) suggested that coarse tracers were preferentially pushed above the normal strand line. Significant differences in tracer c-axis length have been found in both the cross-shore and longshore orientations (Carr, 1974; Caldwell, 1983). In contrast to the work carried out to investigate particle size influences, particle shape has received relatively little attention. Carr (1971) found that shape sorting did not appear to play an important role on Chesil Beach, and Nordström and Jackson (1993) failed to link particle shape to cross-shore position. In contrast, Wright *et al.* (1978a) identified a tendency for more rounded tracers to be recovered on the lower foreshore and angular tracers further towards the crest of the beach. Rather than considering particle size and shape separately (a division which it can be argued is artificial), Lee *et al.* (see Chapter 22) considered the effect of particle form (size and shape) on displacement. Their analysis failed to find a link between longshore displacement and form but it did suggest that a tracer's injection position had an effect on displacement; likewise injection position did intermittently appear to have an influence on the vertical mixing depth of tracers while form did not.

Within fluvial environments, one of the suspected primary controls on displacement is the size or shape of the particle. Early studies were inconclusive, and again the review conducted by Church and Hassan (1992) confirmed the complexity of particle displacements. Absolute particle size does not exert a strong control on displacement, although Church and Hassan (1992) did manage to develop a nonlinear relationship between scaled particle size and scaled transport distance (Figure 2.6). Wilcock (1997) subsequently explored the theoretical basis for the relationship using field and flume data on particle displacements and concluded that the observed increase in the rate of decrease in displacement with increasing particle size results from two processes: first, for coarser particles, transport is mostly partial and therefore relates to relatively short movements of surficial grains over a largely static bed, whilst for smaller grains, mobility is almost equal and full transport conditions pertain, including particles being mobilised below the bed surface. The second effect relates to the higher degree of relative exposure of large grains and the higher opportunity for trapping of fine

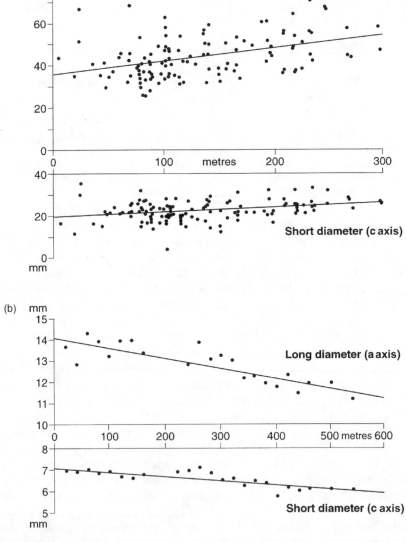

Figure 2.5 Example relationships between tracer size and displacement derived from littoral coarse sediment tracing studies (after Carr, 1974)

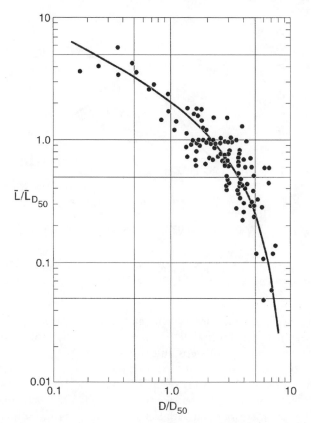

Figure 2.6 Example of the relationship between tracer size and distance of movement derived from the review of tracer studies conducted by Hassan and Church (1992) (after The movement of individual grains on the streambed by M.A. Hassan and M. Church. © Copyright John Wiley & Sons Limited. Reproduced with permission.)

grains which, together with the control exerted by bed structure on entrainment, tends to reduce the distinction between large and small particles in terms of overall displacement. Considerable scatter exists around the relationship developed in Figure 2.6, some of which will relate to the effects of particle shape on transport distance. Research into the effects of particle shape on tracer displacement has not been as prevalent as that into the effects of particle size. Schmidt and Ergenzinger (1992) and Schmidt and Gintz (1995) have conducted the main research programme in a relatively high-energy, boulder/cobble bed step-pool stream, with controlled flume experiments undertaken using the same tracers as used by Carling *et al.* (1992) in support of the field research. Using magnetic tracer technology, they observed similar displacement frequencies and distance travelled for rods, ellipses and spheres, but much lower values for disc-shaped particles. Instead of particle size, this study recorded a decrease in displacement length according to grain weight.

With the advent of tracking technology applied to particle movements, progress has been made towards understanding the characteristics of intra-event displacement.

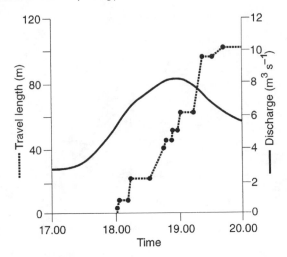

Figure 2.7   Example of the information recorded by radio-tracking particles (after Ergenzinger *et al.*, 1989). Reproduced by permission of E. Schweizerbart'sche Verlagsbuchhardlung.

Using relatively small populations (seven maximum) of large particles (grain size represented $\gg D_{50}$), Ergenzinger (1989) confirmed that the motion of the particles was erratic, moving in a series of steps, interspersed with periods of rest (Figure 2.7). The distribution of step lengths and rest periods was found to follow a Poisson model, providing confirmation of the validity of the Einstein (1937) model assumptions. The step-length distance was found to be strongly controlled by event magnitude, such that larger discharges corresponded with longer step lengths and total displacements. No comparative data exist for the littoral zone, largely due to the lack of tracking technology available for such research. Radio technology, using standard 150 MHz frequencies, does not work in saline water, the medium rapidly absorbing the radio waves, making detection impossible. Low-frequency radio technology has been deployed for the tracking of crustacea, but, to date, intra-event particle motion measurements in the swash, surf and breaker zones remain to be undertaken.

The Use of Coarse Sediment Tracing Technology for Measuring Coarse Sediment Transport

The measurement of coarse sediment transport in fluvial and littoral environments has traditionally utilised similar technologies, although the specific methods of deployment differ according to environment. The technique strongly depends on what aspect of the transport process the study is attempting to investigate. For example, entrainment data are typically derived from trapping bedload in conjunction with some measures of hydrodynamic forces (Wilcock, 1993) or deploying a range of tracers and inferring flow competence from the maximum size mobilised during a given event (Carling, 1983), hydrodynamic conditions being simultaneously measured using electromagnetic current meters (EMCMs). Transport of coarse bed material is typically quantified during floods or tidal events using a range of techniques including trapping, topographic re-survey and tracing (see Table 2.1). Trapping is undertaken using either a fixed structure such as a pit trap (Reid *et al.*, 1985) or a surface trap (Workman, 1997) or through deployment of portable samplers (Emmett, 1980; Van Wellen *et al.*, 1997). The former are confined to specific cross-sections whilst the latter may also be deployed at varying locations to obtain spatially distributed data on sediment transport rates. Spatial and temporal variability in sediment transport rates within and between discrete events makes the acquisition of representative data difficult. Nevertheless, some extensive data sets have been collected using these techniques and these appear to show some correlation between sediment transport rate and hydrodynamic conditions (Chadwick, 1989; Ferguson and Wathen, 1998). Where no net sediment transport occurs beyond a point in space such as a groyne or sediment trap, estimates of transport rate may be derived from the topographic survey of accumulations of coarse sediments (Madej and Ozaki, 1996; Wilson, 1996). Assumptions of zero or a known value of sediment transport at the downstream end of a reach or beach are difficult to substantiate; recent studies have indicated that significant transfer can occur over and around groynes (Coates *et al.*, 1999). The presence of sedimentological contrasts such as gravel–sand transitions can be used to infer zero transfer of coarse sediments, but to date the mobility of these transition fronts is unknown. Problems with this technique include the scale of topographic survey required to generate a given accuracy of net transport estimate and the duration of time it is necessary to wait to allow a measurable quantity of sediment to accumulate. The latter problem often results in measurements straddling a number of different sets of hydrodynamic conditions and consequently makes their use in modelling difficult. Tracers can be used in conjunction with topographic surveys to provide estimates of net sediment transport rates by providing information on the virtual rate of travel of a sediment slug, or establishing confidence in the route taken by particles in transit (Ferguson and Ashworth, 1992).

The use of tracers themselves for determining sediment transport rates or volumes, is considered cost-effective, and in littoral environments it is often a favoured method of estimating coarse sediment transport. The advantages are outlined in Table 2.1, relative to other techniques, and in recent publications the method has been shown to provide estimates comparable to those derived from portable samplers (Haschenburger and Church, 1998). To estimate sediment transport (Q_b) using tracers, information is required on the basis of the following formula:

$$Q_b = v_b \cdot d_s \cdot w_s \cdot (1 - p)g(\rho_s - \rho) \tag{2.1}$$

v_b is the virtual rate of travel (L/t) of moved particles where L is the mean transport distance of moved particles and t is the total time (including rest periods in most instances as no intra-event data are available) over which the movement occurred; w_s and d_s are the width and depth of the active bed layer; p is the porosity of the sediment and ρ_s is the density of sediment, which is commonly taken to be $2650 \, \text{kg m}^{-3}$; ρ is the density of water and g is the gravitational constant. Technologies for determining these parameters must permit detection of buried particles. Studies of the burial and vertical exchange of tracers are necessary in order to determine the thickness of the active layer and the downstream and vertical distance over which full mixing of tracers occurs. If tracers are known to be fully mixed, then transport rates may be computed from surface samples of tracers only. The problem is that few tracer studies are of sufficient duration to ensure full mixing occurs. Instead, burial depths are known to be distributed exponentially, with relatively few tracers recorded at depth. Vertical exchange is profoundly affected by the duration and strength of the flow/tidal event, and simple exponential burial depth models begin to break down under these conditions (Hassan and Church, 1994).

Although tracer estimates of sediment transport have been undertaken in both fluvial and littoral environments, little research has been conducted on the reliability of the estimates of mean width, depth and length of the tracer cloud. Haschenburger and Church (1998) have conducted error analysis on a study of magnetic tracer dispersion in three subreaches of a wandering gravel-bed river. They report that uncertainty in estimates of transport depends largely on errors associated with the derivation of virtual velocity. This arises from the relatively high standard errors associated with downstream dispersion, using small populations of tracers, together with the difficulty in specifying actual tracer dispersion when recovery rates become small. The estimates of uncertainty do not include the general bias arising from assumptions of uniform porosity.

Wilcock (1997) presents a different approach to the determination of sediment transport that uses a direct measure of the rate of entrainment of tracers from a discrete patch of tracers placed in (or on) the streambed. A general formula for estimating sediment transport Q_b using this approach takes the following form:

$$Q_b = M \cdot (N/t) \cdot L \tag{2.2}$$

where M is the mass of particles entrained per unit bed area, N is the number of times an individual particle is entrained during time t where t is the event duration, and L is the length of a single displacement. This general formula can be applied to individual size fractions to provide fractional transport rates. The deployment of tracers, which in this instance can be painted or passive, is different to that in other fluvial studies where lines of tracers are deployed across the channel, and depends on emplacing a known number (and mass) of particles in a patch, and recovering the number that remain in situ following a transport event(s). The values given by equation (2.2) are averages, and again depend on the correct determination of transport duration (t), and are valid only for the area of the patch. The effects of over-loose tracer gravels may be a problem with this method, resulting in the prediction of relatively high rates of entrainment, and thus transport.

Early tracer studies in coarse littoral environments (e.g. Russell, 1960) used what has since become known as the continuous injection method (CIM) to measure longshore transport rate. The basis of the method was to inject tracers at a known steady rate and then to measure the concentration of the tracer at a point downdrift of the injection site. The concentration of the tracer at the sampling site would in theory be q/Q, where q is the tracer injection rate and Q is the rate at which sediment is transported past the injection point. The method was found to give encouraging results when tracer-derived drift rates were compared with those from topographic measurements (Russell, 1960). However, the necessity to assume, when using the method, that the sediment drift repeats itself precisely during each period between injections seems to have restricted its further use.

Since the work of Russell (1960), equation (2.1) (or a variation of it) has almost exclusively been used to calculate transport rates (e.g. Nicholls and Wright, 1991; Bray *et al.*, 1996; Lee *et al.*, see Chapter 22). This method is sometimes referred to as the spatial integration method (SIM). On coarse-grained beaches the SIM usually involves releasing fairly small numbers of tracers (often < 100) at the beach surface during low tide, and allowing these to be transported before recovering them on the next tide (or several tides later). From the longshore and vertical distributions of the tracers, v_b and d_s (see equation (2.1)) are quantified, while w_s is usually taken from the width of the beach covered by the tide. Until the early 1990s a major problem associated with the use of the method was the low recovery rates achieved during high-energy conditions (Bray *et al.*, 1996). With high percentages of the tracer not being recovered (up to 73 percent in the case of Wright *et al.*, 1978b), and thus the location of this material not being known, large uncertainties were associated with the drift rates calculated. To overcome these recovery problems, the electronic pebble was developed (Workman *et al.*, 1994). This system has a much greater detection range than the previously used aluminium system ($\sim 100\,\text{cm}$ as opposed to 35 cm) and it has allowed recovery rates consistently > 70 percent to be achieved (Bray *et al.*, 1996; Lee *et al.*, see Chapter 22).

Coarse sediment tracing studies have made a major contribution to the understanding of sediment transport. In addition to those areas discussed above, tracers have the advantage of providing information on the connectivity between different areas of river or beach. These are important when attempting to understand the spatial development of landforms and particularly when attempting to identify the role of storage in sediment transfer. Further information on the application of different tracing technologies and the advances these have made in the understanding of coarse sediment transport is contained in Tables 2.4 and 2.5, located at the end of this chapter. Although these tables may not contain details of every field study appearing in the literature to date, they are thought to be representative of the coarse tracer literature published to date and provide a useful introduction to others entering or already working with coarse sediment tracers.

TECHNIQUES FOR COARSE SEDIMENT TRACING

The main factors determining the selection of the most appropriate tracing techniques fall into four categories:

(1) The nature of the information one desires to collect (e.g. estimates of transport rates, critical entrainment thresholds, intra-event data on particle motion, etc.).
(2) The character of the environment in which the information is to be collected (e.g. littoral versus fluvial, and the grain size of the material present).
(3) The level of statistical confidence one wishes to achieve with respect to the measurement one takes. This is dependent on the sample size and recovery rate and as such is influenced by the previous two factors.
(4) The cost of the system to be deployed.

Over the history of coarse sediment tracing experiments, a range of technologies have been deployed, with an underlying increase in performance as the aims of tracing research have changed, e.g. the requirement for 3-D tracer location. Figure 2.8 illustrates both the change over time and the differences between the tracing technology deployed in littoral and fluvial environments. The differences can mostly be explained in terms of individual choice and the continuity of techniques once they have become established, e.g. the distinction between the dominance of aluminium in the littoral tracer deployments versus magnetic tracers in fluvial environments. These systems perform similarly, the main difference being the nature of the construction process. In littoral environments a fundamental difference in tracing is the short time (*c.* 6 hours) available for recovery if one needs to determine tracer dispersion for a single tide. In this instance, technologies that minimise recovery time are more important, and aluminium offers the possibility of hiring metal-detecting clubs to increase the numbers of detectors. In contrast, magnetic detection requires the purchase of relatively specialist detectors, but this is compensated for in fluvial environments by the generally longer time available for recovery of dispersed tracers between events. Tracer technologies fall into three main groups: 'visual' tracers which depend

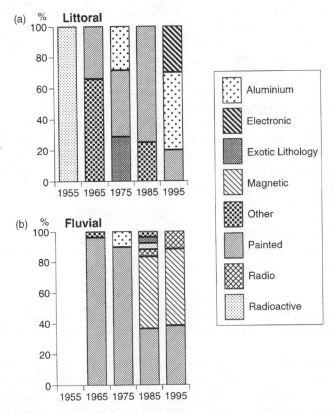

Figure 2.8   The history of tracing technology deployed in (a) littoral and (b) fluvial environments

upon visual identification for relocation (e.g. painted particles, exotic lithology); 'passive' tracers which can be detected remotely, but not tracked (radioactive, magnetic, aluminium, electronic); and active or so-called 'smart' tracers which transmit a signal and can be tracked during events. The last group includes the acoustic or radio-tagged pebbles.

'Visual' tracers account for the bulk of coarse tracer deployments in both littoral and fluvial environments and in earlier deployments were dominated by painted pebbles, artificial tracers and exotic lithologies. Table 2.2 summarises the advantages and disadvantages of existing visual and passive types of tracer. With simple painted or exotic lithology tracers, grain-size representation can be excellent but detection is limited to visual identification of surface particles. For finer grain sizes and with large numbers of particles it may be possible to adopt bulk sampling methods such as have been used in tracer studies of sand-sized particles (Kraus et al., 1982), although the quantities required are likely to be prohibitive and attempts to date have proved unsuccessful (Nordström and Jackson, 1993). The recovery rates of painted and exotic lithology are generally low, although in short distance, partial transport conditions (sensu Wilcock, 1997) performance can match that of the passive technologies (see Tables 2.2 and 2.3). As with any tracing system, performance varies according to the aims of the experiment; for example, Billi (1988) used painted tracer technology to successfully investigate the destruction of pebble clusters, whilst others have recorded useful information on entrainment (Carling, 1983; Wilcock, 1997). The use of painted tracer technology is now generally considered inappropriate for the determination of sediment transport rates unless supported by other methods for estimating the depth of active layer. Even so, the use of surface distribution characteristics to define virtual rates of transport and width of active layer is questionable, the main advantages being that one can generate, with relatively limited funds, information on the movement of smaller particles, and in larger numbers than with other technologies. Cavazza (1981) reviews best practice for the deployment of painted tracers, including choice of site and the type of measurements required to provide information on transport and entrainment. The major work carried out in the littoral zone using visual techniques was in the 1960s and 1970s (e.g. Russell, 1960; Jolliffe, 1964; Carr, 1971, 1974). Considering the limitations of the tracer technologies used (recovery rates often < 10 percent) these studies provided a great deal of useful information. Often drift rates were not measured during such studies; instead travel distances and velocities of tracers were given and the controls on tracer movement were investigated. For example, Carr (1971) found that daily longshore transport could be 343 m and that relationships between pebble size and longshore movement did exist. Carr (1971) also noted that recovered material was not necessarily typical of that injected. The latter finding was supported by the work of Caldwell (1981) who also noted that few tracers recovered at the beach surface had parameters that were similar to the indigenous material surrounding them. Caldwell's (1981) findings suggested that 'visual' tracer techniques may yield misleading results due to tracers that are unrepresentative of the indigenous material being concentrated at the beach surface. Studies carried out since that of Caldwell (1981) have, as a consequence, mainly used 'passive' tracer technologies, with visual techniques only being used for qualitative assessments (e.g. Van Wellen et al., 1999).

Table 2.2 Performance of visual and passive tracer technologies

Technique (environment)	Typical recovery rates	Grain-size representation	Spatial data acquisition	Lifespan	Future potential
Nuclear tagging (both)	c. 5%	> 19 mm	Unknown	Limited by half-life of material	Limited. Environmental hazards. Limited ID potential.
Electronic pebble (littoral)	> 70%	> 16 mm	After-event coverage good owing to high numbers and high detection rates. 3-D detection	c. 1.5 years	High – miniaturisation of circuitry offers greater grain-size range. Improved detection system makes it possible to increase resolution of burial depth prediction and more rapid recovery of data. Battery life means particle active for > 1.5 years
Magnetically tagged particles (fluvial)	25–100%, typically 70–85%	> 10 mm if magnetic inserts used. Below this value and particles attract and aggregate	After-event coverage moderate–good owing to high numbers and detection rates. Burial depth in coastal zone may reduce recovery rates. 3-D detection	Permanent (although once paint abraded, becomes difficult to detect)	Limited – improved detection system needed to increase speed of recovery
Aluminium particles (both)	40–50% littoral, 35% fluvial	> 7 mm although difficult to locate much smaller than 10 mm	After-event coverage moderate–good owing to high numbers and detection rates. Burial depth in coastal zone may reduce recovery rates. 3-D detection	Permanent	Limited – improved detection system needed to increase speed of recovery
Painted particles (both)	4–100%, typically 30–60%	Full range > 4 mm although difficult to locate much smaller than 10 mm	Poor – depends on burial rates and numbers used. 2-D detection	< 5 years depending on abrasion	Limited

Table 2.3 Performance of active tracer technologies

Technique	Intra-event data	Grain-size representation	Multi-particle tracking	Spatial data acquisition	Lifespan	Future potential
Acoustic tracking system	Yes (limited) – unknown in littoral. Limited success in fluvial at low flows. Background noise problematic	> 50 mm	Yes, but currently limited to 10 (costly)	Yes – resolution unknown 2-D detection	< 1 year	Moderate – depends on ability to filter signal/noise. Size reduction needed to represent higher percentage of grain-size populations
Radio tracking system	Yes – some success in fluvial systems. Resolution poor (>2 m) but could be increased. Unsuitable in coastal environments at current frequencies	> 12 mm	Yes, but currently limited to 10 (costly)	Yes – resolution limited. 2-D detection	< 1 year	Moderate – microelectronics offer smaller size. Digital processing makes it possible to increase number of individually tracked particles and improve spatial resolution
Magnetic detector system	Yes – limited to location of detector(s). Cannot distinguish between individual particles	Full range if naturally or enhanced magnetic susceptibility used. > 10 mm if magnetic inserts used	Yes, but no individual particle ID (costly)	Poor – limited to location of detectors. 1-D detection	Permanent	Limited – multiple detector arrays and higher frequency of cross-section detection offer enhanced spatial resolution. Lack of ID remains a limiting factor

Passive (radio, aluminium, magnetic, iron and electromagnetic) tags have been employed to track coarse sediments in rivers since the early 1980s, although Nir (1964) conducted a campaign with a limited number of passive iron tagged tracers in ephemeral channels. In the littoral environment, passive tracing was undertaken using aluminium tracers in the 1980s (Nicholls and Wright, 1991) although wider application really started in the early 1990s with the deployment of the electronic pebble. Passive tracers have an advantage over the 'visual' tracers in that they permit relocation after burial. Recovery rates are, however, still highly variable, being also affected by the local site environment, the prevailing hydrodynamic conditions and the nature of the experiment. In fluvial environments, recovery rates as low as 17 percent for iron tags (Schmidt and Ergenzinger, 1992) and as high as 100 percent for magnetic tags (Gintz et al., 1996) have been recorded. However, recovery rates for these earlier studies are typically around 65 percent in fluvial environments (Reid et al., 1984; Ashworth and Ferguson, 1989; Hassan et al., 1991; Schmidt and Ergenzinger, 1992; Hassan and Church, 1992, 1994; Gintz et al., 1996; Ferguson and Wathen, 1998) and > 70 percent for the electronic pebble in littoral environments (Bray et al., 1996). Perhaps the most reliable indication of potential tracer retrieval rates using such tracers (and in this case magnetic) is from a recent study conducted by Ferguson and Wathen (1998). They monitored displacements of tracer grains on an event-by-event basis over a period of two years and recorded an overall recovery rate of 61 percent (Ferguson and Wathen, 1998). This is in line with the recovery rates recorded by others (as discussed); however, overall recovery rates still fall a good deal short of the idealised 100%. This shortfall in tracer recovery highlights the importance, to the science of sediment transport, of the development of real-time tracking systems, using so-called 'smart tracers'. Theoretically the recovery rate of such systems (excluding sensor malfunction) would be 100%, although the low numbers typically deployed during early studies limit their usefulness.

The problem of short recovery time in littoral environments, coupled with the need to recover high percentages (generally > 70 percent) of tracers, led to the development of the electronic pebble system (Figure 2.9). The principle of the system is fairly simple: essentially a small (presently 23 mm diameter, although it is now considered possible to miniaturise to 10 mm) battery-powered circuit is used to transmit electromagnetic signals (at an optimum frequency) which can be detected at a range of 1 m using a specially designed receiver. The system's performance is the best presently available for use in the littoral environment and this is reflected by the high tracer recoveries achieved to date (Bray et al., 1996; Van Wellen et al., 1999; Lee et al., see Chapter 22). The system is the first to allow high recovery rates to be achieved in high-energy conditions and thus it is contributing greatly to our knowledge of shingle beach behaviour.

Active Tracing Technology

Some attempts have been made in recent years to develop coarse sediment tracing systems that automatically gather grain positional data in the longitudinal downstream plane (i.e. in one dimension) (Ergenzinger and Custer, 1983; Reid et al., 1984; Ergenzinger et al., 1989; Busskamp and Hasholt, 1996). An active real-time system

Figure 2.9 A cut-away of an electronic pebble tracer illustrating the use of resin for construction, circuit board and battery

that logs downstream and cross-stream displacements (i.e. in two dimensions) has yet to be successfully field-tested. This is largely due to the fact that the river in flood is a hostile environment for sensitive electronic equipment. Such information on grain position during flood events would enlighten questions of step length, rest period and particle velocity.

Early experiments on the intra-event detection of bedload transport in fluvial environments were conducted by Ergenzinger and Custer (1983) using the Faraday principle of electromagnetic induction to generate an electrical pulse as magnetic (in this case naturally magnetic) particles passed across a coil buried in the streambed of the Squaw Creek. The same principle was subsequently used by Reid *et al.* (1984) to determine the differential entrainment of particles occupying open and structured positions on the streambed. In this latter case, the ceramic magnets were inserted within a limited number (50) of particles. The Squaw Creek experiments provided real-time logging of the frequency of transport of cobble/large gravel sized material. Later the detectors were modified and two were emplaced in the channel, providing input and output from a reach, whilst cross-stream variation was assessed by mounting four or five individual detectors across the river bed. Figure 2.10 illustrates the type of detector deployed, together with an example of the information that can be generated on transport and entrainment. Problems with this system relate to the lack of individual particle IDs, the fixed position of the detector (and therefore the information), and the interpretation of the signal which may lead to multiple detection as the detector and particle fields interact. Improvements in these designs are possible utilising signal-processing software and more sophisticated electromagnetic detectors; however, problems in terms of the generation of false signals during the

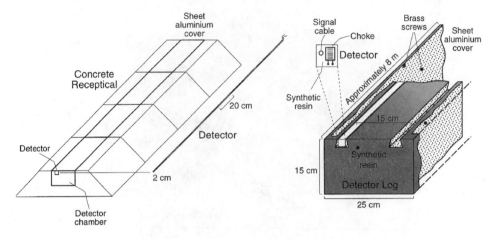

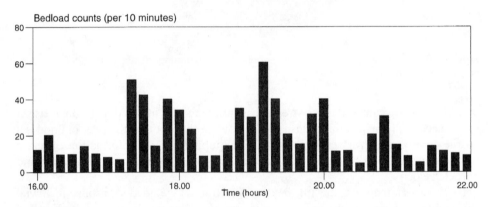

Figure 2.10 A typical active magnetic tracing detection system with an example of output (after Spieker and Ergenzinger, 1990)

passage of magnets over a detector still exist. If issues such as step length, rest period and particle velocity as well as improvements to the recovery rate statistic are to be addressed then real-time acquisition of grain transport data in two dimensions will be required. A fundamental problem exists, however, with the adaptation of the above-mentioned technologies to coarse sediment tracing. At the root of this is the initial question, asked by most (not all) biologists who are interested in the patterns of migratory animals, this being, 'how many individuals passed a particular point?' The question of 'exactly when and where were the individuals?' has not been widely asked and therefore existing animal-tracking systems are not readily adaptable to the real-time tracking of coarse sediments. However, some biologists (e.g. Cole *et al.*, 1998) have used technologies which may be useful for gravel tracking, e.g. the MAP-500 Acoustic Systems (developed by Lotek Marine Inc.) in the coastal environment tracking juvenile Atlantic cod (*Gadus morhua*); close to 1 m resolution was achieved in this environment for 2-D and 3-D positioning. Although promising, it is thought that such a system would have some difficulty functioning correctly in a shallow gravel bed river. Existing radio- and acoustic-tracking technologies are discussed below and the performance of 'active' technology is outlined in Table 2.3. Fundamentally the problems with existing technologies relate to the size of the electronics, the power requirements (more power = larger battery, more data = more power), the absorption of VHF radio waves by salt water, and the current inability to track large numbers of particles at spatial resolutions that are meaningful (i.e. less than a step length). The absorption problem effectively limits radio-tracking technologies in littoral environments to low-frequency radio systems. Tests by the authors, using an electromagnetic system that emits low-frequency radio waves, have demonstrated the possibility of detecting signals in the breaker zone.

The acoustic system of tracking a particle or animal in open water is readily used by biologists interested in the movements of fish or oceanographers plotting the movement of ocean currents. The system involves a transmitter (or pinger) of acoustic energy and three or more detectors or hydrophones. The travel time of the ultrasonic pulse along the direct path to each of the hydrophones is calculated and the position can be determined using the method of triangulation. Caution is necessary, however, if reflection or refraction of the direct path pulse is detected and used in the calculation, as false positioning will result. For correct positioning, a clear 'line of sight' is required between the pinger and the three hydrophones. Further to this, the source of the pulse should be within a near-equilateral hydrophone array in order for all three hydrophones to detect the acoustic pulse. This constraint is very limiting in a narrow river, effectively restricting the detection reach to no more than 8 m in many cases (see Figure 2.11) whilst acoustic noise and multipath problems compound this still further in the littoral environment.

Figure 2.11 illustrates how the conditions mentioned above would be difficult to meet in a typical shallow river with a gravel bed, although this would be less of a problem in deeper wider channels. In Figure 2.11 a typical hydrophone array (H1, H2, H3) is considered with a tracer in the furthest downstream position before leaving the detection area. For a given river width of 7 m (Figure 2.11(b)), it can be seen (without making any trigonometric calculation) that the maximum usable hydrophone (H1) to pinger displacement is about 8 m. This illustration clearly shows that the reach covered by the hydrophone array is relatively small in comparsion to a typical study area, which might be 100 m long. Here (Figure 2.11(a)), where a water depth of 2 m is considered, the signal's direct path is marked as a dotted line and shows that stones with diameters of approximately 200 mm within 0.5 m of the tracer and along the 'line of sight' would interfere with the ultrasonic pulse. This would cause obscuration or refraction of the sound wave resulting in incorrect positioning. Similarly it can also be seen that pebbles of around 400 mm diameter and within 1 m of the tracer grain would have the same effect. The likelihood of signal obscuration is considered to be a significant problem. The situation described in Figure 2.11 is in one sense 'a worst case scenario' in that a displacement of 8 m is likely to be about the furthest the tracer grain would be from any one hydrophone. Therefore, the angle of the direct path to the horizontal (α in Figure 2.11(a)) is at its most acute and is most likely to intersect with stones on the bed. However, in Figure 2.11(b) the tracer grain is characterised as sitting on the surface of the gravel bed, whereas more naturally it would be part of a

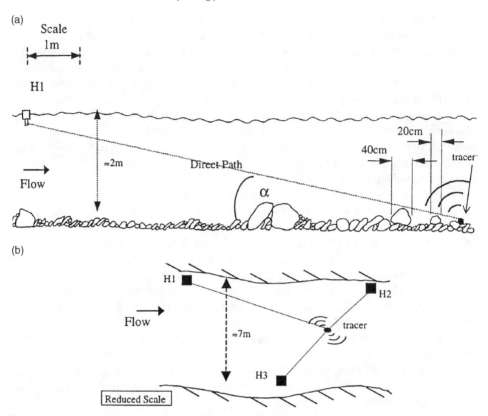

Figure 2.11 (a) A typical acoustic tracking set-up in a gravel-bed river scenario. (b) A reduced-scale plan view showing hydrophone placements and the tracer about to leave the detection array

mobile bed and would be partially or completely buried for a proportion of the time. Therefore, accurate positioning of the tracer would only be possible for a fraction of the total time when the tracer was proud of the gravel bed.

Radio-tracking systems developed for animal studies detect the presence or passage of the transmitter but do not exactly position it in two or three dimensions. These systems have been developed principally to monitor animal migration patterns and 2D and 3D positioning were not a priority. However, positioning with radio systems can be achieved by adjusting the gain at the antenna until the signal's direction can be identified. Then, by manually narrowing the search area, positioning can be achieved. This method, while useful, is far from the ideal of real-time automatic tracking, although some success has been achieved using this type of system but with arrays of fixed antennae (Ergenzinger *et al.*, 1989). Typical detection resolutions are generally poorer than $\pm 2.0\,\mathrm{m}$, limiting the correlation between particle movements to section-averaged values of hydrodynamic forces. Figure 2.12 illustrates a detector array such as deployed by Ergenzinger *et al.* in Germany. The fixed antennae allows detection of tracers as they pass, but not their specific position.

Pebble-Transmitter System

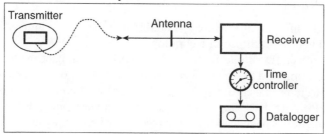

Antenna System

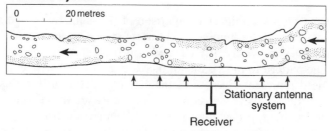

Figure 2.12 Typical radio detection array for tracking sediment tracers (after Ergenzinger *et al.*, 1989). Reproduced by permission of E. Schweizevbart'sche Verlagsbuchhardlung.

Radio direction finding (RDF) is not a new subject and much has been written about it (American Radio Relay League, 1994). In RDF work a compass bearing is aligned with the antenna that is honed in on the radio signal's null position, which is an indication of source direction (American Radio Relay League, 1994). A null can be within half a degree or less of the signal's directivity. Generally, sharper nulls are exhibited with phased arrays (American Radio Relay League, 1994), therefore a 2-D 'real-time' positioning system using radio transmitters and phased aerial arrays will be considered and developed further provided initial investigations are encouraging. The question of refraction at the water's surface is a key consideration and needs to be tested in the field.

DISCUSSION

It is useful to consider finally 'what makes a good coarse sediment tracer'. In reality this will be dependent on the aims of the experiment and the spatial and temporal scale of interest. Nevertheless, there are at least seven properties that should be considered prior to embarking on a a tracer-based research programme:

- *Representativeness*: wide range of sizes, shapes, densities and deployment positions are possible and should be characterised for each site.

- *Recoverability*: the highest rates possible should be sought, including detection when buried.
- *Longevity*: the lifespan of the tracer should match the chosen aims of the research, and will influence the choice of technology.
- *Durability*: tracer ID should be resistant to abrasion, impact, water penetration, algal colonisation and thermal effects, particularly if exposed on bars/beaches.
- *Identity*: tracers need clear unequivocal identities, with colour and codes typically employed. Remote detection is now feasible with electronically tagged particles. Remote detection is beneficial for rapid recovery when deeply buried or recovering from 'deep' water.
- *Reproduction*: tracers should be simple to construct and replace.
- *Economy*: tracers should be cost-effective relative to the information they provide.

Figure 2.13 provides a summary comparison of the main coarse sediment tracing technologies in terms of the numbers of tracers typically deployed, the grain-size range for which they may currently be used and the approximate cost of a typical deployment. The pattern to date is one of increasing cost with active tracing technology, and reduced representativeness of the actual population in terms of size and numbers. In contrast, passive detection systems, such as those using magnetic or aluminium tracers, are approaching the numbers and grain-size range that painted or exotic lithology can provide, but with substantially improved recovery rates and 3-D post-event detection.

Despite several decades of research using coarse sediment tracers, there is a very real need to improve the theory supporting deployment of tracers. It is no longer sufficient just to deploy a 'large number' of tracers, but rather to be explicit about the assumptions made regarding tracer representativeness, the effects of deployment strategy on the results, and the limitations of the tracing technology used. At present, very little attention has been given to describing the form of the tracer distributions in littoral environments. Similarly, in fluvial environments, although estimates of mean displacement length, and simple models of the downstream and vertical distributions exist, very little is known about the validity of the tracer populations deployed in order to generate the data sets. Studies which have considered such factors seem to indicate that the number of tracers required to represent reliably the distribution of movement amounts to $n > 1000$ (Hassan *et al.*, 1991), although much fewer may be required to provide reliable estimates of the mean displacement (Ferguson, 1992). The test used by Hassan *et al.* (1991) was based on the assumption that modelled distributions (compound Poisson and gamma) described the nature of the true distribution of downstream particle frequency, but the validity of this assumption breaks down as reach-scale storage results in more complex distributions. Similarly, no systematic research has been conducted to investigate the effects of tracer recovery rate and deployment strategy on the distributions of lateral, vertical and longitudinal displacements, the latter being of particular concern in the littoral zone where cross-shore variations in longshore transport may be significant (Lee *et al.*, see Chapter 22). The method of tracer deployment and the location of tracers within a river/beach profoundly alter the results of a tracing experiment. Typical deployments in coastal environments include emplacing bulk patches of tracers at three cross-shore locations, whilst fluvial campaigns tend to string tracers across the full-width surface of the channel. The specific nature of the deployment methodology is again dependent on the aims of the experiment, but to date little consideration has been given to its effects on the end results. More research is clearly needed to investigate the assumptions made in earlier tracing studies and to build robust tracing methodologies for the future.

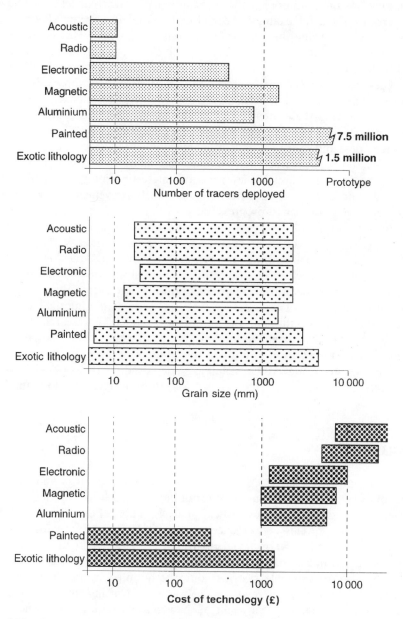

Figure 2.13 Comparison of coarse sediment tracing technologies in terms of the number, grain-size representation and approximate start-up costs

Whilst much attention has been given to exploring the relationship between particle size and displacement, relatively little has been given to the effect of particle shape or bed structure. Those studies that have explored shape effects tend to focus on extremes of shape, but little information exists on the relative importance of shape versus size versus bed structure. There is also an increasing awareness of the role that morphology has on conditioning the distribution of tracer particles, although the signals are confused; for example, Sear (1996) identified a strong effect on tracer dispersal resulting from the pool–riffle topography, whilst Haschenburger and Church (1998) did not, although earlier experiments noted the important controls exerted by bars (Hassan *et al.*, 1991). Much more research is required to investigate the feedback that exists between the creation of morphology, its subsequent influence on hydrodynamics and sediment transport. Active tracing technology offers a means of directly observing these interactions whilst passive tracer deployments provide valuable information on the role of storage in sediment routing.

The theoretical and statistical improvements alluded to above extend to the active technologies for particle tracking. To date, populations of tracked stones have been very small, and although they generate interesting information, the validity of the models developed are statistically uncertain. Improvements in tracking technology are needed in order to progress the work undertaken in the 1980s and early 1990s. Phased arrays and the use of digital signal processing offer the possibility of improved tracking resolution and larger numbers of particles, although existing systems would still not operate in the littoral environment. Tracking technologies that utilise low-frequency radio waves offer the best solution to tracking coarse sediment transport in the littoral environment during tidal events although problems arise from the presence of boundary waves that arise at the interface of different conductive mediums. To this end, the authors are involved in research to develop a system that inverts the current technology by making the pebble the receiver and logging signals emitted from a grid of transmitters to give the position in 2-D and possibly 3-D. Whatever the technology deployed, particle tracking will, for the foreseeable future, be an expensive option, utilising relatively low numbers of particles.

CONCLUSIONS

Coarse sediment tracing has been used as a tool in geomorphology and engineering science for over a century, although it was not until the latter half of the 20th century that the technology received any critical analysis. This review of coarse sediment tracing technology reveals convergence on standard tracing technologies that are individually suited to littoral or fluvial environments. Technology transfer is possible and clearly the advantages of remote access of particle burial depth and identification afforded by electronic pebble systems could be adopted by fluvial research pro-grammes. Particle tracking is relatively well advanced in fluvial environments but unknown in littoral, largely as a result of the problems arising from conduction in seawater, prohibiting signal transmission. This being said, particle tracking in fluvial environments to date is expensive, unrepresentative of most of the grain-size population, and utilises limited numbers of particles. Technological advances are possible however, and could fruitfully be exploited to improve numbers and size range.

In addition to improvements in tracing technology, there is a clear need for research into tracer theory in order to build more statistically rigorous experiments that utilise multi-technology tracing to provide high populations, high recovery rates and intra-event data on the movement of individual grains on coasts and within rivers. Improvements are necessary in terms of defining the minimum numbers of particles needed to provide a given level of accuracy on the basis of the research questions being addressed. Similarly, there has to date been no published work on the effects of deployment methodology on the subsequent movement of tracer particles in fluvial environments. In littoral studies, where depth of disturbance is relatively high, tracer deployments are often made in patches extending down into the beach. In contrast, fluvial experiments largely take the form of surface emplacement, often in lines across a cross-section. M. W. E. Lee (pers. comm.) reports that in littoral environments, such deployments may lead to considerable overestimates of the transport distance and rate. As part of the process of improving tracer experiments and the inter-comparison between them, authors should be encouraged to publish standard information on the numbers of tracers used, the representation of the indigenous grain-size range, the shape of the tracers relative to indigenous population, and the method of tracer deployment.

ACKNOWLEDGEMENTS

This research is made possible by funding under the Engineering and Physical Sciences Research Council Grant GR/L94987 for which the authors are very grateful. In addition, the authors would like to thank the following people for their contributions towards the production of this work: Dr Mark Workman, Dr Malcolm Bray, Jim Smith, Peter Boyce, Alison Houghton, John Cross, Bob Smith, Andy Vowels and Linda Hall.

Table 2.4　Littoral coarse sediment tracing experiments

Study	Tracing technique
Lee *et al.* (see Chapter 22) 　Shoreham Beach, West Sussex, UK	Artificial (electronic and aluminium)
Van Wellen *et al.* (1999) 　Shoreham Beach, West Sussex, UK	Artificial (electronic and aluminium) and natural (foreign painted)
Cooper *et al.* (1996) 　Elmer Beach, West Sussex, UK	Artificial (coarse and fine aluminium) and natural (painted indigenous)
Bray *et al.* (1996) 　Shoreham Beach, West Sussex, UK	Artificial (electronic and aluminium) tracers
Nordström and Jackson (1993) 　Delaware Bay, NJ	Natural (indigenous material) coated with fluorescent paint
Nicholls and Wright (1991) 　Hengistbury Long Beach, Dorset, UK 　Hurst Castle Spit, Hants, UK	Artificial (aluminium tracers)

continues overleaf

Table 2.4 (*continued*)

Study	Tracing technique
Williams (1987) Gileston Beach, South Wales, UK	Natural (indigenous material coated with marine paint)
Caldwell (1983) Gileston Beach, South Wales, UK	Natural (indigenous material coated with marine paint)
Caldwell (1981) Gileston, South Wales, UK	Natural (indigenous material coated with marine paint)

Matthews (1980)
  Wellington Harbour (Eastern),
  New Zealand
	Artificial foreign (fragments of brick roofing tiles)

Hattori and Suzuki (1978)
  Suruga Bay, Shizuoka Prefecture,
  Central Japan
	Natural ('foreign' material – Dactite blocks)

Wright *et al.* (1978a) Hengistbury Long Beach, Dorset, UK	Artificial (aluminium tracers)
Wright *et al.* (1978a) Hengistbury Long Beach, Dorset, UK	Artificial (aluminium tracers)

Gleason *et al.* (1975)
  Describes experiments at Slapton
  already referred to by Carr (1974)	

Carr (1974) Slapton Beach, Devon, UK	Natural (painted 'foreign' pebbles)

Novak (1972)
  Broad Cove Beach, Appledore Island,
  Maine, USA
	Natural (indigenous material painted)

Carr (1971) Chesil Beach, Dorset, UK	Natural (foreign, quartz granulites (QG), quartzite–jasper conglomerates (QJC) and basalt (B)
Yasso (1966) Sandy Hook, New Jersey, USA	Natural (flint treated with fluorescent coating)
Jolliffe (1964) Deal and Rye, Kent, UK	Artificial (concrete, inpregnated with fluorescent plastic pellets) and natural (indigenous pebbles painted with marine paint)

Jolliffe (1961)
  Describes the same experiments as
  Russell (1960)	

Russell (1960) Deal, Rye and Dungeness, Kent, UK	Artificial (crushed concrete containing one of anthracene, rhodamine or fluorescent plastic)
Kidson *et al.* (1958) Orfordness, Suffolk, UK	Indigenous (radioactivity tagged – barium 140)
Richardson (1902) Chesil Beach, Dorset, UK	Artificial (brickbats)

Table 2.5 Fluvial coarse sediment tracing experiments

Study	Tracing technique
Haschenburger and Church (1998) Carnation Creek	Magnetic inserts in natural particles
Ferguson and Wathen (1998) Allt Dubhaig	Magnetic and painted natural particles

Study	Tracing technique
Wilcock (1997) Trinity River	Painted natural particles
Wathen *et al.* (1997) Allt Dubhaig	Magnetic inserts in natural particles
Gintz *et al.* (1996) Lainbach	Magnetic inserts in natural and artificial particles
Busskamp and Hasholt (1996) Arctic Greenland	Radio (150 MHz) inserts in natural particles
Sear (1996) North Tyne, Dry Creek, River Severn, Allt Dubhaig	Painted and magnetically enhanced natural particles
Schmidt and Gintz (1995) Lainbach Stream	Concrete and plastic artificial particles with magnetic inserts
Michalik and Bartnik (1994) Wiloka River, Raba River, Danejec River	Natural particles with radioactive inserts
Hassan and Church (1994) Carnation Creek, Harris Creek, Nahal Hebron, Nahal Og	Magnetic inserts in natural particles
Hassan (1993) Nahal Hebron, Nahal Og	Magnetic inserts in natural particles
Schmidt and Ergenzinger (1992) Lainbach Stream	Iron/magnetic and radio (150 MHz) inserts in natural particles
Larrone and Duncan (1992) Ashburton River	Magnetic inserts in natural particles
Reid *et al.* (1992) Turkey Brook	Painted natural and magnetic particles
Church and Hassan (1992) Compilation of several tracer experiments	Painted and magnetic particles
Hassan and Church (1992) Compilation of several tracer experiments	Painted and magnetic particles
Hassan *et al.* (1992) Compilation of several tracer experiments	Painted and magnetic particles
Hassan *et al.* (1991) Nahal Hebron, Nahal Og	Magnetic particles
Hassan (1990) Nahal Hebron	Magnetic particles
Emmett *et al.* (1990) Toklat River, Phelan Creek	Radio (150 MHz)
Tacconi *et al.* (1990) Virgino Creek	Painted natural particles
Ergenzinger *et al.* (1989)	Radio (150 MHz)
Ashworth and Ferguson (1989) Allt Dubhaig, Glen Feshie, Lyngsdalselva	Painted natural particles
Billi (1988) Farma River	Painted natural particles
Carling (1987) Gt Eggleshope Beck, Carl Beck	Painted natural particles
Petit (1987) La Rulles Stream	Painted natural particles
Schick *et al.* (1987) Nahal Hebron	Magnetic particles

continues overleaf

Table 2.5 (*continued*)

Study	Tracing technique
Custer *et al.* (1987) Squaw Creek	Naturally magnetic particles passing over a detector
Kondolf and Matthews (1986) Carmel River	Exotic lithology (dolerite rip-rap)
Brayshaw (1985) Turkey Brook	Magnetic inserts
Hassan *et al.* (1984) Nahal Hebron	Magnetic inserts
Reid *et al.* (1984) Turkey Brook	Magnetic inserts
Carling (1983) Gt Eggleshope Beck, Carl Beck	Painted natural particles
Arkell *et al.* (1983) Nant Tannlwyth (Forest Drain)	Magnetically enhanced natural material
Ergenzinger and Custer (1983) Squaw Creek	Naturally magnetic particles passing over static detector
Ergenzinger and Conrady (1982) Squaw Creek	Naturally magnetic particles passing over static detector
Oldfield *et al.* (1981)	Magnetically enhanced natural material
Cavazza (1981) River Frigido	Painted natural particles
Tazioli (1981) (unspecified – Tuscany)	Natural particles with radioactive inserts
Thorne (1978) River Severn	Painted natural particles
Mosley (1978) Tameki River	Exotic lithology (limestone)
Butler (1977)	Natural particles tagged with aluminium strip
Larrone and Carson (1976) Seales Brook	Painted natural particles
Hey (1975) River Severn	Painted natural particles
Slaymaker (1972) Nant y Grader 9A, Nant y Grader 9B	Painted natural particles
Wilcock (1971) River Hodder	Painted natural particles
Keller (1970) Dry Creek	Painted natural particles
Helley (1969) Blue River	Painted natural particles
Leopold *et al.* (1966) Arroyo de los Frijoles	Painted natural particles
Takayama (1965) Fukogawa, Hayakawa, Okawa	Painted natural particles
Nir (1964) Nahal Zin	Iron oxide coated artificial cobbles

REFERENCES

American Radio Relay League 1994. *The Antenna Book* (17th edition). 15th Publication of the ARRL, Newington, Connecticut.

Arkell, P. L., Leeks, G. J. L., Newson, M. D. and Oldfield, F. 1983. Trapping & tracing: some recent observations on supply and transport of coarse sediment from Wales. In Collinson, J. D. and Lewin, J. (eds) *Modern & Ancient Fluvial Systems*. International Association of Sedimentologists, Special Publication 6, 107–119.

Ashworth, P. J. and Ferguson, R. I. 1989. Size-selective entrainment of bed-load in gravel-bed streams. *Water Resources Research*, **25**(4), 627–634.

Billi, P. 1988. A note on cluster bedform behaviour in a gravel-bed river. *Catena*, **15**, 473–481.

Bray, M. J., Workman, M., Smith, J. and Pope, D. 1996. Field measurements of shingle transport using electronic tracers. In *Proceedings of the 31st MAFF Conference of River and Coastal Engineers*. Ministry of Agriculture, Fisheries and Food, London, 10.4.1–10.4.13.

Brayshaw, A. C. 1985. Bed microtopography and entrainment thresholds in gravel-bed streams. *Geological Society of America Bulletin*, **96**, 218–223.

Busskamp, R. and Hasholt, B. 1996. Coarse bedload transport in a glacial valley, Sermilik, south-east Greenland. *Zeitschrift für Geomorphologie*, **40**, 349–358.

Butler, P. R. 1977. Movement of cobbles in a gravel-bed stream. *Geological Society of America Bulletin*, **88**, 1072–1234.

Caldwell, N. E. 1981. Relationship between tracers and background beach material. *Journal of Sedimentary Petrology*, **51**, 1163–1168.

Caldwell, N. E. 1983. Using tracers to assess size and shape sorting of pebbles. *Proceedings of the Geologists' Association*, **94**, 86–90.

Carling, P. A. 1983. The threshold of coarse sediment transport in broad and narrow natural streams. *Earth Surface Processes & Landforms*, **8**, 1–18.

Carling, P. A. 1987. Bed stability in gravel streams with reference to stream regulation and ecology. In Richards, K. S. (ed.) *River Channels: Environment & Process*. Institute of British Geographers Special Publication 17, London, 321–347.

Carling, P. A., Kelsey, A. and Glaister, M. S. 1992. Effect of bed roughness, particle shape and orientation on initial motion criteria. In Billi, P., Hey, R. D., Thorne, C. R. and Tacconi, P. (eds) *Dynamics of Gravel-Bed Rivers*, John Wiley, Chichester, 23–37.

Carr, A. P. 1971. Experiments on longshore transport and sorting of pebbles: Chesil Beach, England. *Journal of Sedimentary Petrology*, **41**(4), 1084–1104.

Carr, A. P. 1974. Differential movement of coarse sediment particles. *Proceedings of the 14th Conference on Coastal Engineering*, Copenhagen, ASCE, 851–870.

Cavazza, S. 1981. Experimental investigations on the initiation of bedload transport in gravel rivers. In *Erosion & Sediment Transport Measurement*, IAHS Publication 133, IAHS Press, Wallingford. 53–61.

Chadwick, A. J. 1989. Field measurements and numerical model verification of coastal shingle transport. In Palmer, M. H. (ed.) *Advances in Water Modelling and Measurement 2*. BHRA Conference, Harrogate, UK, 1988, 231–245.

Church, M. and Hassan, M. A. 1992. Size and distance of travel of unconstrained clasts on a streambed. *Water Resources Research*, **28**, 299–303.

Coates, T., Bray, M., Stapleton, K., Van Wellen, E. and Lee, M. 1999. Advances in shingle beach management. In *Proceedings of the 34th MAFF Conference of River and Coastal Engineers*. Ministry of Agriculture, Fisheries and Food Publication PB 4714, 9.4.1–9.4.13.

Cole, D., Scruton, D. A., Niezgode, G. H., McKinely, R. S., Rowsell, D. F., Lindström, R. T., Ollerhead, L. M. N. and Whitt, C. J. 1998. A coded acoustic telemetry system for high precision monitoring of fish location and movement: application to the study of nearshore nursery habitat of juvenile Atlantic cod (*Gadus morhua*). *MTS Journal*, **32**(1), 54–62.

Cooper, N. J. 1996. *Evaluation of the impacts of shoreline management at contrasting sites in Southern England*. Unpublished PhD Thesis, Department of Geography, University of Portsmouth.

Cooper, N., Bray, M. and King, D. 1996. Field measurements of fine shingle transport. Presented at *Tidal '96, Symposium for Practising Engineers*, University of Brighton, 12–13 November 1996.

Custer, S. G., Bugosh, N., Ergenzinger, P. J. and Anderson, B. C. 1987. Electromagnetic detection of pebble transport in streams: a method for measurement of sediment transport waves. In: Etheridge/Flores (Eds) *Recent developments in Fluvial Sedimentology*. Society of Paleontologists and Mineralogists, pp. 21–26.

Einstein, H. A. 1937. Bedload transport as a probability problem. PhD Thesis, published as Appendix C in Shen, H. W. (ed.) *Sedimentation*, 1972, Colorado State University, Fort Collins, Colorado.

Emmett, W. W. 1980. A field calibration of the sediment trapping characteristics of the Helley-Smith bedload sampler. *USGS Professional Paper 1139*.

Emmett, W. W., Burrows, R. L. and Chacho, E. F. 1990. Coarse particle transport in a gravel-bed river. Paper presented at Third International Workshop on Gravel-bed Rivers: Dynamics of Gravel-bed Rivers, 25–29 September 1990, Florence, Italy.

Ergenzinger, P. J. and Conrady, J. 1982. A new tracer technique for measuring bedload in natural channels. *Catena*, **9**, 77–80.

Ergenzinger, P. J. and Custer, S. G. 1983. Determination of bedload transport using naturally magnetic tracers: Squaw Creek, Gallatin County, Montana. *Water Resources Research*, **19**, 187–193.

Ergenzinger, P. J., Schmidt, K.-H. and Busskamp, R. 1989. The pebble transmitter system (PETS) – 1st results of a technique for studying coarse material erosion, transport and deposition. *Zeitschrift für Geomorphologie*, **33**(4), 503–508.

Ferguson, R. I. 1992. Discussion of Hassan & Church. In Billi, P., Hey, R. D., Thorne, C. R. and Tacconi, P. (eds) *Dynamics of Gravel-Bed Rivers*, John Wiley, Chichester, 174.

Ferguson, R. I. and Ashworth, P. J. 1992. Spatial patterns of bedload transport and channel change in braided and near-braided rivers. In Billi, P., Hey, R. D., Thorne, C. R. and Tacconi, P. (eds) *Dynamics of Gravel-Bed Rivers*, John Wiley, Chichester, 477–492.

Ferguson, R. I. and Wathen, S. J. 1998. Tracer-pebble movement along a concave river profile: virtual velocity in relation to grain size and shear stress. *Water Resources Research*, **34**(8), 2031–2038.

Gintz, D., Hassan, M. A. and Schmidt, K.-H. 1996. Frequency and magnitude of bedload transport in a mountain river. *Earth Surface Processes and Landforms*, **21**(5), 433–445.

Gleason, R., Blackley, M. W. L. and Carr, A. P. 1975. Beach stability and particle size distribution, Start Bay. *Journal of the Geological Society of London*, **131**, 83–101.

Haschenburger, J. K. and Church, M. 1998. Bed material transport estimated from the virtual velocity of sediment. *Earth Surface Processes & Landforms*, **23**, 791–808.

Hassan, M. A. 1990. Scour, fill and burial depth of coarse material in gravel-bed streams. *Earth Surface Processes & Landforms*, **15**, 341–356.

Hassan, M. A. 1993. Bed material and bedload movement in two ephemeral streams. In Marzo, M. and Puigdefabregas, C. (eds) *Alluvial Sedimentation*. IAS Special Publication 17, Blackwell Scientific, Boston, MA, 37–49.

Hassan, M. A. and Church, M. 1992. The movement of individual grains on the streambed. In Billi, P., Hey, R. D., Thorne, C. R. and Tacconi, P. (eds) *Dynamics of Gravel-Bed Rivers*, John Wiley, Chichester, 159–175.

Hassan, M. A. and Church, M. 1994. Vertical mixing of coarse particles in gravel-bed rivers – a kinematic model. *Water Resources Research*, **30**(4), 1173–1185.

Hassan, M. A., Schick, A. P. and Laronne, J. B. 1984. The recovery of flood dispersed coarse sediment particles, a three-dimensional magnetic tracing method. *Catena Suppl.*, **5**, 153–162.

Hassan, M. A., Church, M. and Schick, A. P. 1991. Distance of movement of coarse particles in gravel bed streams. *Water Resources Research*, **27**(4), 503–511.

Hassan, M. A., Church, M. and Ashworth, P. J. 1992. Virtual rate and mean distance of travel of individual clasts in gravel-bed channels. *Earth Surface Processes & Landforms*, **17**, 617–627.

Hattori, M. and Suzuki, T. 1978. Field experiment on beach gravel transport. *Proceedings of the 16th International Conference on Coastal Engineering*, vol. 2, ASCE, Hamburg, 1688–1704.

Helley, E. J. 1969. Field measurement of the initiation of large bed particle motion in Blue Creek/Klamath, California. *US Geological Survey Professional Paper 562-G.*

Hey, R. D. 1975. Response of the upper Severn to river regulation. Unpublished report to Severn-Trent Water Authority, 45 pp.

Hjulström, F. 1935. Studies of the morphological activities of rivers as illustrated by the River Fyris. *Bulletin of the Geological Institute,* University of Uppsala, **25**, 221–257.

Jolliffe, I. P. 1961. The use of tracers to study beach movements: and the measurement of littoral drift by a fluorescent technique. *Revue de Geomorphologie Dynamique,* **2**, 81–98, University of Strasbourg.

Jolliffe, I. P. 1964. An experiment designed to compare the relative rates of movement of different sizes of beach pebbles. *Proceedings of the Geologists' Association,* **75**, 67–86.

Keller, E. A. 1970. Bed material movement experiments, Dry Creek, California. *Journal of Sedimentary Petrology,* **40**, 1339–1344.

Kidson, C., Carr, A. P. and Smith, D. B. 1958. Further experiments using radioactive methods to detect movement of shingle over the sea bed and alongshore. *Geographical Journal,* **124**, 210–218.

Kondolf, M. G. and Mathews, G. W. V. 1986. Transport of tracer gravels in a coastal California river. *Journal of Hydrology,* **85**, 265–280.

Kraus, N. C., Isobe, M., Igarashi, H., Sasaki, T. O. and Horikawa, K. 1982. Field experiments on longshore transport in the surf zone. In *Proceedings of the 18th Coastal Engineering Conference,* New York, ASCE, 969–988.

Larrone, J. B. and Carson, M. A. 1976. Interrelationship between bed morphology and bed material transport for a small gravelbed channel. *Sedimentology,* **23**, 67–85.

Larrone, J. B. and Duncan, M. J. 1992. Bedload transport paths and gravel bar formation. In Billi, P., Hey, R. D., Thorne, C. R. and Tacconi, P. (eds) *Dynamics Of Gravel-Bed Rivers,* John Wiley, Chichester, 177–200.

Leopold, L. B. 1992. Sediment size that determines channel morphology. In Billi, P., Hey, R. D., Thorne, C. R. and Tacconi, P. (eds) *Dynamics of Gravel-Bed Rivers.* John Wiley, Chichester, 297–307.

Leopold, L. B., Emmett, W. W. and Myrick, R. M. 1966. Chanel & hillslope processes in semi-arid areas, New Mexico. *US Geological Survey Professional Paper 352–G,* 192–253.

Madej, M. A. and Ozaki, V. 1996. Channel response to sediment wave propagation and movement, Redwood Creek, California, USA. *Earth Surface Processes & Landforms,* **21**, 911–927.

Matthews, E. R. 1980. Observations of beach gravel transport, Wellington Harbour entrance, New Zealand. *New Zealand Journal of Geology and Geophysics,* **23**(2), 209–222.

Mickalik, A. and Bartnik, W. 1994. An attempt at determination of incipient bedload motion in mountain streams. In Ergenzinger, P. and Schmidt, K.-H. (eds) *Dynamics and Geomorphology of Mountain Rivers.* Lecture Notes in Earth Sciences No. 52, 291–299.

Mosley, M. P. 1978. Bed material transport in the Tamaki River near Dannevirke, North Island, New Zealand. *New Zealand Journal of Science,* **21**, 619–626.

Nicholls, R. J. and Wright, P. 1991. Longshore transport of pebbles: experimental estimates of K. In Kraus, N. C. *et al.* (eds) *Coastal Sediments '91.* American Society of Civil Engineers, New York, 920–933.

Nir, D. 1964. Les processus erosifs dans la Nahal Zine (Neguev setentrional) pendant les saisons pluvieuses. *Annales de Geographie,* **73**, 8–20.

Nordstrom, K. F. and Jackson, N. L. 1993. Distribution of surface pebbles with changes in wave energy on a sandy estuarine beach. *Journal of Sedimentary Petrology,* **63**(6), 1152–1159.

Novak, I. D. 1972. Swash zone competency of gravel size sediment. *Marine Geology,* **13**, 335–345.

Oldfield, F., Thompson, R. and Dickson, D. P. E. 1981. Artificial magnetic enhancement of stream bedload: a hydrological application of superparamagnetism. *Physics of the Earth & Planetary Interiors,* **26**, 107–124.

Petit, F. 1987. The relationship between shear stress and the shaping of the bed of a pebble-loaded river, La Rulles, Ardennes. *Catena,* **14**, 453–468.

Reid, I., Brayshaw, A. C. and Frostick, L. E. 1984. An electromagnetic device for automatic detection of bedload motion and its field applications. *Sedimentology*, **31**(2), 269–276.

Reid, I., Frostick, L. E. and Layman, J. T. 1985. The incidence and nature of bedload transport during flood flows in coarse grained alluvial channels. *Earth Surface Processes & Landforms*, **10**, 33–44.

Reid, I., Frostick, L. E. and Brayshaw, A. C. 1992. Microform roughness elements and the selective entrainment and entrapment of particles in gravel-bed rivers. In Billi, P., Hey, R. D., Thorne, C. R. and Tacconi, P. (eds) *Dynamics of Gravel-Bed Rivers*, John Wiley, Chichester, 188–205.

Richardson, N. M. 1902. An experiment on the movements of a load of brickbats deposited on the Chesil Beach. *Proceedings Dorset Natural History and Antiquarian Field Club*, **23**, 123–133.

Russell, R. C. H. 1960. Use of fluorescent tracers for the measurement of littoral drift. *Proceedings of the 7th Conference on Coastal Engineering*, University of California, 418–444.

Sear, D. A. 1996. Sediment transport processes in pool–riffle sequences. *Earth Surface Processes & Landforms*, **21**, 241–262.

Sear, D. A., Newson, M. D. and Brookes, A. 1995. Sediment related rivers maintenance: the role of fluvial geomorphology. *Earth Surface Processes & Landforms*, **20**, 629–649.

Schick, A. P., Hassan, M. A. and Lekach, J. 1987. A vertical exchange model for coarse bedload movement: numerical considerations. *Catena, Suppl.*, **10**, 7–83.

Schmidt, K.-H. and Ergenzinger, P. 1992. Bedload entrainment, travel lengths, step lengths, rest periods – studied with passive (iron, magnetic) and active (radio) tracer techniques. *Earth Surface Processes and Landforms*, **17**(2), 147–165.

Schmidt, K.-H. and Gintz, D. 1995. Results of bedload tracer experiments in a mountain river. In Hickin, E. J. (ed.) *River Geomorphology*, John Wiley, Chichester, 145–158.

Slaymaker, H. O. 1972. Patterns of present sub-aerial erosion and landforms in mid-Wales. *Transactions of the Institute of British Geographers*, **55**, 47–68.

Spieker, R. and Ergenzinger, P. 1990. New developments in measuring bedload by the magnetic tracer technique, erosion, transport and deposition processes. IAHS Publication **189**, IAHS Press, Wallingford. 169–178.

Tacconi, P., Rinaldi, M., Moretti, S. and Matteini, M. 1990. Monitoring of particle movement on Virginio gravel bed stream. Paper presented at the *3rd International Workshop on Gravel-bed Rivers: Dynamics of Gravel-bed Rivers*, 25–29 September 1990, Florence, Italy (unpublished).

Takayama, S. 1965. Bedload movement in torrential mountain streams (in Japanese). *Tokyo Geographical Papers*, **9**, 169–188.

Tazioli, G. S. 1981. Nuclear techniques for measuring sediment transport in natural streams – examples from instrumented basins. IAHS Publication **133**, *Erosion and Sediment Transport Measurement*, IAHS Press, Wallingford. 63–81.

Thorne, C. R. 1978. Processes and mechanics of bank erosion. PhD Thesis, University of East Anglia, 450pp.

Van Wellen, E., Chadwick, A. J., Bird, P. A. D., Bray, M., Lee, M. W. E. and Morfett, J. 1997. Coastal sediment transport on shingle beaches. *Proceedings of Coastal Dynamics '97*, Plymouth, ASCE, 38–47.

Van Wellen, E., Chadwick, A. J., Lee, M. W. E., Bailey, B. and Morfett, J. (1999) Evaluation of longshore sediment transport models on coarse grained beaches using field data: a preliminary investigation. *Proceedings of 26th International Conference on Coastal Engineering*, Vol. 3, Copenhagen, ASCE, 2640–2653.

Wathen, S. J., Hoey, T. B. and Werritty, A. 1997. Quantitative determination of the activity of within-reach sediment storage in a small gravel-bed river using transit time and response time. *Geomorphology*, **20**, 113–134.

Wilcock, D. N. 1971. Investigation into the relations between bedload transport and channel shape. *Geological Society of America Bulletin*, **82**, 2159–2176.

Wilcock, P. R. 1993. The critical shear stress of natural sediments. *Journal of Hydraulic Engineering*, **119**(4), 491–505.

Wilcock, P. R. 1997. Entrainment, displacement and transport of tracer gravels. *Earth Surface Processes & Landforms*, **22**, 1125–1138.

Williams, A. T. 1987. A new model for pebble beach tracer dispersal. *Acta Oceanologica Sinica*, **6**(2), 229–234.

Wilson, S. F. 1996. *Shoreham and Lancing Sea Defence Strategy Plan*. Final report to National Rivers Authority, Southern Region, 45pp.

Workman, M. 1997. The field investigation, using tracers, of meso-scale shingle beach behaviour. PhD Thesis, University of Southampton, 344pp.

Workman, M., Smith, J., Boyce, P., Collins, M. B. and Coates, T. T. 1994. *Development of the Electronic Pebble System*. HR Report SR 405, HR Wallingford Ltd, 70pp.

Wright, P., Cross, J. and Webber, N. (1978a) Shingle tracing by a new technique. *Proceedings of the 16th Coastal Engineering Conference*, American Society for Civil Engineers, 1707–1714.

Wright, P., Cross, J. S. and Webber, N. B. 1978b. Aluminium pebbles: a new type of tracer for flint and chert pebble beaches. *Marine Geology*, **27**, M9–M17.

Yasso, W. E. 1966. Formulation and use of fluorescent tracer coatings in sediment transport studies. *Sedimentology*, **6**, 287–301.

3 Natural Magnetic Tracers in Fluvial Geomorphology

J. A. DEARING

Environmental Magnetism Laboratory, Department of Geography, University of Liverpool, UK

INTRODUCTION

Mineral magnetic techniques (Thompson and Oldfield, 1986) measure the type and concentrations of naturally occurring Fe-bearing minerals. Their ability to detect concentrations of minerals and mineral phases with a higher precision than many other techniques within the ubiquitous and highly reactive Fe system brings many advantages to geomorphological studies (Dearing *et al.*, 1985). The past two decades have seen an increase in their use in characterising the mineralogy of environmental materials and in tracing their movement through fluvial and limnic systems – from hillslopes and channels to floodplains and the bottom-sediments in lakes and reservoirs. The objective of this chapter is to review the present scope of magnetic measurements in different geomorphological contexts and the types of environmental questions that may be addressed. The influences of geology, erosion and transport processes, soil type and processes, particle size, and post-depositional authigenesis and diagenesis on magnetic minerals are exemplified from published works. Different case studies are used to show how the 'magnetic method' needs to be defined and modified according to the type of environmental conditions that control the formation, transport and preservation of magnetic minerals.

Minerals and Magnetic Measurements

All matter can be defined magnetically, but for most tracing purposes it is appropriate to think in terms of a small set of naturally occurring and common Fe-bearing minerals, and how each mineral may be detected using magnetic measurements. Magnetic properties that allow mineral detection and definition are affected by both the chemical and physical states of the mineral. This extends the classification of Fe-bearing minerals beyond their geochemical composition to include size and shape-related properties, embodied in the concept of magnetic domains, which may be more diagnostic of environmental conditions than the mineral itself. The parallel development of rock magnetism theory and magnetic measuring equipment over the past

Tracers in Geomorphology. Edited by Ian D. L. Foster. © 2000 John Wiley & Sons Ltd.

three decades has on the one hand led to a massive expansion in the number of non-specialists able to make basic measurements, but has also led to improved diagnostic tests requiring specialist facilities. Full details of magnetic theory, measurements and their interpretation are provided by Thompson and Oldfield (1986), Dunlop and Özdemir (1997), Dearing (1999a), and several chapters in Walden *et al.* (1999). Table 3.1 summarises the important minerals, domains and mineral origins, and some of the key measurements presently used for their identification. Isothermal (non-varying and usually room temperature) and low-temperature magnetisation and remanence curves showing schematic features diagnostic of minerals and domains are shown in Figure 3.1. One important aspect of magnetic measurements, especially basic room-temperature measurements (Figure 3.1(a) and (b)), that may lead to confusion or difficulties in interpretation is that many are compound. A single measurement value may be found for a range of combinations of minerals or domains, analogous to measurements of total gamma activity or conductivity. Increasingly, measurements using thermal properties are now providing a stronger basis for unambiguous identification of specific minerals and domains but it is still not always possible to express concentrations of specific minerals quantitatively. Low-temperature measurements of remanence (Figure 3.1(c) and (d)) and magnetic susceptibility (Figure 3.1(e)) using liquid nitrogen are relatively straightforward and are preferred to high-temperature measurements because there are fewer irreversible mineral transformations. A combination of low temperature and high field magnetisation measurements can now be used to extend identification to paramagnetic (P) and specific canted antiferromagnetic (CA) minerals. These minerals are usually

Table 3.1   Magnetic states, minerals, domains, mineral origins and common measurements

	Primary	Secondary	Bacterial	Combustion	Room temperature	Low temperature
1. FERRIMAGNETIC (F)						
magnetite-maghemite					χ_{LF}, M_s	
superparamagnetic (SP)		x		x^2	χ_{FD}, $\chi_{FD\%}$	curve
stable single domain (SSD)	x	x	x^1		χ_{ARM}	curve
pseudo-single domain (PSD)	x		x^1	x^3	IRM_{IT} (SIRM)	
multidomain (MD)	x			x^3	IRM_{20mT}	transition
greigite (GR)		x	x		$IRM_{40-80mT}$ $SIRM/\chi_{LF}$	curve
2. CANTED ANTIFERROMAGNETIC (CA)						
haematite (H)	x	x		x^3	$IRM_{300-1000mT}$	transition
goethite (G)	x	x			IRM_{1-7T}	curve
3. PARAMAGNETIC (P)						
Fe-bearing minerals e.g. olivine, biotites, lepidocrocite, ferrihydrite, siderite	x	x			$\chi_{800-1000mT}$	curve
4. DIAMAGNETIC (D) e.g. water, quartz, organic matter, calcium carbonate						

[1]Magnetotactic bacteria. [2]Fire. [3]Fossil fuels.
IRM measurements are normally made in demagnetisation fields after magnetisation at 1T (SIRM).

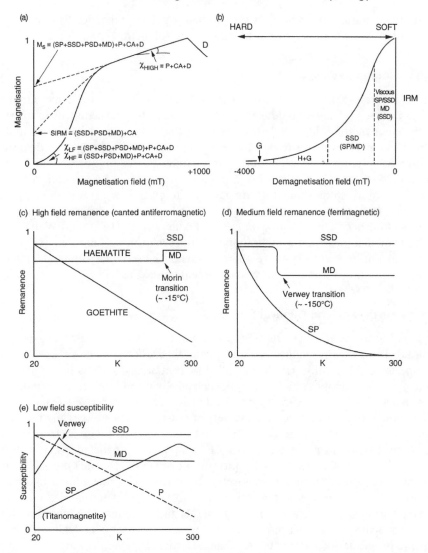

Figure 3.1 Schematic magnetisation and remanence curves for magnetic states, minerals and domains showing key measurement parameters with abbreviations as in Table 3.1.(a) Isothermal magnetisation curve showing low-frequency susceptibility (χ_{LF}), high-frequency susceptibility (χ_{HF}), high field susceptibility ($\chi_{HIGH} = \chi_{800mT}$), saturation isothermal remanent magnetisation (SIRM) and saturation magnetisation (M_s). (b) Isothermal demagnetisation remanence showing the controls on curve shape by magnetic hardness interpreted in terms of different minerals and domains (bracketed = minor control). (c) Low-temperature (20–300 K) high field (300 mT) remanence warming curves showing discrimination of canted antiferromagnetic minerals and domains. (d) Low-temperature (20–300 K) medium field (100 mT) remanence warming curves showing discrimination of ferrimagnetic (magnetite) domains. (e) Low-temperature (20–300 K) susceptibility (χ_{LF}) warming curves showing discrimination of ferrimagnetic, superparamagnetic and paramagnetic components. Compiled and redrawn from Thompson and Oldfield (1986) (Reproduced with kind permission of Kluwer Academic Publishers), Cullity (1972), Dekkers (1989) and France (1997)

quantitatively more important in a sample than the ferrimagnetic (F) minerals and much is often known about their responses to environmental conditions.

Methodologies and Magnetic Variability

In general terms, tracing the movement of sediment by magnetic measurements requires a clear and unambiguous magnetic signature for the sediment in question. The choice of methodology is dependent upon the existing mineral system and the geomorphological questions to be addressed; in practice, the former often determines the latter. Some studies use a signature that is diagnostic of a uniquely defined specific mineral or domain, while others use a signature representing a mineral mixture or assemblage. In all studies minerals are assumed to exist as conservative mixtures of source end-members. Contrasting magnetic mineral systems are exemplified by the early case studies of Thompson *et al.* (1975), Oldfield *et al.* (1979) and Walling *et al.* (1979). They demonstrated a dichotomy of approach, based on either the *primary* mineral content in essentially unweathered clasts or the *secondary* mineral content of soil, which still holds true for many environments today.

Primary Ferrimagnetic Mineral Systems

Values of mass specific χ_{LF} for unweathered rocks vary by five to six orders of magnitude from negative values in diamagnetic (D) chalk to $\sim 10^{-5}\,m^3\,kg^{-1}$ in ultrabasic specimens. Typically, χ_{LF} is controlled by F minerals, such as titanomagnetite, in igneous rocks, and more by P minerals, such as olivine and biotite, in metamorphic and sedimentary rocks (Dearing, 1999a). Tracing sediment derived from different geologies therefore depends upon the variety of potential rock sources and the success of magnetically classifying each one. In a mixture of sediment derived from multiple rock sources, single measurements of the F component, for instance χ_{LF}, will only detect the presence of a specific rock contribution where the contrast in rock magnetism is large, say ultrabasic and metamorphic. Otherwise the whole mineral assemblage of source material requires a magnetic signature, usually in terms of a number of magnetic parameters detecting different minerals and domains. The question of how to unmix a sediment sample into different source components is dealt with below. Magnetic properties of sediments and sediment sources generally vary with particle size (e.g. Thompson and Morton, 1979; Dearing *et al.*, 1981; Björck *et al.*, 1982; Bradshaw and Thompson, 1985; Snowball, 1993), and hence differences in particle size distribution within a sediment sample may be a major control on bulk sample magnetic properties (Figure 3.2). The sorting effects of erosion, transport and sedimentation processes may be expected to produce as much as a fivefold difference between the magnetic concentrations of different textured sediment samples. Where a significant proportion of the sediment comprises primary minerals from weathered igneous sources, peak magnetic concentrations are often associated with the silt or fine sand-sized fraction, equivalent to the actual mineral/crystal dimensions (Thompson and Morton, 1979; Dearing *et al.*, 1981; Thompson and Edwards, 1982; Bradshaw and Thompson, 1985; Colman *et al.*, 1990; Yu *et al.*, 1990; Walden and Addison, 1995). Where magnetically enhanced soil (see below) forms a significant part of a

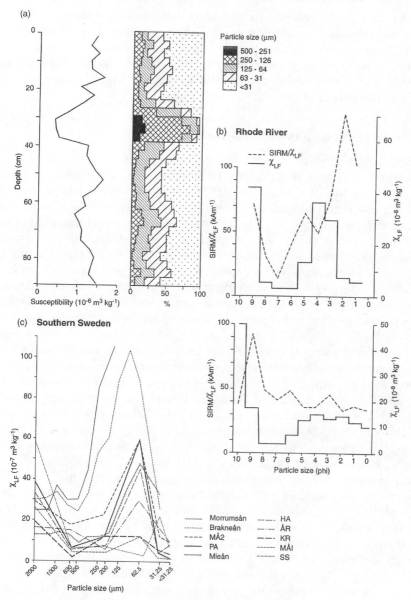

Figure 3.2 Variations in magnetic properties of sediments and sources with particle size. (a) Bulk χ_{LF} and particle size distribution of recent sediments from Loch Lomond showing low χ_{LF} correlating with coarse ($> 125\,\mu m$) sediments (Thompson and Morton, 1979). (b) Magnetic measurements of different particle fractions (10–0 ϕ; 1–1000 μm) from two soil samples in the Rhode River catchment showing bimodal distributions of ferrimagnetic (χ_{LF}) concentrations (fine pedogenic and coarser primary minerals) and wide variability of magnetic mineralogy (SIRM/χ_{LF}) (Thompson and Oldfield, 1986). (c) χ_{LF} values for fractions of stream sediments from five south Swedish rivers showing bimodal distributions with peak ferrimagnetic concentrations transported in the coarse silt/fine sand range (Björck *et al.*, 1982). Reproduced by permission of Scandinavian University Press

sediment sample, the presence of secondary ferrimagnetic minerals (SFMs – see below) may raise the F concentrations of the clay and (through aggregation) the silt fraction. Hence, in many landscapes the variability of F concentrations is controlled as much by particle size as different sediment sources to the extent that magnetic properties may in some situations be useful in providing hydrodynamic information about the transport of different sized clasts.

Secondary Ferrimagnetic Mineral Systems

A unique feature used in magnetic-based tracing is the phenomenon of magnetic enhancement, originally described by Le Borgne (1955) and further examined by Mullins (1977), Maher (1984, 1986, 1998) and Maher and Taylor (1988), where SFMs are produced pedogenically in surface soil. SFMs comprise magnetite and maghemite, formed largely as a result of biogeochemical transformations of weathered Fe, and exist as populations of poorly crystalline grains $< 0.05\,\mu m$ straddling the magnetically important boundary between SP and SSD states (Dearing et al., 1996a, 1997). Measurements of χ_{LF} and χ_{FD} percent of soil samples (Dann, R. J. L., unpublished; Hay, 1998) across England and Wales have helped to identify the environmental controls on the production and accumulation of SFMs (Dearing et al., 1996b) and to provide a regional picture of where enhancement is most likely to lead to successful tracing of surface soil. Soils in which SFMs dominate, as shown by high χ_{FD} percent values (Figure 3.3), are found in soils on rapidly weathering Fe-bearing sedimentary parent materials, especially Carboniferous and Devonian strata found in Devon, Cornwall and south-west Wales, Jurassic limestones (e.g. Cotswold Hills) and Cretaceous chalk (e.g. Salisbury Plain). The χ_{LF} data (not shown) also indicate that surface soils around the major conurbations (Greater London, West Midlands, Merseyside, Greater Manchester and South Yorkshire) may be 'enhanced' through the atmospheric fallout of relatively large magnetite-containing pollution particles or spheres with predominantly MD behaviour (Dearing et al., 1996b). Further magnetic analyses, complemented by scanning electron microscopy, Mössbauer spectroscopy and the national geochemical database for soils (McGrath and Loveland, 1992), suggest that the rate of Fe release to a free-draining soil is one of the important factors in SFM formation (Dearing et al., 1996b, 1997). Crop burning, once thought to be a major factor in causing soil enhancement, appears to be a relatively inefficient enhancement mechanism. The high temperatures required for enhancement (Le Borgne, 1960) are brought about either by intense fires, such as bonfires or hearths, or where the soil organic horizons actually ignite, as in some forest fires. A comparison of the soil magnetic data with regional geology shows that the magnetic properties of surface soils are only rarely inherited from parent materials. Even over igneous and metamorphic rocks (or glacial drift derived from them) which contain primary 'magnetites', the magnetic properties of surface soils are usually affected by the presence of SFM or other secondary mineralogical transformations related to pedogenesis. Soils also vary strongly in terms of non-ferrimagnetic Fe oxides, such as lepidocrocite, goethite and haematite (Table 3.1). Indeed, the contrast between SFM-enhanced topsoil and non-enhanced subsoil is often accentuated by the dominance of CA and P minerals in subsoil horizons, and has led to a common approach of discriminating between sediments derived from topsoil and subsoil sources. Increasingly, with new and improved measurements, there is the opportunity for exploiting the whole range of Fe-bearing minerals to

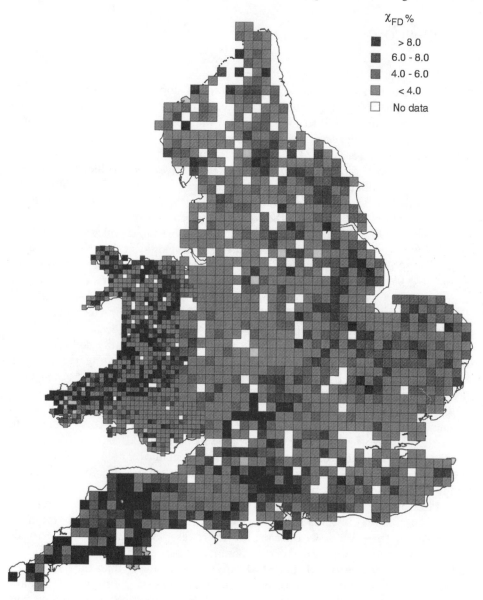

Figure 3.3 Frequency-dependent susceptibility per cent ($\chi_{FD\%}$) of surface soils across England and Wales. Samples were taken by the Soil Survey and Land Resource Centre, Silsoe, UK (McGrath and Loveland, 1992), and subsampled and measured at 10 km (England) and 5 km (Wales) Ordnance Survey grid intersections by Hay (1998) and Rebecca Dann (unpublished data) respectively. Data for England previously published in Dearing *et al.* (1996b). Reproduced by permission of Blackwell Science Ltd

extend the definition of sources from the commonly used 'vertical' divisions of topsoil and subsoil horizons to well-discriminated 'horizontal' divisions of different soil types.

GEOMORPHOLOGICAL CONTEXTS

Hillslopes

A few early studies showed that χ_{LF} varied across a slope sometimes in response to changes in underlying geology, soil type and drainage (e.g. Maher, 1984), but at other times apparently in relation to the slope form only (Dearing, 1979). Consequently, Dearing *et al.* (1986) suggested that the pattern of χ_{LF} along slope transects on Fe-rich Jurassic clays in Oxfordshire, UK, was partly explained by the actions of surface erosional processes. Similarly, Smith *et al.* (1990) described profiles of magnetic measurements at different points on a cultivated slope, over Triassic parent material in Shropshire, UK, which showed features related to the degree to which a previously stable surface horizon had been either eroded or buried by colluvium. Heavy metal analyses showed that the buried surfaces had been exposed to significant atmospheric contamination during the period of local industrial development thus providing some degree of time control. Foster *et al.* (1996), working in the Start Valley, Devon, UK, showed that high χ_{FD} values extend to nearly twice the soil depth at a downslope site compared with two upslope profiles, indicating the redistribution of sediment-associated SP grains from A horizons. Analysis of ^{137}Cs profiles suggests that much of the erosion has taken place within the past 40 years. A recent study (de Jong *et al.*, 1998) combines analyses of ^{137}Cs and magnetic susceptibility to quantify soil erosion on cultivated slopes in the Canadian Prairies, opening up an exciting prospect (cf. Boardman *et al.*, 1990) of making more systematic and widespread use of magnetic measurements in soil redistribution studies.

What are the theoretical difficulties of this would-be method? First, any use of magnetic measurements to study movement of soil on slopes makes assumptions about the nature of enhancement of surface soil by SFMs. If it may be assumed that the production and accumulation of SFMs is constant over a slope then magnetic measurements may be used in an analogous way to inventories of bomb-test [137]Cs (see Chapter 9) – mapping the zones of accumulation and depletion as a spatial expression of surface soil redistribution. Second, there are local factors and internal feedback mechanisms that may invalidate an assumption of constant potential SFM accumulation, especially the rate of weathering and Fe release. For example, rapid erosion might be expected to cause rates of weathering and hence enhancement to increase, and it has been demonstrated that SFMs are destroyed faster than their formation where there is poor drainage, possibly within timescales of $< 10^1$ years (Hannam, J. A., pers. comm.). Lastly, in contrast to ^{137}Cs, the timescale for SFM formation (and for any derived erosional information) is unknown except that recent work suggests that steady-state enhancement may occur over 10^2–10^3 years (Hannam, J. A., pers. comm.).

A recent study was designed explicitly to test the use of magnetic measurements in soil redistribution (Sexton, 1998). Three $100 \, m \times 32 \, m$ agricultural plots on the

Chiltern Hills, UK, were selected and sampled at 4 m grid intersections: a flat culti-vated field as a control, and two plots, cultivated and pasture, with similar concave slope features and maximum gradients. The underlying chalk parent material weath-ers rapidly to produce strongly enhanced rendzina soils and provides the optimum magnetic contrast between surface soil and parent material. Percentage values of $\chi_{FD} > 5$ precent show that the magnetic mineral assemblages across all three plots contain significant concentrations of SFMs. Measurements of χ_{LF}, χ_{FD} and IRM parameters show that the variability of SFM concentrations is higher on the slopes than on the flat surface (Figure 3.4). Both slopes show a similar pattern with the highest values on the flatter upper slope and lower footslopes, and the lowest values coinciding with the steepest mid-slope concave section. The pattern suggests a sign-ificant redistribution of enhanced surface soil by surface erosional processes, a finding supported by field evidence for deeper soils at the base of the slope. The rate of SFM formation appears to be considerably slower than the rate of soil removal. The use of magnetic inventories rather than concentrations within discrete samples might prove to be more appropriate, and the theoretical concerns still require evaluation, but the evidence suggests this may be a useful technique to complement existing radionuclide-based studies.

Suspended Fluvial Sediments

Magnetic measurements of samples of suspended fluvial sediments held on filter papers are straightforward provided that the magnetic properties of the paper are relatively weak and known. Using this approach, magnetic measurements have been used to monitor sediment sources changing through individual storm events, on annual timescales, and to make comparisons of dominant sources in different streams. The earliest studies by Walling et al. (1979) and Oldfield et al. (1979) at Jackmoor Brook, Devon, UK, demonstrated changes in sediment source through storm events. Measurements of soil profiles in the Jackmoor Brook catchment described surface soils enhanced with SFMs in contrast to subsoils and parent materials dominated by CA minerals. Using a combination of IRM demagnetisation curves to match sedi-ments and sources, and measurements of SIRM on discrete suspended sediment samples, the results provided insight into the range of hydrological information that could be gained; direct and lagged relationships between sediment source delivery, rainfall and discharge; the importance of channel scouring and source depletion; soil transport by surface runoff processes; and runoff from snow-melt. Permian rocks and locally derived drift provided a similar degree of magnetic enhancement and potential for sediment-source tracing in Grew's (1990) study of sediment movement in two small instrumented subcatchments in Warwickshire, UK, which were contrasted in terms of arable (S1) and pasture (S2) land uses. He was able to show from measure-ments of daily samples of suspended stream sediment over three years that the concentrations of F minerals ($IRM_{0.8T}$) varied little throughout the year except during storm events when high values indicated topsoil erosion. By using $IRM_{-0.1T}/IRM_{0.8T}$ values he was able to demonstrate a seasonal pattern in sediment source in S2, present in all three years (Figure 3.5). Short-term peaks in topsoil dominance were super-imposed on a seasonal pattern where topsoil components were highest in the winter

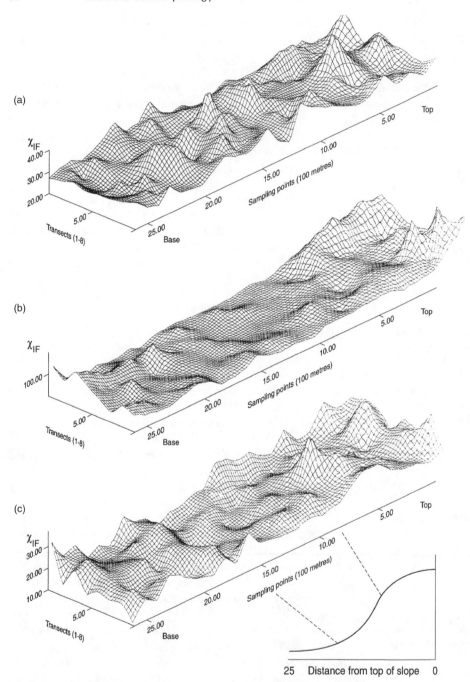

Figure 3.4 Mass specific χ_{LF} ($10^{-8}m^3kg^{-1}$) indicating the distribution of secondary ferrimagnetic minerals on three plots in the Chiltern Hills: (a) flat cultivated; (b) steep concave cultivated; (c) steep concave pasture; inset showing schematic slope profile and approximate position of steepest slopes in (b) and (c) (Sexton, 1998)

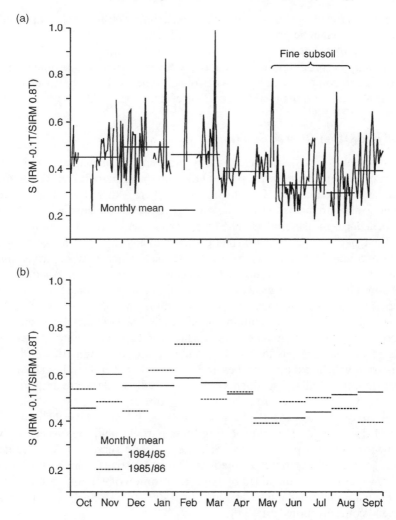

Figure 3.5 Measurements of $IRM_{-0.1T}/IRM_{0.8T}$ on daily suspended sediment samples from stream S2 in the Seeswood Pool catchment showing topsoil-dominated flood events (high values) superimposed upon a seasonal pattern (monthly means) of a greater subsoil component (low values) in summer months for (a) 1986/87 and (b) 1985/86 and 1984/85 (Grew, 1990)

months and subsoil components highest in June–August. He deduced that an energy-limited system led to the erosion of surface soil in winter from intensively cattle-poached areas. Essentially constant magnetic concentrations and sediment sources in S1 pointed to a more sediment-limited system in which buffer strips reduced the connectivity between the cultivated slopes and stream channel. The data provided the key to tracing the historical sources of sediments deposited in the downstream Seeswood Pool and recreating past sediment budgets (e.g. Foster *et al.*, 1990).

Examples of tracing suspended sediment in catchments where primary F minerals dominate are rare. Unless catchment geology (both bedrock and unconsolidated deposits) is heterogeneous and magnetically discriminated for each major sediment source type, the magnetic properties of suspended sediments are likely to be affected by particle size distributions and mineral densities (Thompson and Oldfield, 1986). This effect may be pronounced in upland catchments where the problem of source ascription is also compounded by mosaics of soil types with highly variable distributions of iron oxides and magnetic properties (cf. Dearing, 1992).

Bedload and Other Coarse Sediments

The constitutents of bedload are normally independent of magnetic enhancement processes, possessing magnetic properties linked to the primary mineral content of the sand and larger fractions (e.g. Caitcheon, 1998). There is therefore an opportunity to use magnetic measurements to trace the source of coarse sediments in a similar way to using geochemical measurements. The major experimental consideration is the variability of magnetic properties with fraction size and any sampling programme has to take into account the variability between fractions, which may be larger than the variability between different sources. Caitcheon's (1993) study of Australian rivers showed how χ_{LF} and SIRM measurements could be used to trace the contributions from separate tributaries to the bedload of a main river.

Bedload sediments are often contaminated with a range of particles, some of which are discriminated by their magnetic properties: products of mining, magnetite-rich particles from coal combustion, scale from metalliferous industries and attrition products from vehicle exhaust, brakes and bodywork. Several studies show that the F concentrations in bedload sediments, where the bedload clasts are naturally weakly magnetic, may be used as a proxy of contamination by these particles and in some cases as a proxy of heavy metals. Magnetic surveys of bedload can help identify the location of discharge outlets transporting these particles and the dispersion of particles downstream of a point source. Scholger's (1998) survey of 500 fine bedload sediments from a 190 km stretch of the River Mur, Austria (Figure 3.6), showed that χ_{LF} values were strongly controlled by the concentration of contaminating scale from metal and manufacturing industries, and provided an extremely useful proxy for heavy metal concentrations, especially Cr.

Magnetic measurements of beach and sand bank deposits may provide a view of the spatial differences in mineralogy, the likely geological sources and the apparent contribution by rivers (cf. Setty and Raju, 1988) and offshore sources to local beach material. In the only study of its type, Lees and Pethick (1995) showed how field measurements of magnetic susceptibility (κ) along beach profiles may help map the distribution of heavy minerals related to wave energies. One recent finding of relevance to all studies of coarse fractions where quartz may dominate is that although pure quartz is diamagnetic, magnetic and scanning electron microscopy analyses (Hounslow and Maher, 1996) show that primary magnetite inclusions are common. The extent to which the presence of quartz (and feldspar) grains with and without inclusions could be used in tracing studies has not been evaluated but, in any case, measurable susceptibilities of quartz-rich sand fractions need not be explained just in terms of contamination or 'iron oxide coatings'.

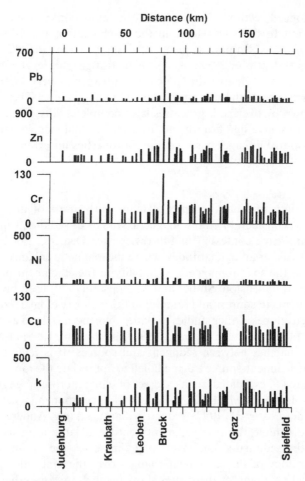

Figure 3.6 Mass specific susceptibilities (k in SI units) and heavy metal concentrations (ppm) of fine bedload sediment along a 190 km (downstream from left to right) stretch of the River Mur, Austria, showing positions of towns and cities (Scholger, 1998)

Floodplains

Despite the recent increase in attempts to understand floodplain dynamics in terms of both accretion rates and sediment sources there are few magnetic studies of floodplain overbank deposits. Foster *et al.* (1996, 1998) have used magnetic methods to trace the source of overbank sediments in the Start Valley, Devon, UK, based on soil enhancement and topsoil–subsoil contrasts expressed by χ_{FD} (SP grains) and HIRM (CA minerals). A decrease in the HIRM/χ_{FD} ratio towards the top of the floodplain core was interpreted as an increase in the rate of supply of eroded topsoil since 1963, based on a ^{137}Cs chronology. One problem with magnetic studies of floodplain deposits is the difficulty in assessing the possible effects of both neo-formation of SFMs in the free-draining surface and reductive diagenetic transformations of Fe-bearing minerals in lower waterlogged sections. Evidence for the latter comes from two floodplain sequences adjacent to the Eau Morte in the catchment of Lac d'Annecy, France (Dearing *et al.*, in press). In both sequences the sediments are mainly silts grading to coarser fractions at depths of 2–3 m. At a similar stratigraphic level in both sequences the magnetic properties change dramatically, with several very high peaks in χ_{LF} caused by the presence of the secondary iron sulphide, greigite. The extent to which minerals laid down by overbank processes have completely dissolved is unclear, but reducing conditions have significantly modified the original magnetostratigraphy and made sediment-source linkages based on the F properties uncertain.

Lake Sediments

The palaeoenvironmental backgrounds of many early users of magnetic measurements have led to considerably more magnetic studies of sediments in lakes (> 100) than in other landscape contexts (for full review, see Dearing, 1999b). Many lakes exist in upland areas, or at high latitudes, where the magnetic enhancement of surface soils is weak, and the spatial magnetic variability in the catchment may be stronger than in the vertical dimension. Stober and Thompson (1979) used bivariate plots to match lake sediments to sources and found that the majority of lake sediments did not fall into source envelopes. Other studies in Wales, Northern Ireland and England (e.g. Thompson and Edwards, 1982; Dearing, 1992; Hutchinson, 1995) also failed to find straightforward matches between sediment and sources. It is tempting to say that early attempts at sediment-source tracing failed to appreciate the range of factors that need to be excluded or evaluated, such as losses of sediment to storage zones upstream of the lake (e.g. Foster *et al.*, 1996); non-detrital F components, such as magnetosomes (e.g. van der Post *et al.*, 1997); reductive diagenesis (e.g. Anderson and Rippey, 1988); failure to sample the true sediment sources; a large atmospheric pollution component in the sediments; and over long timescales the no-analogue problem where the mineralogy of the catchment changes over the timescale of the sediment record. Nevertheless, many of these sites show magnetic records that parallel independent records of environmental change. What are these records showing if not describing a set of common sediment-source linkages?

Previous reviews of European Holocene lake sediment magnetic records (Dearing, 1979, 1991a; Thompson *et al.*, 1980) have noted two distinct types of trend in bulk χ_{LF}. The first type shows highest values in the late-glacial and early Holocene periods with a gradually decreasing trend towards the sediment surface, a trend interrupted only by peaks in the recent sediments. Lakes exhibiting declining trends in χ_{LF} are found in Scandinavia, Iceland, Israel, Russia and North America. A common factor for these lakes is direct pollen evidence or other information for very restricted human activity in the catchment during the Holocene. Most of these lakes exist in high-latitude climatic zones where natural forest, heath or peat covered catchments dominate the present land use, where agricultural production today is extremely limited or, as particularly in the case of North America, where there is a short history of significant human impact.

The second type, exemplified by Thompson *et al.*'s (1975) original magnetic study of lake sediments at Lough Neagh, shows similar high values in the early part of the record but the following declining trend is reversed during the mid to late Holocene. Subsequent increases in χ_{LF} are stepped towards the sediment surface with values in the recent sediments, equivalent to the past 200–300 years, often as high as those measured for late-glacial samples. The earliest rises in the longest European χ_{LF} records appear to cluster in time around periods of major human disturbance: Neolithic (5500–4000 BP); Bronze Age (4200–2500 BP); Iron Age (2500–2000 BP); Historical ($\sim$ 1000 BP). Dearing and Flower (1982) demonstrated that the χ_{LF} of material in modern Lough Neagh sediment traps closely followed the preceding monthly rainfall. They argued that high χ_{LF} was linked to the transport of relatively large (silt and fine sand) and dense particles in flood events which maximised the delivery of titanomagnetite; thus providing a process link between hydrological processes and the lake sediment χ_{LF} record. More recently, Turner (1997) has also shown how storm events recorded in lake sediments, during the time of European settlement in New Zealand, deliver material containing titanomagnetite. Also, Kodama et al. (1997) identified a 50-year period, low-amplitude variation in recent sediments from three lakes in north-eastern Pennsylvania that matched well with regional rainfall variations over the past 120–250 years. It may therefore be proposed that many Holocene χ_{LF} records are proxies for a complex set of hydrodynamic processes acting on particles rather than specific sediment sources. Applying this explanation to long-term decreasing trends in χ_{LF} implies that the availability of sediment or the efficiency of delivery processes in many high-latitude and upland sites has declined during the Holocene and may have been insensitive to climate change (Dearing, 1999b).

In contrast (and as a logical and expected extension of tracing suspended stream sediments), magnetic studies at lowland lake sites show clearer and more direct evidence for the importance of specific sediment sources. In Holocene sediments from lakes in Skåne, southern Sweden, the parameter HIRM/χ_{FD} was used to reconstruct the changing mixture of sources from haematite-rich channels and magnetically enhanced surface soil (Dearing et al., 1987, 1990; Dearing 1991b; Gaillard et al., 1991a,b). Stream channel material was the dominant sediment source during virtually all phases of agricultural expansion since 3000 BP. The exception was in recent times when the stream networks were drastically reduced by artificial drainage, findings which were well-supported by detailed documentary records of agriculture and catchment drainage over the past 200 years. The recent lake sediments are not composed entirely of surface soil, despite evidence in field experiments for soil redistribution under intensive agriculture, possibly indicating considerable sediment storage on the slopes. In contrast, long magnetic records of χ_{FD} and soft IRM (both detecting SP grains) at Lac d'Annecy, France, show that agricultural activities, particularly those related to monastic agricultural activities around AD 1000, have led to the delivery of surface soil to the lake (Higgitt et al., 1991). Subsequent work by Thorndycraft et al. (1998) has shown that the composition of sediment associated with extreme flood events recorded in the sediments covering the past 300 years is also dominated by surface soil, suggesting that the sediment system in this sub-alpine landscape is particularly sensitive to summer storm activity. Tracing sediment sources over the past 250 years in the cultivated catchment of the reservoir Seeswood Pool, Central England (Foster et al., 1985, 1986, 1990; Dearing and Foster, 1986), was also based particularly on measurements of HIRM and X_{FD}, and were later complemented by measurements of ^{137}Cs (Foster et al., 1994). The results from the sediments showed that, as cultivation had become more intensive in the mid-20th century, the topsoil component actually decreased because the main hydrological effect was focused on channel erosion caused by higher runoff. Using a similar approach, Foster and Walling (1994) and Foster et al. (1996, 1998) showed that surface soil from pasture was the dominant sediment source for sediments in the Old Mill Reservoir, Devon, UK. In contrast, recent intensification of cultivation was not linked to an increased delivery of surface soil at nearby Slapton Ley. There, studies of the floodplain sediments show that much of the eroded fine soil has been stored before reaching the lake.

At some sites, a change in sediment source may be linked to non-farming activities. For instance, haematite-rich inwash layers were linked to spoil from iron ore workings in the catchment of Loe Pool, UK (Coard et al., 1983), giving large peaks in X_{LF}. At some sites the erosional response is in terms of low magnetic concentrations, as in the record of 20th century deep ploughing and afforestation around Loch Frisa, Scotland (Appleby et al., 1985) or the effects of slate quarrying around Llyn Peris in north Wales (Dearing et al., 1981). In contrast, mining for Pb and Zn ores around Llyn Geirionydd, north Wales, led to the deposition over a 50-year period of sediments with both low and high F concentrations (Dearing, 1992). In Morocco, an increasing trend and a series of peaks and troughs of X_{LF} in recent sediments from Dayat-er-Roumi (Flower et al., 1984, 1989) were interpreted as the periodic reactivation of subsoil from a single gully system in the catchment.

In theory, the use of magnetic measurements as applied to suspended sediments applies equally to floodplain and lake sediments. In practice, however, an assumption of conservative mineral mixtures may be invalidated by factors and processes that alter the magnetic mineralogy, such as reductive dissolution of F minerals (complete to partial); the formation of bacterial magnetite (magnetosomes); diagenetic formation of iron sulphides (greigite); the formation of SFMs on floodplains; and in recent sediments the accumulation of atmospherically derived 'magnetite' from coal burning. Increasingly there are magnetic and other methods which may identify these effects but it may mean that the detrital F component is difficult to isolate and interpret (Dearing, 1999b). One alternative approach to avoid these problems is to utilise the CA components (haematite and goethite) and P (Fe-bearing minerals) components which may be the least affected by these secondary processes (cf. Dearing et al., 1998; Snowball et al., 1999) and hence, in some situations, the preferred mineral assemblage on which to base sediment-source linkages.

MODELLING SOURCE COMPONENTS

All tracing involves the matching of sediments to sources, through scatterplots for example, in order to confirm that the source is present in a sediment sample. A further stage is to attempt quantitative unmixing of sediments into component sources. There are two main approaches: two-component and multiple-component mixing models. An important issue is whether modelling should be based on the properties of discrete minerals with well-defined magnetic signatures or assemblages of minerals with compounded magnetic signatures. The former has the advantage of working with 'clean' mineral parameters, such as magnetisation curves (e.g. Peters and Thompson, 1998), but the latter may be more realistic or accurate with regards to the true complexity and variability of the source mineral assemblages. It seems that modelling on the basis of mineral end-members is only appropriate where sediment sources are defined and dominated by mutually exclusive mineral properties. One approach that may simplify the problem of using different magnetic parameters is to reduce all measurements to partial susceptibilities or partial saturation magnetisation, which may then be quantitatively compared (Frederichs *et al.*, 1999; Xie *et al.*, 1999).

Simple mixing models can be created using two source components having distinct magnetic properties. Caitcheon (1993) used SIRM and X_{LF} values to classify the sediments carried by two Australian rivers and the mean ratios (the regression line gradient) in simultaneous equations to model the proportions of each to sediments below the confluence. Because the contribution of one component to a bivariate ratio varies nonlinearly with the ratio value, some workers have calibrated the measured ratio values to measurements on artificial mixtures where the component contribution is known. This appears to be particularly successful in lowland agricultural catchments where surface soil and subsoils can often be unambiguously discriminated (Foster and Walling, 1994; Foster *et al.*, 1996). Stott (1986) also used this approach in an upland reservoir catchment in England to show that sediment sources had changed markedly through a 50-year period of afforestation and maturation of forest plantations.

Multicomponent mixing models utilise multivariate statistics, often least-squares regression, to express each magnetic parameter in terms of a weighted combination of source contributions (Yu and Oldfield, 1989, 1993). Obtaining solutions to the equations for the sediment samples may be achieved by numerical linear modelling (Thompson, 1986; Lees, 1994). At Chesapeake Bay, USA (Oldfield *et al.*, 1985, 1989; Yu and Oldfield, 1989), the sources of estuarine sediments were successfully linked to sources that were differentiated on the basis of both magnetic enhancement and geological contrasts. In a study of reservoir sediments in southern Spain, Yu and Oldfield (1993) demonstrated the importance of a detailed knowledge of the relationships between magnetic properties and particle size, and the inclusion of all source types. Shankar *et al.* (1994) utilised a maximum likelihood unmixing algorithm to unmix bedload sediments in the Bhadra River, India, derived in part from mine tailings, on the basis of their combined magnetic and natural radioactivity properties. This approach coped well with high collinearities, a common problem when using magnetic measurements.

Experiments by Lees (1994, 1997) using artificial mixtures with known source contributions highlight a number of particular problems with multiple-source mixing models. First, some magnetic parameters do not show perfect additivity presumably because of magnetic interactions between grains. Second, the accurate prediction of the contribution by sources that are magnetically weak or, in magnetic terms, multiples or compounds of the same minerals, is difficult. Even where sources are classified successfully by multivariate classifications, the unmixing algorithms may be unable to distinguish between, for instance, the wholly paramagnetic behaviour of one component and the partial paramagnetic component of another. Third, the success of unmixing declines markedly with more than four end-member sources. Fourth, simulation of the natural variability of end-member sources in the environment by adding error terms to source parameters showed that the success of the linear modelling methods was severely restricted where the magnetic variability of a source was greater than 10 percent. An interesting aspect of Lees' studies was that optimum discrimination of sources was achieved with just four parameters (X_{LF}, X_{FD}, X_{ARM} and HIRM), representing total F concentrations, SP grains, SSD grains and CA minerals. Shankar *et al.* (1994) were also able to optimise modelling procedures by zero weighting half of the initial magnetic variables.

Recent applications of different modelling approaches to source ascription in stream sediments confirm most of these findings (Walden *et al.*, 1997), but one important issue remains largely unresolved. A high level of source variability may make defining end-members difficult, and it is fair to say that the spatial and vertical variability of sources has not always been properly quantified. In practice, there is a trade-off between simple, but probably also simplistic, definitions of source end-members on the basis of a few highly selective samples, and complex, but more realistic, definitions which take into account the full variability of the 'magnetic landscape'. Prompted by this problem, there have recently been attempts to use systematic approaches to sampling and classifying potential sediment sources. Lees (1994) demonstrated the advantages of rapid field measurements of κ in providing a basis for calculating variograms and identifying preliminary magnetic classifications of the landscape which were later confirmed through multivariate cluster analysis or principal component analysis. Systematic mapping of source magnetic properties is undertaken more rapidly than for many other properties and may provide a strong basis for defining sediment sources. As an example, the mapped data in Figure 3.3 provide information about major differences in the magnetic mineral assemblage of potential sediment sources within the large river catchments of England and Wales. At a larger scale, the magnetic properties of soils sampled on a 1 km grid across the Petit Lac d'Annecy catchment, France, reveal a number of zones of similar magnetic assemblage (Figure 3.7) (Dearing *et al.*, in press). Previous non-systematic sampling of soils in the catchment had shown widespread magnetic enhancement (Dearing, 1979; Higgitt *et al.*, 1991) but the maps now demonstrate a second-order arrangement of spatial magnetic variability related to altitude, parent material and drainage. The high level of coherence of the magnetic properties with the landscape and geology provides a strong basis for establishing a range of sources defined by key mineral associations. At Lac d'Annecy, there is now the possibility of using magnetic-based sediment-source linkages to reconstruct, from the lake and floodplain sediment records, the form and location of past hydrological processes to a high level of detail over the past 5000 years (Figure 3.8). The following hypotheses have recently been suggested (Dearing *et al.*, in press): (i) the erosion of high-altitude substrates has increased within the last 1000 years; (ii) a shift in the proportion of surface soil occurred $\sim$ AD 1000 and may be linked to the accelerated accretion of floodplain overbank deposits; and (iii) the source of sediment transported by high-magnitude, low-frequency flood events shifted at $\sim$ AD 1000 from high-altitude soils to lowland and mid-altitude free-draining soils.

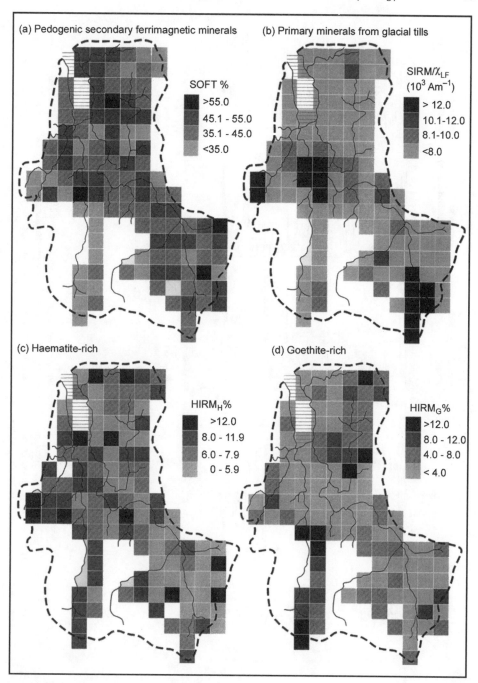

Figure 3.7 Spatial patterns of soils (1 km grid) in the Petit Lac d'Annecy catchment (drainage network and lake basin shown) with magnetic properties dominated by (a) pedogenic secondary ferrimagnetic minerals, (b) primary minerals from glacial tills, (c) haematite and (d) goethite (Dearing *et al.*, in press). Reproduced with kind permission from Kluwer Academic Publishers

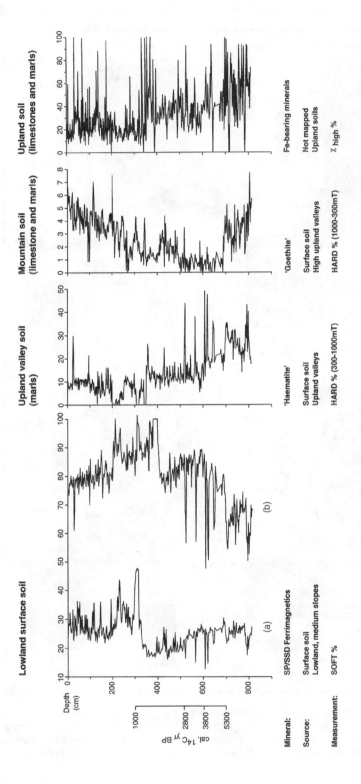

Figure 3.8 Sediment records (~0–6000 cal. years BP) from Petit Lac d'Annecy of magnetically defined sediment sources: lowland surface soil (based on (a) SOFT % data and (b) SOFT % data corrected for the effects of bacterial magnetosomes); upland valley surface soil on marls; mountain soils on limestones and marls; and upland soils on limestones and marls dominated by paramagnetic minerals (Dearing *et al.*, in press). Reproduced with kind permission from Kluwer Academic Publishers

CONCLUSIONS

- There are now many published studies which show convincing evidence that magnetic measurements may be used across the complete range of sediment-source contexts: soil redistribution, linking the slope to the channel, tracing bedload movements, and reconstructing hydrological processes in sediment sequences. The measurements may be used either to complement other tracing techniques or to provide unique diagnostic information about a particular source or process.
- Measurements are generally quick and easy to make, and all but high-temperature measurements are non-destructive. Standard magnetic measurements, such as χ_{LF} and IRMs, may be complemented with high field and low-temperature measurements to measure paramagnetic, goethite and haematite minerals, providing a more complete definition of mineral assemblages, and hence a greater potential for accurate sediment tracing.
- The basis of the methodology is the variability of primary Fe-bearing minerals and the sensitivity of newly formed secondary iron oxides/hydroxides to environmental conditions. The distinctive hydrodynamic properties of magnetite-rich primary minerals lend themselves to tracing hydrological responses, especially in lake and floodplain sediment sequences. The phenomenon of magnetic enhancement of well-drained soil allows the tracing of sediment delivered by surface erosion processes.
- The choice of magnetic measurements to define sediment sources should be based on the causes and nature of magnetic variability – and these may be highly complex. However, many studies show that the first-order controls on magnetic variability may usefully be represented by a simple dichotomy of geology/particle size and surface/subsoil contrasts. Full advantage of field and laboratory measurements should be taken in order to produce detailed magnetic maps of spatial variability, in addition to the vertical variability detected in soil and stream channel profiles.
- There is a hitherto untapped potential to use magnetically enhanced soil profiles, normally found on weakly magnetic substrates, to complement other methods, such as artificial radionuclides, in defining soil redistribution patterns on hillslopes and identifying erosional sequences in colluvium.
- Magnetic-based sediment-source linkages that include either lake sediments or floodplain overbank deposits need to take into account the likelihood of secondary effects on the detrital mineral assemblage: bacterial magnetosomes, reductive dissolution, authigenic iron sulphide formation, recent accumulation of atmospheric pollution spheres, and, in floodplains, the possibility of SFM enhancement.

ACKNOWLEDGEMENTS

I would like to thank Rebecca Dann, Dr Robert Grew, Dr Peter Loveland and the SSLRC, Silsoe, for permission to use unpublished magnetic data; Dr Robert Scholger for permission to use Figure 3.6; Philip Sexton for allowing the use of data from his undergraduate dissertation; Jack Hannam for useful discussions; and Sandra Mather for drafting the figures. Parts of the sections 'Lake sediments' and 'Modelling source components' are published in Dearing (1999b).

REFERENCES

Anderson, N. J. and Rippey, B. 1988. Diagenesis of magnetic minerals in the recent sediments of a eutrophic lake. *Limnology and Oceanography*, **33**, 1476–1492.

Appleby, P. G., Dearing, J. A. and Oldfield, F. 1985. Magnetic studies in a Scottish lake catchment. I Core chronology and correlation. *Limnology and Oceanography*, **30**, 1144–1153.

Björck, S., Dearing, J. A. and Jonsson, A. 1982. Magnetic susceptibility of Late-Weichselian deposits in southeastern Sweden. *Boreas*, **11**, 99–111.

Boardman, J., Dearing, J. A. and Foster, I. D. L. 1990. Soil erosion studies: some assessments. In Boardman, J., Foster, I. D. L. and Dearing, J. A. (eds) *Soil Erosion on Agricultural Land*, John Wiley, Chichester, 659–672.

Bradshaw, R. and Thompson, R. 1985. The use of magnetic measurements to investigate the mineralogy of Icelandic lake sediments and to study catchment processes. *Boreas*, **14**, 203–215.

Caitcheon, G. G. 1993. Sediment source tracing using environmental magnetism: a new approach with examples from Australia. *Hydrological Processes*, **7**, 349–358.

Caitcheon, G. G. 1998. The significance of various sediment magnetic mineral fractions for tracing sediment sources in Killimicat Creek. *Catena*, **32**, 131–142.

Coard, M. A., Cousen, S. M., Cuttler, A. H., Dean, H. J., Dearing, J. A., Eglinton, T. I., Greaves, A. M., Lacey, K. P., O'Sullivan, P. E., Pickering, D. A., Rhead, M. M., Rodwell, J. K. and Simola, H. 1983. Palaeolimnological studies of annually-laminated sediments in Loe Pool, Cornwall, UK. *Hydrobiologia*, **103**, 185–191.

Colman, S. M., Jones, G. A., Forester, R. M. and Foster, D. S. 1990. Holocene palaeoclimatic evidence and sedimentation rates from a core in South western Lake Michigan. *Journal of Paleolimnology*, **4**, 269–284.

Cullity, B. D. 1972. *Introduction to Magnetic Materials*. Addison-Wesley, Reading, MA, 666.

Dearing, J. A. 1979. The use of magnetic measurements to study particulate flux in lake-watershed ecosystems. Unpublished PhD Thesis, University of Liverpool.

Dearing, J. A. 1991a. Lake sediment records of erosional processes. *Hydrobiologia*, **214**, 99–106.

Dearing, J. A. 1991b. Erosion and land use. In Berglund, B. E. (ed.) The cultural landscape during 6000 years in southern Sweden. *Ecological Bulletins* (Special Issue), **41**, 283–292.

Dearing, J. A. 1992. Sediment yields and sources in a Welsh lake catchment during the past 800 years. *Earth Surface Processes and Landforms*, **17**, 1–22.

Dearing, J. A. 1999a. *Environmental Magnetic Susceptibility: Using the Bartington MS2 System* (2nd edition). Chi Publishing, Kenilworth.

Dearing, J. A. 1999b. Holocene environmental change from magnetic proxies in lake sediments. In Maher, B. A. and Thompson, R. (eds) *Quaternary Climates, Environments and Magnetism*. Cambridge University Press, Cambridge.

Dearing, J. A. and Flower, R. J. 1982. The magnetic susceptibility of sedimenting material trapped in Lough Neagh, Northern Ireland, and its erosional significance. *Limnology and Oceanography*, **27**, 969–975.

Dearing, J. A. and Foster, I. D. L. 1986. Limnic sediments used to reconstruct sediment yields and sources in the English Midlands since 1765. In Gardiner, V. (ed.) *International Geomorphology*, Vol. I, John Wiley, Chichester, 853–868.

Dearing, J. A., Elner, J. K. and Happey-Wood, C. M. 1981. Recent sediment flux and erosional processes in a Welsh upland lake catchment based on magnetic susceptibility measurements. *Quaternary Research*, **16**, 356–372.

Dearing, J. A., Maher, B. A. and Oldfield, F. 1985. Geomorphological linkages between soils and sediments: the role of magnetic measurements. In Richards, K. S., Arnett, R. R. and Ellis, S. (eds) *Geomorphology and Soils*. George Allen and Unwin, London, 245–266.

Dearing, J. A., Morton, R. I., Price, T. W. and Foster, I. D. L. 1986. Tracing movements of topsoil by magnetic measurements: two case studies. *Physics of the Earth and Planetary Interiors*, **42**, 93–104.

Dearing, J. A., Håkansson, H., Liedberg-Jönsson, B., Persson, A., Skansjö, S., Widholm, D. and El-Daoushy, F. 1987. Lake sediments used to quantify the erosional response to land use change in southern Sweden. *Oikos*, **50**, 60–78.

Dearing, J. A., Alström, K., Bergman, A., Regnell, J. and Sandgren, P. 1990. Past and present erosion in southern Sweden. In Boardman, J., Foster, I. D. L. and Dearing, J. A. (eds) *Soil Erosion on Agricultural Land*. John Wiley, Chichester, 687.

Dearing, J. A., Dann, R. J. L., Hay, K., Lees, J. A., Loveland, P. J., Maher, B. A. and O'Grady, K. 1996a. Frequency-dependent susceptibility measurements of environmental materials. *Geophysics Journal International*, **124**, 228–240.

Dearing, J. A., Hay, K., Baban, S., Huddleston, A. S., Wellington, E. M. H. and Loveland, P. J. 1996b. Magnetic susceptibility of topsoils: a test of conflicting theories using a national database. *Geophysics Journal International*, **127**, 728–734.

Dearing, J. A., Bird, P. M., Dann, R. J. L. and Benjamin, S. F. 1997. Secondary ferrimagnetic minerals in Welsh soils: a comparison of mineral magnetic detection methods and implications for mineral formation. *Geophysics Journal International*, **130**, 727–736.

Dearing, J. A., Boyle, J. F., Appleby, P. G., Mackay, A. W. and Flower, R. J. 1998. Magnetic properties of recent sediments in Lake Baikal, Siberia. *Journal of Paleolimnology*, **20**, 163–173.

Dearing, J. A., Hu, Y., Doody, P., James, P. A. and Brauer, A. in press. Preliminary reconstruction of sediment-source linkages for the past 6000 years at The Petit Lac d'Annecy, based on mineral magnetic data. Journal of Paleolimnology.

De Jong, E., Nestor, P. A. and Pennock, D. J. 1998. The use of magnetic susceptibility to measure long-term soil redistribution. *Catena*, **32**, 23–36.

Dekkers, M. J. 1989. Magnetic properties of natural goethite – II. TRM behaviour during thermal and alternating field demagnetisation and low-temperature treatment. *Geophysics Journal International*, **97**, 341–355.

Dunlop, D. J. and Özdemir, Ö. 1997. *Rock Magnetism: Fundamentals and Frontiers*. Cambridge University Press, Cambridge, 573.

Flower, R. J., Nawas, R. and Dearing, J. A. 1984. Sediment supply and accumulation in a small Moroccan lake: an historical perspective. *Hydrobiologia*, **112**, 81–92.

Flower, R. J., Stevenson, T., Dearing, J. A., Foster, I. D. L., Airey, A. A., Orend, K., Rippey, B. and Appleby, P. G. 1989. Palaeolimnological studies from three contrasting environments in Morocco. *Journal of Paleolimnology*, **1**, 293–322.

Foster, I. D. L. and Walling, D. E. 1994. Using reservoir deposits to reconstruct changing sediment yields and sources in the catchment of the Old Mill Reservoir, South Devon, UK, over the past 50 years. *Hydrological Sciences Journal*, **39**, 347–368.

Foster, I. D. L., Dearing, J. A., Simpson, A., Carter, A. D. and Appleby, P. G. 1985. Lake catchment based studies of erosion and denudation in the Merevale catchment, Warwickshire, UK. *Earth Surface Processes and Landforms*, **10**, 45–68.

Foster, I. D. L., Dearing, J. A. and Appleby, P. G. 1986. Historical trends in catchment sediment yields: a case study in reconstruction from lake-sediment records in Warwickshire, UK. *Hydrological Sciences Journal*, **31**, 427–443.

Foster, I. D. L., Dearing, J. A., Grew, R. and Orend, K. 1990. The sedimentary data base: an appraisal of lake and reservoir based studies of sediment yield. In *Erosion Transport and Deposition Processes*. IAHS Publication, **189**, IAHS Press, Wallingford. 119–143.

Foster, I. D. L., Dalgleish, H., Dearing, J. A. and Jones, E. D. 1994. Quantifying soil erosion and sediment transport in drainage basins; some observations on the use of Cs-137. In *Variability in Stream Erosion and Sediment Transport*. IAHS Publication **224**, IAHS Press, Wallingford. 55–64.

Foster, I. D. L., Owens, P. N. and Walling, D. E. 1996. Sediment yields and sediment delivery in the catchments of Slapton Lower Ley, South Devon, UK. *Field Studies*, **8**, 629–661.

Foster, I. D. L., Lees, J. A., Owens, P. N. and Walling, D. E. 1998. Mineral magnetic characterisation of sediment sources from an analysis of lake and floodplain sediments in the catchments of the Old Mill reservoir and Slapton Ley, South Devon, UK. *Earth Surface Processes and Landforms*, **23**, 685–704.

France, D. E. 1997. The mineral magnetic characterisation of goethite and haematite in soils and sediments. Unpublished PhD Thesis, University of Liverpool.

Frederichs, T., Bleil, U., Däumler, K., von Doboneck, T. and Schmidt, A. 1999. The magnetic view on the palaeoenvironment: parameters, techniques and potentials of rock magnetic studies as a key to paleoclimatic and palaeoceanographic changes. In Fischer, G. and Wefer, G. (eds) *Use of Proxies in Paleoceanography: Examples from the South Atlantic*. Springer Verlag, Berlin and Heidelberg, 575–599.

Gaillard, M. J., Dearing, J. A., El-Daoushy, F., Enell, M. and Håkansson, H. 1991a. A late Holocene record of land use history, lake trophy and lake-level fluctuations at lake Bjäresjö (south Sweden). *Journal of Paleolimnology*, **6**, 51–81.

Gaillard, M. J., Dearing, J. A., El-Daoushy, F., Enell, M. and Håkansson, H. 1991b. A multi-disciplinary study of Lake Bjäresjösjon (S. Sweden): land use history, soil erosion, lake trophy, and lake-level fluctuations during the last 3000 years. *Hydrobiologia*, **214**, 107–114.

Grew, R. G. 1990. Sediment yields and sources over short and medium timescales in a small agricultural catchment in N. Warwickshire, UK. Unpublished PhD Thesis, Coventry University.

Hay, K. L. 1998. The magnetic properties of English topsoils. Unpublished PhD Thesis, Coventry University.

Higgitt, S. R., Oldfield, F. and Appleby, P. G. 1991. The records of land use change and soil erosion in the late-Holocene sediments in the Petit Lac d'Annecy, eastern France. *The Holocene*, **1**, 14–28.

Hounslow, M. W. and Maher, B. A. 1996. Quantitative extraction and analysis of carriers of magnetisation in sediments. *Geophysics Journal International*, **124**, 57–74.

Hutchinson, S. M. 1995. Use of magnetic and radiometric measurements to investigate erosion and sedimentation in a British upland catchment. *Earth Surface Processes and Landforms*, **20**, 293–314.

Kodama, K. P., Lyons, J. C., Siver, P. A. and Lott, A. M. 1997. A mineral magnetic and scaled-chrysophyte paleolimnological study of two northeastern Pennsylvania lakes: records of fly ash deposition, land-use change, and paleorainfall variation. *Journal of Paleolimnology*, **17**, 173–189.

Le Borgne, E. 1955. Susceptibilité magnétique anormale du sol superficiel. *Annals of Géophysics*, **11**, 399–419.

Le Borgne, E. 1960. Influence du feu sur les propriétés magnétique du sol et sur celles du schiste et du granit. *Annals of Géophysics*, **16**, 159–195.

Lees, J. A. 1994. Modelling the magnetic properties of natural and environmental materials. Unpublished PhD Thesis, Coventry University.

Lees, J. A. 1997. Mineral magnetic properties of mixtures of environmental and synthetic materials: linear additivity and interaction effects. *Geophysics Journal International*, **131**, 335–346.

Lees, J. A. and Pethick, J. S. 1995. Problems associated with quantitative magnetic sourcing of sediments of the Scarborough to Mablethorpe coast, northeast England. *Earth Surface Processes and Landforms*, **20**, 795–806.

Maher, B. A. 1984. Origins and transformations of magnetic minerals in soils. Unpublished PhD Thesis, University of Liverpool.

Maher, B. A. 1986. Characterisation of soils by mineral magnetic measurements. *Physics of the Earth and Planetary Interiors*, **42**, 76–92.

Maher, B. A. 1998. Magnetic properties of modern soils and Quaternary loessic palaeosols: palaeoclimatic implications. *Palaeogeography, Palaeoclimatology and Palaeoecology*, **137**, 25–54.

Maher, B. A. and Taylor, R. M. 1988. Formation of ultrafine-grained magnetite in soils. *Nature*, **336**, 368–370.

McGrath, S. P. and Loveland, P. J. 1992. *The Soil Geochemical Atlas of England and Wales*. Blackie, London.

Mullins, C. E. 1977. Magnetic susceptibility of the soil and its significance in soil science: a review. *Journal of Soil Science*, **28**, 223–246.

Oldfield, F., Rummery, R., Thompson, R. and Walling, D. E. 1979. Identification of suspended sediment sources by means of magnetic measurements: some preliminary results. *Water Resources Research*, **15**, 211–218.

Oldfield, F., Maher, B. A., Donoghue, J. and Pierce, J. 1985. Particle-size related, mineral magnetic source sediment linkages in the Rhode River catchment, Maryland, USA. *Journal of the Geological Society of London*, **142**, 1035–1046.

Oldfield, F., Maher, B. A. and Appleby, P. G. 1989. Sediment source variations and lead-210 inventories in recent Potomac Estuary sediment cores. *Journal of Quaternary Science*, **4**, 189–200.

Peters, C. and Thompson, R. 1998. Magnetic identification of selected natural iron oxides and sulphides. *Journal of Magnetism and Magnetic Materials*, **183**, 365–374.

Scholger, R. 1998. Heavy metal pollution monitoring by magnetic susceptibility measurements applied to sediments of the River Mur (Styria, Austria). *European Journal of Environmental and Engineering Geophysics*, **3**, 25–37.

Setty, K. B. and Raju, D. R. 1988. Magnetite content as a basis to estimate other heavy mineral content in the sand deposit along the Nizampatnam Coast, Guntur District, Andhra Pradesh. *Journal of the Geological Society of India*, **31**, 491–494.

Sexton, P. F. 1998. The use of mineral magnetic measurements to investigate patterns of soil redistribution: a methodological case-study. Unpublished BSc Dissertation, University of Liverpool.

Shankar, R., Thompson, R. and Galloway, R. B. 1994. Sediment source modelling: unmixing of artifical magnetisation and natural radioactivity measurements. *Earth and Planetary Science Letters*, **126**, 411–420.

Smith, J. P., Fullen, M. A. and Tavner, S. 1990. Some magnetic and geochemical properties of soils developed on Triassic substrates and their use in characterisation of colluvium. In Boardman, J., Foster, I. D. L. and Dearing, J. A. (eds) *Soil Erosion on Agricultural Land*. John Wiley, Chichester, 255–272.

Snowball, I. F. 1993. Geochemical control of magnetite dissolution in subarctic lake sediments and the implications for environmental magnetism. *Journal of Quaternary Science*, **8**, 339–346.

Snowball, I., Sandgren, P. and Petterson, G. 1999. The mineral magnetic properties of an annually laminated Holocene lake-sediment sequence in northern Sweden. *The Holocene*, **9**, 353–362.

Stober, J. A. and Thompson, R. 1979. An investigation into the source of magnetic materials in some Finnish lake sediments. *Earth and Planetary Science Letters*, **45**, 464–474.

Stott, A. P. 1986. Sediment tracing in a reservoir-catchment system using a magnetic mixing model. *Physics of the Earth and Planetary Interiors*, **42**, 105–112.

Thompson, R. 1986. Modelling magnetization data using SIMPLEX. *Physics of the Earth and Planetary Interiors*, **42**, 113–127.

Thompson, R. and Edwards K. J. 1982. A Holocene palaeomagnetic record and geomagnetic master curve from Iceland. *Boreas*, **11**, 335–349.

Thompson, R. and Morton, D. J. 1979. Magnetic susceptibility and particle-size distribution in recent sediments of the Loch Lomond drainage basin, Scotland. *Journal of Sedimentary Petrology*, **49**, 801–811.

Thompson, R. and Oldfield, F. 1986. *Environmental Magnetism*. George Allen and Unwin, London.

Thompson, R., Bloemendal, J., Dearing, J. A., Oldfield, F., Rummery, T. A., Stober, J. C. and Turner, G. M. 1980. Environmental applications of magnetic measurements. *Science*, **207**, 481–486.

Thompson, R., Battarbee, R. W., O'Sullivan, P. E. and Oldfield, F. 1975. Magnetic susceptibility of lake sediments. *Limnology and Oceanography*, **20**, 687–698.
Thorndycraft, V., Hu, Y., Oldfield, F., Crooks, P. R. J. and Appleby, P. G. 1998. A high
resolution flood event stratigraphy in the sediments of the Petit Lac d'Annecy. *The Holo-
cene*, **8**, 741–746.
Turner, G. M. 1997. Environmental magnetism and magnetic correlation of high resolution
lake sediment records from northern Hawke's Bay, New Zealand. *Journal of Geology and
Geophysics*, **40**, 287–298.
Van der Post, K. D., Oldfield, F., Haworth, E. Y., Crooks, P. R. J. and Appleby, P. G. 1997. A
record of accelerated erosion in the recent sediments of Blelham Tarn in the English Lake
District. *Journal of Paleolimnology*, **18**, 103–120.
Walden, J. and Addison, K. 1995. Mineral magnetic analysis of a 'weathering' surface within
glacigenic sediments at Glallynnau, North Wales. *Journal of Quaternary Science*, **10**, 367–
378.
Walden, J., Slattery, M. C. and Burt, T. P. 1997. Use of mineral magnetic measurements to
fingerprint suspended sediment sources: approaches and techniques for data analysis. *Jour-
nal of Hydrology*, **202**, 353–372.
Walden, J., Oldfield, F. and Smith, J. P. (eds) 1999. *Environmental Magnetism: A Practical
Guide*. Technical Guide No. 6, Quaternary Research Association, London.
Walling, D. E., Peart, M. R., Oldfield, F. and Thompson, R. 1979. Suspended sediment sources
identified by magnetic measurements. *Nature*, **281**, 110–113.
Yu, L. and Oldfield, F. 1989. A multivariate mixing model for identifying sediment source from
magnetic measurements. *Quaternary Research*, **32**, 168–181.
Yu, L. and Oldfield, F. 1993. Quantitative sediment source ascription using magnetic measure-
ments in a reservoir-catchment system near Nijar, S.E. Spain. *Earth Surface Processes and
Landforms*, **18**, 441–454.
Yu, L., Oldfield, F., Yushu, W., Sufu, Z. and Jiayi, X. 1990. Palaeolimnological implications of
magnetic measurements on sediment core from Kunming Basin, Southwest China. *Journal
of Paleolimnology*, **3**, 95–111.
Xie, S., Bloemendal, J. A. and Dearing, J. A. 1999. A partial susceptibility approach to
analysing the magnetic properties of environmental materials: a case study. *Geophysics
Journal International*, **138**, 851–856.

Section 2

ATMOSPHERIC AND HYDROLOGICAL TRACERS

4 Fingerprinting Airborne Particles for Identifying Provenance

J. R. MEREFIELD,[1] I. M. STONE,[1] J. ROBERTS,[2] J. JONES,[2] J. BARRON[2] and A. DEAN[2]

[1]Advance Environmental, University of Exeter, UK
[2]Earth Resources Centre, University of Exeter, UK

INTRODUCTION

Particles from sub-micrometres (μm) to 100 μm in size, carried by the wind, are in many aspects the aeolian equivalent of suspended clays and silty sediments transported by watercourses. The growing emphasis on air quality world-wide has driven the need to apportion the source of such complex wind-blown composites. As a direct response to this requirement, the 1990s saw significant advances in the technology and methodology employed in, and the applications applied to, dust fingerprinting for identification of provenance.

The work reported here describes the evolution of an increasingly important topic in geomorphology. It illustrates progress by reference to research on fugitive dust from opencast coal mining in South Wales and explores the opportunities for further progress. Essentially, the current state-of-the-art research relies on tried and tested sedimentological and petrological techniques formerly applied to mineral exploration and sedimentology. The novel aspect now links this with analytical work employing X-ray diffraction and scanning electron microscopy with energy dispersive X-ray spectrometry. Particle species (e.g. minerals, soot, organics, aerosol-derived salts) are identified, as are chemistry, particle number, size, shape and roundness characteristics. Together, the results present a comprehensive particulate data set. Judicious sampling and analysis can then be employed to assess the spatial and temporal aspects of provenance.

The chapter concludes with a discussion on the untapped potential of airborne dust fingerprinting for geomorphologists and others concerned with the provenance of actively transported sediments.

NATURE OF AIRBORNE PARTICLES

Airborne dusts comprise a wide range of particle sizes and species in varying concentrations, which makes them especially difficult to examine. The coarsest dust that can

Tracers in Geomorphology. Edited by Ian D. L. Foster. © 2000 John Wiley & Sons Ltd.

be mobilised by wind action is generally considered to reach a mean particle diameter of 1 mm (1000 μm). British Standard 6069 defines the term 'dust' as comprising particles from 1 to 75 μm, but with construction work, this definition is extended to include particles of 2000 μm diameter (Anon., 1987; Arup, 1995). The finest dust mobilised is around 0.02 μm in size (Tsoar and Pye, 1987). However, for practical purposes airborne particulate matter is generally defined as material in the range of sub-micrometre to 100 μm. Recent definitions further divide the particle size range into nuisance dust visible to the naked eye as particles up to 1000 μm (Mark, 1994; Godby and Sharples, 1997), and health-related dust of mean aerodynamic diameters of less than 10 μm or PM_{10} (Anon., 1996).

In general terms, these aerosols can be summarised into two dominant categories which vary in their origin, chemical composition and residence time in the atmosphere:

(1) *Primary*: combustion sources, industrial emissions including minerals and soils, and soots (chiefly traffic in urban areas).
(2) *Secondary*: salts (mainly sulphate and nitrate) formed from chemical reactions in the atmosphere (often from fuel combustion sources) and sea salts (mainly chloride).

Although research into apportionment of airborne particulate matter is at an early stage, it has recently been suggested that three source types – a primary combustion-generated component, a secondary component, and 'other' particles including mineral dusts and primary non-combustion emissions – each form around one-third of total long-term PM_{10} concentrations in the UK (Harrison, 1999). This assessment is necessarily over-simplified as it relies upon annual mean PM_{10} mass measurements.

Whilst much attention has been paid previously to the study of the coarser nuisance dusts, concerns regarding airborne pollution and links with health now focus research on these finest airborne particles (Seaton *et al.*, 1995; Holgate, 1995, 1998; Pless-Mulloli *et al.*, 2000). Consideration to particle numbers is also urged (Merefield and Stone, 1997). Both of these aspects fuel the need for the development of methods in particle characterisation and counting.

HISTORICAL PERSPECTIVE

Compared with other areas of environmental sedimentation, the study of airborne particulates is relatively new and is still in its early stages of development. As yet, therefore, very little standardisation of procedures has been introduced in dust characterisation.

Whilst dust has been examined scientifically in the field since the 1700s (Dobson, 1781; Darwin, 1846), most early work on airborne particles centred on studies of relatively coarse aeolian sands and silt particles through laboratory experiments (Udden, 1894, 1914). Procedures were later developed to describe this material in terms of roundness and sphericity (Russell and Taylor, 1937) and these procedures were later improved (Powers, 1953). Such particles in the size range of 63–2000 μm were suitable for examination of dust dispersion and re-entrainment by gravimetric analysis and conventional optical microscopy (Bagnold, 1937, 1960). The contribution of airborne salts and dusts to soils was first recognised in the 1960s (Yaalon, 1964), during such work in Israel. Advances in scanning electron microscopy (SEM) from then onwards have enabled research on the finest particles and long-range transport of dusts (reviewed by Pye, 1989).

SAMPLING

Sampling of airborne particles for fingerprinting depends on two factors:

(1)   whether dust deposits or the dust flux are to be measured, and
(2)   if nuisance dusts or potentially health-related particles (PM_{10} and finer) are to be provenanced.

Traditionally, relatively low-technology (passive) gauges have been used for the collection of nuisance dusts. Normally, a one-month period of sampling is necessary in order to provide enough material for reproducible analysis. These passive 'wet' samplers require any rainwater also collected to be removed prior to dust analysis. They are, therefore, not well suited to the collection of secondary particulate matter or the finest particles such as sub-micrometre soot deposits important in 'health-dust' investigations, where forced (pumped) samplers prove essential.

Dust Gauges

Deposit sampling gauges include the British Standard Deposit Gauge (Anon., 1969) which consists of a relatively deep collection bowl. This is not considered as aerodynamically sound as the inverted frisbee design first proposed by Hall and Upton (1988), which has performed well in wind tunnes, computer modelling and field trials (Chiu et al., 1995). A foam insert prevents dust blow-off and the sample is washed by the action of rain into a suitable collecting bottle of 1–5 l capacity. This needs to be masked with black tape and doped with a chlorination tablet in order to prevent algal growth during warm periods which can inhibit particle identification and introduce erroneous gravimetric results.

Directional Gauges

The British Standard Directional Gauge (Anon., 1972), consisting of four slotted pipe collectors, is particularly suitable for the sampling of total flux dust for provenance investigations. Its collection efficiency has been criticised (Ralph and Hall, 1989) in part because vortices and eddies can introduce dust into opposing wind direction slots at some periods, and equally can reject dust from them under other weather conditions. However, use of heavy gauge plastic in the construction of these four-way collectors, angular collecting bottles and a large array of samplers has yielded valuable data during provenance investigations in the South Wales Coalfield (Merefield et al., 1994a,b, 1995a,b).

Pumped Samplers

'Active' samplers range in pumping capacity from a few litres per minute to high volume versions of around $1000 \, l \, min^{-1}$. Total sampling heads can be used for nuisance studies, whilst the finest fractions of dust can be collected using PM_{10}, $PM_{2.5}$ and $PM_{1.0}$ heads (Mark, 1994). The Tapered Element Oscillating Microbalance (TEOM) adds a dimension of real-time gravimetric analysis in the field and is central to the UK's Automated Network (AUN) strategy for air-quality monitoring (ETSU, 1997). With an Automated Cartridge Collection Unit (ACCU^{TM}) system, the TEOM can be turned into a directional sampler collecting particles into different ports allocated for specific wind directions (Hooper *et al.*, 1999). Dedicated samplers are also available using an Andersen head for selection of particle size. All of these 'dry' samplers deposit onto a filter, which makes them eminently suitable for particle characterisation and fingerprinting for provenance (Stone *et al.*, 1995).

SAMPLE PREPARATION

Samples collected using 'passive' deposit and directional samples require preparation to remove them from the rainwater before they can be employed for analysis (Merefield *et al.*, 1995a). Evaporation in porcelain dishes under infra-red lamps at 60 °C has been used. Although this presents a combined soluble and insoluble sample for gravimetric analysis, the development of evaporites from ions in the rainwater inhibits accurate identification of the insoluble components (Merefield *et al.*, 1992). The preferred method is thus filtration. Normally, use of $0.1 \, \mu m$ cellulose nitrate filters located in a multi-port vacuum filtration unit enables reliable batch preparation. Filters supplied from the manufacturer will vary in weight from batch to batch, which makes it necessary to weigh representative filters from each series prior to filtration, when the sample weight is needed. If required, once air-dried, these filters can be weighed and used directly without further disturbance for examination of the individual entrained particles (Merefield and Stone, 1998; Merefield *et al.*, 1998).

However, frequently, overloading makes the counting of individual grains difficult if not impossible and the sample requires dispersion. This is achieved by cutting a representative piece of the filter and shaking it in a vial containing 10 ml of deionised water. Once the individual grains are free, the suspension is re-filtered onto a poly-carbonate filter media using the vacuum filtration system. After air-drying, this subsample can be further prepared for instrumental analysis. Pumped (filter) samplers produce a relatively dry deposit free of rainwater which does not normally require secondary filtration.

Both approaches enable samples to be divided for preparation before X-ray diffraction (XRD) of the crystalline species, and/or physical and chemical examination by scanning electron microscopy with energy dispersive analysis (SEM/EDAX) (Figure 4.1). X-ray diffraction requires 'wetting' of the sample under alcohol in an agate pestle and mortar and smearing onto a half (3 cm) glass slide (Merefield *et al.*, 1999a). Subsamples of all types of filter can be mounted on aluminium pin-type stubs that are prepared with non-conducting adhesive film. Once carbon or gold-coated, they are suitable for SEM/EDS examination.

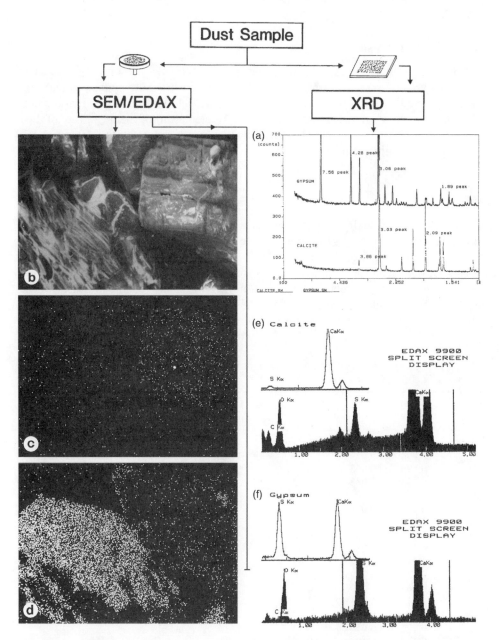

Figure 4.1 Integration of XRD and SEM/EDAX techniques: (a) XRD diffractogram of prepared dust sample indicating the presence of gypsum and calcite; (b) uncoated SEM photomicrograph of the sample in (a); (c) carbon 'element map' calcite particle; (d) sulphur 'element map' delineating a gypsum particle (note the presence of minute gypsum particles adhering to the calcite specimen); (e) EDAX spectrum of a calcite particle (note the presence of a sulphur peak); (f) EDAX spectrum of a gypsum particle

ANALYSIS

Gravimetric Analysis

For both research and regulatory purposes it will be necessary to obtain gravimetric analyses by some means. Samples obtained using frisbee dust deposit gauges are vacuum filtered through pre-weighed 47 mm 0.2 μm cellulose nitrate or nylon membrane filters, which are dried and re-weighed. A calculation of the mean rate of dust deposition of undissolved solids has been devised by Vallach (1995) to determine the dust deposits in milligrams per square metre per day as follows:

$$\frac{(W2 - W1) \times 24.7}{T} \tag{4.1}$$

where $W1$ is the initial dry weight of the filter (in mg), $W2$ is the final dry weight of the filter plus dust (in mg), T is the length of the exposure period, and 24.7 is the surface area of the deposit gauge.

This sample weight is subject to errors from bird strikes, biological insect debris, particle size and weight relationships. It is not a completely reliable indicator of the level of dust deposits and does not in itself indicate particle provenance. However, when repeat measurements are made and the data are used in conjunction with other

Table 4.1 Clay to non-clay ratios and gravimetric data from dust deposit gauges at Nant Helen Opencast Coal Site, South Wales

Month	Site 10 ratio	Off-site 1 ratio	Off-site/site ratio	Site 10 ($\mathrm{mg\,m^{-2}\,day^{-1}}$)	Off-site 1 ($\mathrm{mg\,m^{-2}\,day^{-1}}$)	Off-site/site ($\mathrm{mg\,m^{-2}\,day^{-1}}$)
Jan. 97	0.248	0.178	0.72	145	54	0.37
Feb.	0.094	0.079	0.84	102	47	0.46
Mar.	0.196	0.068	0.35	23	19	0.83
Apr.	0.133	0.083	0.62	144	28	0.19
May	0.218	0.104	0.48	95	22	0.23
June	0.244	0.063	0.26	279	39	0.14
July	0.063	0.107	1.70	585	21	0.04
Aug.	0.131	0.086	0.66	86	39	0.45
Sep.	0.174	0.280	1.61	162	13	0.08
Oct.	0.213	0.122	0.57	164	21	0.13
Nov.	0.248	0.263	1.06	289	42	0.15
Dec.	0.263	0.119	0.45	263	39	0.15
Jan. 98	0.212	0.158	0.75	114	19	0.17
Feb.	0.301	0.202	0.67	52	43	0.83
Mar.	0.122	0.054	0.44	174	29	0.17
Apr.	0.139	0.122	0.88	76	27	0.36
May	0.190	0.168	0.88	121	41	0.34
June	0.124	0.086	0.69	53	32	0.60
July	0.077	0.077	1.00	44	16	0.36
Aug.	0.134	0.119	0.89	61	20	0.33
Mean	0.170	0.120	0.71	148	30	0.20

Table 4.2 Comparison of dust deposit data from Summer 1997 and 1998 at Nant Helen Opencast Coal Site, South Wales

Summer	Site 10 ratio	Off-site 1 ratio	Off-site/site 10/1	Site 10 $(\mathrm{mg\,m^{-2}\,day^{-1}})$	Off-site 1 $(\mathrm{mg\,m^{-2}\,day^{-1}})$	Off-site/site $(\mathrm{mg\,m^{-2}\,day^{-1}})$
1997	0.146	0.085	0.58	317	33	0.10
1998	0.117	0.094	0.80	53	23	0.43
97/98	1.25	0.90	0.72	5.98	1.43	0.23

types of information this form of gravimetric analysis can provide a valuable guide to dust levels. Examples of this are given in Tables 4.1 and 4.2 from data obtained at Nant Helen Opencast Coal Site near Glynneath, South Wales (National Grid Reference, SN810115). A continuous monitoring programme was run there employing frisbee deposit gauges along two transects from the site into two adjacent communities (Gwern-y-gilfach and Penrhos) from November 1995 until March 1999. This enabled any fugitive dust migration to be established, as well as any seasonal influences. Recent work there indicates that concentrations fall to around $25\,\mathrm{mg\,m^{-2}\,day^{-1}}$ in the community (400 m from the workings), a level now established as ambient for the area (Merefield *et al.*, 1999b).

Particle Identification by X-ray Diffraction

Standard XRD procedures prove suitable for identification of the crystalline components of airborne particles. Copper K_α fine focus excitation has been used successfully with the receiving slit set at $0.1°$ and scatter and divergence slits set at $1°$. For general mineralogy ($4-70°\,2\theta$), count rates were adjusted to 1.0 with a step size of 0.02. A PA2000 APD Philips software system was employed to set parameters, to assist with the interpretation of the patterns and to store them on hard and/or floppy disks. For more detailed clay mineral investigations and semi-quantitative analysis ($4-15°\,2\theta$), the count rates were increased to 2.0 seconds and the step size maintained at 0.02 (Merefield *et al.*, 1995a). Manual and automated identification procedures were based on Brindley and Brown (1980). Data from this approach were used to provide time series plots (Figure 4.2) in the vicinity of the former British Coal Opencast coal site at Ffos Las in South Wales (National Grid Reference SN455055) by Merefield *et al.* (1992, 1994a).

In order to provide a robust indication of material disturbed by working of geological materials and soils, a comparative method using XRD data has been devised during studies of these opencast coal sites (Merefield *et al.*, 1994b). Similar approaches could be used for other applications such as areas with a limestone geology. Essentially this method provides information on the relative amounts of point source to ambient dust. In the South Wales Coalfield, clay minerals were used to represent point source emissions and the non-clays to highlight the ambient particles. The clays are usually represented by the clay minerals kaolinite or illite and the non-clays by the ubiquitous mineral quartz. Elevated ratios obtained from XRD patterns usually indicate dust derived from anthropogenic activities.

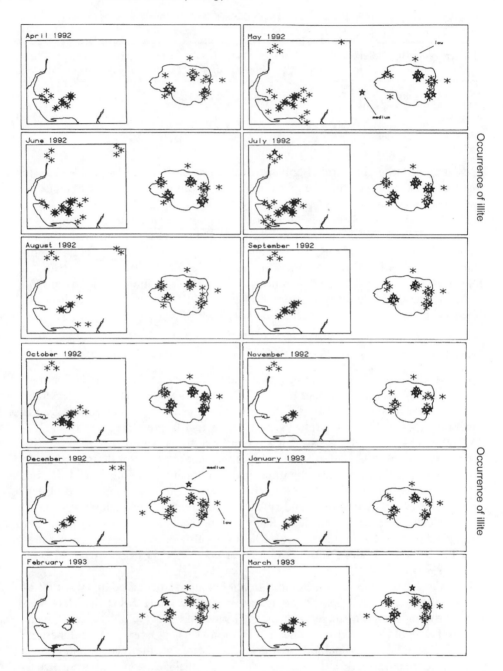

Occurrence of illite

Occurrence of illite

Figure 4.2   Occurrences of illite collected using British Standard directional dust gauges in the vicinity of and on the former Ffos Las opencast Coal Site, Llanelli, South Wales. Stars indicate elevated levels, which occur almost exclusively on the site itself

A revised version which takes into consideration secondary quartz is given in the following equation:

$$35/(Q_{100}/Q_{101} \times 100/1)\, (K_{001}/Q_{101}) = K \text{ ratio} \tag{4.2}$$

where K_{001} is 7.2 Å peak intensity, Q_{100} is 4.26 Å (35 l/l_1) quartz peak intensity, Q_{101} is 3.342 Å (100 l/l_1) quartz peak intensity, and 35 is theoretical 4.26 Å quartz l/l_1.

A further refinement known as a set weight precipitate smear is used to adjust for any variations in the sample weight of the defined area presented to the X-ray diffractometer:

$$0.185/\text{corrected wt} = X \tag{4.3}$$

$$X \times K_{\text{ratio}} = \text{corrected } K_{\text{ratio}} \tag{4.4}$$

where 0.185 is the baseline weight ratio value of weight-corrected standards; the corrected weight is from the weight ratio calibration curve obtained empirically, and K_{ratio} is the weight-corrected kaolinite/quartz ratio.

In order to provide a quantitative estimate of total airborne dust deposits, a gravimetric approach is generally adopted.

Particle Identification by Scanning Electron Microscopy

Physical and chemical examination of airborne particles is readily achieved by scanning electron microscopy with energy dispersive analysis (SEM/EDS). The short wavelength and electromagnetic properties of electrons present several advantages over ordinary light as an image source, enabling very high magnifications and a comparatively great depth of field, which is essential for airborne dust work (Figure 4.3). The X-rays generated by the bombardment of the high-energy electrons incident upon the specimen and their excitation of the electron shells of its constituent elements give rise to a characteristic spectrum (Goldstein, 1992). Use of an advanced polymer super ultra-thin window (S-UTW) detector further enables analysis of these elements down to atomic number 5 (boron). A standardless software package can be employed to analyse the spectra collected by point or area methods (Figure 4.4). The SEM has been used regularly in the UK and internationally to identify such material (Pye, 1989), but usually for qualitative purposes.

Particle Speciation and Counting

New research developments have been designed to describe the characteristics of airborne mineral particles escaping in dust clouds associated with extractive industry workings (Merefield et al., 1999a). These also have applications wherever source apportionment is required (King et al., 1998; Hooper et al., 1999). Analytical work employing scanning electron microscopy with energy dispersive X-ray spectrometry is aided by classical sedimentary petrological techniques in the assessment of particle speciation, number, size, shape and roundness characteristics. Size-selected samples such as PM_{10} can also be analysed for their soluble components (salts and heavy metals) after leaching with warmed (40 °C) de-ionised water.

Figure 4.3 SEM photomicrographs of representative mineral particles from dust gauge samples: (a) general view; (b) coal; (c) quartz; (d) clays; (e) halite; (f) feldspar; (g) calcite; (h) gypsum; (i) dolomite. Note in most samples, the appearance of adhering clay particles. Scale bar is in micrometres

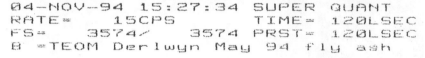

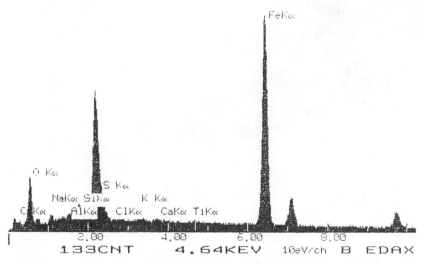

Figure 4.4 PM_{10} particles collected from a TEOM in the Glynneath area, South Wales, showing iron flyash (5 µm), shale particles and secondary calcium sulphate ($CaSO_4$) salts

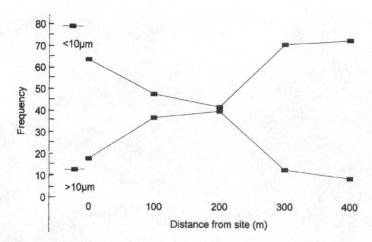

Figure 4.5   Comparison of ≤ 10 μm (PM10) and >10 μm particles of shale deposited along Gwern-y-gilfach transect NH10 to NH1 at Nant Helen OCCS in October 1997. Note the site impact up to 200 m (sampling station NH3)

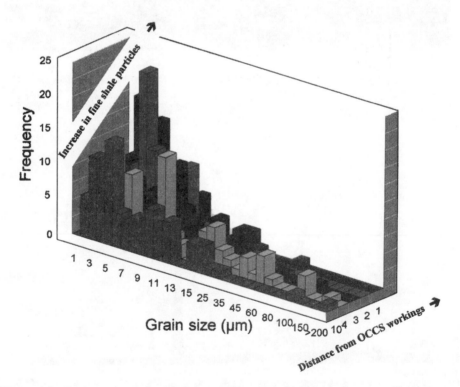

Figure 4.6   Frequency and size of shale particles deposited along Gwern-y-gilfach transect NH10 to NH1 at Nant Helen OCCS in October 1997. Coarse particles drop off with distance from the opencast workings whilst the finest particles increase relatively

In this way airborne minerals from $< 1 \mu m$ to $100 \mu m$ can be characterised spatially and through time series analysis. Examination of the total dust deposits and PM_{10} provides valuable information on the relative persistence of the nuisance and health-risk components of the dust flux (Figures 4.5 and 4.6).

FUTURE DEVELOPMENTS

This chapter has concentrated on the information to be gained from fingerprinting airborne particles. As a result of the implementation of Part IV of the UK Environment Act 1995 on 23 December 1997, all district and unitary authorities in England, Scotland and Wales are under duty to review and assess air quality in their area. From 1996 to 1997 local authorities in 14 areas were funded by the (now) Department of the Environment, Transport and the Regions (DETR) to test various elements of the review and assessment procedures. These findings are to be incorporated into local air quality management (LAQM). Most authorities have had recourse to particle characterisation in some form, such as the calcium, sodium, magnesium, aluminium, silicon, sulphur and potassium analyses undertaken during the High Peak and Derbyshire Dales study; an area with industrial processes including lime and cement production (Nicholls, 1998). These groups will need further support from geomorphologists engaged in dust fingerprinting for provenance as the science develops.

The level of particle counts of individual species is already seen as increasingly important in view of associated health risks (Seaton et al., 1995; Tate et al., 1995; Merefield and Stone, 1997; Pless-Mulloli et al., 2000). Large tracts of land contaminated by former industrial and mining activities have the potential to release airborne dusts to the surrounding urban and agricultural areas (Davies, 1983; Abrahams and Thornton, 1987). As a result of UK regulation for PM_{10} levels (50 $\mu g\,m^{-3}$), research on the status of secondary particles is necessary. Increasing the knowledge base on the soluble (available) components of airborne particles is especially needed in view of the lack of speciation data on health-related 10, 2.5 and sub-micrometre particulates.

CONCLUSIONS

Concern over the potential environmental effects of fugitive particles from opencast mining and quarrying has driven the need to study the provenance of nuisance dusts. It has also provided an opportunity to develop methods for dust fingerprinting. Such work in the vicinity of point sources of airborne particles has since led to effective combinations of sampling and analytical technology. Individual minerals are readily identified and time series plots of atmospheric particles is possible. Minerals such as quartz and calcite prove regional, whilst elevated levels of clays and shale particles can signal dust generation from point sources such as opencast mining and quarrying activity. A combination of this approach coupled with particle size investigations has enabled dust profiling, where sampling has been undertaken along transects away from the point source. When counts are made of key indicator species such as shale, quartz, biological material, flyash and others using a combination of traditional

(1930s) point-counting and modern SEM/EDS techniques, apportionment is achievable. This could prove particularly relevant to investigations concerned with effects of the finest airborne particles on health.

ACKNOWLEDGEMENTS

Grateful thanks are due to Powys, Carmarthenshire and Torfaen County Councils for funding development work of dust fingerprinting at opencast coal sites in South Wales, and for sample collection. Celtic Energy is also gratefully acknowledged for enabling access.

REFERENCES

Abrahams, P. W. and Thornton, I. 1987. Distribution and extent of land contaminated by arsenic and associated metals in mining regions of southwest England. *Transactions of the Institution of Mining and Metallurgy*, **96**, B1–8.

Anon. 1969. *Deposition Gauges*. British Standard 1747, Part 1. HMSO, London.

Anon. 1972. *Directional Dust Gauge*. British Standard 1747, Part 5. HMSO, London.

Anon. 1987 *Glossary of Terms*. British Standard 6069, Part 2. HMSO, London.

Anon. 1996. *Airborne Particulate Matter in the United Kingdom*. Third Report of the Quality of Urban Air Review Group (QUARG), Department of Environment, London.

Arup Environmental, Ove Arup & Partners 1995. *The Environmental Effects of Dust from Surface Mineral Workings*. Report to DoE, Arup Environmental, Ove Arup & Partners, HMSO, London.

Bagnold, R. A. 1937. The transport of sand by wind. *Geological Journal*, **89**, 409–438.

Bagnold, R. A. 1960. The re-entrainment of settled dusts. *International Journal of Air Pollution*, **2**, 357–363.

Brindley, G. W. and Brown, G. 1980. *Crystal Structures of Clay Minerals and Their X-ray Identification*. Mineralogical Society, London.

Chiu, T. W., Grainger, P., Merefield, J. R. and Stone, I. M. 1995. The aerodynamics of the Frisbee dust collectors. In *Proceedings of International Symposium on Air Pollution by Particulates*, Prague, Czech Republic, October 1995. Czech Geological Survey, Prague, 95–115.

Darwin, C. 1846. An account of the fine dust which often falls on vessels in the Atlantic Ocean. *Quarterly Journal of the Geological Society of London*, **2**, 26–30.

Davies, B. E. 1983. Heavy metal contamination from base metal mining and smelting: implications for man and his environment. In Thornton, I. (ed.) *Applied Environmental Geochemistry*. Academic Press, London, 425–462.

Dobson, M. 1781. An account of the Harmattan, a singular African wind. *Philosophical Transactions of the Royal Society of London*, **71**, 46–57.

ETSU 1997. Comparisons of non-urban levels of PM$_{10}$ particulate matter, Urban Network and meteorological data. ETSU Report No N/01/00033/REP.

Godby, S. P. and Sharples, S. 1997. Laboratory technique for evaluation of methods of suppressing nuisance dust. *Transactions of the Institution of Mining and Metallurgy (Section A: Min. Industry)*, **106**, A51–94.

Goldstein, J. I. (ed.) 1992. *Scanning Electron Microscopy and X-ray Analysis*. Plenum, New York.

Hall, D. J. and Upton, S. L. 1988. A wind tunnel study of the particle collection efficiency of an inverted frisbee used as a dust deposit gauge. *Atmospheric Environment*, **22**, 1383–1394.

Harrison, R. M. 1999. *Source Apportionment of Airborne Particulate Matter in the United Kingdom.* Airborne Particles Expert Group (APEG) DETR (Department of the Environment and Transport for the Regions).

Holgate, S. T. 1995. *Non-biological Particles and Health.* Committee on the Medical Effects of Air Pollutants (COMEAP), Department of Health, HMSO, London.

Holgate, S. T. 1998. *Handbook on Air Pollution and Health.* Committee on the Medical Effects of Air Pollutants (COMEAP), Department of Health, HMSO, London.

Hooper, M. J., Hollingsworth, P., Stone, I. M. and Merefield, J. R. 1999. Source apportionment of PM_{10} utilising automatic cartridge collection unit (ACCU) and TEOM systems in Neath Port Talbot. *Clean Air*, **29**, 127–130.

King, A. M., Stone, I. M., Merefield, J. R., Pless-Mulloli, T. and Howel, D. 1998. Separation of local sources of PM_{10} particulate matter from the regional background. In Brebbia, C. A., Ratto, C. F. and Power, H. (eds) *Air Pollution VI, Aerosols and Particles.* CMP, Southampton, 911–917.

Mark, D. 1994. *The Sampling of Aerosols in the Ambient Atmosphere.* Valid Analytical Measurement Initiative: Project 14, Annex D: the development of sampling guidelines, AEA-TPD-353.

Merefield, J. R. and Stone, I. M. 1997. Dishing the dirt on pollution. *New Scientist*, **2100** (September), 58.

Merefield, J. R. and Stone, I. M. 1998. Environmental issues in opencast mining. In Singhal, R. K. (ed.) *Proceedings of the 7th International Symposium on Mine Planning and Equipment Selection (MPES)*, Calgary, October 1998. AA Balkema, Rotterdam/Brookfield, 667–672.

Merefield, J. R., Rees, G., Stone, I., Roberts, J., Parkes, C. and Jones, J. 1992. Mineralogical characterisation of atmospheric dust within and adjacent to opencast coal sites in South Wales. *Proceedings of the Ussher Society*, **8**, 67–69.

Merefield, J. R., Stone, I., Rees, G., Roberts, J., Dean, A. and Jones, J. 1994a. Mineralogy and provenance of airborne dust in opencast coal mining areas of South Wales. *Proceedings of the Ussher Society*, **8**, 313–316.

Merefield, J. R., Stone, I., Jarman, P. J., Roberts, J., Jones, J. and Dean, A. 1994b. Fugitive dust characterisation in opencast mining areas. In Paithankar, A. G. (ed.) *Proceedings of the International Symposium on the Impact of Mining on the Environment: Problems and Solutions*, Nagpur, India, January 1994. AA Balkema, Oxford & IBH, New Delhi, 3–10.

Merefield, J. R., Stone, I. M., Jarman, P., Rees, G., Roberts, J., Jones, J. and Dean, A. 1995a. Environmental dust analysis in opencast mining areas. In Whately, M. K. G. and Spears, D. A. (eds) *European Coal Geology*, Geological Society Special Publication No. **82**, Geological Society Publishing House, Bristol, 181–188.

Merefield, J. R., Stone, I., Roberts, J., Dean, A. and Jones, J. 1995b. Monitoring airborne dust from quarrying and surface mining operations. *Transactions of the Institution of Mining and Metallurgy (Section A: Min. Industry)*, **104**, A76–78.

Merefield, J. R., Stone, I., Roberts, J., Jones, J. and Barron, J. 1998. Airborne particulate characterisation for environmental regulation. In Bennett, M. R. and Doyle, P. (eds) *Issues in Environmental Geology: A British Perspective.* Geological Society Special Publication, Geological Society Publishing House, Bristol, 277–289.

Merefield, J. R., Stone, I., Barron, J. and Jones, J. 1999a. Techniques for tracing fugitive mineral dusts for nuisance control and health risk. *Transactions of the Institution of Mining and Metallurgy (Section A: Min. Industry)*, **108**, A77–81.

Merefield, J. R., Stone, I. and Camp, J. 1999b. Dust characterisation at Nant Helen Glynneath, South Wales. Report to Powys Count Council, No. ERC/99/109, 11pp.

Nicholls, I. 1998. High Peak and Derbyshire Dales. First Phase Authorities Report. *Clean Air*, **28**, 51–52.

Pless-Mulloli, T., King, A., Howel, D., Stone, I. and Merefield, J. R. 2000. PM_{10} levels in communities close to and away from opencast coal mining sites. *Atmospheric Environment.* in press.

Powers, M. C. 1953. A new roundness scale for sedimentary particles. *Journal of Sedimentary Petrology*, **23**, 117–119.

Pye, K. 1989. *Aeolian Dust and Dust Deposition*. Academic Press, London.

Ralph, M. O. and Hall, D. J. 1989. Performance of the BS directional dust gauge. Paper No. W89001 (PA). Presented at the Aerosol Society Annual Conference, March 1989.

Russell, R. D. and Taylor, R. E. 1937. Roundness and shape of Mississippi River sands. *Journal of Geology*, **45**, 225–267.

Seaton, A., MacNee, W., Donaldson, K. and Godden, D. 1995. Particulate air pollution and acute health effects. *Lancet*, **345**, 176–178.

Stone, I. M., Merefield, J. R., Roberts, J., Dean, A. and Jones, J. 1995. A particulate characterisation approach to TEOM, PM10 and TSP samplers. In *Proceedings of International Symposium on Air Pollution by Particulates*, Prague, Czech Republic, October 1995. Czech Geological Survey, Prague, 116–124.

Tate, J., Pless-Mulloli, T., Howel, D., Stone, I. M. and Merefield, J. R. 1995. Measuring the association between dust exposure and health in children living near to opencast coalmining sites – lessons from a pilot study. In *Proceedings of International Symposium on Air Pollution by Particulates*, Prague, Czech Republic, October 1995. Czech Geological Survey, Prague, 211–219.

Tsoar, H. and Pye, K. 1987. Dust transport and the question of desert loess formation. *Sedimentology*, **34**, 139–153.

Udden, J. A. 1894. Erosion, transportation and sedimentation performed by the wind. *Journal of Geology*, **2**, 318–331.

Udden, J. A. 1914. Mechanical composition of clastic sediments. *Geological Society of America Bulletin*, **25**, 655–744.

Vallack, H. W. 1995. A field evaluation of frisbee-type dust deposit gauges. *Atmospheric Environment*, **29**, 1465–1469.

Yaalon, D. H. 1964. Airborne salts as an active agent in pedogenetic processes. In *Transactions 8th International Congress, Soil Science*, Bucharest, V, 997–1000.

5 Tracing Atmospheric Metal Mining Pollution in Blanket Peat

T. M. MIGHALL,[1] J. P. GRATTAN,[2] S. FORSYTH[1] and S. TIMBERLAKE[3]
[1]Centre for Quaternary Science, School of Natural and Environmental Sciences, Coventry University, UK
[2]Institute of Geography and Earth Sciences, University of Wales, Aberystwyth, UK
[3]Early Mines Research Group, Cambridge, UK

INTRODUCTION

Ombrotrophic[1] peat deposits and lake sediments are commonly used to reconstruct atmospheric pollution histories. These deposits collect atmospheric particulate matter released by mining and metalworking and a plethora of studies appear to have successfully identified pollution derived from the use of metals during Roman and historical times (e.g. Lee and Tallis, 1973; Livett et al., 1979; Martin et al., 1979; Jones et al., 1991; Williams, 1991; West et al., 1997). These studies, for example, have interpreted peaks of elements, such as lead (Pb), zinc (Zn) and copper (Cu), as evidence for past historical mining and smelting in the British Isles (Livett, 1988).

Most of these studies have relied on total metal extraction methods which provide limited insight into the behaviour of elements once deposited on the peat surface (Jones, 1987) and make little or no attempt to consider the effect of physico-chemical processes on elemental redistribution. Numerous natural processes can influence element mobility in peat including the position and movement of the water table, pH, redox status, surface vegetation, the degree of humification, ash content and peat accumulation rates, all of which can modify the resultant chemical profile (Van Geel et al., 1989; Stewart and Fergusson, 1994; Shotyk, 1996a). Damman (1978), for example, showed that Pb, Zn and manganese (Mn) are mobilised and removed in permanently anaerobic zones of bogs but are immobile in aerobic zones, whilst Stewart and Fergusson (1994) show that the pattern of Pb, Zn, Mn, calcium (Ca), cadmium (Cd) and Cu profiles in Sphagnum peat at sites in New Zealand can be explained in part by a range of parameters including pH, redox potential, and organic matter content. They conclude that Pb, Zn, Ca and Mn are less suitable for reconstructing atmospheric pollution histories because these elements are more

---

[1] Ombrotrophic peat bogs are also known as raised bogs and the plants which grow on their surface receive their minerals exclusively from the atmosphere as the surface peat layers are hydrologically isolated from the influence of ground and surface waters (Shotyk, 1996a).

Tracers in Geomorphology. Edited by Ian D. L. Foster. © 2000 John Wiley & Sons Ltd.

mobile than Cu and Cd. According to Elomaa (in Shotyk, 1988), interpreting the Zn distribution in peat profiles is problematic because it is affected by numerous processes and therefore its distribution is variable. Despite the fact that few studies have attempted to evaluate the extent to which internal processes modify the distribution of elements, the available results do suggest that elements can be redistributed by internal processes. Because the physico-chemical processes operating in peat are poorly understood, the reconstruction of pollution histories is not straightforward (Jones and Hao, 1993). Given the uncertainty of element behaviour in peat, the reliability of reconstructed atmospheric pollution histories using peats must be questioned.

To overcome the problems of interpreting geochemical records from peat profiles, several authors have suggested that the integration of linked, complementary techniques can provide more insight into element behaviour in the peat ecosystem (Jones, 1987; Stewart and Fergusson, 1994). Jones (1987), for example, used chemical fractionation techniques to demonstrate that elements can be preferentially bound in a range of different forms which may affect their mobility. Van Geel et al. (1989: 471) used pollen analysis as an independent method to provide evidence for human activity and radiocarbon dating to determine peat accumulation rates to compare with geochemical records, although they still conclude that 'at the present state of knowledge the geochemical behaviour of many of the recorded elements cannot be interpreted satisfactorily'.

The studies briefly outlined above fall into two categories. First, there are those which assess the effect of physico-chemical processes on elemental distribution in peat (e.g. Damman, 1978; Jones, 1987; Malmer, 1988; Stewart and Fergusson, 1994); and, secondly, there are studies which reconstruct regional pollution histories (e.g. Lee and Tallis, 1973; Martin et al., 1979). Very few studies have attempted to understand the role of physico-chemical processes in areas with well-known mining histories (Jones and Hao, 1993; Görres and Frenzel, 1997) or on blanket peats in close proximity to industrial sites with independently dated stratigraphies.

The developing interest in early and historical extractive metallurgy and mining in the British Isles and Europe (see Crew and Crew, 1990; Craddock, 1995) provides a wealth of archaeological sites with known mining histories (e.g. Timberlake, 1990a,b; O'Brien, 1994, 1995). Several of these sites are located in areas with suitable peat deposits and therefore provide ideal sites at which to investigate the effect of anthropogenic and post-accumulation redistribution processes on peat chemistry.

In order to assess the effect of peat humification, pH, redox potential, position of the water table and surface vegetation on the pattern of a suite of chemical elements (Cu, Pb, Zn, Ca, Mg, Na, K, Fe, Mn) held within peat, a study was conducted on a profile taken from blanket peat located to the north of the prehistoric and historic mineworkings at Copa Hill, Cwmystwyth, Dyfed, mid-Wales. This chapter discusses the geochemical results in the context of the known physico-chemical processes that influence element distribution and addresses some of the problems in interpreting geochemical evidence from blanket peats by comparing the results with archaeological evidence for prehistoric and historic mining at the site. In conclusion the value of the peat heavy metal record as a reliable tracer for atmospheric pollution records will be evaluated.

Site Description

The Ystwyth Valley is situated 30 km ENE of Aberystwyth. The bedrock of the area comprises Silurian gritstones and shales which constitute part of the Upper Llandovery series. A series of discontinuous ore bodies are associated with a number of major E–W faults where these cut the beds of the Upper Frongoch and basal Cwmystwyth Formations along the north side of the Ystwyth Valley. These ore bodies are predominantly Pb and Zn rich with only minor evidence of Cu mineralisation.

However, the eastern end of the Comet Lode, which strikes the top of Copa Hill on the northern slopes of the Ystwyth Valley, is much richer in copper (Timberlake and Switsur, 1988). Excavations here have shown that a prehistoric opencast or trench mine, at approximately 420 m OD (National Grid Reference SN 811751) (Figure 5.1), was worked during the early Bronze Age (Timberlake, 1990a,b, 1994; Craddock, 1995). Archaeological excavations carried out since 1986 have included a programme of radiocarbon dating. Charcoal, antler and wood recovered from either mine sediments or mineworking spoil located outside the mine, have produced radiocarbon dates to the early to mid Bronze Age (Timberlake, 1987, 1990a,b; 1995; Craddock, 1994). Metal mining continued in the Ystwyth Valley, possibly during Roman and Medieval times, but from at least the 15th century up until the early 20th century for Pb and latterly Zn (Hughes, 1981, 1994; Armfield, 1989). Thus, this location provides an ideal site to reconstruct atmospheric pollution histories for mining spanning four millennia and to trace the movement of metals within the peat.

An extensive area of blanket peat occupies the northern plateau of the Ystwyth Valley. The present-day vegetation consists of a strong moorland element with *Nardus*, *Sphagnum*, and *Calluna* heaths as well as *Eriophorum* blanket peat. In peripheral areas there are grassy heathlands of *Festuca*, *Molinia*, *Nardus*, *Vaccinium* and *Agrostis* (Ball *et al.*, 1981).

METHODS

Field Measurements

Sampling

Samples were extracted from an area of blanket peat located on an upland plateau on the northern side of the Ystwyth valley. A monolith, using monolith boxes of 15 cm × 15 cm cross-section, was taken for analysis from a cleanly exposed section of peat located close to the mine (site CH2), to undertake geochemical, organic carbon, peat humification and pollen analyses. Site CH2 is also the location for investigation into the impact of prehistoric mining on vegetation (Mighall and Chambers, 1993). Selected pollen data from that study are presented in this chapter.

pH and Redox Potential

pH and redox potential were recorded *in situ* every 4 cm down the CH2 peat profile using a ELE field meter (accuracy ±0.02 pH and ±1 mV) and EB202-BNc glass

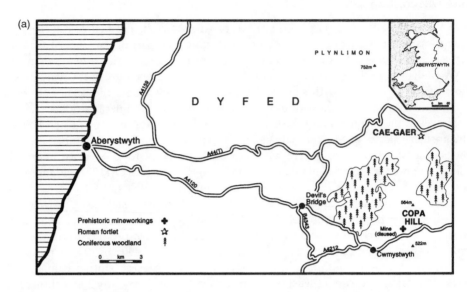

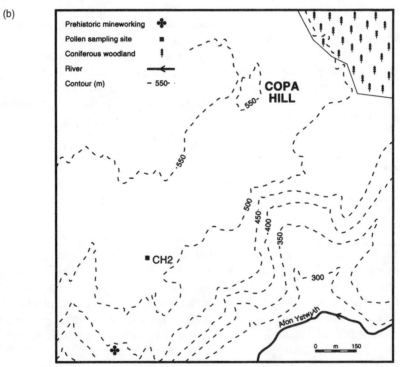

Figure 5.1 Location of (a) the Ystwyth Valley and (b) the pollen and geochemical sampling site

pointed connector. The field meter was first calibrated to pH 4 and pH 7 buffer solutions and to air temperature. A fresh peat section was cut using a spade before the probe was inserted into the peat section.

Laboratory Methods

Samples of 0.5 cm vertical thickness and c. 2 g wet weight were prepared for pollen analysis after Barber (1976). A sum of 500 total land pollen grains were employed, excluding spores and obligate aquatic taxa. Two peat samples from the CH2 profile were submitted to the Centre for Isotope Research, Groningen, for radiocarbon dating.

Contiguous samples of 1 cm thickness were cut for peat humification analyses between 0 and 20 cm and from 110 to 140 cm for the CH2 profile. Between 20 and 110 cm, samples of 1 cm thickness were cut every 2 cm. Peat humification was measured using the procedure outlined by Blackford and Chambers (1993). The data are expressed as percentage light transmission at 540 nm wavelength through an alkali-soluble extract.

Samples of 1 cm thickness were also cut from the monolith for chemical analysis. Contiguous samples were analysed between 0 and 21 cm and in the basal 20 cm. Between 20 and 120 cm, samples were measured at 5 cm increments. Samples were oven-dried at 40 °C before the addition of 10 ml perchloric acid, 2 ml of sulphuric acid and 2 ml of nitric acid to dried peat samples of approximately 0.6 g. A ratio of 4:6:4 ml was added to inorganic samples. Samples were heated on a Kjeldatherm digest unit and each sample was filtered and made up to a 100 ml solution after the addition of 3 ml of Lanthanum chloride. Elements were measured using a Varian model 1472 atomic absorption spectrophotometer.

Contiguous samples from 0 to 21 cm, 121 to 141 cm, and in increments of 5 cm between 21 and 120 cm, were analysed for organic carbon by loss-on-ignition at 925 °C for 4 hours.

RESULTS

Peat Chemistry

The chemical profiles for copper (Cu), lead (Pb), zinc (Zn), potassium (K), calcium (Ca), sodium (Na), magnesium (Mg), iron (Fe) and maganese (Mn) at site CH2 are shown in Figure 5.2.

Copper (Cu), Lead (Pb) and Zinc (Zn)

Cu values (Figure 5.2(a)) show a surface peak of 32 μg g^{-1} but do not exceed 20 μg g^{-1} between 9 and 119 cm. Cu peaks between 134 and 127 cm with values ranging between 83 and 110 μg g^{-1}.

Zn and Pb values (Figure 5.2(b) and (c)) both peak in the top 10 cm of the peat profile (93 μg g^{-1} for Zn and 200 μg g^{-1} between 5 and 7 cm for Pb). Values for both

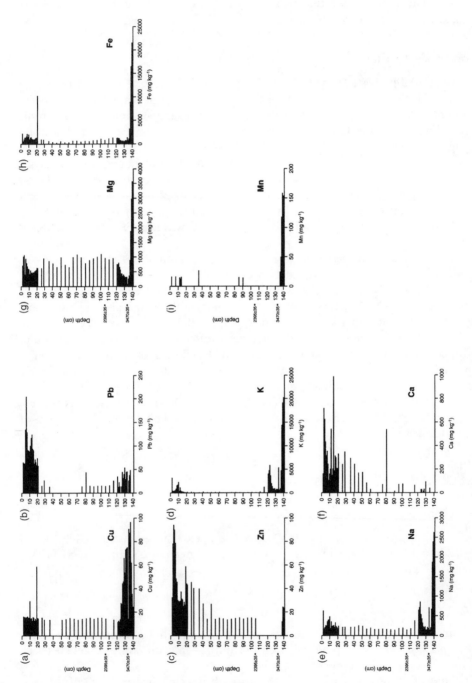

Figure 5.2 Chemical profiles for selected elements from site CH2 located on Copa Hill

elements then gradually fall. Pb reaches zero by 41 cm and remains undetected until between 76 and 125 cm when values range between 14 and 45 μg g^{-1}. Zn averages between 28 and 37 μg g^{-1} from 9 to 18 cm, reaches a peak of 58.73 μg g^{-1} at 19 cm before gradually declining to the lowest limit of detection by 111 cm. A small Zn peak is also recorded at the base of the profile.

Potassium (K), Sodium (Na), Calcium (Ca) and Magnesium (Mg)

The patterns of K and Na are very similar (Figure 5.2(d) and (e)). Both elements have slightly elevated values in the top 12 cm. Their values are consistently low between 13 and 120 cm. Two basal peaks are evident: one small peak between 121 and 125 cm and a second at the base of the profile where K and Na values exceed 20 000 and 2500 μg g^{-1} respectively. Mg values remain relatively constant down-profile until 136 cm when values increase dramatically to over 2700 μg g^{-1} (Figure 5.2(g)). Ca values are highest in the top 90 cm, but gradually decline down profile from 4 cm, with occasional reversals at 12, 15 and 76 cm (Figure 5.2(f)).

Iron (Fe) and Manganese (Mn)

Fe and Mn profiles are very similar (Figure 5.2(h) and (i)). Both elements occur in relatively low concentrations except for the occasional reversal (at 7 cm, between 12 and 14 cm, at 36 cm for Mn and at 21 cm for Fe) before they increase dramatically in the basal 6 cm: Mn to over 100 μg g^{-1} and over 20 000 μg g^{-1} for Fe.

Pollen Diagram

A pollen diagram for selected taxa from site CH2 is shown in Figure 5.3. Only major taxa and those known to affect the accumulation of chemical elements are presented. The pollen data are expressed as percentages of total land pollen (TLP) as described by Mighall and Chambers (1993).

Radiocarbon Dating

A basal sample extracted from the CH2 peat profile, between 134 and 133 cm depth, produced a date of 3470 ± 35 years BP. A second radiocarbon date of 2395 ± 35 years BP was derived from a sample between 105 and 106 cm depth. Dates are expressed in uncalibrated years BP.

Peat Humification

A peat humification curve for site CH2 is presented in Figure 5.4(a). The humification method measures the amount of humic acid present in a peat sample. As peat decomposes it releases humic acid as a breakdown product. Humic acids are dark brown in solution and can be extracted from peat using sodium hydroxide. The colour of the extract is assumed to be indicative of the degree of humification and, therefore, the extent of decomposition. Data are presented as percentage transmission 'with

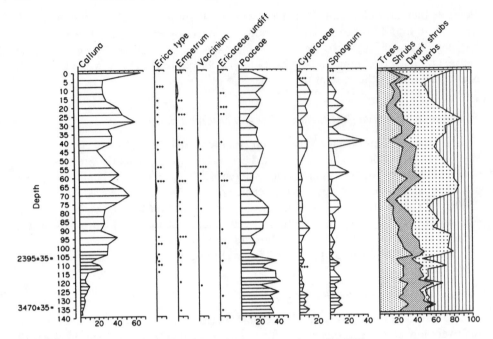

Figure 5.3   Percentage pollen diagram for selected taxa from site CH2

higher transmission values being indicative of and proportional to, but not an exact measure of, lower degrees of humification' (Blackford and Chambers, 1993: 17). The highest transmission values are recorded in the upper part of the peat profile between 0 and 23 cm at site CH2. Values range between 40 and 60%, suggesting a low degree of decomposition. Below 23 cm, transmission values remain relatively constant at between 20 and 25 percent which is indicative of a more humified peat down-profile. At the base of the profile, transmission values increase to 55%. This result is distorted by the inclusion of inorganic matter from the basal clay. Clay particles are thought to prevent humic acids from solubilising in an alkali solution (Blackford and Chambers, 1993).

Redox Potential and pH

Redox potential measurements are shown in Figure 5.4(b). Values range from 150 mV at the base of the profile and gradually rise up core to 205 mV at 70 cm. From 70 cm to the peat surface, values fluctuate between 205 and 210 mV. It is noticeable that redox potential falls from over 170 mV to 155 mV between 124 cm and 128 cm and this probably represents the transition between the acrotelm and the catotelm. In order to locate the position of the water table at Copa Hill, an exposed section cleanly cut down to the clay (to collect the monolith) was left and water was allowed to settle. Water settled at approximately 125 cm depth. Notwithstanding the probability that the water table will fluctuate, this measurement provides an indication of the position of the water table.

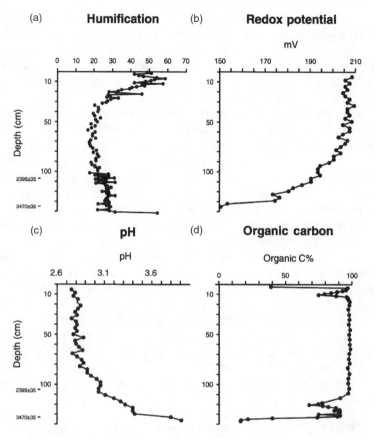

Figure 5.4   (a) Peat humification curve for site CH2; (b) redox potential curve for site CH2; (c) pH profile for site CH2; (d) Organic carbon percentage curve for site CH2

pH values for the CH2 profile are shown in Figure 5.4(c). pH values range from 3.91 and gradually increase in acidity up-profile to between 2.75 and 2.85. The profile is acidic which is most likely due to the production of organic acids released from the humification of peat (Stewart and Fergusson, 1994).

Organic Carbon

Measurements for organic carbon are shown in Figure 5.4(d). Organic carbon dominates the peat profile, with values generally exceeding 98%. Occasional lower values are recorded in the upper 12 cm and in the basal 20 cm. The low values at the base mark the transition from peat into clay.

DISCUSSION

The archaeological evidence confirms that Cu was mined during the Bronze Age on the northern slopes of the Ystwyth Valley. Based on the radiocarbon dates obtained from organic deposits, charcoal and wood fragments collected from the Copa Hill mine and mineworking debris, it is reasonable to assume that mining was taking place for one or more periods between 3690 ± 50 and 3210 ± 80 years BP. Although the exact timespan of mining cannot be determined, the acme of activity appears to have been between 3500 ± 50 and 3405 ± 70 years BP. This is suggested by the dating of in-situ mine deposits buried beneath several metres of infill inside the opencast. In calender years, the most likely range of dates is 1900 to 1600 Cal BC (Timberlake, 1990a,b, 1995). A basal radiocarbon date of the CH2 peat profile, from 134 to 133 cm depth, places the time of peat initiation slightly before 3470 ± 35 years BP. A second radiocarbon date of 2395 ± 35 years BP was derived from a sample between 105 and 106 cm. Using an age–depth curve, the known period of Bronze Age mining corresponds in age with the vegetational record between 138 and 126 cm, an estimate revised since that previously quoted by Mighall and Chambers (1993).

The Cu peak in the basal part of the peat profile (Figure 5.2(a)) corresponds with a period of Bronze Age mining at Copa Hill. Notwithstanding the problems of making direct comparisons between concentrations of chemical elements in soils and peats, it is interesting to note that Cu values peak at over $100 \, \mu g \, g^{-1}$, exceeding the natural background level for Cu in soils, as suggested by Macklin (1992), of $25.8 \, \mu g \, g^{-1}$. Furthermore, the Spearman rank correlation coefficients for each element (Table 5.1) show that Cu does not strongly correlate with any other measured element, suggesting that its distribution within the profile is controlled by different process(es). The correspondence between the Cu curve and the archaeological evidence, combined with the relatively high Cu values and its lack of association with other metal patterns, supports the hypothesis that atmospheric pollution from prehistoric Cu mining is the most likely cause for this peak.

Table 5.1 Correlation coefficients for element against element (r values)

	Fe	K	Mg	Mn	Na	Pb	Zn	Cu
Ca	−0.1554 p=0.236	−0.2821 p=0.029	−0.1447 p=0.270	−0.1797 p=0.170	−0.2425 p=0.062	0.5236 p=0.000	0.697 p=0.000	−0.2873 p=0.030
Fe		0.9084 p=0.000	0.8842 p=0.000	0.9415 p=0.000	0.9156 p=0.000	−0.1832 p=0.161	0.0089 p=0.946	−0.0058 p=0.966
K			0.8577 p=0.000	0.9159 p=0.000	0.9956 p=0.000	−0.2774 p=0.032	−0.1634 p=0.212	0.1228 p=0.363
Mg				0.9172 p=0.000	0.8682 p=0.000	−0.3071 p=0.017	0.048 p=0.716	−0.2235 p=0.095
Mn					0.9205 p=0.000	−0.2121 p=0.104	−0.0205 p=0.876	−0.0171 p=0.900
Na						−0.2489 p=0.055	−0.1216 p=0.355	0.069 p=0.610
Pb							0.6014 p=0.000	−0.0441 p=0.745
Zn								−0.3551 p=0.007

A second known period of Pb–Zn mining occurred during historical times in the Ystwyth Valley, with production peaking in the 18th and 19th centuries (Hughes, 1981). The values for Pb and Zn are highest in the upper 45 cm of the core, with both elements peaking in the top 7 cm (Figure 5.2(b,c)). Pb values are consistently above the natural background level for soils ($29.2 \mu g\, g^{-1}$) whereas Zn only exceeds natural background levels ($59.8 \mu g\, g^{-1}$) in the top 7 cm. The correlation coefficient for Pb against Zn (Table 5.1) suggests that there is a strong, positive correlation between the two elements, indicating that they are influenced by the same process(es). Given that the values occur in the upper part of the profile, it is plausible that they represent atmospheric pollution fallout from mining.

However, the results from previous work suggest that the use of peat heavy metal records to provide this interpretation may not be straightforward. Livett *et al.* (1979), Jones (1987), Livett (1988), Shotyk (1988, 1996a,b), Van Geel *et al.* (1989), Jones and Hao (1993) and Stewart and Fergusson (1994) have all demonstrated that elements deposited in peat can be remobilised after deposition. They suggest that several processes/factors, including the degree of peat humification, plant bioaccumulation, the position of the water table, pH and redox potential of the peat, can affect element movement. If this is the case, the elemental peaks in peat might not accurately reflect a chronological sequence of pollution caused by mining and/or metalworking.

It is interesting to note that the high positive correlation coefficients (Table 5.1) suggest that K, Mg, Na, Fe and Mn are influenced by similar process(es). The correlation coefficients (Table 5.2) show that element distribution appears to be

Table 5.2 Correlation coefficients for elements against pollen percentages

	Depth	Calluna	Cyperaceae	Sphagnum	Poaceae
Ca	−0.6963 p=0.000	0.4543 p=0.058	0.2238 p=0.372	0.0107 p=0.966	−0.2529 p=0.311
Cu	0.4021 p=0.000	−0.5901 p=0.000	0.1984 p=0.220	−0.002 p=0.990	0.4196 p=0.007
Fe	0.2760 p=0.034	−0.2488 p=0.319	0.1265 p=0.617	−0.159 p=0.529	0.2389 p=0.340
K	0.4461 p=0.000	−0.4619 p=0.054	−0.0151 p=0.953	−0.0896 p=0.724	0.3274 p=0.185
Mg	0.2688 p=0.040	0.2375 p=0.343	−0.2693 p=0.280	−0.4766 p=0.046	−0.2027 p=0.420
Mn	0.3002 p=0.021	−0.016 p=0.95	0.0767 p=0.762	0.0091 p=0.971	−0.0023 p=0.993
Na	0.3885 p=0.002	−0.395 p=0.105	0.0182 p=0.943	−0.1065 p=0.674	0.2778 p=0.264
Pb	−0.6242 p=0.000	0.283 p=0.255	0.6746 p=0.002	0.1695 p=0.501	−0.123 p=0.627
Zn	−0.7976 p=0.000	0.6321 p=0.005	−0.0562 p=0.825	−0.1258 p=0.619	−0.3114 p=0.208

significantly correlated with depth. All the metals either peak in the upper and/or basal parts of the peat profile. With the exception of Ca, the alkali metals, Fe, Mn and Cu all have moderate to weak positive r values. These correlation coefficients can be explained by the high values of K, Na, Mg, Fe and Mn at the base of the profile. The basal peaks are likely to have been caused by either bedrock weathering or mineral leaching down-profile. Bedrock weathering provides the main natural source for elements and it is interesting to note that the basal layers become more minerogenic with organic carbon percentages decreasing (Figure 5.4(d)). The r values for these elements are, however, weakened by an occasional peak up-profile (the cause of these peaks is considered below). Cu also has a significant, positive r value (0.4021) with depth caused by the peak between 134 and 124 cm and it is found within peat rather than clay. Cu values actually decline in the clay which suggests that bedrock weathering cannot solely account for its accumulation.

Ca, Pb and Zn have strong negative correlation coefficients when compared with depth (Table 5.2). The strong correlation coefficients for these three elements can be explained by relatively high values at the top of the profile which gradually decline with depth. Two processes can explain this phenomenon. Aaby and Jacobsen (1979) and Pakarinen et al. (1983) have analysed metal deposition rates on the peat surface and have shown that many elements, including Ca, Cd, Cu, Fe, Ni and V, are retained in the surface layer reflecting atmospheric deposition. Alternatively surface peaks can be caused by plant bioaccumulation (Shoytk, 1988). Livett et al. (1979), Stewart and Fergusson (1994) and Espi et al. (1997) have suggested that surface vegetation can adsorb and retain elements, although peat-forming plants have different retention capabilities. Görres and Frenzel (1997) also suggest that elements such as Ca, Mn and K are strongly influenced by the nutrient cycling of mosses (Sphagnaceae) but more immobile elements such as Cu, Pb, Si and Ti could also be affected. Sphagnum mosses are very efficient at accumulating elements passively as they have a porous structure, high content of polygalacturonic acids and high cation exchange capacities. Livett et al. (1979) also suggest that members of the Cyperaceae may help retain elements whilst other species, especially members of the Ericaceae family and Eriophorum, are poor element accumulators although Ericaceae can take up considerable amounts of Mn in their leaves (Van Geel et al., 1989). According to Van Geel et al., (1989), chemical profiles in peat may be diluted by high accumulation rates of plants with poor metal-adsorbing capabilities. Livett et al. (1979) showed that this effect may be exacerbated by plant species, age and state of decomposition. For example, they showed that Zn was more susceptible to movement when compared with Pb. Thus, the ability of vegetation to retain metals could account for surface and near-surface peaks of Cu, Fe, Ca and, to a lesser extent, K, Na and Mn at CH2, especially as Zn, K and Cu are bioessential elements. Those elements with moderate to weak positive r values tend to have both surface and basal peaks (Table 5.2).

In the absence of plant macrofossils in well-decomposed peats, pollen data may be the only type of data available to provide an indication of the type and abundance of mire vegetation growing at, or close to, a sampling site. Although it cannot provide direct evidence of processes such as plant bioaccumulation, individual species composition or the ability of plants to capture metal-bearing particles from the atmosphere, a comparison of metal concentrations with pollen percentages provides an opportunity to investigate whether vegetation might have influenced metal distribution as the peat has accumulated. Notwithstanding the limitations of using pollen data to reconstruct vegetation growing at a site (Moore *et al.*, 1991), the pollen record at CH2 provides an indication of the type and abundance of mire vegetation growing at, or close to, the sampling site during the Late Holocene (Figure 5.3). *Calluna* is the most dominant of these taxa, averaging over 30 percent from 112 to 0 cm, and peaking at over 60 percent at the top of the profile. Cyperaceae pollen are represented at values below 10 percent in the lower part of the profile but increase from 46 to 0 cm, whilst *Sphagnum* peaks at 40 cm before declining to very low values in the top 5 cm. Poaceae values are highest between 106 and 140 cm although grasses are consistently represented in the top 105 cm.

The pollen values for Poaceae, *Sphagnum*, Cyperaceae and *Calluna* were correlated with the elements measured from the Copa Hill peat profile (Table 5.2). The results show that there are only four significant correlations between pollen taxa and chemical values, suggesting that the type of vegetation growing on the mire surface as the peat accumulates has little influence on the distribution of the majority of the metals. There are, however, a number of significant correlations. A negative correlation exists between Mg and *Sphagnum* which suggests that clubmosses have not aided the retention of Mg. The heavy metals also correlate with plant type. Cu appears to be poorly retained in peat with high amounts of *Calluna*, suggesting that Cu values in the upper 120 cm may have been affected by heather growing on the peat. Van Geel *et al.* (1989) note that plants such as *Eriophorum* (a member of the Cyperaceae family) and Ericaceae have poor retention capacities and if they accumulate rapidly can dilute the concentration of the chemicals. A positive correlation between Cu and Poaceae pollen suggests that some grasses may have helped to retain Cu in the basal part of the profile.

If the assertion of Van Geel *et al.* (1989) that Cyperaceae and *Calluna* are poor element adsorbers is correct, the positive correlation between *Calluna* and Zn, and Cyperaceae and Pb, must be treated with caution. The results presented here appear to challenge the findings of Van Geel *et al.* (1989) but a closer examination of the patterns does suggest that the relationship between the pollen taxa and metals is not straightforward. For example, the peak in *Calluna* and the simultaneous drop in Zn and Pb concentrations in the top 5 cm supports Van Geel *et al.*'s (1989) idea that heather is a poor element accumulator. Although a cause and effect relationship of the patterns described above cannot be proved, and notwithstanding the inconsistencies within the patterns, these results suggest that heavy metal values might be, in part, an artefact of vegetation type. An alternative explanation is that the high sub-surface and surface metal concentrations are also the result of increased atmospheric metal deposition. Espi *et al.* (1997) suggest that plant bioaccumulation and/or increased atmospheric metal deposition are responsible for elevated values of Cd, Zn, and Cu in peat profiles in the Ovejuyo Valley, Bolivia. The high concentrations of Pb and Zn in the subsurface layers at Copa Hill could be the result of atmospheric pollution from metal mining and the lack of Pb and Zn at 5 cm could be interpreted as a result of the cessation of metal mining in the Ystwyth Valley. This explanation is more likely for Pb, as Espi *et al.* (1997) suggest that Pb is not required by plants and therefore should not be enriched relative to other elements as a result of plant bioaccumulation.

Many researchers have argued that metals are subject to down-profile translocation and this process could alter the chemical record substantially. For example, Aaby and Jacobsen (1979) note that Zn, Mg, Na, K and Mn can be translocated down-profile whilst Malmer (1988) suggests that Fe, Zn and Pb are amongst the elements with low retention rates. An examination of the metal profiles at Copa Hill supports the idea that some of the metals may have been translocated. It is noticeable that Ca, Zn and Pb values gradually decrease down-profile whilst concentrations of K, Na, Fe, Mn and, to a lesser extent, Mg and Cu are lower throughout the upper and/or middle parts of the peat column. Whilst these lower values may simply reflect a low atmospheric input into the peat, previous work has shown that these metals are prone to translocation. Shotyk (1988), Jones and Hao (1993) and Stewart and Fergusson (1994) all suggest that element mobility in peat is controlled by one or a combination of redox potential, pH and the position of the water table. Redox and pH measurements are presented in Figure 5.4(b,c) and they have been correlated with the metals data (Table 5.3).

The lower concentrations of Na and K within the peat between 3 and 115 cm is not surprising as both form highly soluble monovalent cations in aqueous solutions which have little affinity to organic matter (Shotyk, 1988). Jones and Hao (1993) suggest that K is likely to be responsive to changing pH and redox status. Ca and Mg are also very soluble, with little affinity for organic matter and, combined with the low pH, are likely to be leached below the surface vegetation layer. The r values for K, Mg and Na

Table 5.3 Correlation coefficients for elements against Eh, pH, peat humification and organic carbon content

	Eh	pH	Humification	Organic carbon
Ca	0.6372 p=0.006	−0.6293 p=0.007	0.444 p=0.002	0.1535 p=0.25
Cu	−0.6868 p=0.000	0.6906 p=0.000	0.2122 p=0.035	−0.194 p=0.148
Fe	0.1704 p=0.513	−0.1755 p=0.500	0.2920 p=0.044	−0.7715 p=0.000
K	−0.3214 p=0.208	0.3104 p=0.225	0.2453 p=0.093	−0.8815 p=0.000
Mg	0.3181 p=0.213	−0.3127 p=0.222	0.0800 p=0.589	−0.7303 p=0.000
Mn	0.2417 p=0.350	−0.2130 p=0.412	0.4246 p=0.003	−0.7852 p=0.000
Na	−0.0295 p=0.911	0.0232 p=0.929	0.3379 p=0.019	−0.8758 p=0.000
Pb	0.3672 p=0.147	−0.3725 p=0.141	0.7707 p=0.000	0.1745 p=0.190
Zn	0.7137 p=0.001	−0.7122 p=0.001	0.5365 p=0.000	0.0649 p=0.629

against Eh and pH reveals that there is no significant correlation. This may be partly explained by the relatively low metal values in the peat or by either the pH and/or Eh having reached a critical threshold which has resulted in the translocation of the metal. Ca concentrations gradually decrease with depth (Figure 5.2(f)) and both Damman (1978) and Stewart and Fergusson (1994) suggest that this gradual reduction may be the result of Ca ions being removed by water movement. This may be aided by pH which has a significant negative correlation with Ca (−0.6293) and possibly by redox potential. The poor correlation between most of the elements measured and redox potential possibly results from the relatively high redox potential measurements in the top 100 cm of the peat profile. Positive redox measurements are best explained as the result of oxygenated groundwater moving through the peat and this input may be sufficient to negate the effect of redox potential observed in more anaerobic peat bogs.

Mattson and Koutler-Andersson (1955), Jones and Hao (1993) and Stewart and Fergusson (1994) suggest that Fe and Mn have a strong relationship with redox status and pH and they are readily removed from the zone below the water table. Low pH values assist this process by stabilising Mn^{2+} ions in solution. Low pH values from Copa Hill support this process; however, neither pH nor redox show any significant correlation with Fe or Mn. This might also be a result of very low Fe and Mn concentrations in the peat or the environmental conditions of the peat are still conducive for Mn and Fe loss irrespective of little variation in pH and redox status.

Pb values do not correlate significantly with either pH or redox (Table 5.3) although both these parameters may have influenced the distribution of Pb in the lower part of the profile. Stewart and Fergusson (1994) suggest that Pb is immobilised by low soluble compounds, such as PbS, in a low pH and Eh environment, whilst Shotyk (1988) suggests that Pb is immobilised under strong oxidising conditions but that it can be mobilised by organic acids in low-pH peat waters regardless of redox potential (Livett et al., 1979). The apparent loss of Pb in the lower part of the peat profile may be explained by either of these processes as the peat profile has a low pH which would provide organic acids. It is also possible that there is insufficient sulphur in the profile below 31 cm to retain Pb above the recorded values. The lack of any relationship between Pb and pH or redox potential in the top part of the profile suggests that neither parameter has significantly altered the distribution of Pb. This is supported by research showing that the rate of migration of Pb in ombrotrophic peats is small, if not insignificant (Shotyk, 1996b; Shotyk et al., 1997). The absence of Pb below 31 cm is probably best explained by a lack of either an atmospheric or groundwater input of Pb.

The results for Zn are more difficult to interpret. This is possibly due to the fact that Zn can be affected by numerous processes including pH, redox potential and ash enrichment (Shotyk, 1988, 1996a). Stewart and Fergusson (1994) suggest that Zn behaves in a similar way to Pb whilst Livett et al. (1979) argue that Zn is more prone to leaching. Shotyk (1988) suggests that because Zn has only one oxidation state, its solubility is unaffected by redox potential whilst it can precipitate as a sulphide in neutral–alkaline conditions. The general patterns of Pb and Zn are similar at Copa Hill (a positive r value of 0.6014; Table 5.1), although Zn has a significant positive correlation with Eh and a significant negative correlation with pH (Table 5.3). Whether these are cause and effect relationships is difficult to determine. A closer examination of the Zn profile (Figure 5.2) suggests that there are some inconsistencies. For example, the Zn maxima at the top of the Copa Hill profile and the increase in values between 18 and 46 cm do not correspond with a significant change in either redox and/or pH. Thus, the effect of pH and redox potential on Zn concentrations is unclear.

A number of studies have shown that subsurface peaks of elements such as Al, Fe, Mn, Zn and Pb occur in peat profiles and these are thought to have been caused by the position of the mean water table (Shotyk, 1988; Malmer, 1988; Van Geel et al., 1989). Aaby and Jacobsen (1979) and Pakarinen et al. (1983) observed that Fe, Zn and Pb peaks in subsurface layers of peat coincided with the transition from aerobic to anaerobic conditions although this feature is not always seen for Zn and Pb, whilst Damman (1978) showed that Fe values peaked at the position of the water table before decreasing in value. Thus, the position of the water table is an important environmental parameter which can influence element mobility. At Copa Hill, the water table occurs at approximately 125 cm depth. Notwithstanding the probability that the water table will fluctuate, this measurement provides an indication of the position of the water table. It is noticeable that redox potential falls from over 170 mV to 155 mV between 124 cm and 128 cm and this probably represents the transition between the acrotelm and the catotelm. Compared with other studies (e.g. Stewart and Fergusson, 1994), the position of the water table is located at some depth. This can be explained by the topography and drainage of this area of blanket peat. It covers an area that slopes gently on a N–S axis which allows water to move laterally downslope and drain the peat to create relatively well oxidised conditions. The position of the water table at this depth coincides with the decline in Cu concentrations (Figure 5.2(a)), and a small peak in K and Na and might have been responsible for changes in these metal concentrations.

There are a number of sub-surface peaks in the metal profiles that cannot be explained by a correlation with vegetation, pH, Eh and/or the position of the water table. For example, Ca values peak at 12, 15 and 76 cm, Fe at 21 cm, Zn at 19 cm, and K and Na at 121 cm. It is possible that these peaks are an artefact of past environmental conditions. For example, pH, redox potential and peat decomposition could have varied with time as the peat grew or the peaks could result from the action of an unknown process. Another possibility is that these peaks represent changes in the metal profiles caused by hummock and pool development on the mire surface. Aaby and Jacobsen (1979), Pakarinen et al. (1983) and Van Geel et al. (1989) attribute element movement to variations in the microtopography of the peat bog surface caused by hummock and pool formation. Drier hummocks and wetter hollows create different redox potential, pH, and local hydrological conditions but there is some dispute as to which part of the peat bog is preferable for recording changes in past metal deposition (see Van Geel et al., 1989). Jones and Hao (1993: 70) have demonstrated that the concentration of individual elements varies considerably with pool and hummock formation but that Cu, Pb and Zn 'exhibit considerable consistency'. It is not possible to comment on the effect of microtopography variations across the blanket peat at Copa Hill as there is no well-defined hummock and pool development evident at the site at the present time, but if Jones and Hao (1993) are correct, Pb and Zn should not be affected by hollow and hummock variations.

A statistical comparison of the peat humification (Figure 5.4(a)) and chemical curves suggests that decomposition could determine, in part, metal distribution. Correlation coefficients in Table 5.3 show that those elements, in particular Pb, Zn and Ca, with higher near-surface concentrations are positively correlated with peat humification. This relationship suggests that less humified peat may preferentially hold these elements as peat humification values are highest in the upper 20 cm of the profile. Below 20 cm, they fluctuate between 18 and 30%, suggesting that the degree of decomposition is relatively uniform and is unlikely to explain the changes in the chemical curves in the lower parts of the profile.

Stewart and Fergusson (1994) suggested that Cu showed little down-profile variation from three sites in New Zealand. They suggest that this is explained by its affinity for organic matter, regardless of pH. However, slightly higher values were recorded in the lower half of the peat profile, a feature that the authors considered to result from the slower decay of peat in the catotelm which would produce elevated Cu values. Elevated Cu values are also observed in the profile (Figure 5.2(a)) and may be the result of slow peat decay. Although lower transmission values correspond with elevated Cu concentrations, similar peat humification values occur elsewhere in the profile but do not appear to be associated with higher Cu concentrations. This implies that slow decay of peat cannot solely account for the anomalously high Cu values at the base of the profile. Indeed there is only a weak, but significant, correlation between humification and Cu.

The discussion so far has focused on the potential mobility of metals in the peat profile. There is some evidence to support the role of plant bioaccumulation at the top of the peat profile with elevated concentrations of metals such as K, Na and Ca. However, an examination of the effect of pH, redox status, vegetation type and the position of the water table on the distribution of metals in the Copa Hill peat profile suggests that the effect of these parameters varies with metal type. For example, in this study redox potential and/or pH does not appear to account for the distribution of certain metals such as Fe, Mn, Pb and Mg whilst there are some statistically significant relationships between certain metals (Cu, Pb, Zn and Mg) and specific pollen taxa. The results presented here suggest that the relationship between these parameters and metals is complex.

It seems likely that either anthropogenic causes and/or processes that operate in the peat bog system which render elements immobile help to create the variability in the Copa Hill data set. For example, Jones (1987), Shotyk (1988) and Stewart and Fergusson (1994) have demonstrated that elements behave differently, depending upon the type of sink holding the elements, and suggest that metals, such as Cu and Zn, are usually organically bound and less predisposed to translocation which reduces their solubility in peat. This process would prevent metals from being internally redistributed in the Copa Hill peat profiles irrespective of peat humification, pH and redox potential. The curve for organic carbon at CH2 (Figure 5.4(d)) shows that the profile is dominated by organic matter despite fluctuating surface and basal values. The high organic carbon values support the idea that metals such as Cu, Pb and Zn could be organically bound and not subject to remobilisation. However, none of the heavy metals show a significant positive correlation with organic carbon. This is probably due to the variation in organic carbon values at the surface and base of the peat profile which is best explained by the input of inorganic atmospheric particulate matter at the surface of the profile by historic mining and/or basal weathering and atmospheric pollution at the base of the profile. Shotyk (1996a) has shown that increased mineral matter, measured as ash content, can lead to metal enrichment in peat profiles. Enrichment factors for Cu, Pb and Zn in the peat profiles at Copa Hill suggest that the metal concentrations are enriched relative to natural concentrations and did not significantly alter the pattern of the Cu, Pb and Zn profiles (Mighall, T. M., Grattan, J. P., Timberlake, S. and Forsyth, S., unpublished).

CONCLUSION

Geochemical analysis of the blanket peats located close to an area of known pre-historic and historic mining confirm that comparatively high concentrations of heavy metals have been deposited within the peat and therefore provide the potential to reconstruct the atmospheric pollution history of the Ystwyth Valley. The results presented in this chapter demonstrate that the relationship between metal concentrations and physico-chemical processes is complex. No one single physico-chemical processes can adequately account for the distribution of the elements measured in the blanket peat on Copa Hill. It is evident that the influence of one or more of these physico-chemical processes on element concentration varies from element to element. Certain elements, namely Ca, Na, K, Fe and Mn, appear to undergo post-deposi-tional translocation, which is consistent with findings reported in previous work in that they are affected by one or a combination of internal parameters such as redox potential, pH, the position of the water table, peat humification and plant bio-accumulation. In certain cases it is not possible to make confident assertions about how the physical and chemical processes that operate in peat bogs affect metal concentrations. Some of the results presented in this chapter are also inconsistent with previous research which suggests that the effect of physico-chemical processes varies from site to site.

Notwithstanding the conclusions made in the previous paragraph, the results from this chapter suggest that tracing the atmospheric pollution history of heavy metal mining is still possible and that the geochemical records for Cu, Pb and Zn do not appear to have been altered significantly by physico-chemical processes operating internally within the peat bog to distort this chronological record of mining activities. The metal distributions for Cu, Pb and Zn are consistent with the archaeological and historical evidence for metal mining. In particular, Cu concentrations are still con-sistent with the radiocarbon-dated archaeological evidence for early Bronze Age and therefore appear to be largely unaffected by other processes, suggesting that Cu pollution can be adequately traced using peat records. Although the elevated con-centrations for Zn and Pb that occur in the upper part of the profiles could have been significantly altered by numerous internal parameters, the geochemical pattern for these elements is still consistent with historical metal mining. Both the increase in the concentration of Zn and Pb in the upper part of the profile, and the fall in the concentration of these elements in the top 5 cm, can adequately be explained by the cessation of Pb and Zn mining. However, before these peat metal records can be used as reliable tracers of atmospheric mining pollution, an independent chronology is needed.

ACKNOWLEDGEMENTS

E. Turner and C. White provided assistance with the chemical analyses and fieldwork. Thanks are due to M. Dezulu who helped prepare the samples for humification analyses and permission to include some of his data in this chapter. We are grateful to Professor Ian Foster for helpful discussions on many aspects of this chapter. Figures were drawn by the Cartography Unit, Geography, Coventry University. The authors are grateful to Dr J. M. Jones and Dr F. B. Pyatt who provided useful comments on an earlier draft of the chapter.

REFERENCES

Aaby, B. and Jacobsen, J. 1979. Changes in biotic conditions and metal decomposition in the last millennium as reflected in ombrotrophic peat in Draved Mose, Denmark. *Dansmarks Geologiske Undersogelse*, 1978, 5–43.

Armfield, C. 1989. Dressing floors on the Kingside Lode, Copa Hill, Cwmystwyth. *Archaeology in Wales*, **29**, 27–29.

Ball, D. F., Dale, J., Sheail, J. and Williams, W. M. 1981. *Ecology of Vegetational Change in Upland Landscapes. Part II, Study Areas.* Bangor Occasional Paper 3, Institute of Terrestrial Ecology, Bangor.

Barber, K. E. 1976. History of vegetation. In Chapman, S. B. (ed.) *Methods in Plant Ecology.* Blackwell Scientific, Oxford, 5–83.

Blackford, J. J. and Chambers, F. M. 1993. Determining the degree of peat decomposition for peat-based palaeoclimatic studies. *International Peat Journal*, **5**, 7–24.

Craddock, P. T. 1994. Recent progress in the study of early mining and metallurgy in the British Isles. *Historical Metallurgy*, **28**, 69–84.

Craddock, P. T. 1995. *Early Mining and Metal Production.* Edinburgh University Press, Edinburgh.

Crew, P. and Crew, S. (eds) 1990. *Early Mining in the British Isles.* Plas Tan y Bwlch Occasional Paper 1. Snowdonia Press, Porthmadog.

Damman, A. W. H. 1978. Distribution and movement of elements in ombrotrophic peat bogs. *Oikos*, **30**, 480–495.

Espi, E., Boutron, C. F., Hong, S., Pourchet, M., Ferrari, C., Shotyk, W. and Charlet, L. 1997. Changing concentrations of Cu, Zn, Cd and Pb in a high altitude peat bog from Bolivia during the past three centuries. *Water, Air and Soil Pollution*, **100**, 213–219.

Görres, M. and Frenzel, B. 1997. Ash and metal concentrations in peat bogs as indicators of anthropogenic activity. *Water, Air and Soil Pollution*, **100**, 355–365.

Hughes, S. 1981. *The Cwmystwyth Mines.* British Mining No. 17, Northern Mines Research Society.

Hughes, S. 1994. The hushing leats at Cwmystwyth. In Ford, T. D. and Willies, L. (eds) *Mining before Powder. Peak District Mines Historical Society*, **12**(3) and Historical Metallurgy Society Special Publication, 48–53.

Jones, J. M. 1987. Chemical fractionation of Cu, Pb and Zn in ombrotrophic peat. *Environmental Pollution*, **48**, 131–144.

Jones, J. M. and Hao, J. 1993. Ombrotrophic peat as a medium for historical monitoring of heavy metal pollution. *Environmental Geochemistry and Health*, **15**(2/3), 67–74.

Jones, R., Chambers, F. M. and Benson-Evans, K. 1991. Heavy metals (Cu and Zn) in recent sediments of Llangorse Lake Wales: non-ferrous smelting, Napoleon and the price of wheat – a palaeoecological study. *Hydrobiologia*, **214**, 149–154.

Lee, J. A. and Tallis, J. H. 1973. Regional and historical aspects of Pb pollution in Britain. *Nature*, **245**, 216–218.

Livett, E. A. 1988. Geochemical monitoring of atmospheric heavy metal pollution: theory and applications. *Advances in Ecological Research*, **18**, 65–177.

Livett, E. A., Lee, J. A. and Tallis, J. H. 1979. Lead, zinc and copper analyses of British blanket peats. *Journal of Ecology*, **67**, 865–891.

Macklin, M. G. 1992. Metal pollution of soils and sediments: a geographical perspective. In Newson, M. D. (ed.) *Managing the Human Impact on the Natural Environment: Patterns and Processes*. Belhaven Press, London, 172–195.

Malmer, N. 1988. Patterns in the growth and the accumulation of inorganic constituents in the *Sphagnum* cover on ombrotrophic bogs in Scandinavia. *Oikos*, **53**, 105–120.

Martin, M. N., Coughtrey, P. J. and Ward, P. 1979. Historical aspects of heavy metal pollution in the Gordano Valley. *Proceedings of the Bristol Naturalists Society*, **37**, 91–97.

Mattson, S. and Koutler-Andersson, E. 1955. Geochemistry of a raised bog. *Annals of the Royal Agricultural College of Sweden*, **21**, 321–366.

Mighall, T. M. and Chambers, F. M. 1993. The environmental impact of prehistoric mining at Copa Hill, Cwmystwyth, Wales. *The Holocene*, 3(3), 260–264.

Moore, P. D., Webb, J. A. and Collinson, M. E. 1991. *Pollen Analysis*. Blackwell Scientific Publications, London.

O'Brien, W. F. 1994. *Mount Gabriel: Bronze Age Mining in Ireland*. Galway University Press, Galway.

O'Brien, W. F. 1995. Ross Island and the origins of Irish-British metallurgy. In Waddell, J. and Shee-Twohig, E. (eds) *Ireland in the Bronze Age*. The Stationery Office, Dublin, 38–48.

Pakarinen, P., Tolonen, K., Heikkinen, S. and Nurmi, A. 1983. Accumulation of metals in Finnish raised bogs. *Ecological Bulletins*, **35**, 377–382.

Shotyk, W. 1988. Review of the inorganic geochemistry of peats and peatland waters. *Earth Science Reviews*, **25**, 95–176.

Shotyk, W. 1996a. Peat bog archives of atmospheric metal deposition: geochemical evaluation of peat profiles, natural variations in metal concentrations, and metal enrichment factors. *Environmental Reviews*, **4**, 149–183.

Shotyk, W. 1996b. Natural and anthropogenic enrichments of As, Cu, Pb, Sb and Zn in rainwater-dominated versus groundwater-dominated peat bog profiles, Jura Mountains, Switzerland. *Water, Air and Soil Pollution*, **84**, 1–31.

Shotyk, W., Norton, S. A. and Farmer, J. G. 1997. Summary of the workshop on peat bog archives of atmospheric metal deposition. *Water, Air and Soil Pollution*, **100**, 213–219.

Stewart, C. and Fergusson, J. E. 1994. The use of peat in the historical monitoring of trace metals in the atmosphere. *Environmental Pollution*, **86**, 243–249.

Timberlake, S. 1987. An archaeological investigation of early mineworkings on Copa Hill, Cwmystwyth. *Archaeology in Wales*, **27**, 18–20.

Timberlake, S. 1990a. Excavations at an early mining site on Copa Hill, Cwmystwyth, Dyfed 1989 and 1990. *Archaeology in Wales*, **30**, 7–13.

Timberlake, S. 1990b. Excavations and fieldwork on Copa Hill, Cwmystwyth, 1989. In Crew, P. and Crew, S. (eds) *Early Mining in the British Isles*. Plas Tan y Bwlch Occasional Paper No. 1, 22–29.

Timberlake, S. 1994. Archaeological and circumstantial evidence for early mining in Wales. In Ford, T. D. and Willies, L. (eds) *Mining Before Powder. Peak District Mines Historical Society* **12**(3) and Historical Metallurgy Society Special Publication, 133–143.

Timberlake, S. 1995. Copa Hill, Cwmystwyth. *Archaeology in Wales*, **35**, 40–43.

Timberlake, S. and Switsur, R. 1988. An archaeological excavation of early mineworkings on Copa Hill, Cwmystwyth: new evidence for prehistoric mining. *Proceedings of the Prehistoric Society*, **54**, 329–333.

Van Geel, B., Bregman, R., Van der Molen, P. C., Dupont, L. M. and Van Driel-Murray, C. 1989. Holocene raised bog deposits in the Netherlands as geochemical archives of prehistoric aerosols. *Acta Botanica Neerlandica*, **38**(4), 467–476.

West, S., Charman, D. J., Grattan, J. P. and Cherburkin, A. K. 1997. Heavy metals in Holocene peats from south west England: detecting mining impacts and atmospheric pollution. *Water, Air, and Soil Pollution*, **100**, 343–353.

Williams, T. M. 1991. A sedimentary record of the deposition of heavy metals and magnetic oxides in the Loch Dee Basin, Galloway, Scotland, since AD 1500. *The Holocene*, **1**(2), 142–150.

6 The Use of Tracers to Aid the Understanding of Herbicide Leaching in Clay Soils

C. M. HEPPEL,[1] T. P. BURT[2] and R. J. WILLIAMS[3]

[1]Department of Geography, Queen Mary and Westfield College, University of London, UK
[2]Hatfield College, University of Durham, UK
[3]Institute of Hydrology, Wallingford, UK

INTRODUCTION

The use of pesticides in agriculture has obvious benefits to the farmer, such as an improved yield and a reduced chance of crop failure. Modern pesticides are generally applied at lower application rates and are less persistent in the environment than their predecessors but are still highly active and can pose a threat to non-target organisms in their path. Currently, it is not uncommon for six or seven different pesticides to be applied to a cereal crop during a single growing season (NRA, 1995). Previous studies in the UK have shown that, for a few months after application, some agricultural pesticides are found in streams draining arable land (Harris, 1995). Of this group of compounds, it is the agricultural herbicides that are detected most frequently in freshwaters due to their widespread use and moderate solubility. Isoproturon (IPU) is one such herbicide: this phenyl-urea compound is used for the post-emergent treatment of blackgrass in cereal crops at a recommended application rate of 1.5 kg a.s. ha^{-1} (a.s. = active substance). Since 1994 it has been the herbicide most frequently found in the surface waters of England and Wales at concentrations exceeding the Environment Agency's non-statutory Environmental Quality Standard (EQS) of 20 μg l^{-1} for a maximum allowable concentration (EA, 1997; Williamson and Croxford, 1998).

Due to concerns surrounding the leaching of isoproturon from agricultural land, the Isoproturon UK Task Force was set up in order to provide farmers and advisers with advice on the best ways to prevent isoproturon from reaching surface waters (IPU Task Force, 1993). The Task Force recognised three routes by which IPU enters surface waters: (i) point-source contamination, (ii) movement through soil cracks, and (iii) surface runoff. Studies in Germany have demonstrated that point-source contributions can account for as much as 95 percent of the total isoproturon load in a watershed (Fischer et al., 1996; Seel et al., 1996). In the UK research is currently being

Tracers in Geomorphology. Edited by Ian D. L. Foster. © 2000 John Wiley & Sons Ltd.

undertaken to assess the importance of point-source contributions. Results of studies on the River Thames and River Cherwell in Oxfordshire indicate that a substantial percentage of isoproturon residues detected in freshwater after autumn application may originate from point-source contributions, such as wash water from farm equipment after spraying (Jones *et al.*, in press).

About 45 percent of the cereal crops of England and Wales are grown on clay soil, and isoproturon is applied to over half of the UK cereals acreage annually (Cannell *et al.*, 1978; ENDS, 1995). Close attention has been paid, therefore, to the hydrological mechanisms by which isoproturon leaches from clay soils. Field and catchment studies have been carried out at three main centres in the UK: at ADAS Rosemaund, Brimstone Farm in Oxfordshire and Cockle Park in Northumberland (Harris *et al.*, 1994; Brown *et al.*, 1995; Williams *et al.*, 1995). These studies have concluded that the transport of isoproturon is event-based with high concentrations of herbicide recorded in streams and field drains during rainfall; and that herbicide leaching occurs predominantly due to bypass flow through preferential pathways, macropores and through cracks caused by artificial drainage so that there is little opportunity for readsorption of herbicide within the soil matrix.

The majority of UK-based studies of pesticide leaching have concentrated on data on total losses of herbicide. Little has been reported concerning the losses of herbicide from a field site over the timescale of a single event. A few points concerning the herbicide chemograph have been noted, however. In general, pesticide concentrations rise with the hydrograph and peak at the same time, or slightly prior to it (Kladivko *et al.*, 1991; Brown *et al.*, 1995; Williams *et al.*, 1995). There are two proposed mechanisms in the literature which attempt to explain this phenomenon. The first is that the flow during the early part of an event is dominated by bypass flow through macropores; this results in the rapid transport of water and solutes through the drainage systems to surface waters and correspondingly high concentrations of pesticide. As the event continues, (dilute) matrix flow becomes more important. The solutes in matrix flow have been retarded in the soil profile for longer periods and so concentrations are much lower; an increased proportion of matrix to bypass flow results in a drop in pesticide concentration (Brown *et al.*, 1995). Alternatively, there may be non-equilibrium sorption and desorption in the soil (Kladivko *et al.*, 1991). At the beginning of an event, pesticide resident in the soil solution flushes through large pores into the drainage system, resulting in an initial peak in pesticide concentration. Desorption is not rapid enough, however, to maintain an equilibrium solution and so the levels of pesticide in the macropore flow drop rapidly. When drainage ceases the equilibrium in the large pores can be re-established, ensuring that a new event will result in another high initial maxima. In reality, a combination of these two different effects may occur (Williams *et al.*, 1995).

In summary, there are two scenarios which attempt to explain the observed rise and rapid decline in herbicide concentration in drainflow during rainfall:

(i)   An 'exhaustion of supply' from the soil surface, as the desorption of isoproturon from the surface aggregates does not occur fast enough to maintain herbicide concentrations in the soil solution at the surface. This will cause the isoproturon concentration in the drainage water to drop before the drainflow peaks.

(ii)   An increasing ratio of 'old' to 'new' water through the hydrograph as matrix flow assumes greater importance later in the event.

This chapter reports some of the findings from a two-year field study of the leaching of isoproturon from an underdrained clay soil. An objective of the study was to describe the changes in solute and herbicide data in the field drains at the site, specifically over the time period of individual rainfall events; and to explain these changes in light of the conceptual model of hydrological processes at the site. This would then allow the two hypotheses of herbicide leaching, as described above, to be tested.

METHODS

The fieldwork took place in field 10c at the Oxford University Farm in Wytham, Oxfordshire. The site has been previously described in Kneale (1986), Haria *et al.* (1994) and Johnson *et al.* (1996). The field lies on Oxford Clay of the Denchworth Series, and slopes at an angle between 2° and 5° to a stream. During the study period the stream was 'protected' by a 3 m uncultivated buffer strip (Figure 6.1). Slotted plastic field drains, approximately 40 m apart, run from the top of the slope to the stream. Mole drains are drawn normal to the field drains at 3 m intervals and 50 cm depth. When fieldwork began, the Institute of Hydrology had just completed a preliminary study of the hydrology and pesticide movement from an isolated plot at the site (Figure 6.1). They noted that drainflow at the site began soon after topsoil had become saturated and suggested that drains were filled by a combination of macro-pore and matrix flow within the soil profile (Haria *et al.*, 1994). Water movement in the field as a whole was thought to occur by lateral routes such as lateral interflow and mole drainflow with vertical connections made by macropores.

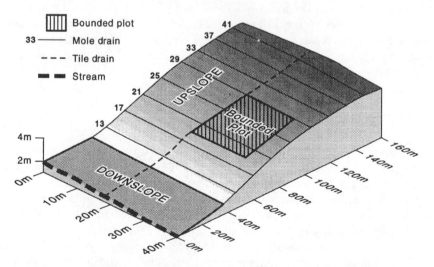

Figure 6.1   A schematic diagram of the tile drain catchment at the Wytham field site

This study attempted to consider the movement of the herbicide over the hillslope scale rather than from a hydrologically isolated plot, therefore, the entire length of a single field drain was instrumented. Only the instrumentation pertaining to the objectives of this chapter are described here. A 90° V-notch weir was installed at the end of the chosen field drain on 15 March 1994. A pressure transducer measured the head of water at the weir plate every 15 min and the resultant data were stored in a Campbell logger (CR 10) so that drainflow could be calculated at a later date. An automatic water sampler was placed by the weir. The sampler was triggered by means of a float switch which responded to slight changes in water height. Once triggered, each sampler took a 500 ml water sample every 30 min for 24 hours. In addition to the samples taken by each water sampler, the Campbell logger recorded the conductivity and temperature of the drainflow every 15 min. An ARG 100 tipping bucket rain-gauge, connected to the logger, measured rainfall at the site. At the base of the slope a saturated layer developed at the base of the A horizon during rainfall. A pressure transducer was placed in a piezometer (35 cm depth) in order to measure changes in the height of this perched water table. Water-table depth was recorded every 15 min by the logger. The depth of water in other piezometers at the base of the slope were recorded manually in order to ascertain whether the automatically logged piezometer was representative of others at the site.

Once collected, water samples were filtered and stored at 4 °C prior to analysis. Anion analysis was carried out by ion exchange chromatography using a Dionex DX 500 system. The eluent consisted of 10.5 mM Na_2CO_3 with 0.5 mM $NaHCO_3$ run at 1.5 ml min^{-1} in isocratic mode. Herbicide analysis was carried out by high-performance liquid chromatography (HPLC) in reverse-phase mode. The eluent system was 35 percent acetonitrile, 65 percent water, and the detection wavelength was 240 nm.

Isoproturon was applied to the field site at a rate of 0.9 kg a.s. ha^{-1} during the first season (12 March 1994) and 2.5 kg a.s. ha^{-1} during the second season (15 November 1994). (This was the recommended dose prior to the review of the product by the Advisory Committee on Pesticides (ACP) in 1995.) Every fortnight after application, 1 kg of soil was collected from the top 0–2 cm of the soil surface in order that the persistence of the herbicide at the site could be measured. Each sample was split into four component samples and isoproturon was extracted in 50 ml of methanol prior to HPLC analysis.

RESULTS AND DISCUSSION

Solute concentrations in drainflow will change in time over a hydrograph in response to the changing contributions of different hydrological pathways through the soil profile to the drains. Longer residence, 'old' water will have had more opportunity to take up ions from the soil and will therefore be higher in solute concentration, hence conductivity. In contrast, 'new' water from rainfall will have had little interaction with the soil and contain few solutes. For the purposes of this study, 'old' water will be considered as matrix flow through the soil and 'new' water as flow that has reached the drainage system through macropore routes, either vertically or laterally. The mixing of 'old' and 'new' water generally leads to an inverse relationship

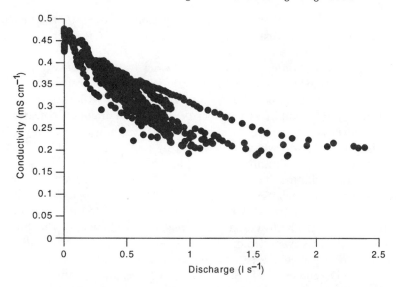

Figure 6.2   The relationship between conductivity and drainflow, January 1995

between solute concentration or conductivity (C) and drainflow (Q). A rating plot of conductivity with discharge indicates that the relationship varies depending on the event (Figure 6.2). Differences between concentrations at similar discharges on the rising and falling limbs of a hydrograph are often observed (Ferguson *et al.*, 1994). For this study the rating plots follow a counter-clockwise path for a simple event, or the first peak of a complex event. However, during complex events the hysteresis becomes less pronounced for subsequent hydrographs. The counter-clockwise motion indicates that there is a greater proportion of (dilute) 'new' water on the rising limb of the hydrograph than on the recession limb.

A two-component mixing model can be used to calculate the changing proportions of chemically dilute new (event) water and more enriched old (pre-event) water during an event as follows (Anderson and Burt, 1982):

$$Q_t = Q_n + Q_o \tag{6.1}$$

$$C_t Q_t = C_n Q_n + C_o Q_o \tag{6.2}$$

$$Q_o = Q_t (C_t - C_n)/(C_o - C_n) \tag{6.3}$$

where Q_t is the tile drain discharge, Q_o is the discharge of 'old' water, Q_n is the discharge of 'new' water, C_t is the conductivity of the total discharge, C_o is the conductivity of 'old' water, and C_n is the conductivity of 'new' water. In this way, the tile drain hydrograph can be split into its 'new' and 'old' water components. This method does not account for solute uptake during passage through the soil and therefore may underestimate the proportion of 'new' water in the hydrograph on the recession limb (Ferguson *et al.*, 1994).

Rainfall events of January 1995 were modelled as described above by equations (6.1)–(6.3). C_o was estimated as $0.474\,\text{mS cm}^{-1}$, the conductance of the water leaving the tile drain on 17 January, after a prolonged period of little rain. C_n, the electrical conductance of rainfall, was estimated as 0.03 mS cm^{-1}, from the weekly averages for Wytham obtained from the Environmental Change Network (ECN).

The overall contribution of 'new' and 'old' water to those events that were monitored during the study are shown in Table 6.1. These data reveal the importance of the matrix water contribution to the drain during an event. With the exception of the high-intensity rainfall event of 5 December 1994, the total drainflow consisted of over 50 percent matrix flow. These data imply that the concentration of any adsorbing solute such as isoproturon which is desorbed by rainfall at the soil surface will be substantially decreased in drainflow by 'old' herbicide-free water.

Figures 6.3 and 6.4 illustrate the results of the hydrograph separation on the data obtained for the rainfall events of 31 March and 5 December 1994. On 31 March 1994 the drainflow consisted of roughly equal proportions of matrix and macropore flow on the rising limb of the hydrograph, whilst matrix flow dominated the drainflow on the recession limb. This event demonstrates the normal response to rainfall at the site: old matrix water assumed a greater importance in drainflow later in the event. In contrast, on 5 December 1994, macropore flow was the dominant contributing hydrological pathway to drainflow throughout the entire event.

Figures 6.3 and 6.4 also present the changes in concentration of the herbicide isoproturon in drainflow during the rainfall events of 31 March and 5 December 1994. The initial rise in isoproturon concentration at the start of the event was a result of rainfall picking up isoproturon from the soil surface and transporting it rapidly through macropores to the drainage system. The concentration and loads of isoproturon in the field drain can be modelled using the hydrograph separation technique described previously, in order to determine whether the change in concentration of isoproturon in the field drain throughout an event is due to some type of 'exhaustion of supply' or is more simply a result of the different contributions of 'new' herbicide-rich and 'old' herbicide-poor water throughout an event. If an instantaneous equilibrium condition is assumed at the surface, then a constant concentration of herbicide (equal to the initial aqueous concentration at the surface) will mix with the 'new'

Table 6.1 The percentage contribution of matrix ('old') and macropore ('new') flow to drainflow for selected rainfall events

Date	Total rainfall (mm)	Max. rainfall intensity (mm h^{-1})	Matrix flow (% of total)	Macropore flow (% of total)
30 Mar. 1994	4.8	8.8	62	38
31 Mar. 1994	7.4	4.8	50	50
18 Nov. 1994	5.5	2	53	47
5 Dec. 1994	9	8	32	68
7 Dec. 1994	5.4	4.2	61	39
18 Dec. 1994	6.4	2	82	18
17 Jan. 1995	8.4	2.8	73	27
19–21 Jan. 1995	16	4.4	65	35
21–23 Jan. 1995	17.6	6.8	60	40
27 Jan. 1995	6.2	1.6	78	22
29 Jan. 1995	9.6	4.2	63	37

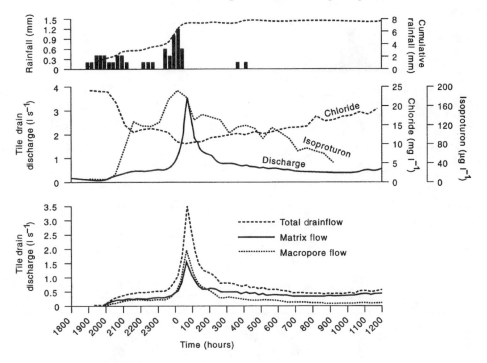

Figure 6.3   Selected hydrological data for the rainfall event of 31 March

water component, Q_n, in the mixing model. If the concentration of herbicide in the 'old' soil water is assumed to be zero (i.e., $I_o = 0$) then using equation (6.3),

$$I_t = I_n \cdot Q_n / Q_t \tag{6.4}$$

Thus the concentration of isoproturon in the tile drain throughout the event can be predicted. Theoretically, if there was no 'exhaustion of supply' from the soil surface, the modelled concentration of herbicide in the tile drain, I_t, and the measured I_m should match. In order to calculate I_t, the amount of isoproturon available for transport at the soil surface on the day of each event was determined from the total isoproturon residues measured at fortnightly intervals, using the following equation:

$$I_n = T/V_s((K_d \cdot \rho) + \theta) \tag{6.5}$$

where T is the total pesticide (g), V_S is the soil volume, K_d is the herbicide partition coefficient ($1\,kg^{-1}$), ρ is the bulk density of the soil ($kg\,l^{-1}$) and θ is the soil moisture content ($cm\,cm^{-1}$) (Johnson *et al.*, 1995). Table 6.2 shows the measured herbicide residues in the soil and associated T values that were calculated for each monitored event. The values of the other parameters were taken as follows:

- K_d of the topsoil at Wytham is $2.5\,l\,kg^{-1}$ (Johnson *et al.*, 1995);
- bulk density of the topsoil is $1.3\,g\,cm^{-3}$ (Johnson *et al.*, 1995);
- soil moisture content is approximately $0.55\,cm\,cm^{-1}$.

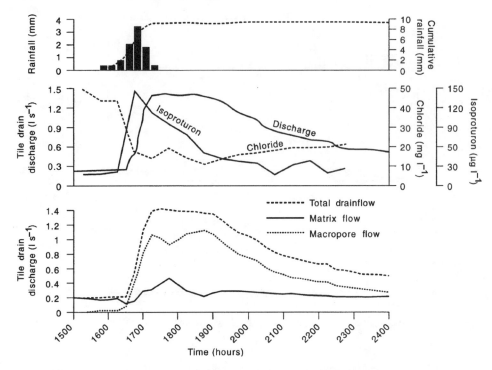

Figure 6.4 Selected hydrological data for the rainfall event of 5 December

Table 6.2 Isoproturon concentrations in the wet soil (t_{wet}) and total mass
of isoproturon in the tile drainage area (T) at the beginning of each event

Date	30 Mar. 1994	31 Mar. 1994	18 Nov. 1994	5 Dec. 1994	7 Dec. 1994	18 Dec. 1994
t_{wet} (mg kg^{-1})	1.36	1.32	8.51	3.0	2.88	1.84
T(g)	254.6	247.1	1593	562	539	345

Figure 6.5(a) shows the graphs of the modelled and actual isoproturon data for the rainfall event of 31 March. Evidently, the model substantially overpredicts the concentration and amount of isoproturon that leaves the tile drain. However, the pattern of release is very similar. In fact, if a concentration of isoproturon in the 'new' water of 165 µg l^{-1} is used, the resultant curve (C_t, fit) fits the measured herbicide quite well. This suggests that the concentration of herbicide in the 'new' water of the tile drain remains at a constant level throughout the event and does not decrease. There does not appear to be an exhaustion of supply and the changes in concentration of isoproturon through an event can in fact be explained by the changing proportions of 'old' and 'new' water in the drain.

During the second season, all the chemographs show that herbicide concentrations rose early in the event, corresponding to the initiation of macropore flow in the

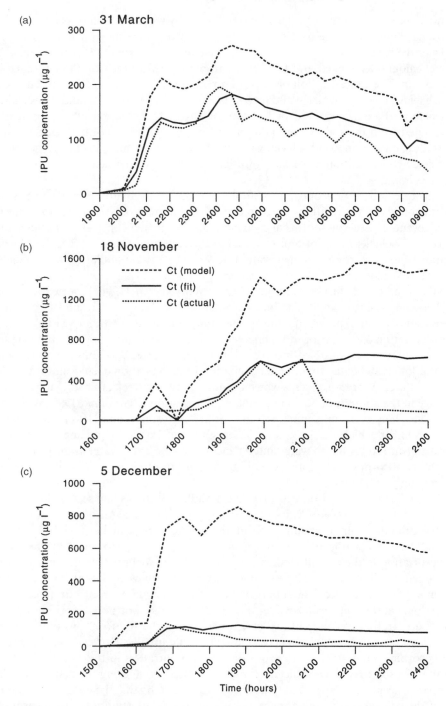

Figure 6.5 Modelled and measured IPU data for (a) the 31 March event, (b) the 18 November event, and (c) the 5 December event

downslope region of the site. This was confirmed by a comparison of perched water-table development at the base of the A horizon (which was indicative of macropore flow to the base of the A horizon) and the time of isoproturon rise for the events. All the chemographs peaked and then began to recede before the drainflow reached its maximum. The concentration of isoproturon in drainflow was modelled and the data are displayed in Figure 6.5(b) and (c) for the events of 18 November and 5 December. As before, the concentration of isoproturon in the soil solution at the surface, as calculated from the soil extraction data, overestimates the loss of herbicide through drainflow during an event. Once again, the model can be made to fit the measured herbicide concentration better if a lower concentration of herbicide than expected is used as input.

The calculation of isoproturon in the topsoil at Wytham from the methanol extraction data substantially overpredicted the amount lost in drainflow during every event. The reason for this discrepancy requires further consideration. If a higher partition coefficient (K_d) of 5 or 6 is used in the equation instead of 2.5, the concentration of isoproturon in the new water is lowered to a value similar to that which fits the data. Walker et $al.$ (1995) have reported a K_d of 2.51 kg^{-1} for isoproturon and Wytham soil under slurry conditions whereas Worrall et $al.$ (1996) reported K_d values of between 4 and 61 kg^{-1}. The linear isotherms calculated as a result of static and slurry adsorption experiments during this study indicated a K_d of 1.951 kg^{-1} (standard deviation 0.03) for topsoil. Differences in either experimental conditions or soil location may be responsible for these differences in reported K_d. Thus the overestimation of loss may be due to an underestimation in the partition coefficient of K_d. This may be the case for those events which occur more than a week after spraying due to a change in the amount of aqueous available isoproturon (Blair et $al.$, 1990; Walker et $al.$, 1995). However, this cannot account for the overestimation in loss in the event of 18 November which occurred soon after spraying. Inefficient mixing between the resident (old) soil solution water and the incoming new (rain) water at the soil surface or the re-adsorption of pesticide within the soil are other possible reasons for the discrepancy.

The concentration of isoproturon in the drainflow of the second season does indeed appear to display a type of 'exhaustion of supply'. The concentration of isoproturon in the drainflow on the rising limb of the chemograph can be modelled using a constant concentration of isoproturon in the 'new' water. However, the herbicide concentration in the drainflow then drops when the model predicts that it should be sustained. In all the events of the second season, the herbicide concentration dropped when rainfall had ceased. The supply of 'new' water containing isoproturon picked up directly from the soil surface will also cease at this time. For non-adsorbing solutes, the concentration of solute in the 'new' water within the soil from this time onward will either remain constant or increase due to solute uptake. However, in the case of an adsorbing solute such as isoproturon, the concentration in the water is likely to drop if the solute is given sufficient contact with the soil. During the first season this drop in concentration appeared to be negligible suggesting that the water was draining from the soil surface through macropores with little opportunity for re-adsorption, even on the recession limb of the hydrograph. However, during the first few months of the second season (November/December) the sharp drop in concentration

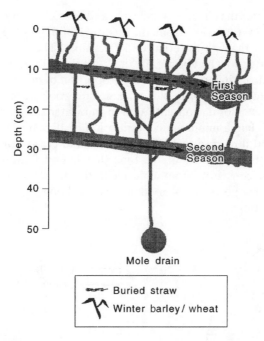

Mole drain

Buried straw

Winter barley / wheat

Figure 6.6 The differences in the hydrological routes at the site over the two studied seasons

over the recession limb suggests that substantial re-adsorption of the herbicide within the soil may have taken place.

Figure 6.6 identifies the main differences in the hydrological routes at the site over the two seasons. Lateral flow through the A horizon was believed to be an important component of the hillslope processes during the early months of the second season, but minimal in extent during the first season due to the closure of inter-pedal cracks on swelling of the clay soil. Germann (1990) noted that a lateral flow component in hillslopes, and the resultant increased length of flowpath, can greatly increase both the residence time of the water and the total volume of water sorbed by the matrix. This increase in both residence time and macropore–matrix interactions will allow re-adsorption of herbicide to occur. As the flow is travelling near the surface where concentrations of chloride in the soil solution are low, there may not be the opportunity to pick up chloride. Hence, for the purposes of the hydrograph separation, the lateral flow contributes a third source of water. This is 'new' water with little chloride, but it has lost much herbicide as it travels through the soil. Initially, before lateral flow commences, the results of the hydrograph separation will correctly predict the concentration of isoproturon in the drainflow, but when lateral flow begins and sustains the drainflow, the model will predict that this 'new' water will be herbicide-rich. Instead, due to its contact with, and residence time in the soil, the water contains little herbicide and so the total concentration of isoproturon in the drainflow drops. In addition, the lateral flow over this period travelled through a layer of buried straw. This straw had a herbicide coefficient of $25 \, l \, kg^{-1}$ and could therefore have acted as a 'sink' which re-adsorbed the herbicide within the soil profile (Heppell, 1997).

Figure 6.7 illustrates the changes in chloride and herbicide concentration in drain-flow and lateral flow (data are from the Institute of Hydrology's bounded plot; Johnson *et al.*, 1995) through the event of 5 December. The first lateral flow sample was taken 20 min after flow had begun. By that time the chloride concentration in the flow had begun to drop due to an increasing proportion of 'new' rainwater. The data indicate that a substantial mixing of old and new water took place between the surface and 30 cm depth; the chloride concentration in rainwater was 1 ppm, but the lateral flow at 30 cm depth had a minimum concentration of 10 ppm. The chloride concentration in the lateral flow rose to 17 ppm after rainfall ceased and remained at this concentration for the remainder of the time that lateral flow was recorded. This corresponded to the concentration of chloride in the drainflow during the period of prolonged flow and supports the theory that the lateral flow was the chief contributor to drainflow at this time.

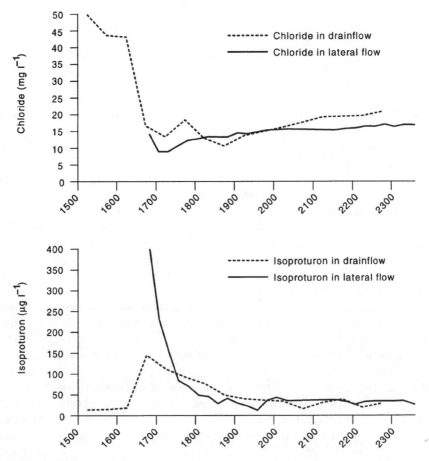

Figure 6.7   (a) Changes in chloride concentration in drainflow and lateral flow (5 December event). (b) Changes in herbicide concentration in drainflow and lateral flow (5 December event)

CONCLUSIONS

This study has demonstrated that a simple two-component mixing model which makes use of changes in indigenous solute concentration during rainfall can help further the current understanding of herbicide transport in clay soil.

The following changes in isoproturon concentrations in drainflow during a rainfall event were observed during both seasons at the site:

- Rainfall picks up high quantities of herbicide from the soil surface and transports it rapidly by macropore flow to the drainage system, so that herbicide concentrations increase on the rising limb of the drainage hydrograph.
- The herbicide concentration reaches a maximum value before the hydrograph peaks and then declines.

During the first season, the changing concentration of herbicide in drainflow could be predicted by calculation of the changing contributions of matrix and macropore flow. There was no evidence for an exhaustion of supply due to a chemical disequilibrium at the surface. In the second season, however, the concentration of isoproturon on the recession limb of the hydrograph dropped at a greater rate than that predicted by the relative contributions of matrix and macropore water. This was thought to be the result of the dominant contribution of lateral flow to the recession limb of the hydrograph. Lateral flow loses substantial amounts of herbicide during its passage through the soil, due to re-adsorption. The presence of a buried straw layer in the A horizon with a high adsorptive capacity may have aided the re-adsorption process.

ACKNOWLEDGEMENTS

This study was supported by a NERC studentship and CASE award from the Institute of Hydrology. The authors would like to thank the following people for their support during the study: John Morgan and Dave Banfield for their help with fieldwork, Chris Jackson for his expertise and advice in the laboratory, and Mike Morecroft of the Environmental Change Network (ECN) for the rainfall chemistry data. The work was carried out whilst Dr Catherine Heppell and Professor Tim Burt were members of the School of Geography at Oxford University.

REFERENCES

Anderson, M. G. and Burt, T. P. 1982. The contribution of throughflow to storm runoff: an evaluation of a chemical mixing model. *Earth Surface Processes and Landforms*, **7**, 565–574.

Blair, A. M., Martin, T. D., Walker, A. and Welch, S. J. 1990. Measurement and prediction of isoproturon movement and persistence in three soils. *Crop Protection*, **9**, 289–294.

Brown, C. D., Hodgkinson, R. A., Rose, D. A., Syers, K. and Wilcockson, S. J. 1995. Movement of pesticides to surface waters from a heavy clay soil. *Pesticide Science*, **43**, 131–140.

Cannell, R. Q., Davies, D. R., Mackney, D. and Pidgeon, J. D. 1978. The suitability of soils for sequential drilling of combine-harvested crops in Britain. A provisional classification. *Outlook on Agriculture*, **9**, 306–316.

EA 1997. *Pesticides in the Aquatic Environment 1995*. Environment Agency, March 1997.

ENDS 1995. Herbicide restricted to protect water supplies. *ENDS Report*, **246**, 33–33.

Ferguson, R. I., Trudgill, S. T. and Ball, J. 1994. Mixing and uptake of solutes in catchments: model development. *Journal of Hydrology*, **159**, 223–233.

Fischer, P., Burhenne, J., Bach, M., Spiteller, M. and Frede, H. G. 1996. Landwirtschaftliche Beratung als Instrument zur Reduzierung von punktuellen PSM-Eintrogen in Fliessgewosser. *Nachrichtenblatt des Deutschen Pflanzenschutzdienstes*, **48**, 265–269.

Germann, P. F. 1990. Macropores and hydrologic hillslope processes. In Anderson, M. G. and Burt, T. P. (eds) *Process Studies in Hillslope Hydrology*. John Wiley, London, 327–364.

Haria, A. H., Johnson, A. C., Bell, J. P. and Batchelor, C. H. 1994. Water movement and isoproturon behaviour in a drained heavy clay soil: 1. Preferential flow processes. *Journal of Hydrology*, **163**, 203–216.

Harris, G. L. 1995. Pesticide loss to water – a review of possible agricultural management opportunities to minimise pesticide movement. *British Crop Protection Council Monograph No. 62: Pesticide Movement to Water*. British Crop Protection Council, Brighton, 371–380.

Harris, G. L., Nicholls, P. H., Bailey, S. W., Howse, K. R. and Mason, D. J. 1994. Factors influencing the loss of pesticides in drainage from a cracking clay soil. *Journal of Hydrology*, **159**, 235–253.

Heppell, C. M. 1997. The fate of pesticide in underdrained clay soil. Unpublished DPhil Thesis, Department of Geography, University of Oxford.

IPU Task Force 1993. *Isoproturon (IPU): A Stewardship Programme to Protect Water Quality*. AgrEvo UK Crop Protection Limited, Ciba Agriculture, Rhone-Poulenc Agriculture Limited, 11 pp.

Johnson, A. C., Haria, A. H., Batchelor, C. H. and Williams, R. J. 1995. *Fate and Behaviour of Pesticides in Structured Clay Soils*. Second Interim Report, Institute of Hydrology, Wallingford.

Johnson, A. C., Haria, A. H., Bhardwaj, C. L., Williams, R. J. and Walker, A. 1996. Preferential flow pathways and their capacity to transport isoproturon in a structured clay soil. *Pesticide Science*, **48**, 225–237.

Jones, R. L., Gatzweiler, E. G., Guyot, C. N., Wicks, R. J., Hardy I. A., Higginbotham, S., Leake, C. R., Arnold, D. J. S., Idstein, H. and Feyerabend, M. in press. Research and monitoring studies on isoproturon movement to surface and groundwater in Europe.

Kladivko, E. J., Van Scoyoc, G. E., Monke, E. J., Oates, K. M. and Pask, W. 1991. Pesticides and nutrient movement into subsurface tile drains on a silt loam soil in Indiana. *Journal of Environmental Quality*, **20**, 264–270.

Kneale, W. R. 1986. The hydrology of a sloping, structured clay soil at Wytham, near Oxford, England, *Journal of Hydrology*, **85** (1–2), 1–14.

NRA 1995. *Pesticides in the Aquatic Environment*. Water Quality Series No. 26, National Rivers Authority.

Seel, P., Knepper, T. P., Gabriel, S., Weber, A. and Haberer, K. 1996. Kloranlagen als Haupteintragspfad fur Pflanzenschutzmittel in ein Fliessgewosser-Bilanzierung der Eintroge. *Vom Wasser*, **86**, 247–262.

Walker, A., Welch, S. J. and Turner, I. J. 1995. Studies of time-dependant sorption processes in soils. *British Crop Protection Council Monograph No. 62: Pesticide Movement to Water*. British Crop Protection Council, Brighton, 13–18.

Williams, R. J., Brooke, D. N., Matthiessen, P., Mills, M., Turnbull, A. and Harrison, R. M. 1995. Pesticide transport to surface waters within an agricultural catchment. *Journal of the Institute of Water and Environmental Management*, **9**, 72–81.

Williamson, A. R. and Croxford, A. C. 1998. Diffuse pollution of environmental waters by pesticides – strategies for monitoring and control. Paper presented at the 3rd International Conference on Diffuse Pollution, International Association of Water Quality, Edinburgh, Scotland, 31 August–4 September 1998.

Worrall, F., Parker, A., Rae, J. E. and Johnson, A. C. 1996. Equilibrium adsorption of isoproturon on soil and pure clays. *European Journal of Soil Science*, **47**, 265–272.

7 Variations in time of Travel in UK River Systems – A Comparative Study

D. A. WILSON,[1] J. C. LABADZ[2] and D. P. BUTCHER[3]
[1]*Environment Agency, Reading, UK*
[2]*Centre for Water and Environmental Management, Department of Geographical and Environmental Sciences, University of Huddersfield, UK*
[3]*Department of Land-Based Studies, Nottingham Trent University, UK*

INTRODUCTION

The use of tracer experiments to simulate the movement of a pollution spillage into fluvial systems has been recognised for many years as a potentially powerful tool for water resource managers. Direct abstraction from rivers currently accounts for approximately one-third of the public water supply in England and Wales and this may increase if inter-regional river transfer schemes are further developed to meet increased demand for water as a result of climate change (Sheriff *et al.*, 1996). Such abstractions are, however, vulnerable to pollution incidents from industrial, waste-water treatment and transport sources. Over 19 000 incidents were substantiated in 1997 (Environment Agency, 1999). Those seeking to take preventative measures cannot target responses effectively if there is no information available about the likely time of travel to a particular point of interest or the concentration of the pollutant as it passes that point. The use of chemical and biological tracers allows an examination of pollutant behaviour and may facilitate rapid remedial action to protect vulnerable water supplies and safeguard human health.

The actual use of travel-time experiments in rivers has been limited in the past by their high cost, although the Environment Agency has commissioned a number of studies in recent years. Even in an area as small as the United Kingdom, it has not been feasible to trace every section of every river because the collection of good quality data is still a laborious and time-consuming exercise. Tracing of reaches that are close to industrial sites or crossed by roads or railways, and are therefore at high risk of being affected by a spillage, has often been neglected. For this reason an approach is desirable which enables prediction of pollutant travel times at a range of flows in previously traced rivers and preferably allows some extension of predictions to untraced rivers.

Tracers in Geomorphology. Edited by Ian D. L. Foster. © 2000 John Wiley & Sons Ltd.

This chapter presents the results of work undertaken in Yorkshire, East Anglia and West Sussex for the Environment Agency and various water companies. Its aim is to review a broad predictive methodology based on an empirical approach, which allows the prediction of travel times on rivers where there have been few or no travel-time studies. Analysis of the travel-time data collected has revealed strong and consistent relationships, both within individual data sets and between data collected at different flow levels. The adequacy of simple empirical equations derived from a number of traces on each river is compared with more sophisticated modelling approaches such as ADZ (Sabol and Nordin, 1978; Beer and Young, 1983; Wallis *et al.*, 1989) and ADE (Taylor, 1954; Fischer, 1967, 1968) which allow a more complete description of the tracer behaviour. The possibility of extension of the empirical approach to untraced rivers by inclusion of map-based descriptors of geomorphology at the reach scale is explored. A major concern was to develop a methodology that was of reasonable accuracy whilst remaining operationally applicable during a pollution incident on a river, allowing flexible prediction of pollutant behaviour from any point along its length.

PREDICTION OF FLOW VELOCITY AND TIMES OF TRAVEL OF POLLUTANTS IN RIVERS

The behaviour of any pollutant within a river depends both upon the nature of that pollution and upon the characteristics of the river. The current work relates only to substances which are carried in solution, which are derived from a point source and which are chemically conservative. In such cases, transport of the pollutant downstream may be described as a solute cloud which disperses as it travels (Leibundgut *et al.*, 1992), with the cloud steadily lengthening and the peak concentration decreasing as the solute is distributed in an ever-increasing volume of water (Young and Wallis, 1993). This is partly a result of processes such as diffusion, which distribute the solute within the flow cross-section, and partly a result of differential advection due to lateral and vertical velocity gradients. These gradients are fundamentally related to the geomorphology of the channel.

Attempts to characterise average velocity at a cross-section may be made by direct measurements or by application of the familiar equations, based upon flow resistance as characterised by slope and a coefficient of friction, such as those of Chezy and Manning. Such 'at-a-point' calculations must be repeated frequently to build up a description of how flow velocity will vary within a complex non-uniform natural river channel. However, Bathurst (1993) noted that whilst the coefficients can be estimated accurately for artificial channels of fixed, regular cross-section, their application to natural channels gives rise to an increased level of uncertainty in the calculated velocity, as a result of the variability in bed slope, plan form and cross-sectional area. Resistance, and therefore velocity, may also be linked to issues such as bedform development, bedload or suspended sediment load and the influence of vegetation in the channel (e.g. Hey and Thorne, 1986).

The use of such methods linking channel form and characteristics to velocity at the cross-section scale does not provide a practical method of obtaining sufficient information to aid the management of a pollution incident. For this purpose a more detailed understanding of the entire velocity distribution within a channel must be developed, necessitating vast amounts of detailed data collection, or an alternative approach must be adopted. One possibility is the concept of hydraulic geometry, first introduced by Leopold and Maddock (1953) and reiterated by Leopold (1994), in which the channel characteristics of natural rivers are seen to constitute an interdependent system that can be described in terms of simple power laws. Of particular interest for pollution management is the possibility that mean velocity is related to discharge at a site:

$$U = kQ^m \tag{7.1}$$

where U is mean velocity and Q is discharge.

The exponent m averages 0.34 at-a-site for rivers in the midwestern USA, but Bathurst (1993) reports values over 0.6 on the rivers Severn and Swale in England. Between sites, however, Hey and Thorne (1986) calculated an exponent of 0.1 when using bankfull discharge (m^3 s^{-1}) to predict velocity ($m\,s^{-1}$) for 62 sites on British gravel and cobble channels. Knighton and Cryer (1990) studied three rivers in Lincolnshire in more detail, comparing values for velocity obtained at cross-sections with those on short reaches and those obtained from tracing longer reaches (up to 7 km length). They discussed only the mean velocity of the tracer, and reported values for m of 0.59 to 0.75 for their longest reaches. It is clear that the relationships, although possibly useful where no alternative is available, are not precise if applied across a variety of rivers.

The use of tracers to record the shape of the pollutant cloud as it travels downstream provides another approach that enables the effect of a section of river on pollutant travel times to be summarised. This method inherently recognises that a range of velocities is experienced over a reach of a river. Changes between the shape of the upstream and downstream tracer concentration curves are influenced by the particular physical properties of the reach even though these are not measured explicitly. Beven and Carling (1992) compared these two approaches for two 1 km reaches on the River Severn. They suggested that good fits for the characterisation of a pollutant cloud could be obtained by recognition of storage within the reach where discharge is effectively zero, as assumed by the Aggregated Dead Zone (ADZ) model (Sabol and Nordin, 1978; Beer and Young, 1983; Wallis et al., 1989).

This varies from the more traditional approach to prediction of the characteristics of the pollutant cloud as it travels downstream, which has been based upon linking advection with Fickian-type diffusion and is exemplified by the work of Fischer (1967, 1968, 1969) in producing the Advection Dispersion Equation (ADE):

$$\frac{\partial \bar{c}}{\partial t} + \bar{u}\frac{\partial \bar{c}}{\partial x} = K\frac{\partial^2 \bar{c}}{\partial x^2} \tag{7.2}$$

where c is concentration, x is distance, t is time, u is velocity and K is the dispersion coefficient.

Strictly speaking this only applies when the concentration distribution attains a Gaussian form, yet many workers report consistent skewness in observed field data.

There also remains the difficulty that values for average longitudinal velocity and the dispersion coefficient are also required before predictions can be made. The velocity is, in practice, usually unknown and the dispersion coefficient varies with discharge. Young and Wallis (1993) summarised several methods of estimating the dispersion coefficient, including both analytical and optimisation techniques, but concluded that there are significant deviations between Fickian theory and observations, particularly with regard to skewness of the solute concentration profiles. Rutherford (1994) stated that formulae to predict the dispersion coefficient were, at best, within a factor of approximately 10 when compared to published observational data.

The Aggregated Dead Zone (ADZ) model (Sabol and Nordin, 1978; Beer and Young, 1983; Wallis et al., 1989; Young and Wallis, 1993) recognises these limitations of the ADE and argues that the advective and dispersive behaviour in real stream channels can be better approached by incorporation of the ideas of dynamic storage zones, in which solute may be temporarily trapped, and of statistical calibration based on observed data from tracer experiments to identify an appropriate model structure. The dead zones may be caused by a combination of features such as turbulent eddies, wakes around roughness elements and reverse flows associated with pools and bends. The model is developed using a transfer function to represent the changes with the passage of the tracer cloud downstream, and its form is determined from the observed data rather than being fixed *a priori* as with the ADE. Wallis et al. (1989) presented examples on river reaches 100–150 m long, and suggested that in some cases the mean travel time for the reach might be inversely related to discharge, rather than requiring the commonly applied power law relationship.

A key concept within the ADZ approach is the dispersive fraction, Df. This is defined as the ratio between the mean time spent by the tracer in the aggregated dead zone (T, calculated from the difference between the time of arrival of the centroid of the tracer cloud and the time of first arrival) and the mean residence time of the tracer in the reach as a whole ($\bar{t}$, the time of arrival of the centroid). Wallis et al. (1989) suggested that all these elements decrease as discharge increases, but that the ratio should remain constant for a given reach, despite changing flows. This would have the advantage of allowing comparisons between different reaches and rivers and can be interpreted as a measure of the fractional volume of the reach which appears to be responsible for dispersion. They suggested that the invariant nature of the dispersive fraction meant that a single tracer experiment could be used to calibrate the ADZ model, and that three or four experiments should be sufficient to characterise relationships between travel time and discharge on a river. They did, however, suggest that the model needed to be applied to a wider range of channel and flow characteristics before widespread practical application. Such a study was reported by Green et al. (1994), who conducted tracer experiments on the River Severn in Shropshire, UK and the River Tonge near Bolton in north-west England. Rather than finding dispersive fraction to be constant, they found that it decreased with discharge. They suggest that this is reasonable, since the proportion of dead zones is likely to decrease at higher flows, and that it is therefore necessary to obtain an empirical expression for its variation with discharge.

The current study aims to undertake a comparison of variations in travel times for conservative pollutants from point sources in a range of rivers within the UK.

Modelling approaches such as ADE and ADZ were applied to the data for the Yorkshire rivers and compared with a simpler empirical approach to predict the major characteristics of the pollution response (see below and Wilson *et al.*, 1997). This approach was then applied to rivers in East Anglia and West Sussex. Also explored is the possibility of the inclusion of geomorphological influence, via the use of simple map-derived channel characteristics, to improve on predictions for untraced rivers made from discharge alone. A subset of the traced rivers have been used for production of simple empirical relationships and the remainder employed in validation of the results.

SCALE CONSIDERATIONS AND REPLICATION IN THE DESIGN OF TRACER EXPERIMENTS

Table 7.1 summarises the scales at which much of the previous work has been completed in the field. Figure 7.1 depicts this review of the available literature, giving details

Table 7.1 Nature and scales of previous studies completed on pollution transport

Publication	Type of channel / situation / name	Maximum length	No. of sites*	No. of runs
Beer and Young (1983)	Semi-natural (part-concrete channel): Tuggeranong Creek, Canberra, Aus.	4.57 km	7	1
	Natural channel: Magela Creek, Northern Terr., Aus.	22.5 km	6	1
Bencala and Walters (1983)	Natural: Uvas Creek, California, US	619 m	5	1
Brady and Johnson (1981)	Natural: River Wear, England			
	Stanhope to Page Bank	24 km	3	1
	Shincliffe to Lumley	40 km	2	1
Buchanan (1964)	Natural:			
	Swatara Creek, Pennsylvania, US	89 km	5	2
	Umpqua River, Oregon, US	222 km	39	1
Carling *et al.* (1994)	Natural: River Severn, UK	500 m	2	1
Day (1975)	Natural (South Island, NZ):			
	Bealey	660 m	9–11	7
	Bruce	775 m	12–21	10
	Craigieburn	780 m	11–19	11
	Porter	825 m	10–18	13
	Thomas	2250 m	11–19	8
Elder (1959)	Lab. flume	3 m	1	133
Fischer (1967)	Lab. flumes	12 m	1	3
		18 m	1–2	154
		40 m	1–6	131
Fischer (1968)	Natural: Copper Creek, Virginia, US	4130 m	6	60
François *et al.* (1997)	Natural: River Doller, France	18 km	4	3
Glover (1964)	Flumes (indoor and outdoor):			
	Fort Collins, Colorado, US	18 m	3	65
	Allenspark, Colorado, US	120 m	5	11
	Natural:			
	South Platte River, Nebraska, US.	8 km	5	1

continued overleaf

Table 7.1 (*continued*)

Publication	Type of channel / situation / name	Maximum length	No. of sites*	No. of runs
	Natural (*contd.*)			
	Mohawk River, New York, US.	21 km	–	1
	Columbia River, Washington, US.	9 km	10	1
Green *et al.* (1994)	Natural: River Severn, UK	1 km	2	7
		1.13 km	2	6
	Natural/concrete walls: River Tonge, Lancashire, UK	330 m	2	8
Harris and Sanderson (1968)	Natural:			
	Long Tom River, Oregon, US	5 km	3	1
	Umpqua River, Oregon, US	222 km	36–38	2
	Willamette River, Oregon, US	280 km	19	3
	Cow Creek, Oregon, US	96 km	16	1
	Calapooya Creek, Oregon, US	53 km	8	1
Hays *et al.* (1966)	Lab. flume	12 m	1	157
	Natural (Tennessee, US):			
	Clinch	–	1	13
	French Broad	–	1	14
	Hiwasse	–	1	18
Kilpatrick and Taylor (1986)	Natural: Shenandoah River, W. Virginia, US	286 km	16	2
Knighton and Cryer (1990)	Natural:			
	Lymn, Lincolnshire UK	6.8 km	1	11
	Witham, Lincolnshire UK	2.4 km	1	4
	Long Eau, Lincolnshire UK	2.9 km	1	10
Leibundgut *et al.* (1992)	Natural: River Rhine, Switzerland/ Germany	*c.*700 km	16–20	4
Nordin and Sabol (1974)	Natural:			
	Wind-Bighorn River, Wyoming, US	181.3 km	7	2
	Bear Creek, Colorado, US	10.9 km	3	1
Sabol and Nordin (1978)	Natural:			
	Conococheague Creek, Maryland, US	–	4	1
	Chatahoochee River, Georgia, US	49 km	4	1
Seo and Maxwell (1991)	Lab. flumes	49 m	2	6
Taylor (1954)	Lab. flumes	4.8 m	2	1
		16.3 m	3	2
Valentine and Wood (1979)	Lab. flume	22 m	6	1
	Natural: water race – Canterbury, NZ	4 km	8	2
Wallis *et al.* (1989)	Natural (Lancashire, England):			
	Brock	128 m	2	32
	Conder	116 m	2	30
	Dunsop	130 m	2	12
	Semi-natural (concrete channel): Ou Beck	127 m	2	16
Yotsukura *et al.* (1970)	Natural: Missouri River	227.4 km	4	1
Zhou Ke-Zhao (1991)	Natural: Tuo River, Sichuan, China	> 200 km	23	7
	Lab. flume	21.6 m	–	111

* Monitoring sites only, not including injection points.

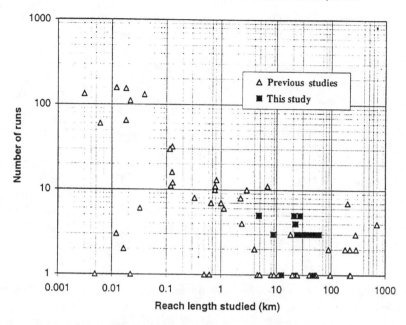

Figure 7.1   Reach length and number of runs in travel-time experiments

of tracer experiments that have been completed over the last 50 years. The number of runs completed has in most cases been inversely related to the reach length, with shorter reaches being studied on a large number of occasions and vice versa. Many previous workers (e.g. Fischer, 1967; Wallis *et al.*, 1989) have undertaken small-scale experiments with a large number of runs under different conditions, whilst others (e.g. Brady and Johnson, 1981; Beer and Young, 1983) have completed large-scale traces with few, if any, repetitions with variables adjusted. Neither of these approaches produces comprehensive information that can be applied to the operational situation.

The diagram also reveals an interesting lacuna in the data, at precisely the scale that is most applicable to pollution incidents on UK rivers. Prior to the studies reported in this chapter, only one of the published experiments for pollution prediction purposes has been undertaken at more than one flow level in a river between 3 and 89 km in length (that reported by François *et al.*, 1997). Knighton and Cryer (1990) conducted 11 replicates over 6.8 km in Lincolnshire, but their concern was not river pollution and they only report centroid time rather than the full concentration distribution or even arrival and through times. Although several other single experiments of this magnitude have been completed, little is known about behaviour under different flow conditions. Figure 7.1 demonstrates how the data reported in the current study go some way to filling this unresearched area.

RIVERS USED FOR DATA COLLECTION

The completion in the current study of a number of long traces over a variety of river systems in the UK (Wilson, 1997; Wilson *et al.*, 1997; Butcher and Labadz, 1999) has allowed an alternative approach to be developed. This has involved the collection of data about the time of travel of pollutants from a number of rivers with different characteristics. By conducting tracer experiments over long river reaches with a number of replications at different discharge levels, it is possible to identify patterns that can be described mathematically.

The rivers used and their basic characteristics are shown in Table 7.2, whilst their locations are indicated in Figure 7.2. The results from 14 rivers are presented. As a result of the considerable length of river to be studied, the Wensum, Waveney, Esk and Western Rother were divided into two sections for some of the actual traces. The same fundamental methodology of data collection, experimental design and field team were employed for all rivers, ensuring that the results are directly comparable.

For each of the rivers studied, a single flow gauging station was selected, although in many instances several were available at different points within the catchment. The decision to use a single reference gauge was based on two factors:

- Using a network of flow gauges, it is possible to predict the flow at any given point within the catchment; however, the uncertainties associated with this process would lead to inaccuracies in the predictions made based upon them.
- During the course of an actual incident the ability to make predictions based on a single flow figure would reduce the complexity of the task and make it more readily understandable.

Table 7.2   Rivers used in the study

Region	River	Gauging station	Catchment area at gauge (km^2)	Q_{50} (m^{3}s^{-1})	FSR Slope (m km^{-1})	Maximum altitude of catchment (m)	Traced length (km)
Yorkshire	Esk	Sleights	308	2.193	4.89	435	29.4
	Hull	Hempholme Lock	378.1	2.182	1.29	246	8.63
	Nidd	Hunsingore	484.3	4.367	2.54	704	21.8
	Swale	Crakehill	1363	11.52	–	713	27
	Ure	Westwick Lock	914.6	10.75	3.24	713	22.3
	Wharfe	Addingham	427	6.38	4.5	704	20.85
	Dibb	Grimwith Res	25.5	0.535	–	550	4.8
	Aire	Armley	691.5	8.986	1.9	594	45.2
East Anglia	Bure	Ingworth	164.7	0.963	2.05	101	33.2
	Gipping	Bramford	298	0.529	1.49	78	23.4
	Waveney	Needham Mill	370	0.769	0.58	65	47.55
	Wensum	Swanton Morley	363	–	–	–	59.25
	Wendling Beck	Swanton Morley					
Southern England	Western Rother	Iping Mill	154	1.318	2.40	271	42

Data derived from National Water Archive, Institute of Hydrology.

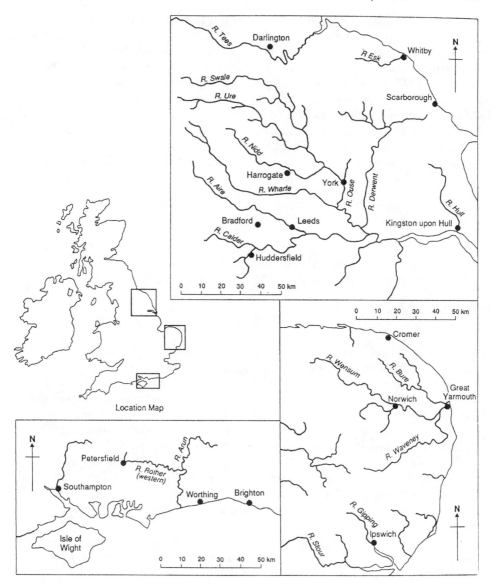

Figure 7.2   Location of study rivers

For the purposes of analysis the mean discharge at the reference gauge for the duration of each trace at the site has been used.

DYE TRACE METHODOLOGY

Rhodamine WT was the tracer selected for use primarily due to its low toxicity, detectability at low concentrations, good resistance to photochemical degradation, and low rates of adsorption onto sediments and vegetation (Smart and Laidlaw, 1977; Smart, 1984). It has been previously widely used in traces of both surface and ground-water (e.g. Beven and Carling, 1992; Rukin et al., 1994) although recently there have been some concerns expressed about its toxicity at high concentrations (Leibundgut and Hadi, 1997). In the current work, therefore, strict limits were set on the maximum dye concentration at any intake for potable supply ($< 0.2\,\mu g\ l^{-1}$). Levels below $0.1\,\mu g\ l^{-1}$ were normally achieved. In the calculation of the dye mass to ensure this limit was not exceeded, but also to make sure enough of the tracer was present to be detected, the method detailed by Church (1974) was used with good success. The equation, set out below, is specifically for use with Rhodamine WT; different coefficients are utilised for other dye types:

$$W = 3.8 \times C_p \left(\frac{QL}{\bar{u}}\right)^{0.93} \tag{7.3}$$

where
    W = mass of 20 percent solution of Rhodamine WT (mg);
    C_p = peak concentration at most downstream monitoring sites; ($\mu g\ l^{-1}$);
    Q = discharge ($m^3\ s^{-1}$);
    L = length of reach (m);
    $\bar{u}$ = mean velocity ($m\ s^{-1}$).

The obvious difficulty is that the equation requires a measure of velocity, which has to be estimated from experience on similar rivers or previous traces.

The required mass of dye was introduced as close to the centre of the river as was feasible, as a 'gulp' injection. Monitoring sites were selected at intervals down-stream, and two Turner 10AU field fluorometers were used to analyse and log concentration data at 20 s intervals. The equipment was moved downstream as the dye plume progressed; with two machines it was normally possible to measure the whole plume as it passed each site. Where necessary, supplementary information was obtained using automatic pump water samplers (Rock and Taylor and Epic) and a field fluorometer in discrete sample mode. Monitoring was continued at each site until the concentration of dye had receded to less than 10 percent of the peak value.

The methodology adopted for this study has been based on the empirical approach derived from the work of Wilson et al. (1997) and Kilpatrick and Taylor (1986). The fundamental basis of the approach has been to identify four facets of the pollution plume. These are as follows:

T_e time from injection to arrival (h)
T_p time from injection to peak concentration (h)
t_d through time (h)
C_{up} concentration unitised peak ($\mu g\ l^{-1}\ g$ injected^{-1})

These facets are expressed as a function of x, the distance from injection (km), and Q, the discharge ($m^3\ s^{-1}$). These particular elements of the concentration–time distribution are shown in Figure 7.3.

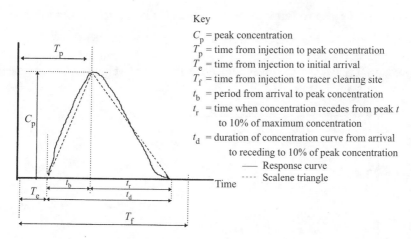

Key

C_p = peak concentration
T_p = time from injection to peak concentration
T_e = time from injection to initial arrival
T_f = time from injection to tracer clearing site
t_b = period from arrival to peak concentration
t_r = time when concentration recedes from peak t
     to 10% of maximum concentration
t_d = duration of concentration curve from arrival
     to receding to 10% of peak concentration
     —— Response curve
     ---- Scalene triangle

Figure 7.3   Features of concentration distribution, with scalene triangle overlaid (based on Kilpatrick and Taylor, 1986)

The characteristic shape of the classic concentration–time distribution is a fairly rapid rise from the leading edge to the peak concentration followed by a more gradual recession to the trailing edge, and this shape is maintained as the plume moves downstream. Kilpatrick and Taylor (1986) used a scalene triangle as an analogy for the response curve. The intention of this was to simplify the more complex response curve and allow predictions to be made based on this simpler form. The main features of the curve (arrival time, peak concentration and through time) form the triangle. Testing of this analogy was undertaken using 184 data sets collected by the US Geological Survey from all over the United States. The area of the triangle was then compared to that of the original response curve. Regression analysis of the two areas yields an expression for the area of the scalene triangle (A_Δ) which is

$$A_\Delta = 1.024(A_c)^{1.006} \tag{7.4}$$

where A_c is the area under the concentration–time distribution.

RESULTS: CHARACTERISTIC RESPONSE CURVES OF THE RIVERS UNDER STUDY

The response curves of all the rivers in the study demonstrate the classic skewed shape that has been described by workers such as Kilpatrick and Taylor (1986). Figure 7.4 shows the response curve at Gypsy Lane in Needham Market following an injection of dye at Stowmarket, 5 km upstream.

The peak at Needham Market is rapidly attenuated as the dye peak moves downstream, as shown in Figure 7.5. This pattern of response is repeated in the dye traces in all the rivers of the study.

Prediction of Time to Peak Concentration for Individual Traced Rivers

The individual dye traces described so far are clearly specific to the observed discharge for each trace. It is essential for operational purposes that some means of interpolation or prediction is available to extend the estimation of travel times to other discharges. The authors (Wilson *et al.*, 1997) have demonstrated a consistent relationship

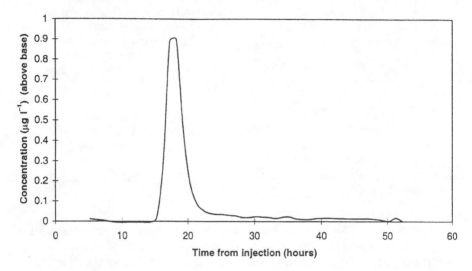

Figure 7.4   Concentration–time distribution for River Gipping at Gypsy Lane (Needham Market) on 30 June 1997

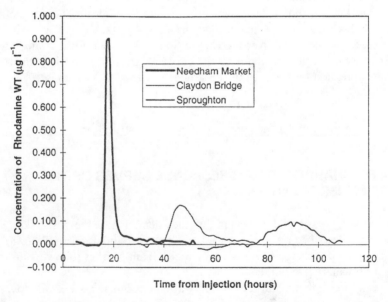

Figure 7.5   Concentration–time distribution for River Gipping for the trace beginning on 30 June 1997

between distance and the time to peak concentration at different discharge levels. This pattern is demonstrated for the River Bure in Figure 7.6, and is composed of three distinct elements:

(a) the time of peak concentration increases as the distance downstream decreases;
(b) the rate of this increase in time to peak decreases as discharge increases;
(c) the time to peak at the initial monitoring site is inversely related to discharge.

The approach adopted to predict Tp from discharge and distance downstream is completed in two stages:

(i)   The data for all of the rivers demonstrate a very high correlation between time of peak concentration and discharge. Behaviour of the pollutant immediately downstream of the injection will, however, be erratic until full transverse mixing occurs. As a result it is inappropriate to attempt to produce simple relationships in the first few kilometres below injection. Therefore an initial relationship is established between discharge and time of peak concentration for most upstream monitoring sites on each river. Thus it is possible to predict the time of peak concentration from discharge for the initial monitoring site.

(ii) A further relationship is then calculated between the discharge level and the gradient of the slope of the relationship between distance from injection and time to peak. A linear relationship is the most widely applicable, since it allows predictions of travel time for a pollution incident originating at any point on the river, although use of curve fitting would give slightly more accurate predictions for one particular injection site. Most rivers in the study demonstrated approximately linear relationships (e.g. the example of the Wharfe in Wilson *et al.*, 1997).

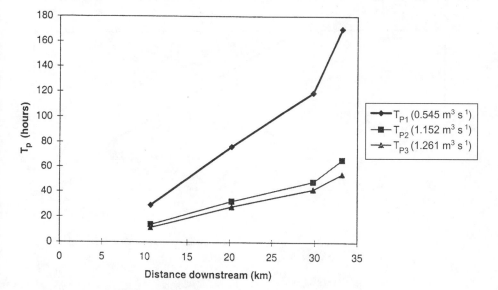

Figure 7.6   River Bure: time of peak concentration (T_p) from Saxthorpe

The final relationship between Tp and the distance from injection (spillage) may be summarised as follows:

$$Tp = A + Bx \tag{7.5}$$

where A is the Tp for the initial monitoring site relative to discharge; and B is the gradient of Tp against distance from injection, relative to discharge.

Other features of the response curve (time of arrival, through time and unitised concentration) may then be predicted from combinations of distance, discharge and time to peak. These further steps are discussed by Wilson *et al.* (1997) but will not be discussed here for the sake of brevity.

This methodology may be exemplified by reference to the River Bure. Three dye traces were undertaken at discharges of 0.534, 1.152 and 1.261 m^3s^{-1} (flow percentiles exceeded 93 percent, 40 percent and 25 percent of the time). The first component of the overall prediction involves the estimation of the time to peak at the first site (Figure 7.7). The resulting regression relationship was then used to predict the arrival time of the dye at any given discharge. Clearly the use of only three data points, the product of three dye traces, represents the absolute minimum necessary for this approach, but the high cost of dye tracing and the high statistical significance of the regression render the outcome more than satisfactory, partly because a wide range of discharge conditions have been sampled.

The second component of the analysis involves the calculation of the gradients of the regression lines fitted to the Tp data for all of the sites. The data for the River Bure are shown in Figure 7.8. The nonlinearity in the lower reaches of the Bure is, at least in part, a result of tidal influences which begin to be apparent close to the drinking water abstraction point. Whilst the inclusion of the most downstream site in the analysis reduces the goodness of fit, it is clearly fundamental in practical terms that this site be included, since the prediction of travel time to the abstraction point is axiomatic to the

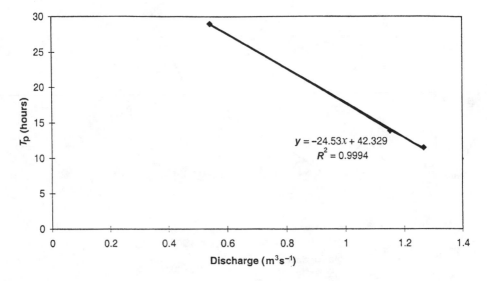

$$y = -24.53x + 42.329$$
$$R^2 = 0.9994$$

Figure 7.7   River Bure at Ingworth: predicted T_p from discharge

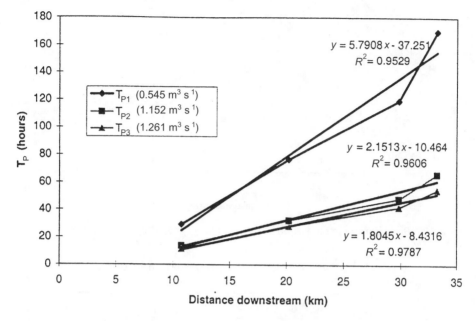

Figure 7.8   River Bure: observed and predicted time to peak concentration from Saxthorpe

research. No other river in the study experienced such conditions within the study reaches, so the Bure is presented here as a worst-case example.

The coefficients of determination for the regression lines indicate how well the points used to calculate the gradient are described by a straight line. The error that does occur in the regression relationship would appear to be a product of a slight increase in the time to peak at the last site on Figure 7.8 (Belaugh). Nevertheless the regression relationship errs on the conservative side, predicting a slightly earlier time to peak than actually occurs. This would appear to be the least harmful error in an operational situation, where the most severe consequences would ensue from a failure to close off the potable water intake before the pollutant arrived on site.

The next stage involves the description of the increase in the gradient of the Tp–distance relationship. It is clear from Figure 7.8 that this increase is a function of the increasing discharge and it is therefore possible to establish a relationship between discharge and the gradient of the Tp–distance relationship. Figure 7.9 shows this relationship.

The equation gives a coefficient of determination of 0.9996 between the observed and predicted values, providing an excellent description of the relationship.

Thus the equation to predict Tp for the Bure is given by

$$Tp = (42.329 - 24.53Q) + (2.8756 - 4.7948(\text{Ln}Q))x \qquad (7.6)$$

The overall relationship gives an excellent prediction of the Tp for the River Bure. A comparison of measured and predicted time to peak values is shown in Table 7.3. The greatest errors in each case occur at Horstead Mill, where the time to peak is over-estimated as a result of the nonlinearity in the original data.

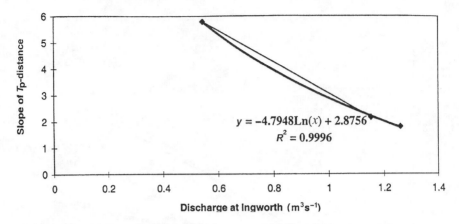

Figure 7.9 River Bure: predicted slope of T_p–distance against discharge at Ingworth

Table 7.3 A comparison of measured and predicted time to peak on the River Bure

Site	Measured T_p (hours)	Predicted T_p (hours)	% Error
Ingworth	29	28.96	0.14
Oxnead	76.5	83.93	−8.85
Horstead Mill	119.25	139.47	−14.50
Belaugh	170	159.14	6.82
Ingworth	13.83	14.07	−1.71
Oxnead	32.24	34.94	−7.73
Horstead Mill	48.08	56.04	−14.20
Belaugh	66	63.51	3.92
Ingworth	11.46	11.40	0.53
Oxnead	27.93	28.15	−0.78
Horstead Mill	41.83	45.08	−7.21
Belaugh	54.5	51.08	6.70

The high cost of each trace precluded the use of further traces on the River Bure, but this was possible on the River Wharfe where five complete traces were carried out. The results for the River Wharfe, shown in Figure 7.10 and Table 7.4, demonstrated a steady improvement in the quality of predictions that were made as a wider range of discharge values were incorporated into the development of the predictive relationships, although the inclusion of intermediate discharges added less to the fit.

Use of data from three dye trace experiments, including those from the highest and lowest discharges available, produced errors in time to peak from −2 to +2.5 hours (−10 to +15 percent of the observed time), whereas incorporation of data from all five dye traces resulted in errors from −1.5 to +3 hours (−7 to +18 percent of the observed). It was apparent from these and similar tests that the number of traces completed made no significant contribution to improving the quality of predictions unless it enhanced the overall flow coverage.

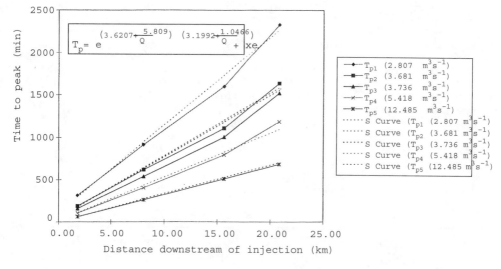

Figure 7.10   River Wharfe: observed and predicted time to peak concentration from Hebden

Table 7.4   Improvement of T_p prediction using selected data sets (River Wharfe data)

Data sets included for prediction (m³s⁻¹)					Errors compared to observed data	
12.485	5.418	3.736	3.681	2.807	Minutes	% of observed
√		√			−184 to 121	−21.3 to 12.1
√		√		√	−117 to 154	−9.9 to 15.4
√	√	√		√	−90 to 175	−7.6 to 17.4
√	√	√	√	√	−88 to 178	−7.4 to 17.7

The necessity to complete tracer experiments at a range of flows must be balanced against the problems associated with working at the extremes of discharge. At high flows the rapid travel times between sites can make it difficult to move monitoring equipment sufficiently quickly to collect full concentration–time distributions. Also, high flow conditions tend to be unstable, making collection of accurate data difficult.

Conversely, experiments at low flows are time-consuming and thus expensive in terms of labour costs. In addition to these drawbacks it is important to realise that, by their very nature, the extremes of flow are only rarely observed. Therefore, if good extrapolations can be made from existing data the necessity to cover the most extreme discharges is reduced. It would appear from the data collected for this project that coverage of half the flow range is necessary, although it would be preferable to collect data from between 60 percent and 70 percent of the flow distribution. The need to collect some data around the median discharge value must not be neglected at the expense of achieving higher and lower flow experimentation, as these still comprise a major part of the distribution.

Table 7.5 Errors produced from modelling of the River Wharfe, from Hebden to Lobwood and Hollins (using three data sets for calibration and four separate observations for validation)

Parameter	Error	ADE	ADZ	Empirical
Initial arrival time	Range (min)	−3 to 3	−201 to −52	−138 to 61
(Te)	Average (%)	0.03	−15.1	−1.7
	Range (%)	−0.22 to 0.34	−22.8 to −5.6	−9.9 to 6.9
Peak concentration time	Range (min)	−3 to −2	−260 to −128	−117 to 73
(Tp)	Average (%)	−0.20	−17.6	−1.7
	Range (%)	−0.25 to −0.12	−26.0 to −10.2	−9.9 to 6.6
Peak concentration	Range (μg l^{-1})	−0.017 to 0.101	−0.071 to 0.051	−0.14 to −0.02
(C_p)	Average (%)	35.46	2.2	−40.6
	Range (%)	−8.4 to 58.4	−35.0 to 28.0	−58.6 to −15.9
Time to 10% of peak	Range (min)	−329 to −50	−363 to −9	−158 to 35
(T_{f10})	Average (%)	−11.3	−12.3	−3.9
	Range (%)	−20.7 to −3.5	−22.9 to −0.63	−7.5 to 3.5

The evaluation for the River Wharfe also demonstrated that the empirical approach produced predictions which were more accurate than the ADE and of comparable or slightly better quality than the ADZ model. As a result of optimisation the ADE produced very small errors in the time domain (less than $\pm$1 percent), whereas the ADZ model produced errors in timing frequently of the order of −10 percent to −26 percent for the River Wharfe between Hebden and Hollins (Table 7.5). Using separate data sets to calibrate and verify, the empirical method produced errors in timing between −10 percent to +7 percent. This represents an operational error of 1–2 h over a trace lasting 15–20 h.

Whilst very strong patterns have been observed in the results obtained from individual rivers, there are very marked differences between rivers. The requirement for some prediction on rivers where tracing has not yet been carried out led to a consideration of the hydrological and geomorphological characteristics of rivers in the study in an attempt to establish some more general relationships.

Generalised Prediction of Time to Peak Concentration for Other Rivers

As discussed previously, the major control on the time of travel of a pollutant is the discharge experienced in the river. Relationships were therefore developed using the rivers in this study to predict the time to peak at an initial site and the gradient of the Tp–distance relationship using discharge per unit area (m^{3}s^{-1}km^{-2}) at a gauging station as the independent variable.

The relationships developed using 187 values are as follows

$$\text{Initial } Tp = 3.2164 * Q^{-1.0041} \qquad (R^2 = 0.5451) \qquad (7.7)$$

$$\text{Gradient } Tp\text{--distance} = 3.4355 * Q^{-0.7104} \qquad (R^2 = 0.7667) \qquad (7.8)$$

For the rivers included in the prediction process, this produces a mean error of just over 4 h, or 14 percent of the observed time for the peak of the dye to arrive. Times were consistently overestimated for the rivers Swale, Wharfe, Dibb and Nidd in Yorkshire, but were underestimated for the Ure and for the Bure and Waveney in

East Anglia. The other rivers were associated with errors of a more variable nature. The worst absolute error was for a low flow on the River Waveney, where time of peak was underpredicted by 7 days in a trace lasting 15 days over a distance of 17.3 km. This was clearly an exceptional circumstance, where the river had an average velocity of the dye peak just over 0.01 m s^{-1} and thus was barely flowing.

The equation was then validated using two traces on rivers not included in its production. On both the Wendling Beck (East Anglia) and the River Aire (Yorkshire) it produced errors of the order of 24 h in the time to peak, equivalent to an under estimation of the actual time by 25–65 percent at a site.

There are clearly several reasons for the relatively poor predictions obtained using this generalised method. These include the following:

- The discharge figure used for a river is that quoted by the Environment Agency at the gauging station. Whilst this is the only operational possibility in a real pollution incident, it is obviously not an adequate description for time of travel to an individual site at a different position along the river. In effect, assuming no exchange with groundwater, discharge is being overestimated for upstream sites and underestimated downstream.
- The method of prediction depends upon prediction of time of the dye peak at an initial monitoring site based upon discharge alone, followed by prediction of its progress downstream on the basis of discharge and distance. For individual rivers this works relatively well, but when several rivers are incorporated into a single model there is no allowance for the variation in distance to the initial site.
- There is no consideration given to the different characteristics of the various rivers. As discussed previously, factors such as the channel size, vegetation, substrate, slope, sinuosity, and degree of management by weirs and sluices are all likely to influence the time of travel of a pollutant. To some extent these will also be reflected in the calculation of discharge per unit area, but the consistency of under-prediction for some rivers and over-prediction for others implies that there are systematic sources of variation not included in the equation.

An attempt was made to address the last point by the incorporation of some simple map-derived geomorphological characteristics into an alternative set of predictive equations. Since the aim was to extend prediction of pollution incident behaviour to untraced rivers in the UK where there is a lack of detailed information on times of travel, it was considered inappropriate to include detailed measurements of channel size or roughness or to require accurate surveying of levels. Two additional variables, derived from the 1:25 000 Ordnance Survey maps, were therefore included for each river reach (Table 7.6).

For calibration purposes, four rivers in East Anglia and seven in Yorkshire were included in the calculations. The Aire, the Wendling Beck and the Rother were reserved for validation, to represent Yorkshire and East Anglia and to simulate the effectiveness of the resulting predictions in a contrasting area, West Sussex. It was considered desirable to predict time to peak from a pollution incident at any point, not just the injection site of the actual trace, so time to dye peak was in each case now converted to a velocity from injection to the monitoring site.

Table 7.6   Calculation of channel characteristics from 1:25 000 maps

Variable	Method
Reach slope $(m\,km^{-1})$	Interpolation of heights between contours; measurement of channel distance from the previous monitoring site
Reach sinuosity	Ratio of channel distance to straight line distance from the previous monitoring site

Correlation coefficients suggested that the velocity of the dye peak (in $km\,h^{-1}$ from the injection) is significantly related to reach slope ($r = 0.5072, n = 178$) although the relationship was not as strong as that with discharge at the gauging station per unit catchment area ($r = 0.8563$). The correlation of reach sinuosity with the velocity of the dye peak was not significant ($r = 0.0376$). In stepwise multiple regression, however, reach slope could not be used because of multicolinearity with the discharge variable. Using discharge at the gauging station alone to predict velocity of the dye peak from injection, the coefficient of determination was 0.7333. The use of a logarithmic–linear relationship improved this to 0.7860 (shown in Figure 7.11) and a cubic function to 0.8040. When discharge and sinuosity were both included in multiple regression, it reached only 0.7517. It would appear, therefore, that the influence of slope is already accounted for in the measurement of river discharge, and that sinuosity at the reach scale adds relatively little explanation to the prediction of velocity of the dye peak. A summary of the residuals for the validation rivers ($n = 32$) is shown in Table 7.7.

The residuals indicate that, on average, velocity of the dye peak was overpredicted by all three equations, and time to peak would thus be underestimated in most cases.

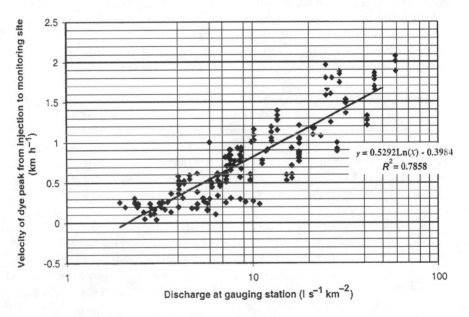

$y = 0.5292Ln(x) - 0.3984$
$R^2 = 0.7858$

Figure 7.11   Velocity of peak: East Anglia and Yorkshire

Table 7.7 Residuals (observed − predicted) on velocity of dye peak from injection to monitoring site (km h^{-1}) for validation rivers (Aire, Wendling Beck and Rother)

	Linear using area-specific discharge	Log–linear using area-specific discharge	Cubic using area-specific discharge
Mean	−0.076	−0.084	−0.054
Minimum	−0.347	−0.479	−0.403
Maximum	0.261	0.074	0.136
Standard deviation	0.155	0.122	0.108

Both the River Aire and the Wendling Beck flowed more slowly during the actual traces than is predicted by these equations. On the Aire, the error was equivalent to an underestimation of time to peak by 15.5 h over the 45.2 km length of the trace, or 15 percent of the total time. On the Wendling Beck the underestimation was more serious, with a predicted time to peak of 28.7 h compared to an observed time of 58.75 h. One explanation for this may be the gravel pits alongside the Beck which appear to be connected to the main channel, increasing the effective storage of water and delaying the passage of any pollutant significantly. On the Rother the predictions were more variable, giving underestimation of the time to peak by around 50 percent in the middle reaches at low flow (an observed time of 137.2 h compared to a prediction of only 68.6 h at the fifth site) but tending to overestimate it in the lower reaches of the river at higher flows (e.g. an observed time to peak of 24.5 h compared to a prediction of 29 h at the last monitoring site).

It would appear that this method works reasonably well on rivers similar to those used for calibration, as is the case with the Aire, but that it fails if extended to other areas or to channels with particular management issues, as was the case with the Wendling Beck.

Attempts to make generalised predictions from readily available characteristics are clearly less reliable than those afforded by a limited number of traces along a river, and do not appear to offer a viable alternative to a more detailed understanding of pollutant behaviour gained by such experiments. In the worst-case scenario, however, where a pollution incident occurs on an untraced river, it may be necessary to employ such a method. From Figure 7.11 it appears that it may be possible to shift the regression line upwards to define a 'fastest known' velocity of peak concentration for a given discharge per unit area in England. It is suggested that by increasing the intercept to zero, as a safety factor, the resultant equation becomes

$$\text{Velocity of } Tp = 0.5292\text{Ln}(Q^*1000/A) \qquad (7.9)$$

where the velocity of Tp is in km h^{-1}; Q is discharge at gauging station (m^{3}s^{-1}) A is the catchment area for the gauging station (km^2).

This suggests that for a discharge of 5 l s^{-1}km^{-2} the velocity of the peak is unlikely to exceed 0.9 km h^{-1} and at 10 l s^{-1}km^{-2} the velocity of the peak is unlikely to exceed 1.2 km h^{-1}, whilst a discharge of 50 l s^{-1}km^{-2} may give rise to a velocity of up to 2.1 km h^{-1}.

It has also been possible to select only those monitoring sites coincident with Environment Agency gauging stations and to calculate a power function relationship between discharge and velocity for those sites. The current data set produced an exponent of 0.87, much higher than those quoted for British rivers by Hey and Thorne (1986) using bankfull discharge on gravel and cobble channels, but more similar to those reported by Bathurst (1993) on the Severn and Swale and by Knighton and Cryer (1990) in Lincolnshire.

CONCLUSIONS

There is a well-recognised requirement for an effective methodology for the prediction of travel times in rivers used for drinking water supply. The high cost of travel-time experimentation has necessitated the use of modelling strategies such as ADZ and ADE, but these approaches have been complex and not transferable to rivers without some travel-time data. Models such as ADZ have not always been calibrated with traces of sufficient reach length or duration, and their complexity does not always lend itself to use in an actual pollution incident.

The empirical procedure to estimate time of travel of the peak of a pollutant from three tracer experiments on an individual river has been shown to provide predictions with a high level of accuracy and confidence. The results are comparable with, if not an improvement upon, those obtained from other approaches such as the ADE and ADZ, although the full response curve is not produced. The empirical technique has been shown to involve a robust and simple methodology for estimating the key points on the curve and to give reliable predictions based upon strong underlying relationships within the data. The consistency and regularity of the relationships observed would also appear to indicate that they are reproducible and can form the basis of predictive systems for river managers such as the Environment Agency and water companies. The simpler method thus forms a significant contribution to the techniques available for use in an operational context.

The use of generalised relationships to predict the time to peak of a pollution incident in a catchment where no dye traces have been carried out, however, is far less satisfactory in its predictive capability. Furthermore the use of explicit descriptors of geomorphology such as reach slope and channel sinuosity do not improve those predictions based on discharge, as a result of the strong multicolinearity between those geomorphological characteristics and discharge. The results do demonstrate, however, that there would appear to be an upper envelope that allows us to identify a fastest arrival time on an ungauged river. Further investigation should include assessment of the influence of channel width, the most popular aspect of channel geometry used in the estimation of discharge (e.g. Wharton and Tomlinson, 1999).

The empirical approach set out by Kilpatrick and Taylor (1986) and applied and extended here would appear to be the most appropriate route towards the management of pollution incidents. It requires a minimum of three traces across as wide a range of discharges as possible, but given this assumption it provides very high quality predictions of river travel time.

ACKNOWLEDGEMENTS

The study on the rivers in Yorkshire formed part of the research for a PhD at the University of Huddersfield (D. A. Wilson). The authors gratefully acknowledge the provision of funding and practical assistance from the Environment Agency, Yorkshire Water Services plc and York Water Co. The East Anglian dye traces were funded by the Environment Agency; thanks to Chris McArthur, Claire Bennett and Andy Baker for their help and support. The Western Rother study was funded by the Environment Agency and completed by Southern Water Technology Group (D. A. Wilson).

The authors would also like to thank the large number of people who helped with the long periods of fieldwork. These included Christine Wilson, John Shacklock, Julie McNish, Adam Potter, Rachel Harding, Sally Barker, Paul Hardwick, Phil Robson, Julia Meaton, Alan Dixon, Rob Johnson, Jonathan Barrett, Jonathan Vann, Sarah Doran, Diana de Nooijer, Liz O'Brien, Richard Coulton, Dennis Sinnott and David Turnbull. The contribution made by others who assisted more briefly is also gratefully acknowledged.

Thanks are due to Marcus Beasant for advice on equipment at various times and to Steve Pratt (Cartographic Unit, University of Huddersfield) who produced Figure 7.2.

REFERENCES

Bathurst, J. C. 1993. Flow resistance through the channel network. In Beven, K. J. and Kirkby, M. J. (eds) *Channel Network Hydrology*. John Wiley, Chichester, 69–98.

Beer, T. and Young, P. C. 1983. Longitudinal dispersion in natural streams. *Journal of Environmental Engineering, ASCE*, **109**, 1049–1067.

Bencala, K. E. and Walters, R. A. 1983. Simulation of solute transport in a mountain pool-and-riffle stream: a transient storage model. *Water Resources Research*, **19**(3), 718–724.

Beven, K. and Carling P. 1992. Velocities, roughness and dispersion in the lowland River Severn. In Carling, P. A. and Petts, G. E. (eds) *Lowland Floodplain Rivers: Geomorphological Perspectives*. BGRG Symposia Series, John Wiley, Chichester, 71–93.

Brady, J. A. and Johnson, P. 1981. Prediction times of travel, dispersion and peak concentrations of pollution incidents in streams. *Journal of Hydrology*, **53**, 135–150.

Buchanan, T. J. 1964. Time of travel of soluble contaminants in streams. *Journal of Sanitary Engineering Division, ASCE*, **90**(SA3), 1–12.

Butcher, D. P. and Labadz, J. C. 1999. Travel times in East Anglian Rivers. Report to Environment Agency, Norwich.

Carling, P. A., Orr, H. G. and Glaister, M. S. 1994. Preliminary observations and significance of dead zone flow structure for solute and fine particle dynamics. In Beven, K. J., Chatwin, P. C. and Millbank, J. H. (eds) *Mixing and Transport in the Environment*. John Wiley, Chichester, 139–157.

Church, M. 1974. Electrochemical and fluorimetric tracer techniques for streamflow measurements. *British Geomorphological Research Group Technical Bulletin No. 12*.

Day, T. J. 1975 Longitudinal dispersion in natural channels. *Water Resources Research*, **11**(6), 909–918.

Elder, J. W. 1959. The dispersion of a marked fluid in turbulent shear flow. *Journal of Fluid Mechanics*, **5**, 544–560.

Environment Agency 1999. *The State of the Environment in England and Wales*. http://www.environment-agency.gov.uk/s-enviro

Fischer, H. B. 1967. The mechanics of dispersion in natural streams. *Journal of Hydraulics Division, ASCE*, **93**(HY6), 187–216.

Fischer, H. B. 1968. Dispersion predictions in natural streams. *Journal of Sanitary Engineering Division, ASCE*, **94**(SA5), 927–943.

Fischer, H. B. 1969. The effects of bends on dispersion in streams. *Water Resources Research*, **5**, 496–506.

François, O., Calmels, P. and Merheb, F. 1997. Field tracer tests for the simulation of pollutant dispersion in the Doller river, France. In Kranjc A. (ed.) *Tracer Hydrology 97*. Balkema, Rotterdam, 121–125.

Glover, R. E. 1964. Dispersion of dissolved or suspended materials in flowing streams. *United States Geological Survey Professional Paper*, **433B**, B1–B32.

Green, H. M., Beven, K. J., Buckley, K. and Young, P. C. 1994. Pollution incident prediction with uncertainty. In Beven, K. J., Chatwin, P. C. and Millbank, J. H. (eds) *Mixing and Transport in the Environment*. John Wiley, Chichester, 113–137.

Harris, D. D. and Sanderson, R. B. 1968. Use of dye tracers to collect hydrologic data in Oregon. *Water Resources Bulletin*, **4**(2), 51–68.

Hays, J. R., Krenkel, P. A. and Schnelle, K. B. 1966. *Mass Transport Mechanisms in Open Channel Flow*. Technical Report No. 8, Vanderbilt University, Tennessee.

Hey, R. D. and Thorne, C. R. 1986. Stable channels with mobile gravel beds. *Proceedings of the American Society of Civil Engineering, Journal of Hydraulic Engineering*, **112**(8), 671–689.

Kilpatrick, F. A. and Taylor, K. R. 1986. Generalization and applications of tracer dispersion data. *Water Resources Bulletin*, **22**(4), 537–548.

Knighton, A. D. and Cryer, R. 1990. Velocity–discharge relationships in three lowland rivers. *Earth Surface Processes and Land Forms*, **15**, 505–512.

Leibundgut, C. and Hadi, S. 1997. A contribution to the toxicity of tracers. In Kranjc, A. (ed.) *Tracer Hydrology 97*. Balkema, Rotterdam, 69–75.

Leibundgut, C., Speidel, U. and Wiesner, H. 1992. Investigation of flow and transport parameters in rivers. In Hotzl, H. and Werner, A. (eds) *Tracer Hydrology*. Balkema, Rotterdam, 379–385.

Leopold, L. B. 1994. *A View of the River*. Harvard University Press, London.

Leopold, L. B. and Maddock, T. 1953. The hydraulic geometry of stream channels and some physiographic implications. *US Geological Survey Professional Paper 252*.

Nordin, C. F. and Sabol, C. V. 1974. Empirical data on longitudinal dispersion in rivers. *Water Resources Investigation 20–74*, United States Geological Survey.

Rukin, N., Hitchcock, M., Streetly, M., al Faihani, M. and Kotoub, S. 1994. The use of fluorescent dyes as tracers in a study of artificial recharge in northern Qatar. In Adar, E. M. and Leibundgut, Ch. (eds) *Application of Tracers in Arid Zone Hydrology, Proceedings of the Vienna Symposium 1994*. IAHS Publication **232**, IAHS Press, Wallingford, 67–78.

Rutherford, J. C. 1994. *River Mixing*. John Wiley, Chichester.

Sabol, G. V. and Nordin, C. F. 1978. Dispersion in rivers as related to storage zones. *Journal of Hydraulics Division, ASCE*, **104**(HY5), 695–708.

Seo, I. W. and Maxwell, W. H. C. 1991. Pollutant transport in open channels with storage zones. In Lee J. H. W. and Cheung Y. K. (eds) *Environmental Hydraulics*. Balkema, Rotterdam, 479–484.

Sheriff, J. D., Lawson, J. D. and Askew, T. E. A. 1996. Strategic resource development options in England and Wales. *Journal of the Chartered Institution of Water and Environmental Management*, **10**, 160–169.

Smart, P. L. 1984. A review of the toxicity of twelve fluorescent dyes used for water tracing. *NSS Bulletin*, **46**, 21–33.

Smart, P. L. and Laidlaw, I. M. S. 1977. An evaluation of some fluorescent dyes for water tracing. *Water Resources Research*, **13**(1), 41–59.

Taylor, G. I. 1954. The dispersion of matter in turbulent flow through a pipe. *Proceedings of the Royal Society of London*, **233A**, 446–468.

Valentine, E. M. and Wood, I. R. 1979. Experiments in longitudinal dispersion with dead zones. *Journal of Hydraulics Division, ASCE*, **105**(HY3), 999–1016.

Wallis, S. G., Young, P. C. and Beven, K. J. 1989. Experimental investigation of the aggregated dead zone model for longitudinal solute transport in stream channels. *Proceedings of the Institution of Civil Engineers*, **87**, 1–22.

Wharton, G. and Tomlinson, J. J. 1999. Flood discharge estimation from river channel dimensions: results of applications in Java, Burundi, Ghana and Tanzania. *Hydrological Sciences Journal*, **44**, 97–111.

Wilson, D. A. 1997. An investigation into travel times and dispersion of pollution incidents into non-tidal river systems and the development of predictive network models. Unpublished PhD Thesis, University of Huddersfield.

Wilson, D. A., Butcher, D. P. and Labadz, J. C. 1997. Prediction of travel times and dispersion of pollutant spillages in non-tidal rivers. *Proceedings of the Salford Symposium*, Institute of Hydrology on behalf of the British Hydrological Society, Wallingford. BHS Annual Conference, Salford 1997.

Yotsukura, N., Fischer, H. B. and Sayre, W. W. 1970. Measurement of mixing characteristics of the Missouri River between Sioux City, Iowa, and Platts Mouth, Nebraska, *United States Geological Survey Water Supply Paper 1899-G*.

Young, P. C. and Wallis, S. G. 1993. Solute transport and dispersion in channels. In Beven, K. J. and Kirkby, M. J. (eds) *Channel Network Hydrology*. John Wiley, Chichester, 129–174.

Zhou Ke-Zhao 1991. A new improvement on Fickian dispersion model. In Lee J. H. K. and Cheung, Y. K. (eds) *Environmental Hydraulics*. Balkema, Rotterdam, 523–528.

Section 3

TRACERS FOR INVESTIGATING SOIL EROSION AND HILLSLOPE PROCESSES

8 Soil Erosion Assessment using ^{137}Cs: Examples from Contrasting Environments in Southern China

D. L. HIGGITT,[1] **X. X. LU**[1] **and L. J. PU**[2]
[1]*Department of Geography, University of Durham, UK*
[2]*Department of Geography, Nanjing University, People's Republic of China*

INTRODUCTION: SOIL EROSION ASSESSMENT USING RADIONUCLIDE TRACERS

The use of caesium-137 as a tracer in soil erosion studies has become increasingly popular over the last two decades. The potential for using ^{137}Cs to estimate soil erosion rates over the medium term is derived from its suitable half-life (30.2 years), its strong retention in surface soils and the relative ease of measurement. Retrospective estimation of average erosion rates for the period since ^{137}Cs accession (1954 to present) from a single site visit offers an efficient means of evaluating erosion status, particularly for areas with limited information on sediment transfer (Ritchie and McHenry, 1990; Walling and Quine, 1992). The reliability of such erosion estimates depends on the validity of certain assumptions inherent in the technique. These include the notions that baseline fallout to the study site is relatively uniform and can be determined within given confidence intervals; that adsorption near the soil surface occurs rapidly, restricting down-profile and downslope movement; and that reliable methods are avilable for translating ^{137}Cs fluxes into soil erosion estimates. This chapter considers the impact of such assumptions on the utility of radionuclide tracing in three contrasting sites, forming a transect across subtropical southern China (Figure 8.1). There has been particular concern that erosion has become widespread and acute in southern China during the last 40 years. Despite comprehensive auditing of erosion hazards there have been relatively few quantitative studies of erosion rates and the use of radionuclide tracers offers an opportunity to assist environmental management in this region. Issues concerning the application of ^{137}Cs in these environments and the estimates of soil loss produced, are discussed.

Measurements of actual soil movement, whether conducted in plot experiments, collected in sediment traps or identified by the exposure of survey markers, require an investment in time. Methods that allow rapid appraisal of erosion status are particularly sought. In this context, determination of rates of soil loss provided by variations in the ^{137}Cs inventory of sampled soil profiles, is useful for research-led investigations

Tracers in Geomorphology. Edited by Ian D. L. Foster. © 2000 John Wiley & Sons Ltd.

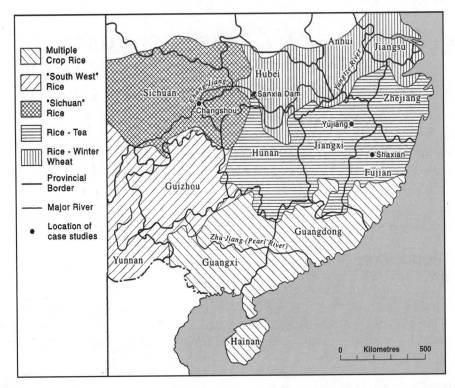

Figure 8.1 Map of traditional agricultural areas of southern China (based on Vermeer, 1977) indicating the distribution of the provinces and the location of the three study sites

of erosion, validation of other assessment methods or in support of conservation planning. Preliminary applications in the Loess Plateau (Zhang *et al.*, 1990, 1994) have supported the potential of the method and Quine *et al.* (1992) report favourably from a purple soil area in Sichuan Province. More recently their studies have demonstrated the significance of tillage erosion in redistributing soil within terraced fields (Quine *et al.*, 1997, 1999a,b). Nevertheless, it might be argued that there has been limited effort to investigate whether the assumptions of the ^{137}Cs technique can be supported in Chinese agricultural environments and the extent to which interpretations based on ^{137}Cs can be corroborated.

CONTEXT: SOIL EROSION IN SOUTHERN CHINA

More commonly recognised as a problem in the semi-arid environments of northern China, soil erosion and land degradation have emerged as significant environmental concerns in the humid south in recent decades. Inventories of land affected by erosion provide stark reading, indicating a dramatic increase in the extent of degradation (Wen, 1993; Zhao, 1994) and leading to speculation about future food security (Smil, 1993). Like most resource inventory estimates they are based upon criteria that may be difficult to replicate and need to be treated with caution. Nevertheless, that the soil erosion problem of southern China has largely emerged in the last three to four decades is indicative of a delicate balance between environmental conditions and land management. Southern China has experienced profound changes in agricultural practice during this period.

The principal challenge to agricultural development in southern China has been a shortage of level land (Murphey, 1982), necessitating the exploitation of sloping land and the construction of terraces in response to population growth. The dominant 'red soils' of southern China have a number of characteristics that render them vulnerable to degradation. They are acidic, and have a low cation exchange capacity and low organic matter content. In addition, poor soil structure results in limited available moisture-holding capacity (Grimshaw and Smyle, 1991). Impacts of degradation include reduction of the naturally low nutrient status, particularly phosphorous, potassium and calcium contents, loss of organic matter, soil pollution from fertiliser, pesticides and industrial discharges and increasing stoniness resulting from the erosion of fine-grained material (Gong and Shi, 1992). Enhanced acidification and aluminium toxicity threatens to become more prominent as an acid rain problem develops in the wake of rapid industrialisation. In addition to the loss of fertility experienced at eroding sites, the fate of the eroded material can cause a variety of off-site sedimentation and pollution problems.

Quantitative information about soil loss dynamics and associated spatial patterns are limited. National erosion inventory estimates represent intensive auditing of land resources by administrative teams at various levels of government, but in practice much of the designation of erosion classes is based on qualitative criteria. Many studies have been based around plot experiments or the application of predictive models, especially the Universal Soil Loss Equation. Advances in modelling can be expected to improve the efficiency and accuracy of erosion hazard evaluation in due course, but frequently there is insufficient information about parameters or accurate representation of field conditions (notably DEM construction in terraced landscapes). In these circumstances the use of a rapid appraisal technique for a quantitative assessment of erosion status has much potential. Though the factors controlling erosion are well known in a general sense, the identification of process dynamics, rates of soil loss and the pathways by which soil is transported into the fluvial system present a challenge to environmental management.

STUDY AREA AND SAMPLING

The present study summarises current work from three investigations of erosion and sedimentation in an attempt to illustrate the constraints facing erosion assessment in a region where land degradation can be severe and to compare sites. Brief descriptions of the study sites are provided below and summarised in Table 8.1.

Shaxian County, Fujian Province

A study on erosion assessment is being carried out in conjunction with a land resource project in Dongxi River Basin (830 km^2), conducted by Nanjing University. The

Table 8.1 Summary of study site characteristics

Site	Shaxian, Fujian	Yujiang, Jiangxi	Changshou, Sichuan
Latitude (°N)	26.4	28.4	29.9
Longitude (°E)	117.7	116.9	107.0
Altitude range (m)	300–1000	150–250	240–380
Mean annual precipitation (mm)	1643	1733	1031
Mean annual temperature (°C)	19.5	17.4	17.6
Dominant geology	Mesozoic shale and sandstone, deep Tertiary regolith	Quaternary red clay	Triassic shale and sandstone
Soils	Red soil (ultisol/oxisol)	Red soil (ultisol)	Purple soil (inceptisol)
Land uses	Rice, oil tea, citrus	Rice, tea, citrus	Rice, wheat, rape, maize

Dongxi River is a left bank tributary of the Shaxi River, which drains the mountainous interior of Fujian Province. The topography is rugged with mountains rising to over 1000 m. Palaeozoic granites and metamorphic rocks dominate the upland areas. Tectonic activity during the Mesozoic resulted in a series of down-faulted basins, up to 20 km in length. The majority of these were infilled by sediments derived from the surrounding uplands. Deep weathering of the various basin infills during the Tertiary generated clay-rich regoliths known generally as Red Beds which are largely capped by Quaternary river deposits. Fluvial dissection of the Red Beds has resulted in a series of low relief hills, typically 100 m in length and 30 m relative relief, whose summits characteristically lie at two distinct terrace levels. The soils of the low relief hills are classified as Red Soils under the Chinese taxonomic system and are equivalent to the transition between ultisols and oxisols in the American taxonomic classification.

Shaxian experiences a humid subtropical climate (Figure 8.2(a)). Precipitation totals are influenced by the arrival of the south-eastern monsoon in mid-March, with May and June being the wettest months. Being relatively close to the coast, the area is sometimes affected by typhoon activity in late summer. The agriculture of Fujian has traditionally been dominated by rice and tea production. Tea prices declined in the 1960s and many tea gardens switched to oil tea production. As the economic viability of oil tea has also declined in recent years, many of the hillsides are now being converted to citrus fruit production and relatively little arable land remains. Although the Red Bed hills are relatively small features, slope gradients are sufficient to promote erosion unless precautions are taken. Preliminary surveys of the catchment (Higgitt, 1995a) identified that the majority of contemporary erosion is associated with arable land, predominantly in the down-faulted basins.

Two phases of field sampling have been conducted. In the first phase (1994) a series of slope transects were surveyed and sampled in three of the down-faulted basins of Fukou, Gaoqiao and Xiamo. Logistical difficulties in transporting samples from the field site restricted the preliminary study to 16 bulk samples for gamma ray analysis plus smaller incremented subsamples for chemical analysis. A second visit in 1997 enabled a more substantial sampling programme to be conducted concentrating on slope transects in Fukou and Xiamo, the former comprising slopes that have recently

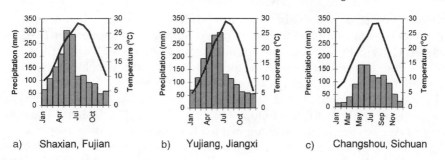

Figure 8.2 Climatographs for the three study sites: (a) Shaxian (1957–1976); (b) Yujiang (precipitation 1955–1994; temperature 1980–1990); and (c) Changshou (1957–1987)

been adapted for citrus production and the latter poor-quality arable land. Soils were sampled at 5 cm increments to variable depths.

Yujiang County, Jiangxi Province

A sampling programme has been conducted in and around the Red Soil Ecological Field Station maintained by the Institute of Soil Science, Academia Sinica, located close to the city of Yingtan in Jiangxi Province. Part of the field station was established on barren land to conduct experiments in rehabilitation on the severely degraded soils developed on Quaternary red clay. Intense erosion of red soils (ultisols) throughout Jiangxi Province following clearance of woodland and scrub has resulted in large expanses of abandoned land where the plinthitic C horizon of the red clay-derived soil is exposed at the surface and vegetation growth is inhibited. Locally these areas are referred to as 'Red Desert'.

In addition to sampling on abandoned land, attention was paid to agricultural soils in the vicinity of the field station. Soils were sampled at 5 cm increments to a depth of 20 cm and at 10 cm increments to 40 cm. The transects concentrated on arable land (15 profiles), but also included a grassland transect, a sequence of paddy fields, a tea garden and an area that had been cleared of tea to plant citrus trees. The Yujiang study site has some similarities to Shaxian, in terms of the distribution of annual precipitation and the dominance of red soils. The temperature regime of Yujiang exhibits enhanced continentality (Figure 8.2(b)) and the topography is much more subdued than the mountainous terrain of Fujian Province. Nevertheless, many parts of Jiangxi, especially those areas on Quaternary red clay, have been subject to severe degradation.

Changshou County, Sichuan Province

As part of a study of erosion and sediment delivery in the area adjacent to the Three Gorges Reservoir Project, on the Yangtze River, a small reservoir catchment, representative of the broader region, was selected for more detailed analysis. The Yiwanshui catchment (0.7 km^2) is located close to the south bank of the Yangtze in Changshou County, downstream of Chongqing (Figure 8.1). The catchment is dominated by

Triassic purple shale and sandstone, which produce soils with high mineral content but a renowned susceptibility to erosion (inceptisols under the American Taxonomic System). Catchment slopes are steep and traditionally divided by terraces. Slope angles across terraced field surfaces, measured during the sampling, ranged from 16° to 36° (mean = 23°). Mean field lengths are only 8 m.

The mean annual rainfall of 1030 mm is highly seasonal with a risk of flooding and severe erosion in May and June followed by risk of drought in the hottest months of July and August. Seasonality is more marked here than at the two other study sites (Figure 8.2(c)). Traditional cultivation practices are therefore geared to both soil and water conservation. The climate enables two main crops per year. Rice paddies used in the summer support wheat or rape in the winter, while the slopes are typically used for a wheat–maize–sweet potato rotation. In the Yiwanshui catchment, arable lands comprise almost 60 percent of the land area.

In order to investigate patterns of soil loss and redistribution throughout the small catchment, a total of 66 bulk samples were collected from arable fields. In all fields, samples were collected from the middle of the terrace surface, with additional samples at the upper and lower boundaries of the terrace surface at 11 sites. More detailed examination of the vertical and spatial distribution of ^{137}Cs was carried out in one field and the study was supplemented by two reservoir cores.

METHODS

The study is concerned with gamma-emitting radionuclides and ^{137}Cs in particular. Determinations of gamma activity were undertaken using HPGe detectors (EG&G Ortec) at the University of Durham (Shaxian and Changshou) and at the Institute of Geography and Limnology, Nanjing (Yujiang). Caesium-137 activity is determined by the photopeak at 662 keV, while ^{210}Pb$_{excess}$ is determined from measurements of ^{210}Pb and ^{214}Pb at 46.5 and 352 keV, respectively. Samples were measured either in 500 ml Marinelli beakers, or as 50 g samples in flat-bottomed pots. The use of small samples is sufficient for determination of ^{137}Cs activity but requires longer count times (between 84 and 172 ks) compared to the 500 ml samples (48–84 ks). The logistic advantage of retrieving smaller samples from remote field locations is therefore tempered by the increased time needed for analysis. Sample activity can be expressed as a concentration (Bq kg^{-1}) which can be translated into an inventory (Bq m^{-2}) using the appropriate bulk density.

RESULTS

Estimating Baseline Fallout

The application of the ^{137}Cs technique requires the comparison of soil core inventories with an estimate of the cumulative atmospheric deposition at that location. Because direct fallout measurements are rarely available, the baseline fallout inventory is usually determined by sampling nearby undisturbed sites. The lack of suitable sites is a major difficulty in the Chinese setting. The ideal reference site is a flat, non-eroding site which has remained under permanent grassland since the onset of fallout acces- sion (normally considered to be 1954). In southern China, where the amount of arable land per capita has declined to about 0.07 ha, gently sloping topography is a com- modity not left unused by human activity. The few exceptions are sacred sites such as hilltop shrines or burial grounds, which pose ethical problems for sampling. Applica- tions of ^{137}Cs tracing have therefore tended to adapt the criteria for a baseline site according to local circumstances (Zhang et al., 1990; Quine et al., 1992; Chappell et al., 1998).

In the present study, sites considered to reflect conditions conducive to the accu- mulation of net fallout deposition included hilltop groves, where a remnant of the original forest is left as agricultural land is created (Shaxian); ridge locations in undisturbed primary forest (Shaxian); replanted secondary woodland on hilltops or ridges (Changshou, Yingtan); and a watershed paddy field (Changshou). As some authors have reported considerable variability between individual soil inventories at reference site locations (e.g. Bachhuber et al., 1987; Fredericks et al., 1988; Suther- land, 1991), the confidence limits associated with the baseline estimate should also be determined. Unfortunately, the lack of suitable sites and the logistic difficulty of transporting a large volume of samples, have sometimes dissuaded researchers from expending effort in collecting replicate cores. Some baseline inventories for previous investigations in China are reported in Table 8.2.

In the Changshou site, a paddy field straddling the watershed boundary on a low col was selected for investigation. Upland paddies in this region are double cropped with rice in the summer and wheat during the winter. Terrace boundaries effectively isolated the paddy from the limited run-on areas on adjacent slopes. Eight cores, each

Table 8.2   Baseline inventories reported in applications of the ^{137}Cs technique in China

Authors	Location	Number of samples	Mean annual precipitation (mm)	Baseline (Bq m^{-2})	Standard error (Bq m^{-2})	Coefficient of variation (%)
Zhang et al. (1990)	Lishi, Shanxi	1	506	2009	–	–
Zhang et al. (1990)	Luochuan, Shaanxi	1	622	2529	–	–
Wang et al. (1991)	Kangxian, Sichuan	1	1285	1731	–	–
Quine et al. (1992)	Yanting, Sichuan	Not reported	900–1100	2600	Not reported	Not reported
Zhang et al. (1992)	Jiangjiagou Yunnan	2	680–1200	1549	84	8
Zhang et al. (1994)	Xifeng, Gansu	10	565	2600	100	12
Pu et al. (1998)	Korla, Xinjiang	1	50	10 292	–	–
This study	Changshou, Sichuan	7	1031	2163	76	9
This study	Yujiang, Jiangxi	3	1733	4063	510	22
This study	Shaxian, Fujian	5	1643	4237	374	20

comprising four increments to a depth of 40 cm were collected. Excluding one outlier, the mean inventory ($\pm$ 1 SD) is 2163 $\pm$ 201 Bq m^{-2}. The inventory is consistent with levels reported elsewhere in Sichuan (Wang et al., 1991; Quine et al., 1992), but the uncertainty associated with the estimate is somewhat lower than other reported studies (Higgitt, 1995b). Attempts to collect baseline inventory samples at the Yingtan study site were fraught with difficulty. A substantial amount of relatively level grass-land with sparse masson pine cover is found in the area, but this represents land that was severely degraded following clearance in the late 1950s and early 1960s. Exposed subsoils developed on Quaternary Red Clay are indurated and have low moisture storage capacities, which inhibits recolonisation by plants. Bulk samples collected under secondary woodland have considerably higher inventories (4063 $\pm$ 854 Bq m^{-2}). In the Shaxian site, potential baseline sites were again confined to sites under woodland. The mean inventory ($\pm$ 1 SD) is 4237 $\pm$ 837 Bq m^{-2}. The higher baseline values at the Yujiang and Shaxian sites are consistent with higher mean annual precipitation.

In the absence of secure reference sites or systematic measurements of direct fallout, an alternative method of estimating baseline fallout is to employ established relation-ships between fallout rate and site location. Previous studies have determined linear relationships between rainfall totals and ^{137}Cs deposition rates for given latitudinal ranges (Davis, 1963; Basher and Matthews, 1993), but these cannot be extrapolated. Instead a model for predicting global distribution of ^{90}Sr fallout (Sarmiento and Gwinn, 1986) has been adapted using monthly precipitation values for the Changshou County meteorological station. The Sarmiento and Gwinn model estimates a fallout deposition of 1683 Bq m^{-2} between 1954 and 1974, while direct measurements of fallout in Japan (Hirose et al., 1987) between 1974 and 1984 indicate that another 200 Bq m^{-2} might be derived from Chinese weapons testing during the decade. Evidence for a modest enrichment of Chernobyl-derived radiocaesium in Chinese lakes (Wan et al., 1990) suggests additional fallout deposition is likely to have occurred during the last decade. In sites where the suitability of reference sites cannot be guaranteed, the use of precipitation data to estimate past fallout deposition offers an alternative means of obtaining the baseline inventory and also the time-dependent trends in fallout accumulation which have relevance to the quantification of erosion rates from ^{137}Cs measurements.

Sediment-associated Redistribution

The essence of a good tracer is that it mimics the behaviour of the process of interest. In the case of ^{137}Cs its utility is dependent on its redistribution being predominantly sediment-associated through the transport of surface materials by sheet wash and rill erosion processes. Investigations into the environmental mobility of ^{137}Cs suggest that losses from the sampling layer via downward diffusion or uptake by vegetation and cropping are insignificant (e.g., Tamura, 1964; Rogowski and Tamura, 1970), with the exception of some organic soils (Livens and Rimmer, 1988). There is no evidence to suspect that soil characteristics in southern China would compromise the rapid adsorption of ^{137}Cs observed elsewhere, but there has been little research on radio-nuclide cycling in subtropical vegetation and associated production systems.

The fixation of ^{137}Cs close to the soil surface can be observed by sampling soil profiles incrementally. On agricultural land, tillage practices are assumed to mix the ^{137}Cs inventory within the plough layer, while undisturbed sites typically exhibit an exponential decline in concentrations from the surface. Sectioned profiles from all agricultural sites in the study sites in southern China exhibit relatively uniform mixing within the plough layer. Profiles from uncultivated sites tend to display the exponential decline with depth that has been observed elsewhere (Figure 8.3). Evidence of the retention of ^{137}Cs close to the soil surface supports the applicability of the technique in soils encountered in southern China. By contrast, a related study in the arid environments of Xinjiang, western China (Pu *et al.*, 1998), has revealed considerable down-profile penetration of ^{137}Cs, which is thought to relate to translocation through desiccation cracks and macropores, making soil redistribution applications difficult.

However, one issue that impinges on the use of ^{137}Cs to decipher patterns of soil loss and gain, is the extent to which farmers, as opposed to the action of erosion processes, affect its redistribution by the deliberate removal of soil. Most notably, the construction of terrace systems will result in the disruption of ^{137}Cs labelling of surface materials, introducing substantial uncertainty into the interpretation of inventories. Many locations in south-east China (including the Yujiang and Shaxian study sites) have experienced profound changes in agricultural practices in the last 40 years, the latest trend being a move towards citrus production. Ground preparation for tree planting, involving the digging of deep pits and terrace reconstruction, destroys the surface labelling properties of ^{137}Cs, preventing its use for soil erosion assessment. However, terrace building also preserves the previous soil surface prior to land-use change. This is useful in the Shaxian site for interpreting pre-agricultural erosion rates on the Red Bed hills.

On arable land, where terrace construction occurred prior to the accession of fallout, erosion-related ^{137}Cs flux may be supplemented by maintenance activity. On the sloping terrace benches of the Changshou site, farmers have been observed adding newly weathered (^{137}Cs-deficient) material from upslope terrace risers and deposited fines (^{137}Cs-enriched) from lower boundary ditches, into the field surface to maintain fertility. It is not possible to distinguish ^{137}Cs redistribution by soil erosion from the action of deliberate removal, though the latter may be suspected to be insignificant compared to the former. The deliberate spreading of soil from the upslope portion of terrace fields towards the lower part to reduce gradients might even be considered a form of tillage erosion. Where intensive grid sampling has been feasible, the patterns of ^{137}Cs in relation to topography have been used to distinguish between tillage and water erosion in China and elsewhere (Zhang *et al.*, 1993; Quine *et al.*, 1997).

Quantification Procedures

The reliable translation of ^{137}Cs inventories into soil redistribution estimates is perhaps the most critical element to the future development of the ^{137}Cs technique. Proposals for quantification procedures are almost as plentiful as examples of applications. Methods have evolved from simple regression relationships (Ritchie *et al.*, 1974; Campbell *et al.*, 1986), to proportional loss models (de Jong *et al.*, 1983; Martz and de Jong, 1987), to mass balance methods (Kachanoski and de Jong, 1984; Higgitt

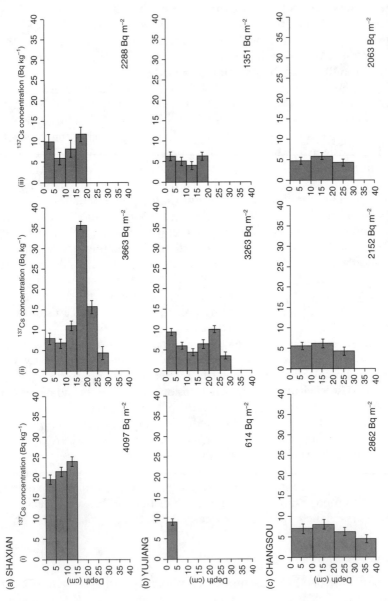

Figure 8.3 Examples of the vertical distribution of ^{137}Cs concentration. (a) Shaxian: (i) hillcrest forest grove, Fukou (baseline site = 4097 Bq m^{-2}); (ii) buried profile in terrace edge following conversion from scrub to citrus production, Fukou. Original soil surface is at 15 cm depth (3663 Bq m^{-2}); (iii) abandoned arable land sloping at 5°, Xiamo (2288 Bq m^{-2}). (b) Yujiang: (i) midslope position on degraded grassland sloping at 15° (614 Bq m^{-2}); (ii) lower slope position on arable land sloping at 2° (3263 Bq m^{-2}); (iii) level secondary masson pine scrub, previously degraded (1351 Bq m^{-2}). (c) Changshou: three baseline profiles from a watershed paddy field. Sample (i) is an outlier from multiple cores at this site and has some evidence of deposition through the deeper accumulation of ^{137}Cs activity

and Walling, 1993; Quine, 1995; Walling and He, 1999). The mass balance procedure partitions ^{137}Cs between input as fallout, storage within the sampling volume and output through radioactive decay, soil erosion flux and other losses. As such, a detailed mass balance procedure requires information on the variable atmospheric fallout rates for successive time increments from 1954 onwards. Additional refinements can be added by considering the fate of ^{137}Cs at the soil surface prior to mixing within the plough layer (Higgitt and Walling, 1993) or scenarios developed for uncultivated soils (Higgitt *et al.*, 1994; Yang *et al.*, 1998). In the absence of data on atmospheric deposition rates in China, Zhang *et al.* (1990) proposed simplified models for cultivated and uncultivated scenarios, which assume that all fallout occurred in 1963. The model for cultivated soils is given by the equation:

$$A_t = Y_t(1 - \Delta H/H)^{[t-1963]} \tag{8.1}$$

where
  A_t = ^{137}Cs inventory in plough layer at time (year), t(Bq m^{-2});
  Y_t = cumulative baseline fallout at year t (Bq m^{-2});
  H = depth of plough layer (cm);
  ΔH = depth of annual soil loss (cm).

The simplified mass balance model has been used to estimate erosion rates in each of the study sites (Table 8.3). In addition, the data derived from the Sarmiento and Gwinn (1986) model of ^{90}Sr deposition, based on detailed precipitation records for Changshou, provide the necessary atmospheric fallout series to apply the Kachanoski and de Jong (1984) time-step model at this site. In comparison to the more detailed time-step simulation, the simplified mass balance model overestimates erosion rates, markedly so, at low ^{137}Cs inventories. Apart from the discrepancy between alternative models, the error associated with soil erosion estimates is not frequently reported in ^{137}Cs papers. The curve relating ^{137}Cs inventory to average annual depth of soil loss for the Shaxian site is displayed in Figure 8.4. Given the uncertainty associated with the estimation of baseline fallout (CV = 21 percent), the envelope marking ± one standard deviation is broad.

Interpretation of Soil Erosion Dynamics

The scarcity of suitable land for agricultural production, compounded by population growth and the spread of urbanisation, has engendered the need for hillslope production systems throughout southern China. The erosion potential of steep land is widely

Table 8.3 Estimates of erosion rates for arable land at the three sites

Site	Erosion rate (t km^{-2}year^{-1})			
	Mean	Standard error	Minimum	Maximum
Shaxian, Fujian	8422	1301	3047	15 600
Yujiang, Jiangxi	6142	302	1571	12 132
Changshou, Sichuan	6532	435	1400	20 920

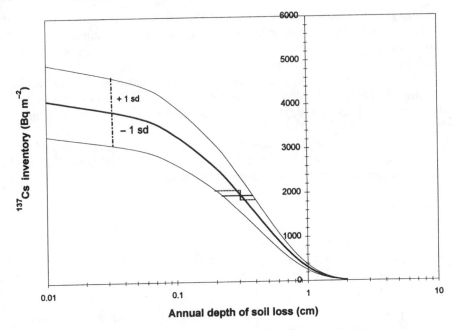

Figure 8.4   Example of a ^{137}Cs–soil loss curve for Shaxian based upon the simple mass balance model. The upper and lower curves indicate one standard deviation for the estimated baseline inventory (4237 Bq m^{-2}). For illustrative purposes, an inventory of 2000 Bq m^{-2} with an analytical precision of 10 percent is indicated

appreciated and is the principal criterion in the semi-quantitative natural resource and erosion surveys that have been undertaken. The use of a tracer to provide quantitative estimates of erosion status enables the relationship between slope gradient and slope position to be analysed in more detail.

In Shaxian, examination of ^{137}Cs inventories across slope transects on Red Bed hills displays a consistent decline with distance downslope. There is a strong statistical relationship with slope gradient ($R^2 = 0.55$), with some sites on steeper slopes having no detectable levels of ^{137}Cs (Figure 8.5). The results point to the lower parts of the convex Red Bed hills as the most intensely eroded locations, in spite of the presence of terraces. In recent years, many of the arable fields on these slopes are being replaced by citrus fruit production and the original terraces are being reinforced or rebuilt. By contrast, the downslope relationship with erosion status is not observed, and to some extent reversed, at Yujiang. The lowest ^{137}Cs inventories tend to occur close to slope crest locations. Here, in northern Jiangxi, slope gradients are considerably less than in the mountainous terrain of Fujian. The relationship between ^{137}Cs inventory and slope gradient is not statistically significant ($R^2 = 0.10$), although the steeper grassland sites have endured greater ^{137}Cs loss than the arable sites. This may reflect the intensity of erosion following clearance of forest–scrub during the Great Leap Forward period, which made these slopes unproductive for arable cultivation.

Further west, the site at Changshou demonstrates a statistically significant relationship between slope gradient and ^{137}Cs inventories, but provides a poor explanation of

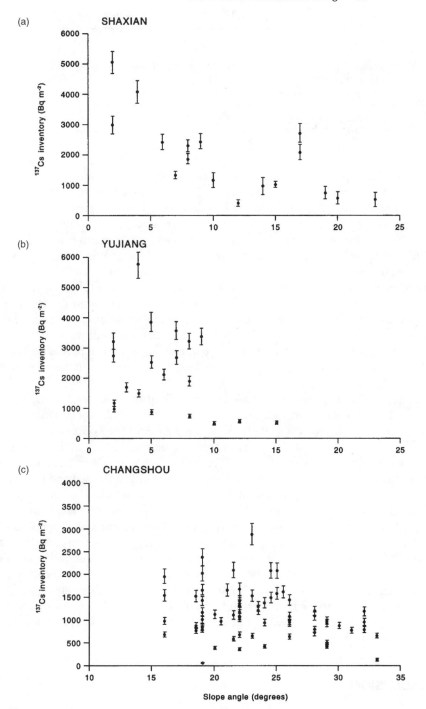

Figure 8.5 The relationship between ^{137}Cs inventory and slope angle at (a) Shaxian, (b) Yujiang, and (c) Changshou

Table 8.4   Correlation between ^{137}Cs inventory, slope angle and position

Site	n	Slope angle	Slope length
Shaxian, Fujian	22	−0.746[a]	−0.280[b]
Yujiang, Jiangxi	21	−0.318	+0.362
Changshou, Sichuan	66	−0.320[a]	−0.066

[a] Correlation significant at $\alpha = 0.01$.
[b] Correlation for Red Bed hills only ($n = 17$) is significant, $r = -0.647$.

the scatter ($R^2 = 0.10$). Three bulk samples were collected from upper, middle and lower positions within some of the individual terrace fields. Generally, the inventories show enhanced depletion at the upper locations, which is consistent with interpretations of tillage translocation across terrace surfaces elsewhere in China (Quine *et al.*, 1997). No relationship between ^{137}Cs inventories and slope length is apparent at any of the sites (Table 8.4). The results suggest that combinations of slope gradient and slope length, such as the LS factor employed within the USLE, are meaningless as a method of predicting soil erosion hazard on terraced land within southern China. More elaborate, physically based models require accurate representation of the topography which is not generally available at suitable resolutions to create DEMs. Furthermore, the angle and length of field plots encountered in the Changshou and Shaxian sites are typically beyond the range of existing empirical plot experiments. It seems logical that terrace configuration and design is a factor that influences observed patterns of ^{137}Cs inventories.

Estimates of erosion rates for arable land in each of the study sites, based upon the simplified mass balance model, are indicated in Table 8.3. The usefulness of an average erosion rate for each of the sites can be questioned on two counts. First, uncertainties associated with the estimates of baseline fallout and the analytical precision of gamma-ray spectrometry affect the translation of a ^{137}Cs residual into a soil erosion estimate. The suitability of the mass balance model used and the choice of parameters for plough depth and bulk density, also constrain the accuracy of soil redistribution estimation. Second, the measurements demonstrate large variations in the incidence of net soil loss over comparatively short distances. Recognition of the worst affected areas is of more practical value for conservation activity. Within China, erosion hazard reporting uses a series of notional magnitude boundaries to indicate increasing levels of severity. A combined frequency distribution of the arable land within each class can be produced (Figure 8.6). According to these preliminary data, the severity of soil erosion is greatest at Shaxian. The modal class of erosion hazard is 'severe' (2500–5000 t km^{-2} year^{-1}) at Changshou, but 'very severe' (8000–13 500 t km^{-2} year^{-1}) at Yujiang and Shaxian.

CONCLUSION

The rapid appraisal of soil erosion status by gamma spectrometry has attracted the attention of scientists interested in examining problems concerned with the study of

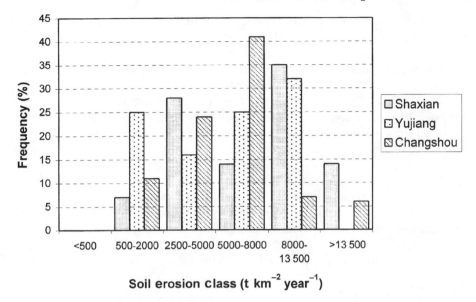

Figure 8.6 Frequency distribution of the relative amount of arable land, at each site, within soil erosion severity classes

erosion processes. As such, most attention has been directed to research investigations of a technical nature and it can be argued that the routine use of ^{137}Cs measurements as a means of informing conservation planning and environmental management has yet to be realised. In order to promote the transition of a technique from a tool for investigating problems in erosion studies to a method to support agencies with soil erosion problems, three areas of development are suggested.

First, and the main theme of the chapter, is the continued attention to the reliability of the assumptions inherent in the technique and their impact on the uncertainty associated with the results. The data presented in this chapter demonstrate that the fundamental tracer properties attributed to ^{137}Cs appear to be broadly supported in a subtropical environment. Land management practices interfere with ^{137}Cs redistribution. Soil spreading on existing terrace fields is considered to be relatively minor compared to the mass soil flux due to tillage and water erosion. Renewed terrace construction or a major land-use change, however, will prevent using ^{137}Cs measurements to infer soil redistribution. Conversion of arable land to citrus production is widespread in south-east China such as that in parts of the Shaxian case study only 'fossilised' [137]Cs profiles, buried within terrace risers, could be used to infer erosion rates prior to the land-use change. Information about the history of land use and terrace construction is crucial for data interpretation. As ^{137}Cs redistribution is derived from an estimate of the total fallout to the study area, the difficulty of obtaining reliable baseline fallout estimates remains a key issue. The development of a model based on temporal variations of ^{90}Sr fallout calibrated with local precipitation data, demonstrated at the Changshou study site, is an alternative to the problems of obtaining baseline fallout information.

Second, the development of corroborative rapid appraisal methods that provide supporting evidence for ^{137}Cs-based interpretations is required. The observation that ^{137}Cs redistribution is consistent with the expected pattern of soil redistribution does not provide independent evidence of validity. Erosion-induced changes in soil properties, such as a reduction in A-horizon thickness or the exposure of tree roots, can be noted during soil surveys (Olsen *et al.*, 1994) but require comparison with uneroded sites on similar materials and slope positions; a requirement that is difficult to meet in southern China. Alternative tracer properties including other gamma-emitting radionuclides and magnetic properties offer some prospect of corroboration. Preliminary data from Shaxian indicate a positive relationship between ^{137}Cs and excess ^{210}Pb, as noted previously by Walling and Woodward (1992). Recent investigations into the environmental behaviour of excess ^{210}Pb have suggested its use in conjunction with ^{137}Cs for refining erosion rate estimates on both uncultivated (Wallbrink and Murray, 1996) and cultivated soils (He and Walling, 1997; Walling and He, 1999). Finally, the ability to derive information on the magnitude and extent of soil erosion rates should be embedded within analysis of the long-term impacts of erosion on soil quality.

REFERENCES

Bachhuber, H., Bunzl, K. and Schimmack, W. 1987. Spatial variability of fallout Cs-137 in the soil of a cultivated field. *Environmental Monitoring and Assessment*, **8**, 93–101.

Basher, L. R. and Matthews, K. M. 1993. Relationship between ^{137}Cs in some undisturbed New Zealand soils and rainfall. *Australian Journal of Soil Research*, **31**, 655–663.

Campbell, B. L., Elliott, G. L. and Loughran, R. J. 1986. Measurement of soil erosion from fallout [137]Cs. *Search*, **17**, 148–149.

Chappell, A. C., Warren, A., Oliver, M. A. and Charlton, M. 1998. The utility of ^{137}Cs for measuring soil redistribution rates in southwest Niger. *Geoderma*, **81**, 313–337.

Davis, J. J. 1963. Cesium and its relationship to potassium in ecology. In Schultz, V. and Klement, A. E. (eds) *Radioecology*. Reinhold, New York, 539–556.

De Jong, E., Begg, C. B. M. and Kachanoski, R. G. 1983. Estimates of soil erosion and deposition for some Saskatchewan soils. *Canadian Journal of Soil Science*, **62**, 673–683.

Fredericks, D. J., Norris, V. and Perrens, S. J. 1988. Estimating erosion using caesium-137 I. Measuring caesium-137 activity in a soil. In *Sediment Budgets*, Proceedings of the Porto Alegre Symposium, December 1988. IAHS Publication **174**, IAHS Press, Wallingford. 225–231.

Gong, Z. T. and Shi, X. Z. 1992. Rational soil utilization and soil degradation control in tropical China. In Gong, Z. T. (ed.) *Proceedings of an International Symposium on the Management and Development of Red Soils in Asia and the Pacific Region*. Science Press, Beijing, 8–12.

Grimshaw, R. G. and Smyle, J. W. 1991. An approach to agricultural development: soil and soil-moisture conservation in red soil areas of South China. In Horne, P. M., Macleod, D. A. and Scott, J. M. (eds) *Forages on Red Soils in China*. ACIAR Proceedings No. 38, Australian Centre for International Agricultural Research, Canberra, 74–81.

He, Q. and Walling, D. E. 1997. The distribution of fallout Cs-137 and Pb-210 in undisturbed and cultivated soils. *Applied Radiation and Isotopes*, **48**, 677–690.

Higgitt, D. L. 1995a. Assessment of erosion and sediment delivery in a mountainous subtropical catchment, Fujian province, China. In Singh, R. B. and Haigh, M. J. (eds) *Sustainable Reconstruction of Highland and Headwater Regions*. Balkema, Rotterdam, 299–306.

Higgitt, D. L. 1995b. The development and application of caesium-137 measurements in erosion investigations. In Foster, I. D. L., Gurnell, A. M. and Webb, B. W. (eds) *Sediment and Water Quality in River Catchments*. John Wiley, Chichester, 287–305.

Higgitt, D. L. and Walling, D. E. 1993. The value of caesium-137 measurements for estimating soil erosion and sediment delivery in an agricultural catchment, Avon, UK. In Wicherek, S. (ed.) *Farm Land Erosion: In Temperate Plains Environment and Hills*. Elsevier, Amsterdam, 301–315.

Higgitt, D. L., Walling, D. E. and Haigh, M. J. 1994. Estimating rates of ground retreat on mining spoils using caesium-137. *Applied Geography*, **14**, 294–307.

Hirose, K., Aoyama, M., Katsuragi, Y. and Sugimura, Y. 1987. Annual deposition of Sr-90, Cs-137 and Pu-239, 240 from the 1961–1980 nuclear explosions: a simple model. *Journal of the Meteorological Society of Japan*, **65**, 259–277.

Kachanoski, R. G. and De Jong, E. 1984. Predicting the temporal relationship between soil caesium-137 and erosion rate. *Journal of Environmental Quality*, **13**, 301–304.

Livens, F. R. and Rimmer, D. L. 1988. Physico-chemical controls on artificial radionuclides in soil. *Soil Use and Management*, **4**, 63–69.

Martz, L. W. and de Jong, E. 1987. Using cesium-137 to assess the variability of net soil erosion and its association with topography in a Canadian Prairie landscape. *Catena*, **14**, 439–451.

Murphey, R. 1982. Natural resources and factor endowments. In Barker, R. and Sinha, R. (eds) *The Chinese Agricultural Economy*. Westview Press, Boulder, CO, 49–63.

Olson, K. R., Norton, L. D., Fenton, T. E. and Lal, R. 1994. Quantification of soil loss from eroded soil phases. *Journal of Soil and Water conservation*, **49**, 591–596

Pu, L. J., Bao, H. S., Peng, B. Z. and Higgitt, D. L. 1998. Preliminary study of using ^{137}Cs to estimate soil erosion rates in wind eroded area, China: Case study on the Korla Area, Xinjiang Autonomous Region. *Acta Pedologica Sinica*, **35**, 441–449 (in Chinese).

Quine, T. A. 1995. Estimation of erosion rates from caesium-137 data: the calibration question. In Foster, I. D. L., Gurnell, A. M. and Webb, B. W. (eds) *Sediment and Water Quality in River Catchments*. John Wiley, Chichester, 307–329.

Quine, T. A., Walling, D. E., Zhang, X. B. and Wang, Y. 1992. Investigation of soil erosion on terraced fields near Yangting, Sichuan Province, using caesium-137. In *Erosion, Debris Flows and Environment in Mountain Regions*, Proceedings of the Chengdu Symposium, July 1992. IAHS Publication **209**, 155–168.

Quine, T. A., Govers, G., Walling, D. E., Zhang, X. B., Desmet, P. J. J., Zhang, Y. and Vandaele, K. 1997. Erosion processes and landform evolution on agricultural land – new perspectives from caesium-137 measurements and topographic-based erosion modelling. *Earth Surface Processes and Landforms*, **22**, 799–816.

Quine, T. A., Walling, D. E., Chakela, Q. K., Mandiringana, O. T. and Zhang, X. B. 1999a. Rates and patterns of tillage and water erosion on terraces and contour strips: evidence from caesium-137 measurements. *Catena*, **36**, 115–142.

Quine, T. A., Walling, D. E. and Zhang, X. B. 1999b. Tillage erosion, water erosion and soil quality on cultivated terraces near Xifeng in the Loess Plateau, China. *Land Degradation and Development*, **10**, 251–274.

Ritchie, J. C. and McHenry, J. R. 1990. Application of radioactive fallout cesium-137 for measuring soil erosion and sediment accumulation rates and patterns: a review. *Journal of Environmental Quality*, **19**, 215–233.

Ritchie, J. C., Spraberry, J. C. and McHenry, J. R. 1974. Estimating soil erosion from the redistribution of fallout ^{137}Cs. *Proceedings of the Soil Science Society of America*, **38**, 137–139.

Rogowski, A. S. and Tamura, T. 1970. Erosional behavior of cesium-137. *Health Physics*, **18**, 467–477.

Sarmiento, J. L. and Gwinn, E. 1986. Strontium-90 fallout prediction. *Journal of Geophysical Research*, **91**, 7631–7646.

Smil, V. 1993. *China's Environmental Crisis: An Inquiry into the Limits of National Development*. M. E. Sharpe, New York.

Sutherland, R. A. 1991. Examination of caesium-137 areal activities in control (uneroded) locations. *Soil Technology*, **4**, 33–50.

Tamura, T. 1964. Selective sorption reactions of cesium with soil minerals. *Nuclear Safety*, **5**, 262–268.

Vermeer, E. B. 1977. *Water Conservancy and Irrigation in China*. Leiden University Press, The Hague.

Wallbrink, P. J. and Murray, A. S. 1996. Determining soil loss using the inventory ratio of excess lead-210 to cesium-137. *Soil Science Society of America Journal*, **60**, 1201–1208.

Walling, D. E. and He, Q. 1999. Improved models for estimating soil erosion rates from caesium-137 measurements. *Journal of Environmental Quality*, **28**, 611–622.

Walling, D. E. and Quine, T. A. 1992. The use of caesium-137 measurements in soil erosion surveys. In *Erosion and Sediment Transport Monitoring Programmes*, Proceedings of the Oslo Symposium, August 1992. IAHS Publication **210**, IAHS Press, Wallingford. 143–152.

Walling, D. E. and Woodward, J. C. 1992. Use of radiometric fingerprints to derive information on suspended sediment sources. In *Erosion and Sediment Transport Monitoring Programmes in River Basins*, Proceedings of the Oslo Symposium, August 1992. IAHS Publication 210, IAHS Press, Wallingford. 153–164.

Wan, G. J., Lin, W. Z., Huang, G. R. and Cheng, Z. L. 1990. Caesium-137 dating and erosion tracing in Hongfeng Lake. *Science Bulletin*, **19**, 1487–1490 (in Chinese).

Wang, Y. C., Zhang, X. B., Li, S. L., Zhao, Q. C., Jiao, J. J., Zhang, Y. Y., Yan, M. Q. and Bai, L. X. 1991. The quantification of siltation depth of arable land in the Gaoqiao of the Three Gorges area using Cs-137. *Geography*, **4**, 63–64 (in Chinese).

Wen, D. Z. 1993. Soil erosion and conservation in China. In Pimental, D. (ed.) *World Soil Erosion and Conservation*. Cambridge University Press, Cambridge, 63–85.

Yang, H., Chang, Q., Du, M. Y., Minami, K. and Hatta, T. 1998. Quantitative model of soil erosion rates using ^{137}Cs for uncultivated soil. *Soil Science*, **163**, 248–257.

Zhang, X. B., Higgitt, D. L. and Walling, D. E. 1990. A preliminary assessment of the potential for using caesium-137 to estimate soil loss in the Loess Plateau of China. *Hydrological Sciences Journal*, **35**, 243–252.

Zhang, X. B., Wang, Y. C., Li, S. N., Zhang, Y. Y., Zhao, Q. C., Jiang, J. J., Bai, L. X., Yan, M. Q. and Wu, L. P. 1992. Preliminary study of the origin of fine grained sediments from soil erosion and mudflows in Jiangjiagou watershed by ^{137}Cs. *Soil and Water Conservation in China*, **2**, 28–31 (in Chinese).

Zhang, X. B., Li, C. L., Quine, T. A. and Walling, D. E. 1993. Tillage effect on soil erosion quantification of arable land using Cs-137. *Chinese Science Bulletin*, **38**, 2072–2076 (in Chinese).

Zhang, X. B., Quine, T. A., Walling, D. E. and Li, Z. 1994. Application of the caesium-137 technique in a study of soil erosion on gully slopes in a yuan area of the Loess Plateau near Xifeng, Gansu province, China. *Geografiska Annaler*, **76A**, 103–120.

Zhao, S. Q. 1994. *Geography of China: Environment, Resource, Population and Development*. John Wiley, New York.

9 Use of Radiocaesium to Investigate Erosion and Sedimentation in Areas with High Levels of Chernobyl Fallout

D. E. WALLING,[1] V. N. GOLOSOV,[2] A. V. PANIN[2] and Q. HE[1]

[1]*Department of Geography, University of Exeter, UK*
[2]*Department of Geography, Moscow State University, Russia*

INTRODUCTION

In recent years, there have been numerous reports in the literature of studies that have confirmed the potential for using measurements of ^{137}Cs inventories in soils to document rates and patterns of soil loss from agricultural land (cf. Ritchie and McHenry, 1990; Walling and Quine, 1992, 1995). The technique has now been employed in contrasting environments in many different areas of the world (cf. Walling, 1998). Its key advantages are seen to be the potential for obtaining retrospective estimates of medium-term rates of erosion and deposition on the basis of a single site visit and the ability to assemble distributed information on erosion and deposition rates for a large number of points within the landscape. Most studies that have employed ^{137}Cs to obtain information on soil redistribution rates have been undertaken in areas where bomb-derived ^{137}Cs represents the sole, or at least the dominant, source of radio-caesium. Additional inputs of ^{137}Cs associated with Chernobyl fallout are commonly seen as introducing complications, which may compromise the validity of the approach (cf. Walling and Quine, 1991). These complications include the need to take account of two periods of fallout input, uncertainty as to the relative magnitude of the bomb- and Chernobyl-derived fallout inputs, and the likelihood of increased spatial variability of the Chernobyl-derived inputs, as a result of the low-level traject-ory of the fallout plume and the limited period of fallout deposition. Although it was possible to use measurements of ^{134}Cs activity to separate the bomb- and Chernobyl-derived components of the ^{137}Cs inventory during the period shortly after the Cher-nobyl accident, the short half-life of ^{134}Cs ($t_{0.5} = 2.2$ years) means that this is generally no longer possible and the problems associated with taking account of two radio-caesium inputs occurring at different times and with markedly different temporal distributions are greatly increased, to the extent that the approach may be unviable. However, in areas that received very high levels of Chernobyl-derived ^{137}Cs fallout,

Tracers in Geomorphology. Edited by Ian D. L. Foster. © 2000 John Wiley & Sons Ltd.

some of these problems will be reduced, since the Chernobyl-derived component of the total ^{137}Cs inventory may be several orders of magnitude greater than the bomb-derived component and the latter can be effectively ignored when interpreting measurements of ^{137}Cs inventories.

In areas characterised by high levels of Chernobyl-derived ^{137}Cs fallout, the use of ^{137}Cs measurements to investigate rates and patterns of soil redistribution could involve less uncertainty than with the use of bomb fallout and such areas could therefore afford greater scope for application of the ^{137}Cs approach. In this context, possible advantages of Chernobyl-derived ^{137}Cs include the fact that the Chernobyl input was limited to a very short and well-defined time period, the greater availability of information regarding land use, land management and soil conditions during the period from 1986 through to the present, and the existence of empirical observations of the initial behaviour and fate of Chernobyl fallout. In addition, in view of the high levels of radiocaesium activity associated with such areas, measurement problems linked to low activities and long count-times essentially disappear and there may be potential to use in-situ measurements of ^{137}Cs activity as an alternative to the collection of soil cores for subsequent laboratory analysis. Possible disadvantages, nevertheless, include the high levels of ^{137}Cs activity, which could be less rapidly and strongly fixed to soils and sediments than the much lower concentrations associated with bomb-derived input, local spatial variability in fallout inputs and the limited time elapsed since 1986 to produce significant changes in ^{137}Cs inventories in response to soil redistribution. In the absence of existing studies aimed at assessing the potential for employing ^{137}Cs measurements to document rates and patterns of soil redistribution in areas receiving high levels of Chernobyl fallout, this contribution aims to consider further this potential, by presenting some initial results from an investigation undertaken in a small agricultural drainage basin located about 250 km south of Moscow.

THE STUDY AREA

The study was undertaken in the 2.18 km^2 Lapki balka catchment, located in the Lokna River basin, 5 km west of Plavsk in the Tula region of central Russia (53°40′ N, 37°09′ E). This region was characterised by high levels of Chernobyl fallout, and ^{137}Cs inventories in excess of 200 kBq m^{-2} were recorded by reconnaissance surveys undertaken shortly after the Chernobyl accident (Figure 9.1). Pre-existing bomb-derived ^{137}Cs inventories were almost two orders of magnitude lower and are therefore unlikely to have had a significant effect on the contemporary spatial distribution of ^{137}Cs inventories within the study area. The study catchment has a relief range of 62 m and the maximum elevation is 236 m a.s.l. The topography is dominated by a relatively flat interfluve area surrounded by gentle slopes, which have been dissected by a balka or ephemeral valley (Figures 9.2 and 9.3). The catchment is underlain by limestone and dolomite of Carboniferous age, mantled by Holocene loess. The soils are primarily typical and leaching chernozems, with a loamy texture. Most of the catchment is occupied by arable land, although about 14 percent of its area, representing the steeper balka sides and the balka bottom, supports permanent pasture. The main

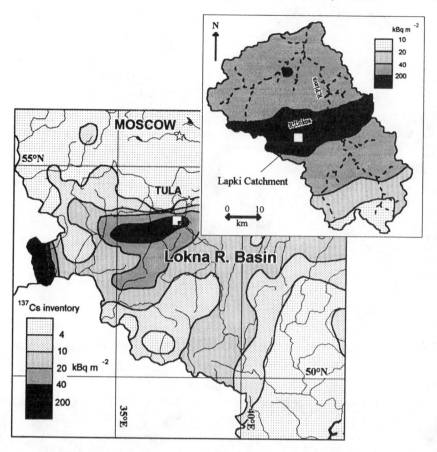

Figure 9.1 The location of the study area in relation to the regional pattern of Chernobyl fallout inventories. (Based in part on Izrael, 1993)

crops include cereals, maize and potatoes. The cultivated slopes are characterised by slope angles between 1° and 7°, and the uncultivated balka sides by slopes of up to 25°. Mean annual precipitation for the study area is estimated to be c. 650 mm, with about half of this falling as snow. Maximum rainfall occurs during the period June to August and the mean water equivalent of the snow before the spring melt, based on the period 1986–1987, is 111 mm. Soil erosion is a significant problem in the local area, and substantial quantities of sediment are deposited in the balka bottoms. Soil erosion occurs mainly during the spring snow-melt, but can also occur as a result of heavy rainfall during the period May to September. Aeolian processes are of limited significance. The mean annual soil loss estimated for the study catchment using a modified version of the Universal Soil Loss Equation (USLE) (Laroniov, 1993) for rainfall erosion and the State Hydrological Institute model for erosion during snow-melt is $6.5 \, \text{t ha}^{-1} \, \text{year}^{-1}$, with values ranging up to $7.5–10 \, \text{t ha}^{-1} \, \text{year}^{-1}$ on the main slopes of the cultivated areas.

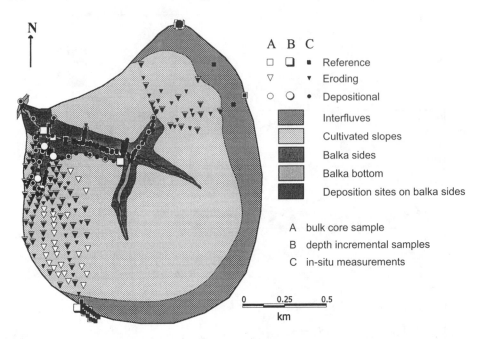

Figure 9.2 The Lapki balka study basin showing the main geomorphological units and the sampling points

Figure 9.3 A view of the central portion of the main balka in the study catchment

METHODS

The primary aim of the sampling programme undertaken within the Lapki balka catchment was to investigate the magnitude, vertical distribution and current spatial pattern of Chernobyl-derived ^{137}Cs inventories within the catchment, in order to assess the potential for using ^{137}Cs measurements to establish rates and patterns of soil redistribution. Initially, the different geomorphological units comprising the catchment (shown in Figure 9.2) were identified. These include cultivated and uncultivated hilltops or interfluves, the main portions of the cultivated slopes, which show no evidence of sediment accumulation, the lower parts of the cultivated slopes characterised by sediment accumulation, the uncultivated balka sides further subdivided according to whether or not they showed evidence of sediment accumulation, and the balka bottoms characterised by sediment throughput or deposition. This subdivision of the catchment was used as a basis for subsequent soil and sediment sampling and in-situ measurement of ^{137}Cs inventories. Sampling was undertaken during the period May–June 1997 and in September 1997.

Particular attention was given to the identification of reference locations where the measured ^{137}Cs inventories should be representative of the total fallout input. Three pits were dug in different undisturbed areas (Figure 9.2) for collection of depth-incremental samples from an area 250×250 mm. A depth increment of 2–3 cm was used in the upper part of the soil profile and this was increased to 5 cm at depth. These three reference sites were located, first, within an uncultivated forest shelterbelt on the relatively flat interfluve area in the southern corner of the basin; secondly, on an uncultivated balka side in the middle section of the balka; and, thirdly, on a terrace in the lower part of the balka bottom. Similar depth incremental samples were collected from three areas selected as representative depositional sites. These were located in the balka bottom at its mouth and in its upper part, and in a depositional zone on the uncultivated balka side. Bulk core samples were collected from a range of other points representative of the different geomorphological units within the study catchment, using a 36.2 cm^2 core tube inserted to a depth of 40 cm. These cores were collected along transects representative of the cultivated slopes, from a transect along the floor of the main balka and from five cross-sections of the balka (Figure 9.2). The balka cross-sections were located in the upper, middle and lower reaches of the main balka and in the upper part and at the outlet of the lower left tributary of the main balka (Figure 9.2). In addition, a more detailed network of coring sites was established within the cultivated area in the subcatchment of the lower left tributary of the main balka (Figure 9.2). In parallel with the collection of the depth-incremental samples and the soil cores, in-situ measurements of ^{137}Cs activity were made adjacent to most sampling points and at additional sites using a Corad field-portable collimated spectrum sensitive NaI detector (cf. Govorun *et al.*, 1994; Chesnokov *et al.*, 1997). Based on field experiments undertaken by the designers of the Corad equipment, the detector is capable of measuring ^{137}Cs inventories in the range 20–20 000 kBq m^{-2} to depths of up to 40 cm with a precision better than $\pm 30\%$. Information on the ^{137}Cs penetration depth is derived from measurements of the count rate in the Compton part of the spectrum as well as in the photopeak. Because of the high levels of ^{137}Cs present in the soils of the study area, problems associated with interference by other radionuclides were minimised and the in-situ measurements required count-times of only a few minutes. The measurements undertaken using this equipment were designed to evaluate the potential for using in-situ measurements, as an alternative to sample collection and subsequent laboratory analysis.

A detailed topographic survey of the entire catchment, including the location of the sampling points, was made using a differential GPS system, which provided measurements of height and position with a maximum error of ±2 cm. The resulting data were used to produce 1:5000 plans of the study catchment and 1:2000 plans of the main balka. Observations of soil morphology were used to estimate long-term rates of soil loss from the cultivated area of the lower left tributary of the main balka. For this component of the study, five downslope transects were established in different parts of the subcatchment and 23 soil pits were dug. The soil pits were used to establish the reduction in the thickness of the humus horizon, relative to sites on the interfluve where erosion could be expected to be negligible. Estimates of the duration of the period of cultivation were derived from existing historical information and the mean annual erosion rate for the entire period of cultivation for individual observation points was determined as follows:

$$E = hT^{-1}$$

where
E = mean annual soil loss (mm year^{-1});
T = period of cultivation (years);
$h = H_f - H_i$, where H_f is the depth (mm) of the humus horizon on the interfluve and H_i is the depth (mm) at the observation point.

All soil samples were dried and sieved to < 2 mm prior to laboratory measurement of their ^{137}Cs content by gamma spectrometry using an HPGe coaxial detector calibrated with Standard Reference Materials and laboratory standards made using standard solutions. Count-times were sufficient to provide a typical analytical precision of ± 4–5 percent. All activities were corrected for radioactive decay to 1 June 1997.

RESULTS

A Comparison of In-situ and Laboratory Measurements

A detailed comparison of the results obtained from the Corad field-portable in-situ detector used in the study with those obtained from core samples analysed in the laboratory has been reported by Golosov et al. (2000). This showed that in-situ measurements afford a cost-effective and rapid means of assembling meaningful information on ^{137}Cs inventories in areas receiving high Chernobyl fallout inputs, such as the study catchment. Table 9.1 provides further details on this comparison by presenting summary statistics for the parallel measurements obtained from different geomorphological situations, including reference sites, cultivated midslope areas, and depositional areas at the bottom of the slope and in the balka bottom. In all cases the mean values of the two sets of measurements are not significantly different ($P = 0.05$), but there are, nevertheless, some differences in the values of standard deviation and

Table 9.1 A comparison of ^{137}Cs inventory values obtained from in-situ measurements and from equivalent laboratory measurements of soil cores for different geomorphological units within the Lapki balka catchment

Position	No. of points	Mean (kBq m^{-2})	SD (kBq m^{-2})	CV (%)	Range (kBq m^{-2})	Correlation, r
Reference sites on interfluves						
and on balka sides	11	479[a]/478[b]	93/44	19/9	283/137	0.4
Cultivated midslopes	54	364/372	53/59	15/16	239/252	0.56
Slope base	16	579/502	97/56	17/11	315/182	0.56
Balka bottom	32	598/576	121/102	20/18	532/517	0.36

[a] Left column = laboratory measurements on cores.
[b] Right column = Corad in-situ measurements.

range for the two data sets which reflect the different nature of the measurements and which result in correlation coefficients of c. 0.5. More particularly, it must be recognised that the measurements provided by the Corad equipment relate to a surface area of 2.1 m^2, which represents the field of view below the detector collimator, whereas those for the cores relate to an area of only 36.2 cm^2. In view of the microscale spatial variability of ^{137}Cs inventories, which has been widely reported in the literature (cf. Sutherland, 1991, 1994; Owens and Walling, 1996), the larger surface area sampled by the in-situ detector could be expected to reduce the effects of this variability and should therefore result in lower values of standard deviation, coefficient of variation and range for the in-situ measurements. This effect is clearly evident for the data presented in Table 9.1 for the undisturbed reference sites and the depositional areas at the bottom of the slopes and in the balka bottom. In the case of the cultivated midslope areas, however, there is much less difference between the two data sets; this reflects the mixing caused by ploughing and cultivation which will reduce the microscale variability of ^{137}Cs inventories and thus the potential contrast between the two sets of measurements. The lower correlation between the two data sets for the undisturbed reference sites and the balka bottom again reflects the increased microscale variability of ^{137}Cs inventories in these uncultivated areas. The lowest level of correlation is associated with the balka bottom where the spatial variability of sediment deposition can be expected to further increase the microscale spatial variability of ^{137}Cs inventories and thus the potential difference between the areal average provided by the in-situ detector and the essentially point values provided by the soil cores.

^{137}Cs Depth Distributions

The ^{137}Cs depth distributions associated with the three depth-incremental sampling points selected to be representative of uncultivated reference sites are presented in Figure 9.4. The associated inventories are listed in Table 9.2. The depth distributions conform closely to those encountered in areas where bomb-derived ^{137}Cs has been used to estimate soil redistribution rates (cf. Walling and Quine, 1995), although both the total inventories involved and the levels of ^{137}Cs activity are several orders of magnitude greater. In these three examples, c. 60 percent or more of the total

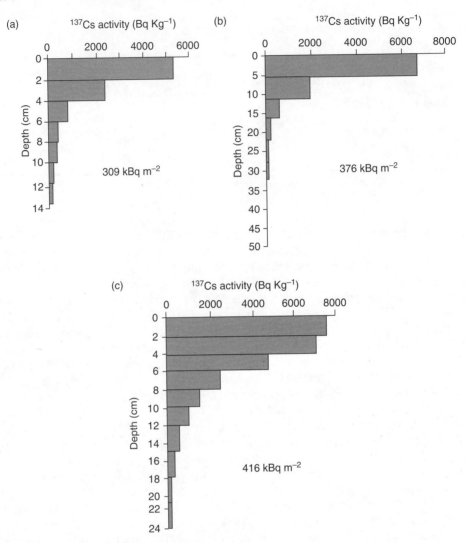

Figure 9.4 Depth-incremental ^{137}Cs profiles recorded for three reference sites within the Lapki balka basin (see Table 9.2 for further details of sites)

inventory is found in the top 5 cm of the profile and comparatively little ^{137}Cs is found below 10 cm, indicating that the fallout input is firmly fixed within the upper horizon of the soil. There is, however, little evidence of the downward displacement of the surface maximum which has frequently been reported for sites receiving bomb fallout, where the peak concentration may occur several centimetres below the surface. This contrast undoubtedly reflects the limited period of time elapsed since the Chernobyl fallout in 1986, so that bioturbation and other biogeochemical processes have had only limited opportunity to operate. It could, however, also reflect the relatively coarse depth increment (2–3 cm) used for documenting the vertical profiles.

Table 9.2 ^{137}Cs inventories associated with the depth incremental ^{137}Cs profiles presented in Figures 9.4 and 9.5

Profile	^{137}Cs inventory (kBq m^{-2})
Reference sites (Figure 9.4)	
A Forest shelter belt on interfluve	309
B Uncultivated balka side	376
C Balka terrace	416
Depositional sites (Figure 9.5)	
A Balka bottom near mouth	554
B Upper balka bottom	659
C Uncultivated balka side with deposition	570

Equivalent ^{137}Cs depth distributions for representative undisturbed depositional sites within the study catchment are presented in Figure 9.5 and the associated total inventories are listed in Table 9.2. These depth profiles are again typical of those reported by studies in areas with bomb-derived fallout. The total inventories are significantly greater than those associated with adjacent undisturbed reference sites and the depth distributions show clear evidence of a subsurface peak in ^{137}Cs activity, which represents the approximate level of the soil surface at the time of Chernobyl fallout in 1986. The ^{137}Cs concentrations found at this level are similar to those found at the surface of the undisturbed reference profiles (Figure 9.4). The reduced concentrations found above the subsurface peak reflect the progressive burial of the 1986 surface by sediment eroded from upslope or upstream areas. Since most of these areas are cultivated and the Chernobyl-derived ^{137}Cs input will have been mixed into the plough layer, the concentrations associated with eroded sediment will be substantially less than those found at or near the surface of uncultivated soils. Since the typical plough depth in the area is *c*. 25–30 cm, concentrations of about 20 percent of the mean value for the upper 5 cm of uncultivated soils are to be expected and the concentrations associated with the upper parts of the profiles shown in Figure 9.5 are close to this level. The reduction in ^{137}Cs concentrations towards the surface is likely to reflect the increased mixing of the plough layer and the progressive reduction in the ^{137}Cs inventory of eroding areas as time proceeds. The clear evidence of sediment deposition provided by the ^{137}Cs depth distributions at these sites confirms that the increased inventories reflect sediment deposition rather than mobility of radiocaesium fallout inputs in runoff prior to adsorption by the soil.

In contrast to the undisturbed areas considered above, ^{137}Cs concentrations associated with cultivated areas indicate that the radiocaesium has been mixed within the plough layer by cultivation and that ^{137}Cs concentrations are therefore relatively uniform within the upper part of the soil profile. Over most of the study catchment, ^{137}Cs concentrations within the plough layer are *c*. 1.4 kBq kg^{-1} and this value is consistent with the mixing of the fallout input within this layer, which extends to a depth of *c*. 25–30 cm. In eroding areas, ^{137}Cs concentrations decline to near zero immediately below the plough depth, whereas at depositional sites appreciable concentrations of ^{137}Cs are found to greater depths, reflecting the progressive accumulation of ^{137}Cs-bearing sediment at these sites.

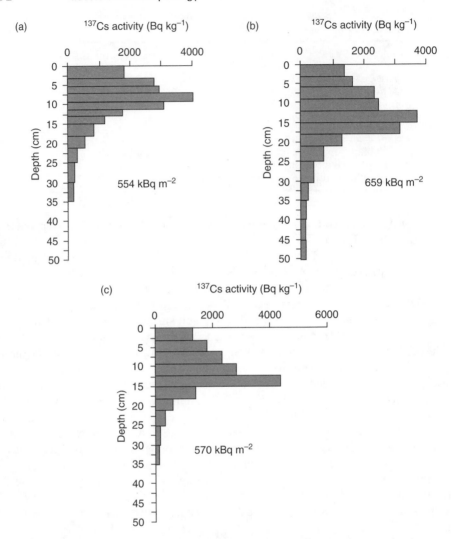

Figure 9.5 Depth-incremental ^{137}Cs profiles recorded for three depositional sites within the Lapki balka basin (see Table 9.2 for further details of sites)

Reference Inventories

The spatial distribution of Chernobyl-derived ^{137}Cs fallout inputs within the Plavsk region derived from reconnaissance measurements and presented in Figure 9.1 evidences strong local gradients. ^{137}Cs fallout inputs are seen to vary by an order of magnitude within a distance of less than 10 km and this marked spatial variability contrasts greatly with that commonly reported for areas receiving bomb-derived fallout, where total inventories are often closely correlated with mean annual precipitation (cf. Cawse and Horrill, 1986). Since most of the Chernobyl-derived fallout will have occurred during a very short period and will have been associated with a small number of rainfall events, the systematic spatial variability evident in Figure 9.1 is likely to have been coupled with greater local-scale variability associated with such factors as topographic influence on rainfall patterns, than reported for bomb-derived fallout. This local variability can, however, be expected to be much more limited in magnitude than the systematic variation shown in Figure 9.1. Against this background, it is necessary to consider whether it is possible to establish meaningful reference inventories in areas receiving high levels of Chernobyl fallout.

The total ^{137}Cs inventories associated with the three reference sites documented in Figure 9.4, and which are listed in Table 9.2, exhibit significant contrasts, since they range from 309 to 416 kBq m^{-2}. The lowest value of 309 kBq m^{-2} was obtained for a site within a forest shelterbelt on the southern watershed of the study catchment. The trees were without leaves at the time of the Chernobyl fallout, and the reduced inventory at this site is unlikely to reflect interception by the tree cover. Figure 9.1 indicates that Chernobyl fallout inputs across the local region evidenced appreciable gradients with a general north–south trend, and the variability of the three reference inventories shown in Table 9.2 is more likely to reflect a systematic increase in fallout receipt from south to north across the watershed. This is confirmed by Figure 9.6, which plots the inventory values for several reference sites within the study catchment and a number of other sampling points on the interfluves where erosion could be expected to be minimal, as a function of latitude expressed in terms of distance from the southern boundary of the catchment. Figure 9.6 indicates that the reference or input inventory varies from *c.* 300 to 500 kBq m^{-2} across the catchment, over a distance of *c.* 1.75 km. It is clear that this systematic variation in the reference inventory would need to be taken into account in any attempt to estimate rates of soil redistribution within the catchment by comparing the inventories existing at eroding

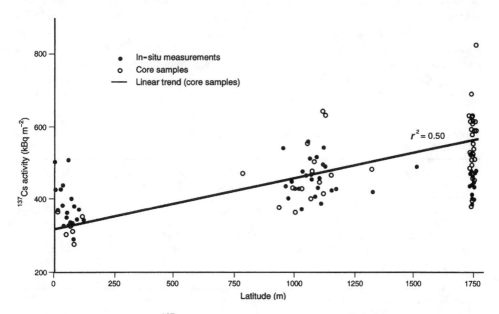

Figure 9.6   The variation of ^{137}Cs reference inventories across the Lapki balka basin

and depositional sites with the reference inventory. This situation differs from that existing in most studies undertaken in areas where the ^{137}Cs is bomb-derived, since reference inventories are generally assumed to be essentially uniform over such distances, unless there are marked changes in altitude.

Relating [137]Cs Inventories to Soil Redistribution

The fundamental premise underlying the use of ^{137}Cs measurements to estimate rates of soil redistribution within the landscape is that eroding areas will be marked by reduced ^{137}Cs inventories, whereas areas of deposition will be marked by increased inventories. The extent to which the ^{137}Cs inventories documented within the study catchment substantiate this premise is considered in Table 9.3, which lists the mean ^{137}Cs inventories recorded for different geomorphological units within the study catchment. Interpretation of the data presented in Table 9.3 is clearly complicated by the south to north gradient of the reference inventory across the study basin described above. Nevertheless, since the sampling points are distributed across the basin, it is possible to undertake a general comparison of the mean ^{137}Cs inventories associated with the different geomorphological units. This comparison reveals significant differences between some geomorphological units, which can in turn be related to soil redistribution within the study basin during the 11 year period since the input of Chernobyl fallout.

If an average reference inventory of c. 400 kBq m^{-2} is assumed for the study catchment, Table 9.3 provides little evidence of significant reduction in ^{137}Cs inventories on the cultivated slopes. Furthermore, the mean Corad-derived inventory for the cultivated slopes is almost identical to that for the interfluve areas, where erosion rates are likely to be minimal. This lack of differentiation between the interfluve and midslope areas is further confirmed by Table 9.4, which presents a comparison of ^{137}Cs inventories from these two geomorphological units for two smaller subcatchments. These subcatchments represent the lower left (southern) tributary and the right (northern) tributary of the main balka. In this case there is no significant difference between the Corad measurements obtained from the interfluve and the midslopes, for either subcatchment. The ^{137}Cs inventory values derived from the core samples show

Table 9.3 ^{137}Cs inventories associated with the main geomorphological units within the Lapki balka catchment

Geomorphological unit	No. of observations	Mean inventory (kBq m^{-2})	SD	Coeff. var. (%)
Interfluves	35[a]/42[b]	515/407	129/54	25/13
Cultivated slopes	96/103	372/403	60/48	16/12
Bottom of cultivated slopes	17/42	567/502	106/74	19/15
Undisturbed Balka sides and terraces	13/23	481/461	85/51	18/11
Balka sides with deposition	4/10	478/570	67/101	14/18
Balka bottom	35/59	601/574	125/98	21/17

[a] Left column = laboratory measurements on cores.
[b] Right column = Corad in-situ measurements.

Table 9.4 ^{137}Cs inventories in the lower left and right subcatchments of the Lapki balka catchment

Geomorphological unit	No. of observations	Mean inventory (kBq m^{-2})	Coeff. Var. (%)
Lower left subcatchment			
Interfluves	7^a/20^b	320/377	9/15
Midslopes	61/56	361/392	17/12
Right subcatchment			
Interfluves	28/22	563/435	16/7
Midslopes	13/28	406/432	12/12

^{a} Left column = laboratory measurements on cores.
^{b} Right column = Corad in-situ measurements.

some contrast between the two morphological units, but in the case of the southern tributary the mean inventory for the midslopes exceeds that for the interfluves. Although there is clear visual evidence of substantial soil erosion within the study catchment, both during the period of snow-melt and as a result of heavy rainfall in the summer, this lack of differentiation between the ^{137}Cs inventories recorded for the eroding midslope sites and the essentially stable interfluve sites is not unexpected in view of the limited period of time that has elapsed since the input of Chernobyl fallout in 1986.

The estimate of mean annual soil loss from the slopes of the study catchment derived using the modified Universal Soil Loss Equation proposed by Laroniov (1993) for rainfall erosion and the State Hydrological Institute model for erosion during snow-melt is 6.5 t ha^{-1} year^{-1}. Assuming a bulk density of *c.* 1 g cm^{-3}, this equates to an annual rate of surface lowering of *c.* 0.65 mm year^{-1}. This value may be compared with the long-term erosion rate derived from the measurements of the reduction in depth of the humus horizon of the soils in the study catchment. These measurements indicated an average reduction in the depth of the humus horizon on the cultivated midslopes of *c.*15 cm over a period of 320 years since the area was first cultivated, which equates to a long-term average erosion rate of 0.47 mm year^{-1}. These two estimates are highly consistent, since the increased erosion rate associated with current conditions can be readily accounted for by the changes in crop rotation after World War Two, which saw an increase in maize, potato and cereal cultivation, and by the increased use of heavy machinery, which has increased the incidence of surface runoff. If the estimate of 0.65 mm year^{-1} is assumed to be representative of recent years, this would mean that during the 11 years that have elapsed since the time of Chernobyl fallout the total soil loss has been equivalent to *c.* 7 mm. If it is assumed that the Chernobyl-derived ^{137}Cs rapidly became mixed within the plough layer by cultivation, erosion of 7 mm from the 25–30 cm deep plough layer could be expected to have reduced the ^{137}Cs inventory by about 3 percent and therefore probably less than 5 percent over most of the eroding area. In view of the imprecision of the laboratory and in-situ measurements, which can itself be expected to be about ±5 percent, the microscale variability of Chernobyl-derived ^{137}Cs inventories, and the systematic variability of fallout inputs outlined above, soil erosion that has occurred in the post-Chernobyl period is unlikely to be clearly reflected by the spatial distribution of ^{137}Cs inventories in the study catchment. This situation explains the lack of differentiation of the ^{137}Cs inventories associated with midslope areas apparent in Tables 9.3 and 9.4. Furthermore, it contrasts with the situation commonly encountered in areas where bomb-derived ^{137}Cs fallout has been successfully used to estimate soil redistribution rates. In the latter case, the period elapsed since the fallout input is substantially greater (e.g. currently *c*. 40 years) and this has permitted the development of significant reductions in ^{137}Cs inventory within eroding areas.

Inspection of Table 9.3 suggests that areas of the study catchment with evidence of deposition are marked by significant increases in ^{137}Cs inventory. Mean inventories for the bottoms of the cultivated slopes and the bottoms of the balka, based on both core and in-situ measurements, are both significantly greater than those for the interfluve areas. In the case of areas of deposition on the balka sides, however, only the inventories based on the Corad measurements are significantly greater. This situation is consistent with that for the eroding areas described above, since deposition is spatially limited and concentrated into small areas, and rates of deposition in these zones are likely to be substantially greater than the erosion rates existing across the eroding areas. Areas of deposition are therefore commonly marked by significant increases in ^{137}Cs inventory.

Figures 9.7 and 9.8 provide more specific evidence of the close relationship between areas of deposition and increased ^{137}Cs inventories. Figure 9.7 depicts the values of ^{137}Cs inventory associated with a series of cores collected along the thalweg of the main balka in its middle reaches. In this case, virtually all of the inventories are in excess of 500 kBq m^{-2} and since the reference inventory for the east–west zone in which the main balka lies can be estimated to be approximately 400 kBq m^{-2}, they provide clear evidence of sediment deposition in the balka bottom. Figure 9.8 plots the ^{137}Cs inventory values associated with cores collected along a north–south cross-

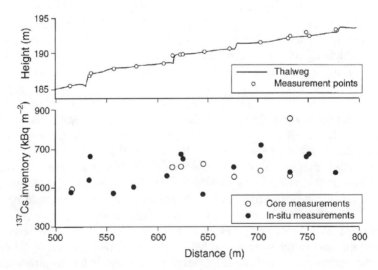

Figure 9.7 ^{137}Cs inventories within the balka bottom in the central reach of the main balka

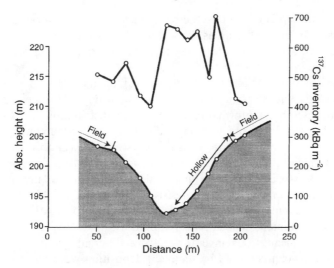

Figure 9.8 ^{137}Cs inventories associated with a cross-section of the central reach of the main balka

section of the main balka in its middle reaches. The northern portion of this cross-section passes through a hollow which connects to the bottom of the cultivated field and which serves as a route for the transport of sediment from the field to the balka bottom. There is visual evidence of deposition in this hollow and this is further confirmed by the increased inventories which reach 700 kBq m^{-2} at the top of the hollow. High inventories are also found in the bottom of the balka, where values exceed 650 kBq m^{-2}. There is no hollow on the southern side of the balka, but the core collected from the bottom of the field in a site where deposition might be expected shows an increased inventory (*c*. 500 kBq m^{-2}), relative to the local inventory within this zone of the study catchment, which, as noted above, is expected to be *c*. 400 kBq m^{-2}.

Perspective

The results presented above confirm that the general behaviour of the high levels of Chernobyl-derived ^{137}Cs fallout input received by the Lapki balka is similar to that reported by studies involving bomb-derived ^{137}Cs fallout. In undisturbed areas, the fallout input is found concentrated near the surface, such that *c*. 60 percent of the inventory is found within the upper 5 cm of the soil profile. In cultivated areas, the fallout input is mixed within the plough layer, providing near-uniform concentrations within the upper part of the soil profile. Concentrations decline to near zero immediately below the plough layer. At undisturbed sites experiencing deposition, the vertical distribution of ^{137}Cs again conforms to that documented in areas with bomb fallout, with a subsurface peak marking the approximate position of the surface at the time of the fallout maximum in 1986 and with a decline in ^{137}Cs concentration towards the surface reflecting the progressive reduction in the radiocaesium content of eroded sediment as the fallout is mixed into the plough layer and inventories are reduced by erosional losses. In cultivated areas, depositional sites are marked by an extended ^{137}Cs profile, with appreciable radiocaesium concentrations being found below the depth of the plough layer.

The comparison of in-situ and laboratory measurements of ^{137}Cs inventories undertaken in this study has confirmed the potential for using in-situ measurements as a rapid means of assembling information on the spatial distribution of ^{137}Cs inventories in areas receiving high levels of Chernobyl fallout. As well as overcoming the need for the collection of soil cores, in-situ measurements also appear to offer a significant advantage in terms of providing values representative of a larger area, which in turn reduces problems associated with the microscale variability of soil inventories and the representativeness of small-diameter cores. However, collection of soil cores is likely to provide more meaningful data in depositional areas, if information on the detailed spatial pattern of sediment deposition is required.

Comparison of mean ^{137}Cs inventories from different geomorphological units indicated that units characterised by sediment deposition were generally marked by increased inventories. More detailed investigation of depositional sites (cf. Figures 9.7 and 9.8) indicated that ^{137}Cs inventories afford a sensitive indicator of sediment deposition within the balka bottom and in hollows on the balka sides. However, because of the limited length of time (11 years) that has elapsed since the input of Chernobyl fallout and the relatively low magnitude of the soil erosion rates in the study catchment ($c.$ 6.5 t ha^{-1} year^{-1}), eroding areas are not yet characterised by significant reductions in ^{137}Cs inventories. This represents a limitation in terms of the potential for using Chernobyl fallout as a basis for estimating rates of soil loss, but, as the period elapsed since 1986 increases towards that associated with bomb fallout (i.e. 30–40 years) this limitation should disappear. In view of the high activities found in areas receiving high levels of Chernobyl fallout, the precision of both laboratory and in-situ measurements should be improved relative to areas with bomb fallout and this will assist the future differentiation of erosional areas.

The scope of this study has been restricted to consideration of ^{137}Cs inventories and inferences regarding rates of soil redistribution within the Lapki balka catchment. Further work is required to modify and extend existing procedures for deriving quantitative estimates of soil redistribution rates from ^{137}Cs measurements developed for areas with bomb fallout (cf. Walling and Quine, 1990; Walling and He, 1999). In view of the essentially instantaneous nature of Chernobyl fallout, compared with the extended period of bomb fallout, and improved knowledge of land-use activities since the period of Chernobyl fallout, it should prove easier to develop reliable calibration relationships. For example, in many areas receiving high levels of Chernobyl fallout, cultivation will have occurred shortly after the fallout input and the uncertainties associated with the fate of freshly deposited ^{137}Cs prior to its incorporation into the plough layer by cultivation are substantially reduced. However, based on the findings of this study, it seems likely that in many areas the increased local variability of fallout inventories associated with Chernobyl fallout will mean that there is an increased need to take account of spatially variable reference inventories. In this study area, the existence of a systematic south to north increase in ^{137}Cs inventories would appear to afford a relatively simple basis for taking account of this variability. As indicated above, the current differentiation of eroding areas in terms of reduced ^{137}Cs inventories may be insufficient to provide a basis for estimating erosion rates in some areas receiving high levels of Chernobyl fallout, where erosion intensity is relatively low. However, this limitation should be overcome as the length of time elapsed since 1986 increases. Areas characterised by sediment deposition are already clearly differentiated and this discrimination should increase further with time. Furthermore, scope clearly exists to undertake periodic repeat measurements in such areas in order to obtain estimates of deposition rates associated with relative short periods (e.g. *c.* 5 years). This approach could have considerable potential in areas where land use is changing and there is a need to document the effects of such changes on rates of sediment redistribution. When it is possible to use Chernobyl-derived ^{137}Cs inventories to estimate both erosion and deposition rates in areas receiving high levels of Chernobyl fallout, considerable potential should exist for establishing detailed sediment budgets (cf. Golosov *et al.*, 1992). It is important that this potential should be recognised and that information on land use, land management, the incidence of erosional events and related aspects should now be assembled for the period since 1986, with a view to assisting the interpretation of the resulting sediment budgets. Coming at a time when there is increasing concern for problems of soil loss and the off-site impact of soil erosion, such data could prove of very great value in guiding the development of future land management strategies.

ACKNOWLEDGEMENT

The authors gratefully acknowledge financial support from INTAS-RFBR (Grant no. 95-0734) and from the IAEA Co-ordinated Research Project on the Assessment of Soil Erosion through the Use of ^{137}Cs and Related Techniques (IAEA Contract no. 9044). The manuscript benefited from the constructive comments provided by two referees.

REFERENCES

Cawse, P. A. and Horrill, A. D. 1986. A survey of caesium-137 and plutonium in British soils in 1977. UK Atomic Energy Authority Report AERE R-10155.

Chesnokov, A. V., Fedin, V. I., Govorun, A. P., Ivanov, O. P., Liksonov, V. I., Potapov, V. N., Smirnov, S. V., Scherbak, S. B. and Urutskoev, V. I. 1997. Collimated detector technique for measuring a ^{137}Cs deposit in soil under a clean protected layer. *Applied Radiation and Isotopes*, **48**, 1265–1272.

Golosov, V. N., Ivanova, N. N., Litvin, L. F. and Sidorchuk, A. Yu. 1992. Sediment budget in river basins and small river aggradation. *Geomorphologiya*, **4**, 62–71 (in Russian).

Golosov, V. N., Walling, D. E., Kvasnikova, E. V., Stukin, E. D., Nikolaev, A. N. and Panin, A. V. 2000. Application of a field-portable scintillation detector for studying the distribution of Cs-137 inventories in a small basin in Central Russia. *Journal of Environmental Radioactivity*, **48**, 79–94.

Govorun, A. P., Liksonov, V. I., Romasko, V. P., Fedin, V. I., Urutskoev, L. I. and Chesnokov, A. V. 1994. Spectrum sensitive portable collimated gamma-Radiometer CORAD. *Pribory i Tekhnika Experimenta*, **5**, 207–208 (in Russian).

Izrael, Yu. A. (ed.) 1993. *Map of the Caesium-137 Radioactive Contamination of the European Part and the Ural Region of Russia, 1:500000*. Russian State Cartographic Service (in Russian).

Laroniov, G. A. 1993. *Soil and Wind Erosion: Main Regularities and Quantitative Estimation*. Izd-vo MSU, Moscow (in Russian).

Owens, P. N. and Walling, D. E. 1996. Spatial variability of caesium-137 inventories at reference sites: an example from two contrasting sites in England and Zimbabwe. *Applied Radiation and Isotopes*, **47**, 699–707.

Ritchie, J. C. and McHenry, J. R. 1990. Application of radioactive fallout cesium-137 for measuring soil erosion and sediment accumulation rates and patterns: a review. *Journal of Environmental Quality*, **19**, 215–233.

Sutherland, R. A. 1991. Examination of caesium-137 areal activities in control (uneroded) locations. *Soil Technology*, **4**, 33–50.

Sutherland, R. A. 1994. Spatial variability of ^{137}Cs and the influence of sampling on estimates of sediment redistribution. *Catena*, **21**, 57–71.

Walling, D. E. 1998. Use of ^{137}Cs and other fallout radionuclides in soil erosion investigations: progress, problems and prospects. In *Use of ^{137}Cs in the Study of Soil Erosion and Sedimentation*. International Atomic Energy Agency TECDOC-1028, 39–62.

Walling, D. E. and He, Q. 1999. Improved models for estimating soil erosion rates from caesium-137 measurements. *Journal of Environmental Quality*, **28**, 611–622.

Walling, D. E. and Quine, T. A. 1990. Calibration of caesium-137 measurements to provide quantitative erosion rate data. *Land Degradation and Rehabilitation*, **2**, 161–175.

Walling, D. E. and Quine, T. A. 1991. The use of ^{137}Cs measurements to investigate soil erosion on arable fields in the UK: potential applications and limitations. *Journal of Soil Science*, **42**, 147–165.

Walling, D. E. and Quine, T. A. 1992. The use of caesium-137 measurements in soil erosion surveys. In *Erosion and Sediment Transport Measurement Programmes in River Basins*. International Association of Hydrological Sciences Publication **210**, 143–142.

Walling, D. E. and Quine, T. A. 1995. The use of fallout radionuclides in soil erosion investigations. In *Nuclear Techniques in Soil–Plant Studies for Sustainable Agriculture and Environmental Preservation*. International Atomic Energy Agency Publication ST1/PUB/947, 597–619.

10 The Use of Marked Rock Fragments as Tracers to Assess Rock Fragments Transported by Sheep Trampling on Lesvos, Greece

D. J. OOSTWOUD WIJDENES,[1] J. POESEN,[1,2] L. VANDEKERCKHOVE[1] and C. KOSMAS[3]

[1]*Laboratory for Experimental Geomorphology, K. U. Leuven, Belgium*
[2]*Fund for Scientific Research-Flanders.*
[3]*Laboratory of Soils and Agricultural Chemistry, Agricultural University of Athens, Greece*

INTRODUCTION

On many arid and semi-arid slopes rock fragments protect the soil against erosion and increase infiltration rates (Poesen and Lavee, 1994; Poesen *et al.*, 1994). The removal of the rock fragments from such slopes therefore results in increasing exposure of the fine earth fraction to erosion, accumulation of stony colluvium at the footslopes and increasing sediment loads in (ephemeral) streams. Thus rock fragment transport can be an important land-degradation process. Downslope movement of rock fragments, however, is widely ignored in soil erosion assessment and models. Within the framework of the Mediterranean Desertification and Land Use Project (MEDALUS) research was initiated to study this process (Poesen and Bunte, 1996).

Downslope transport of rock fragments has been attributed to a variety of processes such as rock fall (Kirkby and Statham, 1975), surface wash (Kirkby and Kirkby, 1974), creep and frost heaving (Schumm, 1967), rill flow (Poesen, 1987) or a combination of mass wasting and overland flow (Frostick and Reid, 1982; Abrahams *et al.*, 1984). Recently, Butler (1995) has focused the attention on animals as geomorphological agents, and Govers and Poesen (1998) have shown the importance of animal trampling on the transport of rock fragments on scree slopes in southern Turkey. Trampling by sheep is also a major cause of the collapse of stone-walled terraces on the Greek island of Naxos (Lehmann, 1994) and contributes to severe soil erosion.

The number of sheep on the Greek island of Lesvos increased sharply (267 percent) during the 20th century (Kosmas *et al.*, 1996). A particularly strong increase occurred between 1976 and 1981 (47 percent). The number of goats increased by 61 percent between 1981 and 1987. Field observations of steep degraded slopes on the western

Tracers in Geomorphology. Edited by Ian D. L. Foster. © 2000 John Wiley & Sons Ltd.

part of the island suggest that trampling by sheep has caused accelerated movement of rock fragments. Many rock fragments on the slopes were not embedded in the soil surface but were resting on it (and often on the grassy vegetation), indicating recent movement. Rock fragments also accumulated behind small shrubs (*Sarcopoterium spinosum*) which prevented the fragments from further movement. In addition, rock fragments concentrated in convergent hillslope sections such as small depressions and hollows.

This chapter presents the results of a one-year field experiment to measure rates of downslope movement of rock fragments using marked rock fragments as tracers. The main objective of this study was to determine the effect of grazing animals on the downslope transport of rock fragments on degraded rangelands and to assess the effect of slope gradient, vegetation cover, surface roughness and rock fragment characteristics on the distances moved. Monitoring of rock fragments continued over a one-year period, i.e. May 1997–May 1998.

THE USE OF MARKED ROCK FRAGMENTS AS TRACERS

The main advantage of using marked (here painted) rock fragments as tracers is that the measured rate of tracer movements can be directly compared to the movement of the rock fragments that are present on the surrounding soil surface. In addition, it is a cheap and simple method which can be easily repeated (Caine, 1981). The two main disadvantages of using marked rock fragments are (i) that they may attract unwanted attention from passers-by (Frostick and Reid, 1982); and (ii), as Caine (1981) explained, there is a lag time (settling time) before they reflect real downslope movement because rock fragments need some time to settle into a natural position (partly embedded into the soil surface). On the selected slopes the only form of land use was grazing by sheep. Shepherds do not stay with their sheep during the day but bring them to their grazing area in the morning and collect them at the end of the day. The grazing areas are fenced and grazed by the sheep of one shepherd only, all year round. No other people normally enter the fenced rangeland parcels. Therefore, it is unlikely that passers-by will move the rock fragments. Since only the rock fragments that are resting on the surface are studied, it is not necessary to account for a lag time. Another potential problem may be that other geomorphological processes, such as water erosion, might move the tracers. During the observation period, 12 runoff events (as recorded on a nearby runoff plot) were observed, ranging from 0.05 to 3.4 mm. However, the selected sites did not show any sign of concentrated flow erosion such as rills or gullies. Therefore, it was assumed that if sheet flow occurred, flow power was not high enough to transport the rock fragments as indicated by field experiments conducted elsewhere (Poesen, 1987). Thus the only process by which the tracers could move was through trampling by sheep.

RESEARCH AREA AND EXPERIMENTAL DESIGN

The western part of the island of Lesvos is characterised by steep slopes (up to 80 percent), shallow, rocky soils (Leptosols) and the absence of forest vegetation.

Average annual rainfall amounts to 414 mm, which indicates semi-arid conditions (Kosmas *et al.*, 1996). The dominant form of land use on these volcanic slopes is grazing by sheep and goats. The grazed slopes carry a herbaceous vegetation which is invaded in many places by *Sarcopoterium spinosum*, a thorny, unpalatable perennial shrub. A representative study area near the small town of Antissa was selected to monitor the effect of trampling (Figure 10.1). The selected area of land belongs to one shepherd. His grazing area covers approximately 150 ha, which supports about 350 sheep all year round. According to the shepherd the entire area is grazed about equally, but sheep may initially prefer flatter areas where the vegetation is denser and less physical effort, particular during warm periods, is required to reach these places. The *Sarcopoterium* shrubs encroach very rapidly and are burned every 2–3 years. Sheep tend to avoid the areas with dense *Sarcopoterium*. Another shepherd in the area grazed his 100 sheep on an area of about 40 ha. According to this shepherd, sites near the farm are more intensively grazed than the most uphill sites, but this depends on the supply of fodder. He burned the *Sarcopoterium* after 3 years. This information indicates that local stocking rates are about 2.3 to 2.5 sheep per hectare. Moreover, grazing pressure may not be evenly distributed over the area on a very short-term basis, but becomes more so in the longer term (one year). The grazing is also temporally variable, with a peak during the months of April, May and June when the grasses and herbs are most abundant. In order to account for these variations as far as possible the experiments were extended over a one-year period.

In May 1997, 21 sites with a range of representative slope gradients, aspect and surface characteristics were selected (Figure 10.2). At each site a row of 30 rock fragments was laid out parallel to the contour lines over a distance of 5 m (the mean distance between the centres of the rock fragments was 16.5 cm). A cord connecting two fixed iron rods parallel to the contour was used as a reference line when laying out the rock fragments. The iron rods also marked the baseline from which the displacement distances were measured. The site characteristics are shown in Table 10.1. In order to include a representative sample of rock sizes and shapes in the rows, a large number of rock fragments (> 600) were randomly collected from the slopes. Each rock fragment was painted, numbered and weighed. In addition, its size was determined by measuring the three length axes (a, b and c). Each row consisted of one-third small (average b axis 3.5 cm), one-third intermediate (4.9 cm) and one-third large (7.3 cm) rock fragments. Each row of rock fragments was painted the same colour (white, blue or green), which helps to relocate them easily in the field.

Surface roughness was determined using a 120 cm long metal chain (chain size 7 mm) which was laid out over the surface perpendicular to the contour in such a way that it followed the microtopography of the soil. The difference between the down-slope projected length of the chain and its full length (120 cm) was used as a roughness index. This roughness index for a site was taken as the mean of three measurements. Each site was photographed after the rows were laid out. The photos were analysed for surface characteristics such as rock fragment and vegetation cover by the point count method. Three photos from each site were analysed to produce one average value for the vegetation and rock fragment cover.

The displacement distance of the rock fragments was measured after 7 months (January 1998) and after 12 months (May 1998). At each site the perpendicular

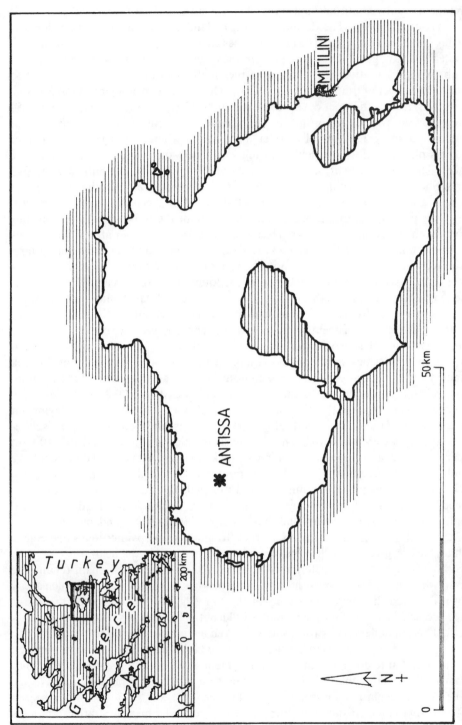

Figure 10.1 Location of the study area on the island of Lesvos (Greece)

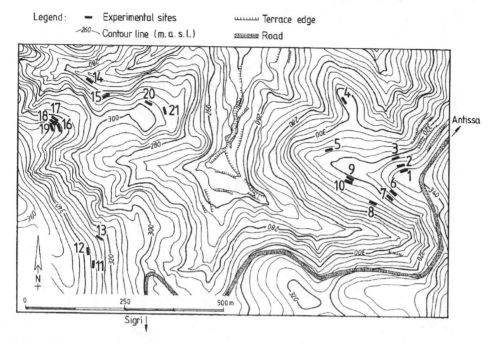

Figure 10.2   Location of experimental sites near Antissa

distance of each rock fragment to the baseline was measured with a measuring tape. In addition, the condition of the vegetation and the surface roughness were re-assessed.

The factors that control the displacement distances are slope gradient (SLOPE); vegetation cover, i.e., all annual types (VEG) and the *Sarcopoterium* cover (SARCO); rock fragment cover (RFC); surface roughness (ROUGH); and rock fragment characteristics such as mass (MASS), intermediate or b-axis (SIZE) and flatness index (FI). The flatness index is calculated as $(L + I)/(2S)$, where L is the long axis, I is the intermediate axis and S is the short axis (Poesen, 1987). The displacement distance perpendicular to the contour and parallel to the soil surface, is referred to as Y (cm).

RESULTS

Frequency Distribution of Rock Fragment Displacement Distances

Of the original 627 marked rock fragments, 623 (99%) were recovered after 7 months and 605 (97 percent) after 12 months. Table 10.2 shows the recovery rate per site. Rock fragments that were not recovered are believed to be embedded in the soil layer and overgrown with vegetation. Only on site 2, near the road, were there indications of human interference: three rock fragments were recovered outside the possible transport area.

Most of the rock fragments moved a distance of 0–1 m over the 7 or 12 month measuring interval. The most extreme downslope displacement distances recorded

Table 10.1 Site characteristics

Site	Colour of rock fragments	No. of rock fragments	Rock fragments cover (%)	SLOPE gradient (%)	Slope aspect (°)	UTM-coordinates	ROUGH[a] (cm)	SARCO7[b] (%)	SARCO12[b] (%)	VEG[c] (%)	Vegetation type and characteristics
1	Green	30	10	28	337	410986 4342870	8	0	5	60	Grasses and thistles
2	White	30	10	44	332	410986 4342870	8	0	5	80	Grass, many thistles
3	Blue	30	24	37	328	410874 4343015	12	25	25	70	Many small *Sarcopoterium* sp.
4	Blue	30	17	13	245	410790 4343115	6	0	0	20–50	Short vegetation, few shrubs
5	White	30	6	25	334	410796 4342961	3	0	2	90	Short grasses
6	Green	30	17	50	214	410960 4342884	7	0	1	60	Short grasses, few shrubs
7	Blue	30	17	23	224	410917 4342844	8	20	20	70	Short grasses, some shrubs
8	Green	30	14	43	213	410932 4342718	6	0	0	90	Short grasses
9	Green	30	4	4	190	4110814 4342817	2	0	0	100	Short grasses
10	Blue	30	7	25	196	4110814 4342817	14	0	0	100	Short grasses
11	Blue	29	45	29	90	410173 4342619	6	0	1	20	Recently burned *Sarcopoterium*, grasses
12	Green	29	30	31	84	410125 4342559	8	0	5	20–25	Short grasses, burned *Sarcopoterium*
13	White	30	9	19	40	410151 4342679	8	0	2	50–60	Recently burned
14	White	30	29	23	32	410195 4343188	8	15	20	30	Burned *Sarcopoterium*

15	White	30	28	53	345	410195 4343188	8	10	10	40	Recently burned *Sarcopoterium*
16	Green	28	29	16	80	410149 4342985	11	20	22.5	40–50	Newly growing *Sarcopoterium*
17	White	30	18	25	27	410107 4342984	9	33	55.5	50–60	Short grasses, newly growing *Sarcopoterium*
18	Blue	30	23	44	38	410107 4342984	10	27.5	45.5	40–50	Many short *Sarcopoterium*
19	Green	29	19	24	64	410104 4342985	10	22.5	50	50	Short *Sarcopoterium*
20	Blue	30	6	30	10	41302 4343048	4	0	1	100	Short grasses
21	White	30	9	10	65	410418 4342951	7	25	27.5	75	Grasses

[a] This number indicates the length (cm) by which the horizontally projected length of the chain (120 cm) is shortened through the effect of roughness elements. The higher numbers indicate greater surface roughness.
[b] SARCO7 is the surface cover of *Sarcopoterium spinosum* after 7 or 12 (SARCO12) months.
[c] VEG includes the surface cover by annuals.

Table 10.2 Recovery rate per site and of the total number of marked rock fragments

Site	Initial no. of rock fragments	Recovered after 7 months	Recovered after 12 months	Recovery rate, 7 months (%)	Recovery rate, 12 months (%)
1	30	30	29	100	97
2	30	30	25	100	83
3	30	30	29	100	97
4	30	30	30	100	100
5	30	30	29	100	97
6	30	30	24	100	80
7	30	30	29	100	97
8	30	30	29	100	97
9	30	30	30	100	100
10	30	30	30	100	100
11	29	28	28	97	97
12	29	29	29	100	100
13	30	29	28	97	93
14	30	30	30	100	100
15	30	30	29	100	97
16	28	28	28	100	100
17	31	31	30	100	100
18	30	30	30	100	100
19	29	29	29	100	100
20	30	28	28	93	93
21	31	31	30	100	100
Total	627	623	605	99	97

were 982 cm (site 6) after 7 months, and 1390 cm (site 10) after 12 months. Some rock fragments had even moved upslope over small distances. This is possible because sheep can walk in any direction over the rock fragments. The occurrence of upslope movements of rock fragments distinguishes animal trampling from other hillslope processes such as creep and flow, which are only driven by gravity. The furthest upslope movement occurred at site 8 where a displacement distance was measured for the same rock fragment of −40 cm after 7 months and −36 cm after 12 months. After 7 months, 5 percent of the rock fragments had moved in an upslope direction, while after 12 months this percentage was reduced to 3 percent.

Not all rock fragments did move. Table 10.3 shows that after 7 months 15 percent of all rock fragments were still in place, while this was reduced to 7 percent after 12

Table 10.3 Direction of movement of rock fragments as a fraction of the total population after 7 and 12 months

	Fraction of rock fragments after 7 months, $n = 623$	Fraction of rock fragments after 12 months, $n = 605$
Not moved	0.15	0.07
Moved upslope	0.05	0.03
Moved downslope	0.80	0.90

Table 10.4   Mean values and standard deviations for rock fragment displacements per site and for the combined data set (including zero displacement distance)

Site	7 months				12 months			
	Mean (cm)	SD (cm)	Median (cm)	$\dfrac{\sum(mass_i * Y_i)}{\sum(mass_i)}$ (cm)	Mean (cm)	SD (cm)	Median (cm)	$\dfrac{\sum(mass_i * Y_i)}{\sum(mass_i)}$ (cm)
1	67	50	60	80	85	50	96	99
2	146	186	72	230	235	210	153	390
3	9	15	3	10	45	71	14	61
4	1	9	0	2	12	22	9	19
5	93	93	69	85	134	117	114	128
6	277	236	246	229	410	240	388	346
7	63	54	54	65	95	97	70	129
8	103	123	49	137	140	133	93	168
9	15	24	9	10	54	61	34	28
10	11	161	49	126	249	287	135	314
11	35	36	25	26	47	44	36	36
12	50	65	18	64	83	89	48	108
13	40	53	21	29	61	72	29	49
14	10	24	4	12	12	24	5	15
15	115	184	49	118	142	188	56	148
16	19	25	12	13	34	45	18	26
17	2	4	2	3	4	5	4	5
18	25	33	12	21	35	39	20	36
19	19	21	15	18	27	25	19	24
20	313	257	219	328	379	268	314	378
21	7	23	0	3	6	9	0	6
All sites	69	133	18	70	104	169	38	111

Y_i = displacement distance for an individual rock fragment.
$mass_i$ = mass of individual rock fragment.
SD = standard deviation of Y_i.

months. Thus, 80 percent of the rock fragments was transported in a downslope direction after 7 months and 90 percent after 12 months.

Table 10.4 shows that the mean displacement distances are considerable larger than their medians, indicating a positively skewed distribution of the data. Log transformation of the data clearly improves the normality of the distributions when negative (upslope displacement distances) and zero values are excluded or when the whole distribution is made positive by adding a constant value (38 cm; Figure 10.3). However, the data in Table 10.4 also show that the mean displacement distances for each site are very close to the standard deviations (Figure 10.4), which is a property of the exponential distribution. Negative values have to be ignored again, but not zeros. Since the negative values (upslope movement) are only a small percentage of the data set (3 percent after one year) and tend to disappear in time, describing the effect of trampling by sheep on rock fragment movement with a decay function could be justified. Initially, all rock fragments in a row have the same probability of being hit and moved. After some time, t_1, some rock fragments are caught behind a *Sarcopoterium* shrub and cannot move further (until the shrub is burned). After time t_2, some

(a)

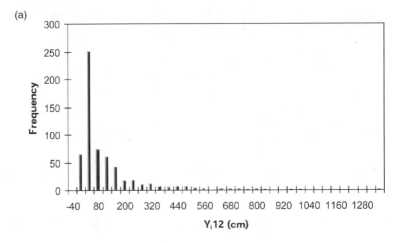

(b)

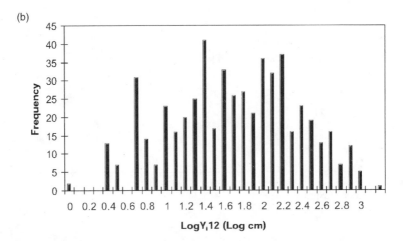

Figure 10.3 (a) Absolute frequency distributions of all individual tracer displacement distances after 12 months ($Y_i 12$). (b) Log-transformed frequency distributions of all individual tracer displacement distances after 12 months (Log $Y_i 12$), $n = 605$

of the remaining rock fragments are fixed by the *Sarcopoterium* shrubs and so on. During the year the proportion of tracers that are fixed by the *Sarcopoterium* shrubs increases. The tracers that are not fixed are dispersed downslope. Because of this dispersion, the concentration of tracers that can still move by trampling is smaller than it initially was and the probability of being hit by a sheep declines. However, this is partly an artefact of the experiment which considers a discrete number of 30 rock fragments. In reality the concentration of rock fragments does not decline downslope. A plot of the cumulative frequency distribution of the relative tracer displacements after 12 months (ratio of displacement distance for an individual rock fragment and the mean displacement distance for all rock fragments at the corresponding site) for each site is shown in Figure 10.5. This cumulative frequency distribution closely

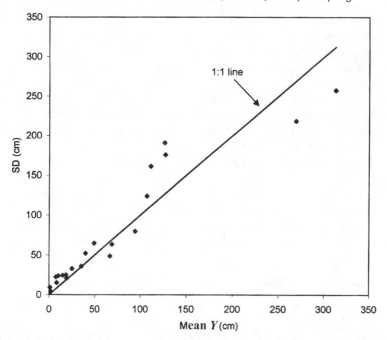

Figure 10.4 Similarity for MEAN *Y* and standard deviation (SD). MEAN *Y* equals mean displacement distance after 12 months of all tracers at a given site

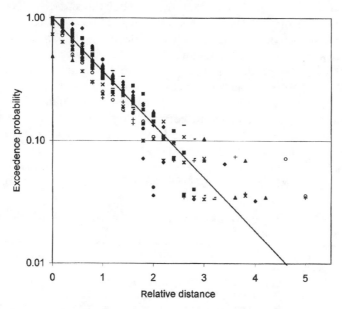

Figure 10.5 Cumulative frequency distribution of the individual tracer displacements after 12 months. The relative displacement distance equals the ratio between the individual tracer displacement distance and the mean displacement distance of all tracers at a given site. A different symbol has been used for each site ($n = 21$)

follows the theoretical exponential distribution of the relative distances up to a relative distance of 2. The increasing scatter after a relative distance of 2 is caused by tracers that moved much further than the mean displacement on a particular site. These tracers may have received many more kicks than others because they rested on a busier pathway. Overall, the exponential distributions appear to describe the process well, which agrees with Kirkby and Stathams' (1975) model for scree development.

Table 10.4 indicates that the mean of the displacement distances of all rock fragments measured is 69 cm after 7 months and 104 cm after 12 months. However, this includes the rock fragments that moved upslope and the rock fragments that did not move. Separate means for up and downslope displacement distances are given in Table 10.5. It is noteworthy that the mean displacement in the uphill direction increases while the absolute number of upslope displaced rock fragments decreases. As stated above, it could be expected that over a longer time period all rock fragments move downslope because a blow from a sheep's hoof into a downslope direction results in a larger displacement distance than a blow in the upslope direction.

Factors that Influence Displacement

Factors that potentially control the displacement distances are SLOPE, VEG, SARCO, ROUGH, MASS, SIZE and FI, but also the 'animal factors', such as the frequency, direction and the impact of the blows of animals' hoofs. The latter factors are a result of the grazing pressure which, according to the shepherd, is about equally distributed over all the sites after one year. However, if, for example, the cover of *Sarcopoterium spinosum* is high so that sheep avoid the site, SARCO will be the controlling variable and not grazing pressure. Thus, frequency, direction and impact of blows are not included in the analyses. Table 10.6 summarises the descriptive statistics of the measured variables and Table 10.7 shows the linear correlation between the variables at the site level. Because the mean tracer characteristics (MASS, SIZE and FI) vary very little between sites, the effect of these variables on the mean displacement distance (MEAN Y) cannot be evaluated at the site level. Therefore, the tracer characteristics are correlated with the displacement distance (Y) on an individual basis ($n = 573$–589; Table 10.8). The other variables are not included in this analysis because they represent mean values per site ($n = 21$).

The mean displacement distance for each site (MEAN Y) is significantly correlated with SLOPE and SARCO (Table 10.7). The latter is negatively correlated, indicating

Table 10.5 Mean and median displacement distance (excluding zero displacement distance) after 7 and 12 months

	Downslope		Upslope		Total movement	
	7 months (cm)	12 months (cm)	7 months (cm)	12 months (cm)	7 months (cm)	12 months (cm)
Mean	87	118	7	10	69	104
Median	29	48	4	7	18	38
	$n = 484$	$n = 540$	$n = 31$	$n = 20$	$n = 605$	$n = 605$

Table 10.6 Summary statistics of the measured variables

	Y_i (cm)	SLOPE (%)	SARCO7 (%)	SARCO12 (%)	VEG (%)	RFC (%)	ROUGH (cm)	$MASS_i$ (g)	$SIZE_i$ (cm)	FI_i
Mean	104	28.0	9.4	14.2	61.6	17.7	7.7	186.3	5.2	1.67
SD	167	13	11.9	17.7	25.3	10.5	2.8	192.3	1.8	0.3
Variance	28438	169	144.0	314.7	643.3	109.4	8.0	36970	3.35	0.12
Kurtosis	2.98	−0.41	−1.25	−0.51	−1.04	0.61	0.56	2.79	−0.51	1.20
Skewness	11.48	0.22	0.68	1.26	−0.09	0.84	0.02	1.72	0.49	1.08
Range	1426	49	33	55.5	80	41	12	956	9.0	1.98
Minimum	−36	4	0.00	0.00	20	4	2	8	1.7	1.04
Maximum	1390	53	33	55.5	100	45	14	964	10.7	3.02
Count	605	21	21	21	21	21	21	605	605	603

Y_i = displacement distance for an individual rock fragment.
$MASS_i$, $SIZE_i$ and FI_i = mass, size and flatness index for an individual rock fragment respectively.

Table 10.7 Linear correlation coefficient between the variables at the site level

	MEAN Y	SLOPE	SARCO7	SARCO12	VEG	RFC	ROUGH	SLP/RFC
MEAN Y	1							
SLOPE	0.50^*	1						
SARCO7	-0.51^*	−0.07	1					
SARCO12	-0.49^*	−0.03	0.93^{**}	1				
VEG	0.47^*	−0.10	−0.20	−0.25	1			
RFC	−0.35	0.25	0.22	0.21	-0.83^{**}	1		
ROUGH	−0.13	0.20	0.47^*	0.44^*	−0.21	0.27	1	
SLP/RFC	0.78^{**}	0.42	-0.46^*	−0.38	0.67^{**}	-0.61^{**}	−0.17	1
SLP/ROUGH	0.63^{**}	0.69^{**}	−0.41	−0.36	0.20	−0.05	-0.46^*	0.67^{**}

MEAN Y is the mean displacement distance after 12 months for all tracers at a given site.
SLP/RFC is the ratio of SLOPE and RFC.
SLP/ROUGH is the ratio of SLOPE and ROUGH.
[**] Significant at 0.01 level.
[*] Significant at 0.05 level.
$n = 21$.

Table 10.8 Linear correlation coefficients between tracer characteristics and displacement distances for all tracers

	Y_i7	Y_i12	$MASS_i$	$SIZE_i$
Y_i7	1			
Y_i12	0.87^{**}	1		
$MASS_i$	−0.01	0.03	1	
$SIZE_i$	0.01	0.05	0.86^{**}	1
FI_i	-0.18^{**}	-0.17^{**}	−0.07	0.14^{**}

Y_i7 and Y_i12 are displacement distances for individual rock fragments after 7 and 12 months respectively.
[**] Significant at 0.01 level.
$n = 573\text{--}589$.

greater displacement distances with less *Sarcopoterium* cover. The variables SLOPE and VEG are both positively correlated with displacement distances. This was expected for the slope gradient, but seems unusual for the vegetation cover. However, the correlation matrix also shows that VEG and RFC are strongly negatively correlated. This is due to the fact that some very smooth surfaces were covered by an almost continuous mat of very short grass (due to grazing), while bare patches showed a rough, stony surface. This explains why displacement distances tend to be larger on well vegetated (smooth) surfaces than on bare (rough) surfaces.

The negative correlation between SARCO and VEG (Table 10.7) implies that *Sarcopoterium* cover reduces the overall annual vegetation cover while it also reduces the displacement distances. This seems somewhat contradictory, but can be explained by two factors. First, the sites with a dense cover of grass have no *Sarcopoterium*, and second, an increase of *Sarcopoterium* cover very often reduces the other vegetation cover. The latter results from the nature of *Sarcopoterium* as an invader species that eventually overruns all annual vegetation. It can establish itself anywhere except on very thin soils, and those are the sites with a dense short grassy vegetation. As might be expected, ROUGH is positively correlated with SARCO.

None of the roughness parameters (RFC and ROUGH) show a significant effect on the mean displacement distance. This might be masked by the effect of slope. The variables SLOPE, RFC and ROUGH are positively (but not significantly) correlated, which means that there is greater a tendency for rougher surfaces to occur on steep slopes than on less steep slopes. When the ratios of the slope gradient and the rock fragment cover (SLP/RFC) or roughness (SLP/ROUGH) are considered, very significant correlations are found with the mean displacement distance (Table 10.7). This indicates that on steeper slopes the displacement distance of the tracers is retarded when the RFC is high and vice versa. The RFC is therefore a key resistance parameter that has an important effect on the tracer displacement in connection with SLOPE. Figure 10.6 shows the regression line and equation of the relationship between MEAN *Y* and SLP/RFC. Since SLP/RFC is more strongly related to MEAN *Y* than SLP/ROUGH is to MEAN *Y* it is given preference over SLP/ROUGH in the analyses.

The variable SLP/RFC may also provide information about the role of *Sarcopoterium*. Figure 10.7 shows that when SLP/RFC exceeds 2, SARCO12 is very low. The greatest displacement distances are found in the same reach (SLP/RFC > 2) (Figure 10.6). This may indicate that the absence of SARCO has a positive effect on the movement of the tracers. The simple correlation of MEAN *Y* and SARCO is not very strong but is significant (Table 10.7). The combined effect of SLP/RFC and SARCO can be assessed in a multiple regression analysis.

The compound variable SLP/RFC explains 61 percent of the variation in the mean travel distance of the tracers (MEAN *Y*; Figure 10.6). When the *Sarcopoterium* cover (SARCO) is included in this multiple regression equation the correlation coefficient is raised by only 4 percent to 65 percent (Table 10.9). The reason for the rather small effect of SARCO 12 might be that the *Sarcopoterium* bushes only become effective after a threshold cover value is exceeded. Figure 10.8 shows that about half of the experimental sites have very low SARCO12 values (open squares). These low cover percentages will not be an effective obstacle for rock fragment movement. When only the

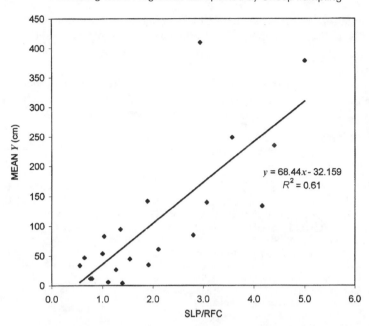

Figure 10.6 Relationship between MEAN *Y* and SLP/RFC. MEAN *Y* equals mean displacement distance after 12 months of all tracers at a given site. SLP/RFC is the ratio between the slope of the soil surface (SLOPE) at a given site and the corresponding rock fragment cover (RFC)

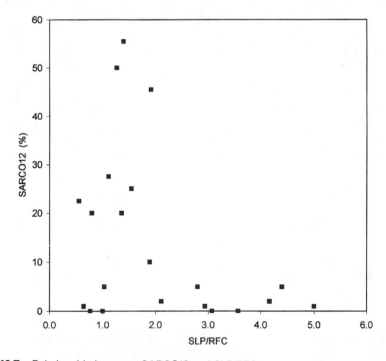

Figure 10.7 Relationship between SARCO12 and SLP/RFC

cover percentages > 10 percent (SARCOHIGH) are used, the correlation coefficient (with MEAN *Y*) rises from 0.49 to 0.60. When we enter SARCOHIGH in the multiple regression equation the r^2 rises to 73 percent (Table 10.9), despite the low number of cases ($n = 10$).

There are several reasons to give preference here to *Sarcopoterium* cover over vegetation cover. First, SARCO is a dynamic variable which may play an important

Table 10.9 Summary of regression models

Regression models	r^2	p	n
MEAN $Y = -32.16 + 68.44$ (SLP/RFC)	0.61	< 0.001	21
MEAN $Y = 155.12 - 3.25$ (SARCO12)	0.24	0.023	21
MEAN $Y = 4.72 + 60.84$ (SLP/RFC) -1.49 (SARCO12)	0.65	< 0.001	21
Partial r^2: (SLP/RFC) 0.61, (SARCO12) 0.40			
MEAN $Y = 26.51 + 61.08$ (SLP/RFC) $- 2.03$ (SARCOHIGH)	0.73	0.02	10
MEAN $Y = -17.04 + 73.87$ (SLP/RFC) $- 0.43$ (VEG)	0.61	< 0.001	21

MEAN Y = mean tracer displacement distance for each site after 12 months.
SLP/RFC = ratio of slope gradient and rock fragment cover.
SARCO12 = *Sarcopoterium* cover after 12 months.
SARCOHIGH = *Sarcopoterium* cover (after 12 months) > 9%.
VEG = vegetation cover after 12 months.

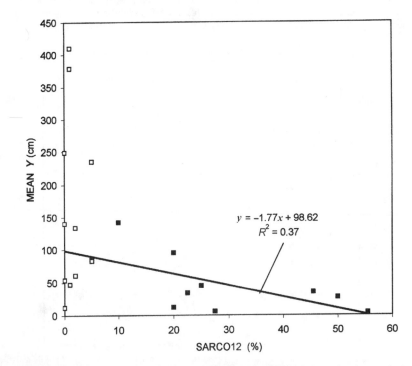

Figure 10.8 Relation between MEAN *Y* and SARCO12. Half of the experimental sites have very low SARCO12 values (< 9%; open squares). The regression line is based on SARCO12 > 9% (solid squares)

role in the transport of rock fragments over the hillslopes in the study area. Management practices such as the burning of *Sarcopoterium* every two to three years reduce the surface roughness, which may have a great effect on the rock fragment displacement. Second, while SARCO12 and VEG both have an effect on ROUGH, the negative correlation of SARCO12 with displacement distance seems more logical than the positive correlation between VEG and downslope displacement. Adding VEG to the regression equation with SLP/RFC does not result in a higher r^2 (Table 10.9).

The tracer (rock fragment) characteristics show remarkably little correlation with the displacement distance (Table 10.7). Only the flatness index (FI) is weakly but significantly correlated with the displacement distance (Y). The negative sign indicates that rounder rock fragments tend to travel further than flatter ones. In other studies it has been reported that large rock fragments moved further (Kirkby and Statham, 1975; Frostick and Reid, 1982; Govers and Poesen, 1998). Statham and Francis (1986) pointed out that the resistance to motion is smaller for larger rock fragments than for smaller ones when they roll or slide over a surface with a given roughness. However, in our experiments the displacement distance is also determined by an impulse force. In order to bring rock fragments into motion a greater critical impulse is needed for a large (heavy) rock fragment than for a small one. Also, the same impetus applied to a large and a small rock fragment should result in a larger horizontal projection for the smaller rock fragment on a frictionless surface. Thus, the impulse factor and the resistance to motion together cancel any size advantage.

The fraction of displaced rock fragments (tracers) is large: 85 percent is moved after 7 months and 93 percent after 12 months. However, as Table 10.10 shows, there are important differences between sites. The fraction may be used as an indicator for the trampling intensity. Small fractions may indicate that the site is not grazed intensively. A regression analysis (Table 10.11) shows that after 7 months, SLOPE and SARCO7 explain 62 percent of the variation in the fraction of displaced rock fragments between the different sites. However, the coefficient of determination has decreased to 27 percent after 12 months, which can be expected because the chance of a hit increases

Table 10.10   Fraction of displaced rock fragments

Site	Df7	Df12	Site	Df7	Df12
1	0.97	1.00			
2	0.92	1.00	12	0.90	0.90
3	0.76	0.90	13	0.86	1.00
4	0.73	0.93	14	0.63	0.73
5	0.97	0.97	15	1.00	1.00
6	1.00	1.00	16	0.86	1.00
7	0.90	0.93	17	0.71	0.84
8	0.97	1.00	18	0.90	0.87
9	0.77	0.93	19	0.83	1.00
10	0.90	0.93	20	1.00	1.00
11	0.89	0.96	21	0.52	0.52
			All sites	0.85	0.93

Df7 and Df12: fraction of displaced tracers after 7 and 12 months respectively.

Table 10.11   Regression models for the fraction of displaced rock fragments

Regression model	r^2	p	n
Df7 = 0.74 + 0.006 (SLOPE) −0.006 (SARCO7)	0.62	0.0001	21
Partial r^2: (SLOPE) 0.38, (SARCO7) 0.24			
Df 12 = 0.87 − 002 (SARCO12) +0.003 (SLOPE)	0.27	0.06	21

over time. Consequently, all fractions approach one and the relationships become less strong. The regression model for the situation after 7 months suggests that sites with high *Sarcopoterium* cover are less frequently grazed. This agrees with the information from the shepherd who stated that sheep tend to avoid areas with high *Sarcopoterium* cover.

CONCLUSIONS

The experiments indicated that after 12 months the marked rock fragments (tracers) had moved a mean distance of 104 cm. This movement was attributed entirely to the effect of trampling of sheep because the rock fragments are relatively large for sheet flow transport, no rills were observed in the plots, no burrowing animal traces were observed and the fields are only used for sheep grazing. The variation in rock fragment movement could be partly explained by the ratio of slope gradient and rock fragment cover, and the *Sarcopoterium* cover. The flatness index probably also has a small influence on the displacement of the tracers. Contrary to other studies of rock fragment movement, no relationship was found between displacement distance and rock fragment size. This is due to the opposing effects of impact and friction forces on small and large rock fragments. A certain impact of a sheep's hoof will project a small (light) rock fragment further downslope than a large (heavy) one. However, friction forces due to soil surface characteristics have a greater retarding effect on small rock fragments than on large ones.

The effect of soil surface roughness on rock fragment movement was best expressed by the rock fragment cover in combination with the slope gradient. Roughness tends to adjust the effect of slope gradient on the downslope movement in these environments.

The effect of *Sarcopoterium* cover on tracer displacement suggests that the cycle of *Sarcopoterium* burning and subsequent encroaching significantly influences rock fragment movement. After burning, the bare surfaces are prone to water erosion, creating erosion pavements. New rock fragments can be detached by trampling from the topsoil, and rock fragment transport will probably be at its highest rate. When the *Sarcopoterium* cover increases again during the subsequent 3 years, the rock fragments will be increasingly trapped by these shrubs and will also receive fewer blows because sheep will avoid these areas.

It is difficult to compare the data with other studies of rock fragment movement because conditions are very different. It was assumed that the rock fragments could only be moved by sheep trampling because no rills were present on the sites. One way to prove this would be to use control sites where no grazing occurs. However, this would result in a more vigorous growth of the vegetation within the enclosure compared to the outside area. The enclosed surface would therefore not be entirely comparable with the experimental sites. Moreover, the fact that some rock fragments moved upslope and that almost all of the transported rock fragments also moved laterally over some distance strongly suggests that the movement was only due to animal trampling.

The mean displacement distance of 104 cm suggests that the movement of rock fragments on the slopes of western Lesvos is a significant form of erosion where many rock fragments outcrop at the soil surface, and should be taken into account in denudation models. The high recovery rate and the ease of detecting the marked rock fragments indicates that they functioned well as tracers. Future research will focus on rock fragment transport rates by means of fluxes and diffusion constants.

ACKNOWLEDGEMENTS

The research for this study was carried out as part of the MEDALUSIII (Mediterranean Desertification and Land Use) collaborative research project. MEDALUSIII was funded by the European Commission Environment and Climate Research Programme (Contract ENV4–CT95–0118, Climatology and Natural Hazards) and this support is gratefully acknowledged. John Babetellis and Naoum are thanked for providing information on the grazing pressure and for allowing us to conduct experiments on their grazing lands. We acknowledge the critical and stimulating comments by Mike Kirkby on an earlier draft of this chapter. Patriek Bleys and Robert Geeraerts are thanked for their technical assistance.

REFERENCES

Abrahams, A. D., Parsons, A. J., Cooke, R. U. and Reeves, W. R. 1984. Stone movement on hillslopes in the Mojave desert, California. *Journal of Geology*, **98**, 264–272.
Butler, D. R. 1995. *Zoogeomorphology: Animals as Geomorphic Agents*. Cambridge University Press, Cambridge.
Caine, N. 1981. A source of bias in rates of surface soil movements as estimated from marked particles. *Earth Surface Processes and Landforms*, **6**, 69–75.
Davis, J. C. 1986. *Statistics and Data Analysis in Geology* (2nd edition). John Wiley, New York.
Frostick, L. E. and Reid, I. 1982. Talluvial processes, mass wasting and slope evolution in arid environments. *Zeitschrift für Geomorphologie N.F.*, **44**, 53–67.
Govers, G. and Poesen, J. 1998. Field experiments on the transport of rock fragments by animal trampling on scree slopes. *Geomorphology*, **23**, 193–203.
Kirkby, A. and Kirkby, M. J. 1974. Surface wash at the semi-arid break of slope. *Zeitschrift für Geomorphologie. N.F. Supp. BD*, **21**, 151–176.
Kirkby, M. J. and Statham, I. 1975. Surface stone movement and scree formation. *Journal of Geology*, **8**, 349–367.
Kosmas, C., Yassoglou, N., Briasouli, H., Danalatos, N., Gerontidis, S. and Dalakou, B. 1996. Lesvos, a review of the geography. Unpublished Medalus Working Paper No. 72.
Lehmann, R. 1994. Landschaftsdegradierung, Bodenerosion und –Konservierung auf der Kykladeninsel Naxos, Griechenland. *Physiogeographica, Basler Beitrage zur Physiogeographie*, Band 21, University of Basel.

Poesen, J. 1987. Transport of rock fragments by rill flow: a field study. *Catena, Supp.*, **8**, 35–54.

Poesen, J. and Bunte, K. 1996. Effects of rock fragments on desertification processes in Mediterranean environments. In Brandt, J. and Thornes J. (eds) *Mediterranean Desertification and Land Use*. John Wiley, Chichester, 257–269.

Poesen, J. and Lavee, H. 1994. Rock fragments in top soils: significance and processes. *Catena*, **23**, 1–28.

Poesen, J., Torri, D. and Bunte, K. 1994. Effect of rock fragments on soil erosion by water at different spatial scales: a review. *Catena*, **23**, 141–166.

Schumm, S. A. 1967. Rates of surficial rock creep on hillslopes in western Colorado. *Science*, **155**, 560–561.

Statham, I. and Francis, S. C. 1986. Influence of scree accumulation and weathering on the development of steep mountain slopes. In Abrahams, A. D. (ed.) *Hillslope Processes*. Allen and Unwin, Boston, 245–267.

11 Using Geostatistics to Improve Estimates of ^{137}Cs-derived Net Soil Flux

A. CHAPPELL[1] and M. A. OLIVER[2]

[1]Department of Geography, University of Salford, Manchester, UK
[2]Department of Soil Science, The University of Reading, UK

INTRODUCTION

The ^{137}Cs technique for estimating soil redistribution has received considerable attention in recent years (Loughran et al., 1993; Walling and Quine, 1993; Basher et al., 1995; Chappell et al., 1996a; Bajracharya et al., 1998) because it enables the net soil flux for the last c. 30 years to be quantified from one field visit (Ritchie and McHenry, 1990; Sutherland and de Jong, 1990; Walling and Quine, 1991). Several ^{137}Cs studies have been conducted for individual fields (Quine and Walling, 1991), along transects (Chappell et al., 1999) or across landscape toposequences (Kulander and Strömquist, 1989; Loughran et al., 1989; Chappell et al., 1998). Recent work (Chappell, 1999) has suggested that ^{137}Cs samples from toposequences do not adequately represent the spatial variation in soil redistribution processes and therefore they are likely to provide unreliable information about soil redistribution rates over a region. There have been few spatially intensive investigations of ^{137}Cs, especially over large areas, because its measurement in the laboratory is time-consuming. To describe its spatial variation and thus soil redistribution accurately many samples are required (Sutherland, 1991, 1994; Chappell, 1995, 1998; Owens and Walling, 1996). The studies that have mapped net soil flux accurately over large areas have used innovative sampling designs and/or simple geostatistical procedures to reduce the number of samples for ^{137}Cs analysis (de Roo, 1991; Chappell et al., 1996b).

Over larger or more complex terrain than a single toposequence there might be distinct geomorphological regions, such as a plateau area and hillslope. Within these units there could be different slope units, different soil types and different processes, each with their own scale of soil redistribution, which could be quite different. In such situations the area should be stratified to coincide with the main factor(s) controlling the spatial variation of net soil flux (McBratney et al., 1991; Stein et al., 1991), and should be sampled accordingly. Sampling the spatial variation accurately within each stratum (Webster and Oliver, 1992) requires more samples than if the region is treated

Tracers in Geomorphology. Edited by Ian D. L. Foster. © 2000 John Wiley & Sons Ltd.

as a single unit. Alternatively the pooled within-strata variance (Voltz and Webster, 1990) may be determined to provide a partial solution (Chappell *et al.*, 1996b).

If expensive-to-measure properties, such as ^{137}Cs, are spatially correlated with a cheap-to-measure property the spatial variation of the sparsely sampled one can be determined more reliably using the auxiliary data. Several techniques for improving the estimates of soil properties using such data have been compared recently by Odeh *et al.* (1994, 1995) and Knotters *et al.* (1995). These techniques include geostatistical methods such as co-kriging, universal kriging, and kriging using intrinsic random functions (IRF-*k*), traditional statistical methods of multiple linear regression, and the combination of kriging and regression. Co-kriging has been widely used to improve the accuracy of the estimates of expensive properties using one or more cheaper properties (McBratney and Webster, 1983; Leenaers *et al.*, 1990; Zhang *et al.*, 1992; Atkinson *et al.*, 1994). In particular, Chappell (1998) used co-kriging to improve the estimates of ^{137}Cs-derived net soil flux using remotely sensed properties.

If there is a trend present, as is likely over large areas with distinct regions, then this must be accounted for in the analysis. Universal kriging or generalised IRFs can be used where there is an underlying trend. Ordinary kriging techniques require the removal of any trend before analysis of the residuals. Methods that combine the benefits of linear regression and kriging use auxiliary data to improve the accuracy of prediction (Delhomme, 1979; Ahmed and De Marsily, 1987). They may enable the ^{137}Cs measurement error to be incorporated into the ^{137}Cs inventory estimated at unsampled locations, which is not done at present. The ^{137}Cs measurement error is not a regionalised variable and cannot be modelled by a variogram. Therefore, in this situation the regression–kriging approach is useful for incorporating this type of data for geostatistical estimation.

The aim of this chapter is to compare the performance of several geostatistical techniques for estimating a sparsely sampled ^{137}Cs inventory and its measurement error using more intensively sampled elevation measurements. The estimates from the different procedures will be compared with several samples of ^{137}Cs activity that were removed from the data before the analyses. The methods are summarised and their value for practical application is discussed. Finally, the results of the mapped ^{137}Cs estimates are interpreted geomorphologically.

MATERIALS AND METHODS

Climate and Geomorphology

The study area (13°33.63′N, 2°34.22′E) in south-west Niger is about 60 km east of Niamey. It has a semi-arid climate with an average rainfall of 560 mm year^{-1} (measured over the period 1905–1989). The prevailing winds in summer are southwesterly, but they are replaced at times by strong easterly winds. The latter precede squalls which bring high-intensity rainfall of short duration which causes soil crusting and surface wash. During the dry winter the northeasterly Harmattan winds dominate, and they transport material from the southern Sahara. However, the strong easterly winds in summer appear to account for more of the aeolian deposits than the Harmattan (Drees *et al.*, 1993).

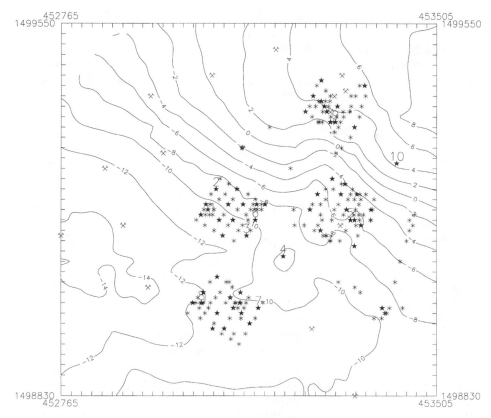

Figure 11.1 The study area, using UTM co-ordinates (m) and orientated north, showing isarithmic lines of ordinary block (20 m) kriging estimates for elevation relative to an arbitrary datum, the locations where ^{137}Cs activity (Bq m^{-2}) and elevation were sampled ($\star$) and where [137]Cs ($\times$) and elevation ($\ast$) were sampled separately. The large numbers indicate the samples that were removed for estimation performance statistics

The physiographic regions of the study area are distinct: they comprise a plateau (> 0 m) and a hillslope (< 0 m) (Figure 11.1). The topography, soil and vegetation of the two regions are quite different, which results in different processes of soil redistribution and interactions with the vegetation in each region (Chappell, 1995; Chappell *et al.*, 1996a). The spatial variation in the soil of the plateau results from the characteristic bands of vegetation or *brousse tigrée* (Chappell *et al.*, 1999). These are orientated with their long axis orthogonal to the slight slope towards the plateau edge. Between the vegetation bands the topsoil is a gravelly loam over cemented ironstone gravels (Manu *et al.*, 1991). It has a pH < 5, low nutrient status and small water storage capacity because of the shallow soil depth and large content of the coarse fraction. The soil associated with the vegetation is thicker, approximately 20 cm deep above the ferricrete cobbles, is rich in organic matter and has a pH of 5–6.

The hillslope region starts at the plateau edge with a ferricrete cobble talus slope. Downslope the terrain is hummocky, with pedestals or mounds of acid soil (pH < 5)

capped with vegetation. These mounds suggest that there has been considerable soil loss locally. The surface gradient decreases further downslope to a shelf that is controlled by the underlying ferricrete. The vegetation is sparse in this area where the soil of the exposed B horizon is compacted. Considerable loss of topsoil appears to have occurred in these areas which are also often adjacent to gullies. The latter incise the hillslope and are fed by rills and surface wash from the plateau.

Field and Laboratory Procedures

In the study area, approximately 600 m × 600 m, the soil was sampled between July and October 1992 at the intersections of a nested grid, with sampling intervals of 5 m, 20 m and 100 m. This scheme was adopted because we had few clues about the spatial scale of variation of the properties. The aim was to sample at a range of distances to ensure that we would resolve any variation at the scale of our investigation. The sampling was also stratified according to the two main physiographic units (McBratney et al., 1991; Stein et al., 1991). Several properties of the soil were measured and observed, but the one that we focus on here is ^{137}Cs inventory (Bq m^{-2}) which was measured at only 73 locations. In addition we also consider elevation (RE) which was recorded at 217 sites, a more intensive sample than the full grid survey (Figure 11.1). Elevation was determined relative to an arbitrary datum by ground-based survey from reference locations fixed by Global Positioning System (GPS).

Soil samples for γ-ray energy spectrometry were obtained in one of two ways depending on the site and the depth of the ^{137}Cs profile. On the plateau, a 20 cm deep pit (600 cm^2 area) was dug and a complete soil and ^{137}Cs profile was removed. On the hillslope the soil was sampled to a depth of 51 cm using a Dutch auger (19.6 cm^2 internal area) to obtain a complete ^{137}Cs profile. At each site the soil was bulked, homogenised and subsampled by hand to provide a representative sample (c. 200 g) of the ^{137}Cs activity in the profile. The ^{137}Cs inventory (Bq m^{-2}) was measured by γ-ray energy spectrometry on a horizontally orientated, 20 percent relative efficiency, hyper-pure germanium γ-ray detector (Chappell, 1995). The interference of bismuth-214 (^{214}Bi) was avoided by inserting a 408A biassed amplifier into the standard equipment circuit to spread the signal over a larger range of channels and provide a greater resolution for the MCA. Measurement error was determined using the approach described by Sutherland (1991). The validity of the ^{137}Cs technique for measuring soil redistribution has been established for this region (Chappell et al., 1998). A first approximation of the reference ^{137}Cs inventory for this region is 2066 ± 125 Bq m^{-2}.

Geostatistical Background

The methods embodied in geostatistics provide suitable tools for analysing spatial data such as those of the soil and landscape (Oliver et al., 1989a). Geostatistics is the practical application of the theory of regionalised variables (Matheron, 1971), which regards spatial properties as random functions. The approach is stochastic although spatial data are essentially deterministic, i.e. they are a function of their position in space, but their variation is so irregular that the best way of treating them is as if they were random variables. The basis of the theory is that the variation in an attribute Z can be defined by a stochastic component and a constant. Following Oliver et al. (1989a), this may be written as

$$Z(\mathbf{x}) = m_v + \varepsilon(\mathbf{x}) \tag{11.1}$$

where $\mathbf{x}$ denotes the spatial coordinates, m_v is the mean in an area v, and the quantity $\varepsilon(\mathbf{x})$ is the spatially dependent random variable. This last component has a mean of zero,

$$E[\varepsilon(\mathbf{x})] = 0 \tag{11.2}$$

and a variance defined by

$$\text{var}[\varepsilon(\mathbf{x}) - \varepsilon(\mathbf{x} + \mathbf{h})] = E[\{\varepsilon(\mathbf{x}) - \varepsilon(\mathbf{x} + \mathbf{h})\}^2] = 2\gamma(\mathbf{h}) \tag{11.3}$$

where $\mathbf{h}$ is a vector, the lag, that separates the two places $\mathbf{x}$ and $\mathbf{x} + \mathbf{h}$ in both distance and direction. Thus, the variance of $\varepsilon(\mathbf{x})$ depends on the separation $\mathbf{h}$ and not on the actual position of $\mathbf{x}$. This assumes that the variable is second-order stationary and that the mean, variance, covariance and variogram exist. Matheron (1971) realised that second-order stationarity was too strong for many spatial variables and reduced the assumptions to stationarity of the mean and variance of the differences. This is the intrinsic hypothesis. Equation (11.3) is then equivalent to

$$\text{var}[z(\mathbf{x}) - z(\mathbf{x} + \mathbf{h})] = E[\{z(\mathbf{x}) - z(\mathbf{x} + \mathbf{h})\}^2] = 2\gamma(\mathbf{h}) \tag{11.4}$$

The semi-variance, $\gamma(\mathbf{h})$, is half the expected squared difference between two values separated by $\mathbf{h}$, as above. The function that relates γ to $\mathbf{h}$ is the semi-variogram or more commonly the variogram. In practice, the experimental variogram is constructed by grouping semi-variance according to lag distance. Models are commonly fitted to the experimental variogram to derive parameters for interpreting the structure of spatial variation and for use in estimation at unsampled locations. The nugget variance (Table 11.3) is typically present in samples of continuous data (McBratney and Webster, 1983; Chappell et al., 1996a). It may arise partly from measurement error, although this is usually small in relation to the spatial variation (Webster and Oliver, 1990). The main cause is usually spatially dependent variation that occurs over distances much smaller than the shortest sampling interval. When the dissimilarity of a property reaches a maximum over the average separation distances, the sill variance is reached and the model is bounded (Table 11.3). The lag separation distance at which the variogram reaches its sill is the range (Table 11.3); this is the limit of spatial dependence. Beyond this limit the semi-variance bears no relation to the separation distance.

The intrinsic hypothesis assumes weak local stationarity, i.e. within some neighbourhood v. There are situations where this does not hold. In large regions the mean values of variables vary predictably or deterministically from one area to another. This is evidence of a trend under which second-order stationarity and the intrinsic hypotheses no longer hold. Trend can be expressed in terms of IRFs of order k, where $k > 0$. Delfiner (1976) suggested that trend could be removed by fitting a polynomial up to degree $2k + 1$. The variogram is then estimated from the residuals of the trend, which are a random variable (Olea, 1975, 1977; Oliver, 1987; Lam and Barrett, 1992). These polynomials are not covariance functions of the non-stationary variables themselves but of their stationary increments (Knotters *et al.*, 1995) and are called generalised covariance functions (GCFs) (Matheron, 1973; Kitanidis, 1983; Stein *et al.*, 1991).

Ordinary Kriging and Co-kriging

When data are sparse, as is the case with most survey data, an important aim in subsequent analyses is to estimate the values of a property at unsampled locations. Kriging is the method of geostatistical estimation (Oliver *et al.*, 1989b). It is essentially a moving weighted average, but the weights are derived from the variogram model rather than in an arbitrary way, as with most other methods of interpolation (Webster and Oliver, 1990). Hence, the estimates are based on knowledge of the spatial variation of the property modelled by the variogram. Kriging has been found to be one of the most reliable two-dimensional spatial estimators (Laslett *et al.*, 1987; Laslett and McBratney, 1990). Furthermore, Laslett (1994) has shown that kriging estimates can be improved by multi-stage sampling designs because the variogram can be determined more reliably, especially over the first few lags which are the most important in kriging.

Co-kriging is the logical extension of ordinary kriging to situations where two or more variables are spatially interdependent. Using traditional ordinary co-kriging (Deutsch and Journel, 1992), the estimate is a weighted average of the available data with weights used so that the estimate is unbiased and has minimum variance (McBratney and Webster, 1983). The sum of the weights applied to the primary variable is set to one, and the sum of the weights applied to any other variable is set to zero. Other non-bias conditions are possible with different types of co-kriging (Deutsch and Journel, 1992). This second condition tends to limit severely the influence of the secondary variable and since K^2 covariance functions are required when K auxiliary variables are considered, the reduction in estimation variance is not worth the additional modelling effort unless the primary variable is underestimated relative to the secondary variable(s) (Deutsch and Journel, 1992).

The linear model of co-regionalisation provides a framework for modelling the variograms and covariance functions so that the variance of any possible linear combination of these variables is always positive (Journel and Huijbregts, 1978). The model for each of these sample variograms may consist of one or more conditional negative semi-definite (or authorised) models. However, both properties must have similar variogram models (Isaaks and Srivastava, 1989). For instance, if the variograms are bounded then their ranges must be similar for co-kriging to be sensible.

Regression–Kriging

A technique proposed by Delhomme (1979), and applied by Ahmed and de Marsily (1987) to deal with trend is 'kriging with a guess field' or regression–kriging model B (Odeh *et al.*, 1995). This model uses normal regression followed by the computation of variograms and ordinary kriging of the regressed values and residuals separately. It has the advantage of explicitly accounting for the spatial correlation in the errors from the regression model (Ahmed and de Marsily, 1987). These techniques appear to perform as well as co-kriging and universal kriging techniques despite being unable to account for spatial covariation in several properties (Odeh et al., 1994; Knotters et al., 1995). This is probably due to the flexibility of the kriging system to combine the regression errors or measurement errors.

One advantage of kriging combined with regression is that use can be made of a known and physically interpretable relationship between the target variable and the auxiliary variable, which can have any form (Knotters et al., 1995). Co-kriging does not use physically interpretable relationships between the target and the auxiliary variable, but assumes a linear relationship. Kriging combined with regression requires only a single model of the spatial correlation between the target variable and auxiliary variables.

Regression–kriging model B is used here to combine the ^{137}Cs measurement with the ^{137}Cs inventory (Bq m^{-2}) estimate at unsampled locations which has not been done before in studies using the ^{137}Cs technique. Least-squares regression equations were calculated for (i) the measurement activity of ^{137}Cs (in Bq m^{-2}) and relative elevation (RE, in m); and (ii) ^{137}Cs, RE and the ^{137}Cs measurement error (EBq, in Bq m^{-2}), separately using the Genstat software (Genstat 5 Committee, 1992). The two regression equations, RB1 and RB2, respectively are as follows:

$$^{137}Cs = \alpha_1 RE + c_1 + \varepsilon_1 \tag{11.5}$$

$$^{137}Cs = \alpha_2 RE + \beta_2 EBq + c_2 + \varepsilon_2 \tag{11.6}$$

where α and β are model parameters, c is a constant and ε is the regression residual. Table 11.1 shows the coefficients of these regression equations and their explanation of variance.

Table 11.1 Coefficients and explanation of variance for regression equations

Model parameters	RB1	RB2
% variance explained	46	53
c	2470	1466
α (RE)	143	151
β (EBq)	–	8

RESULTS

Ordinary experimental variograms were computed and models fitted using weighted least-squares in Genstat (Genstat 5 Committee, 1992). Variograms for the ^{137}Cs inventory and relative elevation (Figure 11.2) showed evidence of trend: for the ^{137}Cs inventory, by the sudden increase in the estimation variances after the sill had been reached; and for relative elevation, by its concave downwards shape near to the

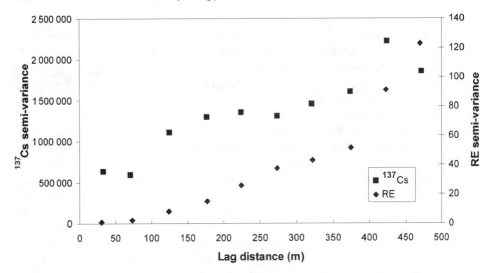

Figure 11.2 Experimental variograms for ^{137}Cs activity (Bq m^{-2}) and elevation (m) relative to an arbitrary datum exhibiting evidence of trend, or systematic change over space, in the data

origin. Trend was removed from the raw data by fitting quadratic polynomials to both variables on their spatial coordinates using least-squares regression. Table 11.2 shows that approximately 50 percent of the variation for ^{137}Cs activity and 96 percent of the variation for relative elevation was accounted for by the quadratic functions. Experimental variograms (Figure 11.3) were then computed on the residuals from the trend as before. In both cases the variograms are distinctly periodic.

Periodic models were not fitted because our aim was to estimate values by kriging for which the periodic model can be unstable, giving rise to negative kriging variances. Therefore, only simple models were fitted (Table 11.3).

A cross-variogram was computed between the residuals of the polynomials fitted to ^{137}Cs and RE using Geopack (Yates and Yates, 1989) and modelled (Table 11.3) 'by eye' (Figure 11.3) using the Variowin software (Pannatier, 1997). Fitting is

Table 11.2 Coefficients and explanation of variance for quadratic polynomials fitted to each property

Model parameters	^{137}Cs(Bq m^{-2})	Relative elevation (m)	Regression RB1 fitted values	Regression RB2 fitted values
% variance explained	49.8	96.0	95.8	79.7
C	+3918.00	−6.70	+1495.00	+1853.00
X	−3.51	−0.00593	−0.81	−2.08
Y	−14.92	−0.04932	−7.13	−7.84
X^2	−0.00165	+0.00000219	+0.00084	+0.00263
Y^2	+0.01694	+0.00007236	+0.01107	+0.01163
XY	+0.01463	+0.00006104	+0.00773	+0.00772

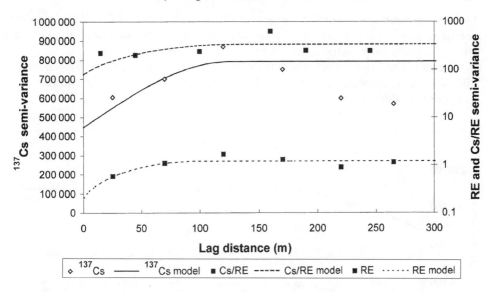

Figure 11.3 Experimental variograms and models of the residuals from quadratic polynomials for ^{137}Cs activity (Bq m^{-2}), elevation (m) relative to an arbitrary datum and the cross-variogram for ^{137}Cs and RE

somewhat imprecise because of adhering to the restrictions of the linear model of co-regionalisation.

As expected, the variograms of the fitted values for both regression equations (RB1 and RB2; see equations (11.5) and (11.6)) exhibited trend (not shown) which was removed by subtracting fitted quadratic polynomials (Table 11.2) as before. The residuals of these fitted polynomials were used to compute the variograms for model fitting (Table 11.3), as before, for both regression equations RB1 and RB2 (Figure 11.4). The variogram of residuals (ε_1) for the regression equation RB1 (Figure 11.5) exhibits a very similar structure to the variogram of ^{137}Cs with trend. Intuitively,

Table 11.3 Model parameters of spherical functions used for kriging and co-kriging

Property	Range	Sill variance	Nugget variance
^{137}Cs(Bq m^{-2})	126	344 000	448 000
Relative elevation (m)	101	1	≈ 0
^{137}Cs / RE[a]	130	259	79
^{137}Cs (restricted)[a]	130	344 000	448 000
RE (restricted)[a]	130	2	0
RB1 fitted values	140	46 570	0
RB2 fitted values	135	52 505	119 200
RB1 ε_1	174	687 174	151 622
RB1 ε_2	301	602 434	285 040

[a] These models are restricted by the linear model of co-regionalisation to the same model type and range.

there should be no trend in the variogram of the regression residuals. To test this assertion, a quadratic polynomial was fitted to the regression residuals and the variogram of the polynomial residuals did not have any spatially dependent structure (not shown). Thus, the variograms of the regression residuals (ε_1 and ε_2) for RB1 and RB2 (Figure 11.5) were modelled (Table 11.3) and used for interpretation and geostatistical estimation.

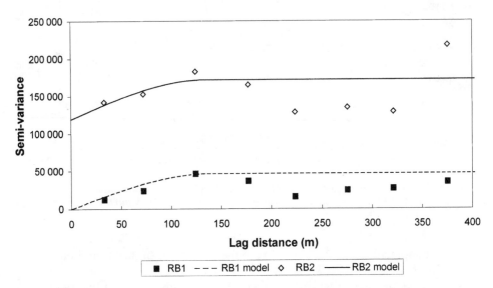

Figure 11.4   Experimental variograms and models of the residuals from quadratic polynomials for fitted values of the regression equations RB1 and RB2 (see equations (11.5) and (11.6))

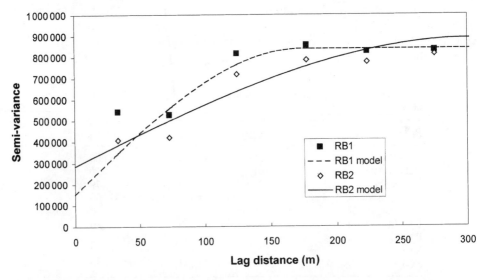

Figure 11.5   Experimental variograms and models for residuals (ε_1 and ε_2) of regression equations RB1 and RB2 (see equations (11.5) and (11.6))

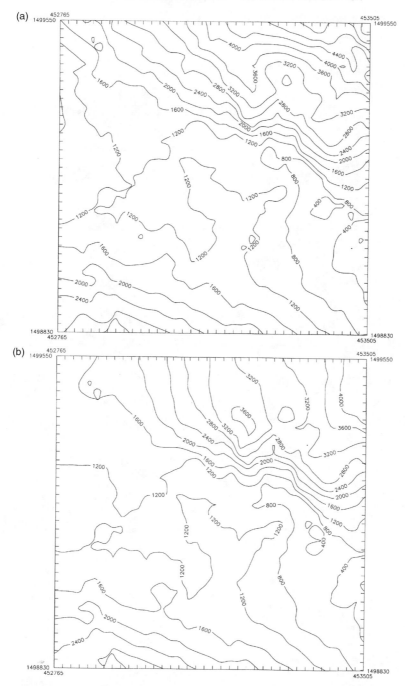

Figure 11.6(a,b)   Isarithmic maps, using UTM co-ordinates (m) and orientated north, of ^{137}Cs (Bq m^{-2}) block (20 m) estimates for (a) ordinary kriging, (b) ordinary co-kriging with relative elevation, (c) regression kriging including relative elevation (RB1: equation (11.1)), and (d) regression kriging including relative elevation and the ^{137}Cs measurement error (RB2: equation (11.2))

(c)

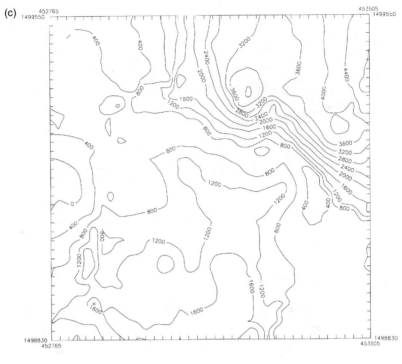

(d)

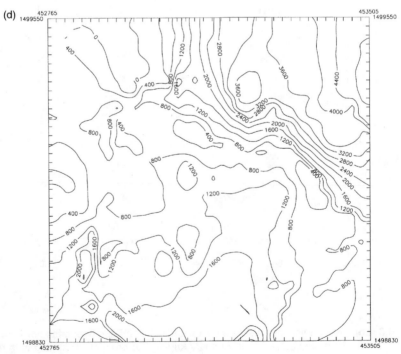

Figure 11.6 (c, d)

The parameters of the variograms (Table 11.3) were used to solve ordinary block kriging and co-kriging equations. The estimates and estimation variances were produced over 20 m blocks using the GSLIB software (Deutsch and Journel, 1992). In the case of regression–kriging, ordinary block (20 m) kriging was performed on the fitted values and the residuals separately for RB1 and RB2. The quadratic polynomials for each property (Table 11.2) were added back to the kriging estimates using a routine in the Genstat software. The fitted values and the residuals of the regression were added together to provide ^{137}Cs estimates for RB1 and RB2. Finally, isarithmic lines were threaded through the kriging estimates (Figure 11.6) for each property using the same isoline frequency with the Surfer package (Golden Software Inc., 1990). Summary statistics for the ^{137}Cs samples and geostatistical estimates are shown in Table 11.4.

Kriging is an exact interpolator, in that when the estimation location coincides with an observation the original sample values are used in the estimation map. Consequently, the performance of the estimation procedures can only be examined by removing some of the sample values prior to estimation. Since the amount of ^{137}Cs data are limited, only very few samples could be removed without having a detrimental

Table 11.4 Summary statistics for the ^{137}Cs inventory (Bq m^{-2}) samples and the geostatistical estimates

^{137}Cs statistics	Samples	Kriging	Co-kriging	Regression–kriging	
				RB1	RB2
Sample size	73	1377	1148	1134	1132
Average	1717.15	1850.79	1871.68	1781.33	1829.73
Standard deviation	1185.88	1120.71	1163.26	1378.41	1378.29
Coefficient of variation	0.69	0.61	0.62	0.77	0.75
Minimum	172.42	48.47	5.50	25.84	5.94
Maximum	4766.03	6360.97	6360.97	5993.54	5915.86

Table 11.5 ^{137}Cs inventory (Bq m^{-2}) samples and the corresponding estimates for each geostatistical technique

^{137}Cs samples and estimates	Sample values (Bq m^{-2})	Kriging estimates	Co-kriging estimates	Regression–kriging estimates	
				RB1	RB2
1	843	1612	1575	1236	1348
2	2091	1507	1550	1741	1843
3	2690	1623	1825	1699	1837
4	1336	1352	1452	1395	1290
5	1250	1254	1242	1290	1163
6	2035	963	964	991	1021
7	2042	1247	1214	1289	1102
8	1174	1247	1232	1205	1154
9	4425	2830	2911	3878	3806
10	3731	2461	2372	2334	2369

Table 11.6   Geostatistical estimation performance statistics

^{137}Cs comparisons	Kriging	Co-kriging	Regression–kriging	
			RB1	RB2
Sample size	10	10	10	10
ME[a]	552.3	528.1	456.1	468.6
RMSE[b]	895.4	869.8	719.4	721.5
MRE[c]	5.5	5.5	4.7	4.8
SDRE[d]	10.7	10.2	8.9	9.0

[a]Mean error.
[b]Root mean square error.
[c]Mean rank error.
[d]Standard deviation of rank error.

effect on the estimation performance. A total of 10 sample locations, where both ^{137}Cs and RE had been measured, were chosen systematically to represent the variation in ^{137}Cs amounts throughout the region (Figure 11.1). The kriging, co-kriging and regression–kriging procedures were repeated, this time with the 10 samples removed so that the estimates could be compared with the removed sample values (Table 11.5). Estimation performance statistics (Table 11.6) were calculated (Laslett *et al.*, 1987; Bourennane *et al.*, 1996) and include the mean error (ME), the root mean squared error (RMSE), the mean rank error (MRE) and the standard deviation of rank error (SDRE) between the sample values and the kriging, co-kriging and regression–kriging (RB1 and RB2) estimates.

DISCUSSION

Variogram Analysis and Geomorphological Significance

The variogram of ^{137}Cs (Figure 11.2) exhibits some periodic structure at medium lag distances and the semi-variances increase sharply at 400 m lag distance which is approximately equivalent to the average size of the valley and the edge of the plateau. The sharp increase of the variogram has been associated with the average separation distance of distinct geomorphological regions (McBratney *et al.*, 1991; Chappell, 1995; Chappell *et al.*, 1996b).

The variogram of residual ^{137}Cs, RE (Figure 11.3), and residual fitted values for RB1 and RB2 (Figure 11.4) exhibited a distinct periodic structure suggesting that there is some cyclical repetition in the variation. The periodic structure is also unbounded, signalling increasing sources of variation with increasing separation distances. The distances at which the minima of cyclical variograms occur, correspond to the wavelengths of pronounced repetitive elements (Olea, 1977). Chappell and Oliver (1997) suggested that periodicity in residual relative elevation for the hillslope was closely related to the proximity to the surface of an underlying ferricrete layer, indicated by badly eroded, compact red exposed subsoil. The range of these variograms (Table 11.3) is very similar to the average distance between these badly eroded areas (150 m) and to the wavelength of the presence or absence of gullying (125 m) found elsewhere (Chappell and Oliver, 1997).

Geomorphological Significance of Mapping [137]Cs Estimates

The general pattern in the maps of the ^{137}Cs estimates is broadly similar for the four geostatistical techniques (Figure 11.6). The north-eastern corner of the study area is a plateau with considerably larger amounts of ^{137}Cs than the hill sloping towards the south-west. The edge of the plateau can be identified from the densely packed ^{137}Cs isolines. The large amount of ^{137}Cs on the plateau is associated with the accumulation of ^{137}Cs-enriched dust in the *brousse tigrée* (Chappell *et al.*, 1998, 1999). The hillslope is less densely vegetated and exposed to wind and water erosion. Notably, the ^{137}Cs isolines are perpendicular to the contours near the base of the hillslope, especially in the south-eastern part of the study area, highlighting the areas affected by wind erosion from easterly squalls (Chappell *et al.*, 1996a). The mid-slopes have less ^{137}Cs than the base of the slopes, and the striking 'V-shaped' pattern in the centre of the slope is the location of the main sediment pathway via two gullies which deposit near the base of the study area. The presence of gullies in the study area has important implications for tracing the sediment transfer process using the ^{137}Cs technique because of the presence of soil unlabelled by ^{137}Cs mixing with topsoil. The limitations of the ^{137}Cs technique are covered elsewhere (Chappell *et al.*, 1998; Chappell, 1999).

A noticeable difference between the ordinary kriging map of [137]Cs (Figure 11.6(a)) and the other maps occurs on the plateau. This difference is probably caused by the co-kriging and regression–kriging techniques making use of the large number of elevation measurements as ancillary information for ^{137}Cs estimation. Although there is little difference between the distribution of ^{137}Cs in these last three maps (Figure 11.6(b)–(d)), there is a greater similarity in the ^{137}Cs estimated by the regression–kriging techniques (Figure 11.6(c) and (d)) than the co-kriging estimates (Figure 11.6(b)). This observation is borne out by the standard deviation of the ^{137}Cs estimates (Table 11.4) which is considerably larger for the regression–kriging estimates than for the kriging and co-kriging estimates.

Comparison of Estimation Procedures

The ME should be close to zero for unbiased methods and the RMSE should be small for an unbiased and precise prediction (Bourennane *et al.*, 1996). None of the estimation techniques have produced ME statistics close to zero because the estimates were over 20 m blocks and the samples were from considerably smaller areas (< 1 m blocks). Thus, the performance statistics are not absolute measures but are valid and consistent for this comparison of estimation techniques. The difference in support between estimates and samples has the advantage of avoiding negative and positive biases that may cancel out with the same support (Bourennane *et al.*, 1996). Performance statistics based on the rank of the results avoid the problem with differences in sample and estimate support. A method that performs well should have a small MRE and SDRE.

All statistics (Table 11.6) indicate that regression–kriging has performed better than either ordinary kriging or co-kriging. The statistics also suggest that regression–kriging including only relative elevation (RB1) performed better than regression–kriging with relative elevation and ^{137}Cs measurement error (RB2). This is slightly surprising since the explanation of variance for the RB2 regression equation (Table 11.1) was larger than that for the RB1 regression equation. This result suggests that the additional explanation of aspatial variation provided by the ^{137}Cs measurement error is less important here than the spatially dependent information modelled by the variogram and utilised in kriging.

Another surprising result is the poor performance of the co-kriging estimates, especially since this technique proved so effective in the estimation of ^{137}Cs-derived net soil flux using remote-sensing data (Chappell, 1998). However, unlike the uniform distribution of remote-sensing pixels, the relative elevation samples are clustered around the main GPS reference sites (Figure 11.1). It is likely that the clustered nature of the auxiliary RE data has reduced the effectiveness of the co-kriging estimation procedure.

CONCLUSIONS

The variograms produced in this study provided information on the structure of spatial variation which was related to known geomorphological processes in the study area. The variograms also provided models of spatial variation which were central to the estimation of ^{137}Cs at unsampled locations using several geostatistical techniques. This approach avoided the use of arbitrary (mathematical) models commonly used for interpolation. The general pattern in the maps of the ^{137}Cs estimates was broadly similar for the geostatistical techniques and was related to known processes of erosion and deposition by wind and water erosion in the study area.

In accordance with other workers (Odeh et al., 1994; Knotters et al., 1995), the regression–kriging technique has performed better than ordinary kriging and co-kriging techniques. This is despite being unable to account for spatial covariation in several properties. In this study the poor estimation performance of the co-kriging technique is related to the limited spread of the auxiliary RE data across the study area. As suggested by Odeh et al. (1994) and Knotters et al. (1995), it is likely that this performance is due to the flexibility of the kriging system in combining the regression residuals. Regression–kriging has the distinct advantage of utilising a single physically interpretable relationship that can be modelled using linear combinations of multi-variate data (regression equations). Co-kriging does not use physically interpretable relationships between the target and the auxiliary variable and requires considerable modelling (K^2 covariance functions when K auxiliary variables are considered). Thus, regression–kriging provides the potential to accurately map a linear combination of properties using the power of kriging and an appropriate model (variogram) of the spatial variation.

It was somewhat surprising to find that the addition of the ^{137}Cs measurement errors did not improve the estimation performance, as intuitively this would serve to reduce the explanation of variance. More work is required to investigate the cause of this apparently anomalous relationship and to investigate the utility of regression–kriging with multivariate data.

ACKNOWLEDGEMENTS

The fieldwork was conducted whilst one of the authors (A.C.) was in receipt of a NERC framework award at the Department of Geography, University College London. The assistance with gamma-ray measurements by M. Charlton and helpful comments from I. Odeh are greatly appreciated. The authors are grateful to P. Owens and R. Loughran for providing useful comments.

REFERENCES

Ahmed, S. and de Marsily, G. 1987. Comparison of geostatistical methods for estimating transmissivity using data on transmissivity and specific capacity. *Water Resources Research*, **23**(9), 1717–1737.

Atkinson, P. M., Webster, R. and Curran, P. J. 1994. Cokriging with airborne MSS imagery. *Remote Sensing Environment*, **50**, 335–345.

Bajracharya, R., Lal, R. and Kimble, J. M. 1998. Use of radioactive fallout cesium-137 to estimate soil erosion on three farms in west central Ohio. *Soil Science*, **163**(2), 133–142.

Basher, L. R., Matthews, K. M. and Zhi, L. 1995. Surface erosion assessment in the South Canterbury downlands, New Zealand using ^{137}Cs distribution. *Soil and Water Management and Conservation*, **33**, 787–803.

Bourennane, H., King, D., Chery, P. and Bruand, A. 1996. Improving the kriging of a soil variable using slope gradient as external drift. *European Journal of Soil Science*, **47**, 473–483.

Chappell, A. 1995. Geostatistical mapping and ordination analyses of ^{137}Cs-derived net soil flux in south-west Niger. PhD Thesis, University of London.

Chappell, A. 1998. Using remote sensing and geostatistics to map ^{137}Cs-derived net soil flux in south-west Niger. *Journal of Arid Environments*, **39**(3), 441–456.

Chappell, A. 1999. The limitations for measuring soil redistribution using ^{137}Cs in semi-arid environments. *Geomorphology*, **29**, 135–152.

Chappell, A. and Oliver, M. A. 1997. Geostatistical analysis of soil redistribution in SW Niger, West Africa. In Baafi, E. Y. and Schofield, N. A. (eds) *Quantitative Geology and Geostatistics*. Kluwer, Dordrecht, 961–972.

Chappell, A., Oliver, M. A., Warren, A., Agnew, C. T. and Charlton, M. 1996a. Examining the factors controlling the spatial scale of variation in soil redistribution processes from south-west Niger. In Anderson, M. G. and Brooks, S. M. (eds) *Advances in Hillslope Processes*. John Wiley, Chichester, 429–449.

Chappell, A., Oliver, M. A. and Warren, A. 1996b. Net soil flux derived from multivariate soil property classification, SW Niger: a quantified approach based on ^{137}Cs. In Buerkert, B., Allison, B. E. and von Oppen, M. (eds) *Wind Erosion in West Africa: Implications and Control Measures in a Millet-based Farming System*. Kluwer, Dordrecht, 69–85.

Chappell, A., Warren, A. Oliver, M. A. and Charlton, M. 1998. The utility of ^{137}Cs for measuring soil redistribution rates in south-west Niger. *Geoderma*, **81**(3–4), 313–338.

Chappell, A., Valentin, C., Warren, A., Noon, P., Charlton, M. and d'Herbes, J.-M. 1999. Testing the validity of upslope migration in banded vegetation from south-west Niger. *Catena*, **37**, 217–229.

Delfiner, P. 1976. Linear estimation of non stationary spatial phenomena. In Guarascio, M., David, M. and Huijbregts, C. (eds) *Advanced Geostatistics in the Mining Industry*. Reidel, Dordrecht, 49–68.

Delhomme, J. P. 1979. Spatial variability and uncertainty in groundwater flow parameters: a geostatistical approach. *Water Resources Research*, **15**, 269–280.

De Roo, A. P. J. 1991. The use of ^{137}Cs as a tracer in an erosion study in south Limburg (The Netherlands) and the influence of Chernobyl fallout. *Hydrological Processes*, **5**, 215–227.

Deutsch, C. V. and Journel, A. G. 1992. *GSLIB Geostatistical Software Library and User's Guide*. Oxford University Press, Oxford.

Drees, L. R., Manu, A. and Wilding, L. P. 1993. Characteristics of aeolian dusts in Niger, West Africa. *Geoderma*, **59**, 213–233.

Genstat 5 Committee 1992. *Genstat 5, Release 3, Reference Manual*. Oxford University Press, Oxford.

Golden Software Inc. 1990. *SURFER 4.0 Reference Manual*. Golden Software Inc., Golden, Colorado.

Isaaks, E. H. and Srivastava, M. R. 1989. *An Introduction to Applied Geostatistics*. Oxford University Press, Oxford.

Journel, A. G. and Huijbregts, Ch. J. 1978. *Mining Geostatistics*. Academic Press, London.

Kitanidis, P. K. 1983. Statistical estimation of polynomial generalized covariance functions and hydrologic applications. *Water Resources Research*, **19**(4), 909–921.

Knotters, M., Brus, D. J. and Oude Voshaar, J. H. 1995. A comparison of kriging, co-kriging and kriging combined with regression for spatial interpolation of horizon depth with censored observations. *Geoderma*, **67**, 227–246.

Kulander, L. and Strömquist, L. 1989. Exploring the use of top-soil ^{137}Cs content as an indicator of sediment transfer rates in a small Lesotho catchment. *Zeitschrift für Geomorphologie*, **33**, 455–462.

Lam, F. and Barrett, J. D. 1992. Modeling lumber strength spatial variation using trend removal and kriging analyses. *Wood Science Technology*, **26**, 369–381.

Laslett, G. M. 1994. Kriging and splines: an empirical comparison of their predictive performance in some applications. *Journal of the American Statistical Association*, **89**, 391–409.

Laslett, G. M. and McBratney, A. B. 1990. Estimation and implications of instrumental drift, random measurement error and nugget variance of soil attributes – a case study for soil pH. *Journal of Soil Science*, **41**, 451–471.

Laslett, G. M., McBratney, A. B., Pahl, P. J. and Hutchinson, M. F. 1987. Comparison of several spatial prediction methods for soil pH. *Journal of Soil Science*, **38**, 325–341.

Leenaers, H., Okx, J. P. and Burrough, P. A. 1990. Employing elevation data for efficient mapping of soil pollution on floodplains. *Soil Use and Management*, **6**(3), 105–114.

Loughran, R. J., Campbell, B. L., Elliott, G. L., Cummings, D. and Shelly, D. J. 1989. A caesium-137 sediment hillslope model with tests from south-eastern Australia. *Zeitschrift für Geomorphologie*, **33**, 233–250.

Loughran, R. J., Elliott, G. L., Campbell, B. L., Curtis, S. J., Cummings, D. and Shelly, D. J. 1993. Estimation of erosion using the radionuclide caesium-137 in three diverse areas in eastern Australia. *Applied Geography*, **13**, 169–188.

Manu, A., Geiger, S. C., Pfordresher, A., Taylor-Powell, E., Mahamane, S., Ouattara, M., Isaaka, M., Salou, M., Juo, A. S. R., Puentes, R. and Wilding, L. P. 1991. *Integrated Management of Agricultural Watersheds (IMAW): Characterisation of a Research Site near Hamdallaye, Niger*. TropSoils Bulletin No. 91–03. Soil Management CRSP North Carolina State University, USA; S & CSD Texas A & M University, USA; INRAN Niamey, Niger; USAID Niamey, Niger.

Matheron, M. A. 1971. *The Theory of Regionalized Variables and its Applications*. Cahiers du Centre de Morphologie Mathématique de Fountainebleau no. 5.

Matheron, M. A. 1973. The intrinsic random functions and their applications. *Advances in Applied Probability*, **5**, 439–468.

McBratney, A. B. and Webster, R. 1983. Optimal interpolation and isarithmic mapping of soil properties. V. Co-regionalisation and multiple sampling strategy. *Journal of Soil Science*, **34**, 137–162.

McBratney, A. B., Hart, G. A. and McGarry, D. 1991. The use of region partitioning to improve the representation of geostatistically mapped soil attributes. *Journal of Soil Science*, **42**, 513–532.

Odeh, I. O. A., McBratney, A. B. and Chittleborough, D. J. 1994. Spatial prediction of soil properties from landform attributes derived from a digital elevation model. *Geoderma*, **63**, 197–214.

Odeh, I. O. A., McBratney, A. B. and Chittleborough, D. J. 1995. Further results on prediction of soil properties from terrain attributes: heterotopic cokriging and regression-kriging. *Geoderma*, **67**, 215–226.

Olea, R. A. 1975. *Optimum Mapping Techniques Using Regionalized Variable Theory*. Series on Spatial Analysis, no. 2. Kansas Geological Survey, Lawrence.

Olea, R. A. 1977. *Measuring Spatial Dependence with Semi-Variograms*. Series on Spatial Analysis no. 3. Kansas Geological Survey, Lawrence.

Oliver, M. A. 1987. Geostatistics and its applications to soil science. *Soil Use and Management*, **3**, 8–20.

Oliver, M., Webster, R. and Gerrard, J. 1989a. Geostatistics in physical geography. Part I: theory. *Transactions of the Institute of British Geographers*, **14**, 259–269.

Oliver, M., Webster, R. and Gerrard, J. 1989b. Geostatistics in physical geography. Part II: applications. *Transactions of the Institute of British Geographers*, **14**, 270–286.

Owens, P. N. and Walling, D. E. 1996. Spatial variability of caesium-137 inventories at reference sites: an example from two contrasting sites in England and Zimbabwe. *Applied Radiation Isotopes*, **47**(7), 699–707.

Pannatier, Y. 1997. MS-Windows programs for exploratory variography and variogram modeling in 2D. *Computers and Geosciences*, **13**, 23–50.

Quine, T. A. and Walling, D. E. 1991. Rates of soil erosion on arable fields in Britain: quantitative data from caesium-137 measurements. *Soil Use and Management*, 7(4), 169–176.

Ritchie, J. C. and McHenry, J. R. 1990. Application of radioactive fallout cesium-137 for measuring soil erosion and sediment accumulation rates and patterns: A review. *Journal of Environmental Quality*, **19**, 215–233.

Stein, A., Statisky, I. G. and Bouma, J. 1991. Simulation of moisture deficits and areal interpolation by Universal Cokriging. *Water Resources Research*, **27**(8), 1963–1973.

Sutherland, R. A. 1991. Examination of caesium-137 areal activities in control (uneroded) location. *Soil Technology*, **4**(1), 33–50.

Sutherland, R. A. 1994. Spatial variability of ^{137}Cs and the influence of sampling on estimates of sediment redistribution. *Catena*, **21**, 57–71.

Sutherland, R. A. and de Jong, E. 1990. Estimation of sediment redistribution within agricultural fields using caesium-137, Crystal Springs, Saskatchewan, Canada. *Applied Geography*, **10**, 205–221.

Voltz, M. and Webster, R. 1990. A comparison of kriging, cubic splines and classification for predicting soil properties from sample information. *Journal of Soil Science*, **41**, 473–490.

Walling, D. E. and Quine, T. A. 1991. The use of caesium-137 to investigate soil erosion on arable fields in the UK – potential applications and limitations. *Journal of Soil Science*, **42**, 146–165.

Walling, D. E. and Quine, T. A. 1993. The use of caesium-137 measurements in soil erosion surveys. In Bogen, J., Walling, D. E. and Day, T. (eds) *Erosion and Sediment Transport Monitoring Programmes in River Basins*. IAHS Publication **210**. IAHS Press, Wallingford. 143–152.

Webster, R. and Oliver, M. A. 1990. *Statistical Methods in Soil and Land Resource Survey*. Oxford University Press, Oxford.

Webster, R. and Oliver, M. A. 1992. Sample adequately to estimate variograms of soil properties. *Journal of Soil Science*, **43**, 177–192.

Yates, S. R. and Yates, M. V. 1989. *Geostatistics for Waste Management: A User's Manual for the GEOPACK (version 1.0) Geostatistical Software System*. EPA US EPA report no. 600/8-90/004.

Zhang, R., Warrick, A. W. and Myers, D. E. 1992. Improvement of the prediction of soil particle size fractions using spectral properties. *Geoderma*, **52**, 223–234.

12 The Origin of Sediment in Field Drainage Water

I. A. J. HARDY,[1] A. D. CARTER,[2] P. B. LEEDS-HARRISON,[3] I. D. L. FOSTER[4] and R. M. SANDERS[4]

[1]Environmental Sciences, Aventis CropScience UK Ltd, Ongar, UK
[2]ADAS Rosemaund, Preston Wynne, UK
[3]School of Agriculture, Food and the Environment, Cranfield University, UK
[4]Centre for Environmental Research and Consultancy, NES (Geography), Coventry University, UK

INTRODUCTION

In order to control the losses of pesticides to surface waters it is necessary to understand all potential mechanisms by which they move from the soil surface to the water body. Compound properties can play an important role in controlling the amount of pesticide lost through leaching, but generic loss mechanisms, such as preferential flow, can be an overriding factor in determining the level and pattern of losses of widely differing compounds in drainage water (Brown et al., 1995; Jones et al., 1995).

In the UK, the installation of field drainage systems on clay soils is required to ensure that the growing of winter crops, especially cereals, is viable (Trafford and Massey, 1975). These drainage systems are designed to remove excess rainwater rapidly from the soil system and thus have a significant potential to transport pesticides (Brown et al., 1995).

Considerable research effort has been put into understanding preferential flow mechanisms and aqueous phase transport of pesticides to field drains (Brown et al., 1995; Jones et al., 1995; Johnson et al., 1996; Williams et al., 1996) but, until recently, little research has been carried out on the importance of transport sorbed to sediments in drainflow. Lysimeter studies on colloidal phase transport have shown that significant amounts (up to 12 percent) of pesticides can be associated with dissolved organic matter as a mechanism for the loss of compounds of widely differing properties, although it was concluded that transport sorbed to particulates was negligible (Worrall et al., 1999).

In a study carried out at Boarded Barns Farm in Essex (Hardy, 1997; Hardy et al., 2000), it was concluded that the major mechanism for the transport of pesticides from the field was sorbed to sediments carried in the field drainage water. Sediments in tile drainage water have also been implicated as a significant loss mechanism for

Tracers in Geomorphology. Edited by Ian D. L. Foster. © 2000 John Wiley & Sons Ltd.

phosphates from agricultural land (Dils and Heathwaite, 1996, 1999; Haygarth *et al.*, 1998). In order to gauge the importance of these mechanisms and losses, and ultimately to develop farm management practices to reduce them, it is essential that the processes involved in the origin and movement of sediment are well understood.

The aim of this chapter is to identify the origins of the sediments observed in drainflow from the field drains during the experiments, with its effect on the transport of pesticides being considered in more detail separately (Hardy *et al.*, 2000).

SITE LOCATION AND DESCRIPTION

The field site was located at Boarded Barns Farm at Ongar, Essex, UK which is managed by Rhône-Poulenc Agriculture Ltd (Figure 12.1). The farm is typical of many cereal-growing farms in East Anglia, with a rotation of winter cereals and oil seed rape or beans as break crops.

The 1994/5 season described in this chapter was the first year of winter wheat following a break crop of oil seed rape. The field, Stocklands North (Figure 12.1(d)), has tile drains (plastic slotted pipe) installed 80 cm below the surface at 40 m spacing (installed 1989/90), backfilled with aggregate to within 40 cm of the soil surface. Lateral mole drains are present 60 cm below the surface at 2.8 m spacing, and drain into the permeable fill above field drains. The mole drains were installed in September 1994. Two major soil types are present (Figure 12.1(c)), which are characterised as Melford Series and Stretham Series, with separate drainage outfalls for the two areas (Figure 12.1(d)). The Melford Series soil (argillic brown earth, subclass 5.71; Hodge *et al.*, 1984) was characterised as having a clay content of 20 percent in the Ap horizon (0–29 cm) rising to 50 percent in the lower Bt horizons (45–82 cm), with 1.2 and 0.5 percent organic carbon in these horizons respectively. The profile was non-calcareous throughout; the calcium carbonate equivalents for the two horizons being 1.0 and 2.4 g kg^{-1} respectively. The Stretham Series soil (brown calcareous earth, subclass 5.11; Hodge *et al.*, 1984) was characterised as having a clay content of 31 percent in the Ap horizon (0–32 cm) and 34 percent in the lower Bw_3 horizon (52–68 cm), with 1.5 and 0.4 percent organic carbon in these horizons respectively. The profile was calcareous throughout, with the calcium carbonate equivalents for the two horizons being 41.7 and 477.8 g kg^{-1} respectively.

Stocklands North field was ploughed to 15 cm depth (20 September 1994), power harrowed (27 September 1994), and drilled with Mercia wheat at a rate of 180 kg ha^{-1} (29 September 1994). The only application of autumn herbicides was on 23 November 1994, when Javelin Gold® was applied delivering 2500 g ha^{-1} isoproturon and 50 g ha^{-1} diflufenican.

Plot Isolation, Designation and Monitoring

The drainage system (tile drains and moles) was installed to hydrologically isolate the two soil types, with separate collected drainage outfalls for the two areas (Figure 12.1(d)). The two separated areas of the field (Stocklands North) are designated Stocklands North West (Stretham Series, calcareous) and Stocklands North East

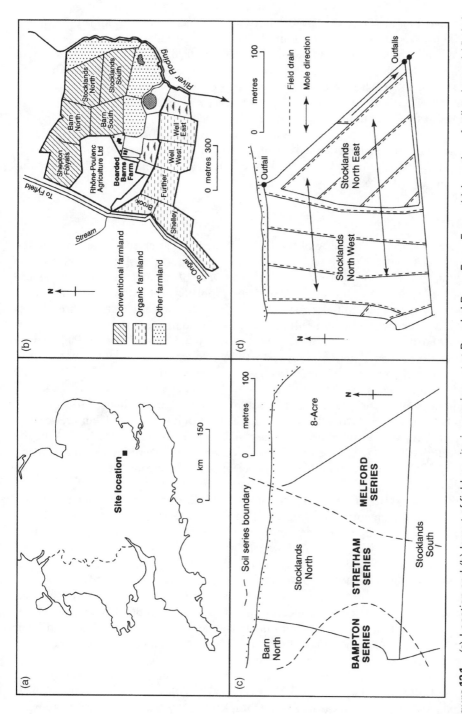

Figure 12.1 (a) Location and (b) layout of field monitoring equipment at Boarded Barns Farm, Essex. (c) Location of soil series and (d) drainage system layout for Stocklands North

(Melford Series, non-calcareous). Mole drains were installed so as not to cross the soil boundary, thus isolating the two drainage areas, which were determined as 1.6 and 1.45 ha for Stocklands North East and North West respectively.

V-notch mini-weirs were fitted to the collector field drainage outfalls for both Stocklands North West and Stocklands North East to record flow during drainage events. A float switch within the weir box triggered autosamplers to start sampling drainflow when the flow rate went above $0.05 l s^{-1}$, these being set to collect 24 samples at fixed 30 minute intervals when triggered.

Following each drainage event, samples were removed from the field as soon as practicable, with pH and conductivity measured immediately on return to the laboratory. All collected samples were subsampled (20 ml) for ion analysis, with the remainder being stored frozen at $-20\,^{\circ}C$.

Sediment Analysis

Problems with reproducibility of the sediment load measurements were found in the early sets of data. A potential source of error occurs in drying the filters prior to weighing due to the small sample volumes used (10 ml) and the low subsequent weights (<0.05 g) of sediment. To overcome problems encountered with the sealed disposable type filters initially tried, a Gelman filter unit in which the disc filter is replaced for each sample was used. This allowed for complete drying of the filter both before and after filtering was carried out. Individual filter discs (0.45 μm PTFE) were placed into numbered glass vials and stored in a desiccator overnight to ensure they were dry prior to initial weighing. Each was then weighed (in the glass vial) and the weight recorded. The filter was placed into the holder and attached to a vacuum manifold. A 10 ml sample was then passed through the filter and allowed to air dry for 15 min before removal from the manifold. The filter unit was carefully unscrewed and the filter returned to its numbered glass vial, which was then placed back in the desiccator to dry (24 h) before being reweighed to determine sediment mass. Twenty-four hours was found to be sufficient, with reweighing after 48 h showing no significant difference from results obtained after 24 h of drying.

Following extraction from the sediment, diflufenican content was determined by gas chromatography/mass spectrometry (Hardy et al., 2000).

Samples of drainflow sediment and topsoil samples from Stocklands North field were analysed for their clay mineralogy. Dried samples were suspended in water, subjected to ultrasonic dispersal and finally centrifuged to separate the clay fraction. Analysis was carried out using a Siemens D-5000 X-ray diffractometer.

A bulked sample of drainflow sediment collected from the weir box and five soil samples, corresponding to units identified in the soil profile at Stocklands North East, were subjected to a range of mineral magnetic measurements. Low (χ_{LF}) and High (χ_{HF}) frequency magnetic susceptibility and a range of remanent magnetic properties were measured (cf. Thompson and Oldfield, 1986; Foster and Walling, 1994; Foster et al., 1998). Parameters are defined, except for ARM, with measurement methods and units in Foster et al. (see Table 12.2 in Chapter 12). Anhysteretic remanent magnetisation (ARM) was also measured on all samples in a Molspin fluxgate magnetometer after an anhysteretic remanence was imparted by smoothly ramping down a mains frequency alternating field of 0.1 T while the samples were subjected to a steady field of 40 μT. ARM units are mAm^{2}kg^{-1}. Selected samples were used to derive isothermal remanent magnetisation (IRM) acquisition curves. These were measured in a Molspin fluxgate magnetometer after subjecting them to an increasing magnetic field, up to a maximum of 0.8 T, in a Molspin pulse magnetiser. All samples were measured after oven drying at 40° C. Corrections for organic matter content were made by measuring weight loss on ignition at 450° C for 12 h. All magnetic measurements were made on soils sieved to < 2 mm and collected drainflow sediments.

In order to determine the impact of particle size on the magnetic signatures of source materials, samples were wet sieved into four particle size classes above 63 μm diameter. For samples below 63 μm diameter, samples were obtained by using settling columns following the methods described by Foster *et al.* (1998).

Rainfall Simulation Experiments

To investigate some of the sediment generation and transport effects within the topsoil (0–20 cm) in more detail, small soil cores were collected from Stocklands North East and North West and subjected to simulated rainfall using a rainfall simulator developed at Cranfield University (Marsh, F. J., pers. comm.). The small-scale simulator consists of a pumped water supply (needle feeds) with a gauze distributor, and has a vertical drop of 4.25 m. Droplets below 1400 μm diameter achieve terminal velocity (Marsh, F. J., pers. comm.). Soil cores were subjected to simulated rainfall at a rate of 3.4 mm h^{-1}, this being a rate known to generate sediment movement to field drains in the field experiments (Figures 12.6 and 12.7). Experiments were conducted on 20 cm diameter × 20 cm deep intact topsoil cores taken from Stocklands North East and Stocklands North West (using PVC tubes inserted into the ground which were then excavated), to which a steel retaining gauze and funnel (for leachate collection) were attached.

The cores were subjected to simulated rainfall at 3.4 mm h^{-1} for 4 h, with the leachate collected over time in 20 ml glass vials.

RESULTS

Hydrology

Hydrological measurements (soil water pressure and water table depth) for Stocklands North West indicated the presence of a water table at 60 cm depth during mid-January–late March 1995 (Hardy, 1997). Drainflow did not commence until the heavy persistent rainfall of late January 1995 and continued until early April 1995, with responses to rainfall events and an almost continual low background baseflow. Significant drainflow events were mainly concentrated in the late January 1995 period (maximum flow rate of 1.31 s^{-1}), with the last rainfall triggered event occurring on 7 March 1995. Recession time constant analysis (Dougherty *et al.*, 1995) of the drainflow hydrographs indicate a slow response, suggesting matrix flow as the dominant mechanism of water movement (Hardy, 1997).

Hydrological measurements (soil water pressure and water table depth) for Stock-lands North East showed a freely draining profile, with only very transient perched water tables during January 1995 (Hardy, 1997). Significant drainflow did not com-mence until the heavy persistent rainfall of late January 1995 (except for three small events in December 1994) and continued until late March 1995, with drainflow being very responsive to rainfall events. A maximum flow rate of $1.8\,l\,s^{-1}$ was recorded, with the last drainflow event occurring on 29 March 1995. Recession time constant analysis (Dougherty *et al.*, 1995) of the drainflow hydrographs indicates a rapid response, suggesting preferential flow as the dominant mechanism of water movement (Hardy, 1997).

A drainflow autosample event for Stocklands North West (maximum flow $0.61\,s^{-1}$) was triggered on day 60 (22 January 1995) following a rainfall event of 9.7 mm, with samples collected at 30 min intervals. Figure 12.2 shows the 24 samples collected during the flow event, illustrating that drainflow remained sediment-free. Drainflow from Stocklands North West remained sediment-free throughout the season.

A drainflow autosample event for Stocklands North East (maximum flow $0.721\,s^{-1}$) was triggered on day 57 (19 January 1995) following a rainfall event of 17.5 mm, with samples collected at 30 min intervals. A striking feature seen during the event was the appearance of sediment in the drainflow, as illustrated in Figure 12.3, which shows the 24 drainflow samples collected. The contrast with Stocklands North West can be seen by comparing Figures 12.2 and 12.3 and the weir outfall photographs taken on 26 January 1995 for Stocklands North West and Stocklands North East (Figures 12.4

Figure 12.2 Photograph showing autosampler run samples from Stocklands North West event, 22 January 1995

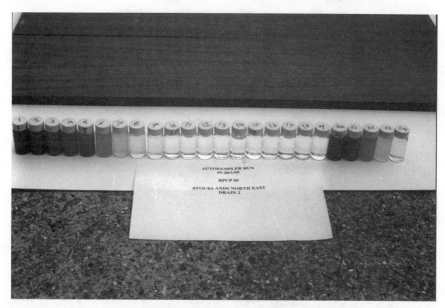

Figure 12.3 Photograph showing autosampler run samples from Stocklands North East event, 19 January 1995

Figure 12.4 Photograph showing drainage outflow from Stocklands North West weir, 26 January 1995 (Stocklands North East weir in background)

Figure 12.5 Photograph showing drainage outflow from Stocklands North East weir, 26 January 1995

and 12.5, respectively). Figure 12.4 shows the outflow from Stocklands North West (with Stocklands North East seen in the background), and Figure 12.5 shows the Stocklands North East outflow with a large sediment load. Figure 12.6 shows the sediment load, flow rate and diflufenican chemograph for the 24 samples collected during the event. For the Stocklands North East event of 19 January 1995, sediment loss was estimated at 6 kg (3.7 kg ha^{-1} equivalent for the 1.6 ha drained area), with a maximum measured sediment concentration of 5.1 g l^{-1}.

Figure 12.7 shows the hourly rainfall data (Meteorological Office) for Stansted Airport (15 km north of Ongar) for the time period of the event, with the first two rainfall events seen in Figure 12.7 being captured in the drainflow autosample pattern and hydrograph (Figure 12.6). The third event was recorded in the hydrograph but not captured by the autosampler. Diflufenican was found to be entirely associated with the sediment, as would be expected for a compound with high adsorption coefficient ($K_{oc} = 2000$ ml g^{-1}), with a maximum concentration of 2.4 μg l^{-1} equivalents. For diflufenican all values are quoted as μg l^{-1} equivalents, this being the apparent concentration in the water although diflufenican remains sorbed to the suspended sediment fraction.

Similar temporal sediment load patterns were found in other events throughout the 1994/5 season (data not presented), with the last event for Stocklands North East on 26 March 1995 still showing sediment in drainflow (Hardy, 1997). Figure 12.5 shows the Stocklands North East drain outfall for an event on 29 January 1995 which was not sampled, but the photograph clearly shows the presence of sediment.

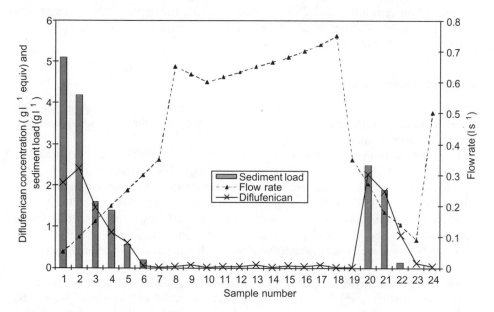

Figure 12.6 Sediment load, diflufenican chemograph and flow rate charts for Stocklands North East event, 19 January 1995

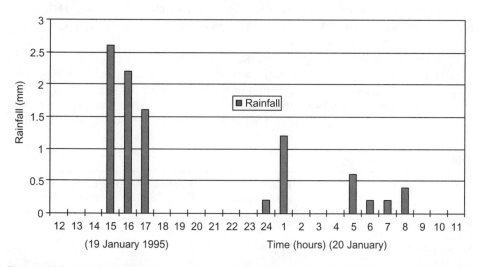

Figure 12.7 Hourly rainfall data at Stansted Airport, 19–20 January 1995. Reproduced by permission of the Meteorological Office

Rainfall Simulation Experiments

Leachate collected from the Stocklands North West cores remained clear throughout the simulated rainfall experiments, which is consistent with the field data. However, leachate from Stocklands North East cores contained significant amounts of sediment, thus confirming that the sediment seen in drainflow could be generated at the soil surface or within the 20 cm topsoil layer. During the experiments evidence was seen of generation of sediment on the soil surface for both soils, but only Stocklands North East showed movement through the core.

Sediment Properties

The clay mineralogy comparisons of the drainflow sediment with topsoil samples from Stocklands North East and Norh West (Table 12.1) showed good agreement although subsoil samples were not analysed.

Selected mineral magnetic signatures of the Stocklands North East soil profile are given in Figure 12.8(a). χ_{LF}, ARM, IRM$_{0.8T}$ and high field remanent magnetisation (HIRM) show clear discrimination between the upper 45 cm and the remainder of the soil profile. The distinction between topsoil and subsoil is less clearcut for the S ratio profile. IRM acquisition curves are plotted for the 0–29 cm (topsoil) and 82–105 cm (subsoil at drain depth) profile units and for the drainflow sediment sample in Figure 12.8(b). There is some evidence from these curves that the drainflow sediment exhibits characteristics which are more similar to topsoils than subsoils.

For χ_{LF}, ARM, IRM$_{0.8T}$ and HIRM, data derived from the mineral magnetic analysis of particle size fractions suggest that the bulk soil signature would be inadequate for modelling the relative contribution from topsoils and subsoils found in the drainflow sediment (Figure 12.9) since 98 percent of the drainflow sediment lies below 31 μm diameter. Below 31 μm (Class 6; Figure 12.9) both χ_{LF} and ARM show significant enhancement in topsoils whilst peak values for IRM$_{0.8T}$ and HIRM are recorded in Classes 7 and 7 & 8 respectively. Less enhancement in the finer particle size fractions is evident for subsoil samples.

In order to overcome the particle size problem in characterising source materials, a particle size enrichment/depletion ratio was calculated for χ_{LF}, ARM, IRM$_{0.8T}$ and HIRM using the particle size distribution of the bulk topsoil and subsoil samples and the mineral magnetic data of Figure 12.9. Adjusted source signatures were then used to model the relative proportion of topsoil and subsoil contained in the drainflow

Table 12.1 Clay mineralogy analysis

	Illite (%)	Interstratified illite–smectite (%)	Kaolinite (%)	Quartz (%)	Calcite (%)	Geothite (%)
Stocklands North West topsoil	33	33	31	2	1	< 1
Stocklands North East topsoil	29	39	30	2	–	< 1
Drain sediment	31	33	34	2	–	< 1

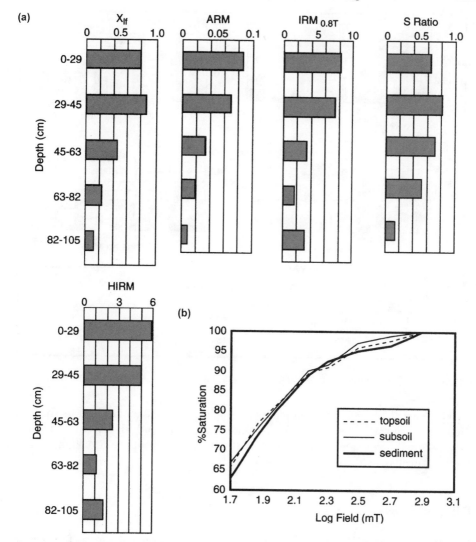

Figure 12.8 Mineral magnetic characterisation of the Stocklands North East soil profile and drainflow sediment. (a) Downprofile properties and (b) IRM acquisition curves

sediment using a two-, three- and four-component mixing model and a SIMPLEX algorithm (Microsoft Excel Solver Routine) as described by Walden *et al.* (1997).

The lowest proportion of topsoil in drainflow sediment (71%) is predicted by a two-component model using χ_{LF} and ARM, whilst the highest proportion of topsoil (91%) is predicted by all four parameters in combination. Whilst there is some uncertainty in the predictions derived by modelling source mixtures, probably relating to the methods used to characterise particle size enrichment/depletion ratios, the results appear to confirm the coincidence between the IRM acquisition curves of topsoils and drainflow sediment plotted in Figure 12.8(b). In the absence of detailed

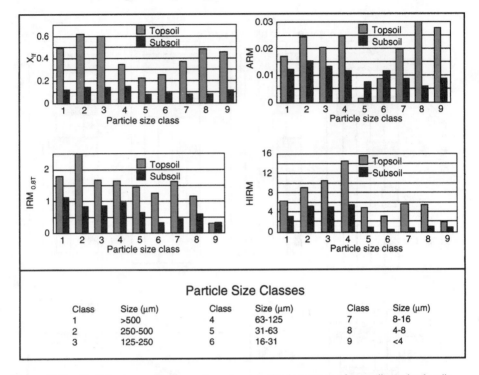

Figure 12.9 Particle size controls on mineral magnetic signatures of topsoils and subsoils

data on particle size controls on magnetic signatures, the data and results from this preliminary modelling exercise suggest that between 71 and 91 percent of the bulk drainflow sediment sample derived from the upper 45 cm of the Stocklands North East soil profile.

DISCUSSION

Evidence to support the hypothesis that the drainflow sediment was surface derived and generated by rainfall impact can be drawn from the above results and is discussed in detail below.

During the analysis of the drainflow samples, the presence of diflufenican sorbed to the sediments was surprising and gave an early indication as to their topsoil origin. Diflufenican is strongly sorbed to soil (K_{oc} 2000 ml g^{-1}), is known not to move vertically in solution and is typically confined to the top few centimetres of the soil profile (Main *et al.*, 1995). The appearance of diflufenican sorbed to drainflow sediment is therefore very strong evidence for the surface origins of the sediment.

The patterns of sediment occurrence in the drainflow appear to be correlated with the hourly rainfall patterns for the event (Figures 12.6 and 12.7), which is evidence that the sediment is generated by the impact of rainfall at the soil surface. However, the rainfall pattern correlation could still be accounted for by subsurface erosion effects. A subsurface origin is very unlikely, as Figure 12.6 shows the appearance of sediment in drainflow even at very low flow rates.

Particle size analysis of the drainflow sediment (Hardy, 1997) also shows it to be predominantly clay and silt fractions (D_{50} 3.4 μm, D_{90} 9.8 μm). If subsurface erosion was occurring, especially from the mole channels, a skew towards larger diameter silt and sand fractions would be expected.

The clay mineralogy of the sediment corresponds with that of the topsoil samples from Stocklands North East and North West (Table 12.1) and is therefore consistent with a topsoil origin. However, as subsoil samples were not taken for comparison no definite conclusions can be drawn.

Analysis of the mineral magnetic signatures of the soil profile and a bulked drainflow sample from Stocklands North East provide further evidence for the likely topsoil origin of the sediment collected in drainflow. Preliminary attempts to quantitatively model the source mixtures suggest that between 71 and 91 percent of the bulked drainflow sediment sample is topsoil in origin.

When subjected to simulated rainfall the cores from Stocklands North East clearly showed the presence of sediment in the leachate, thus indicating that sediment can be produced within the top 20 cm of the soil profile and transported through the core. There was evidence from the simulator studies that sediment was generated at the soil surface during the experiment, thus lending further weight to the surface origin hypothesis. Attempts were made to minimise any edge effects using petroleum jelly added between the tube and core, with the lack of sediment in the Stocklands North West leachates, despite being formed at the soil surface, indicating that this was successful.

The generation of sediment is a physical process which can be caused by the impact of rainfall on the soil surface or the erosion effect of water passing down the soil profile or through the mole channels. In most circumstances the linear velocity of water within mole channels is not sufficient to cause mole channel erosion (Hallard and Armstrong, 1992). Raindrop impact on tilled soil causes breakdown of soil aggregates (Reichert and Norton, 1994), with aggregate stability being probably the single most important property governing erodibility.

Soil erosion in relation to runoff is a two-phase process, consisting of detachment of individual soil particles and then their subsequent transport by water or wind (Kwaad, 1994). Amounts of soil detached by rainsplash and its transport in runoff was investigated, with erosion described as either detachment-limited or transport-limited. The severity of erosion depends on the quantity of material supplied by detachment and the capacity of the eroding agents (water or wind) to transport it. Raindrop impact is considered as the most important agent of soil particle detachment (Young and Wiersma, 1973), with rainfall rates of $> 2 \, \text{mm h}^{-1}$ being found to cause particulate detachment and transport in runoff (Harrod, 1994).

The lack of sediment in leachate from the Stocklands North West core, despite generation at the surface, is thought to be caused by flocculation of the sediment due to the calcareous nature of the soil. The concept of colloidal flocculation has been investigated by several researchers (McBride, 1994; Reichert and Norton, 1994), with the critical coagulation concentration (CCC) for divalent cations such as calcium being in the range 0.0005–0.002 M. For a calcareous soil at pH 7.8 the equilibrium calcium ion concentration would be > 0.003 M (McBride, 1994), which is therefore above the CCC and thus colloidal flocculation could be responsible for preventing the movement of sediment through the calcareous Stocklands North West soil.

CONCLUSIONS AND IMPLICATIONS

Erosion and surface runoff of sediment to surface waters has long been recognised as an important contamination mechanism for both phosphates and pesticides (Boardman and Evans, 1994; Harrod, 1994). However, until recently, the potential of sediment-associated phosphate (Haygarth *et al.*, 1998; Dils and Heathwaite, 1999) and pesticide (Sanders, 1997; Hardy *et al.*, 2000) transport in drainflow has not been fully recognised and researched.

The evidence presented above supports the hypothesis that the sediment found in drainflow is of surface origin and is probably generated by the impact of rainfall on the soil aggregates. Once generated at the surface, the transport of sediments to drainflow is postulated to occur via preferential flow routes (soil macropores). Losses of sediment to drainflow may also be a significant factor in the transport of pesticides (Hardy, 1997; Hardy *et al.*, 2000). Levels of both applied nutrients and pesticides will be potentially highest at the soil surface and so any mechanism capable of moving large amounts of sediment to surface waters may lead to significant concentrations being detected. Furthermore, the concentrations detected in drainflow sediments are likely to be higher than those found in topsoils as a result of the selective transport of only the fine silts and clays.

It is currently not known how widespread and important the phenomenon of drainflow sediment transport is in the UK. Other researchers have observed and commented on the presence of sediment in drainflow for a total of 15 soil series (Table 12.2). To date, there has been no comprehensive inventory of soils capable

Table 12.2 Soil series for which sediment load effects have been reported

Soil series (subclass)	Coverage (%)	Soil type	Reference
Melford (5.71)	0.07	Clay loam over clayey	This study
Ludford (5.71)	0.25	Fine loamy head	Hardy (1997)
Bromyard (5.71)	0.53	Stoneless silty clay loam	Williams *et al.* (1996); Foster and Chapman (pers. comm.)
Denchworth (7.12)	1.79	Fine loamy over clayey	Jones *et al.* (1995)
Dunkeswick (7.11)	1.74	Fine loamy over clayey	Brown *et al.* (1995)
Worcester (4.31)	0.48	Clayey	Sanders (1997)
Whimple (5.72)	0.80	Fine loamy	Sanders (1997)
Efford (5.71)	0.08	Fine loamy	Parsons (pers. comm.)
Bishampton (5.72)	0.14	Fine loamy	Parsons (pers. comm.)
Parkgate (8.41)	0.09	Stoneless silty gley	Parsons (pers. comm.)
Salop (7.11)	0.80	Stony clay loam	Dils and Heathwaite (1999)
Hodnet (5.72)	0.19	Stony clay loam	Dils and Heathwaite (1999)
Compton (8.13)	0.16	Clayey	Foster and Chapman (pers. comm.)
Middleton (5.72)	0.21	Fine silty	Foster and Chapman (pers. comm.)
Hallsworth	0.73	Clayey	Haygarth *et al.* (1998)

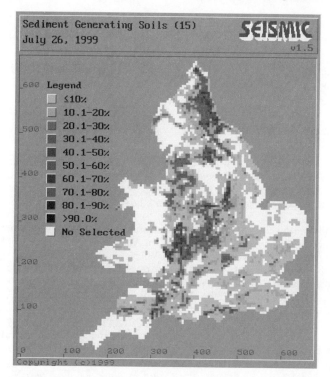

Figure 12.10 UK soil types with the potential for drainflow sediment generation. Reproduced by permission of Cranfield University

of generating drainflow sediment and many factors, such as pH and topsoil calcium carbonate content, need to be considered in defining the likelihood of this process occurring in a variety of UK soils.

Using the SEISMIC database (Hollis *et al.*, 1993), with the 15 soil series in Table 12.2 as input, a distribution map of soils with potential for drainflow sediment generation has been produced as a first approximation of the potential significance of this process in the UK (Figure 12.10). This map suggests that soils covering 8.0 percent of the total land area of England and Wales have this potential. A range of factors are likely to control the appearance of sediment in drainflow, with important requirements being preferential flow routes, soil surface instability (weathering), non-calcareous soils, rainfall intensity and cultivation practice. Further research is required in order to define precisely those conditions promoting sediment movement to land drains.

ACKNOWLEDGEMENTS

The work was funded by and carried out in association with Rhône-Poulenc Agriculture Ltd. The authors would like to thank Fiona Marsh for the development and set-up of the rainfall simulator and for help and advice on its use. The data in Figure 12.10 are derived from the SEISMIC database system, the copyright of Cranfield University and MAFF, and may not be reproduced without their permission. The rainfall data in Figure 12.7 are reproduced with kind permission of the Meteorological Office.

REFERENCES

Boardman, J. and Evans, R. 1994. Soil erosion in Britain: a review. In Rickson, R. J. (ed.) *Conserving Soil Resources – European Perspectives*. CAB International, Wallingford.

Brown, C. D., Hodgkinson, R. A., Rose, D. A., Syers, J. K. and Wilcockson, S. J. 1995. Movement of pesticide to surface waters from a heavy clay soil. *Pesticide Science*, **43**, 131–140.

Dils, R. M. and Heathwaite, A. L. 1996. Phosphorus fractionation in hillslope hydrological pathways contributing to agricultural runoff. In Anderson, M. G. and Brooks, S. M. (eds) *Advances in Hillslope Processes*. Vol. 1, John Wiley, Chichester, 229–251.

Dils, R. M. and Heathwaite, A. L. 1999. The controversial role of tile drainage in phosphorus export from agricultural land. *Water Science Technology*, **39**(12), 55–61.

Dougherty, E., Leeds-Harrison, P. B., Youngs, E. G. and Chamen, W. C. T. 1995. The influence of soil management on drainage hydrographs. *Soil Use and Management*, **11**, 177–182.

Foster, I. D. L. and Walling, D. E. 1994. Sediment yields and sources in the catchment of the Old Mill Reservoir, South Devon, UK over the past 50 years. *Hydrological Sciences Journal*, **39**, 347–368.

Foster, I. D. L., Lees, J. A., Owens, P. N. and Walling, D. E. 1998. Mineral magnetic characterisation of sediment sources from an analysis of lake and floodplain sediments in the catchments of the Old Mill reservoir and Slapton Ley, South Devon, UK. *Earth Surface Processes and Landforms*, **23**, 685–704.

Hallard, M. and Armstrong, A. C. 1992. Observations of water movement to and within mole drainage channels *Journal of Agricultural Engineering Research*, **52**, 309–315.

Hardy, I. A. J. 1997. Water quality from contrasting drained clay soils: the relative importance of sorbed and aqueous phase transport mechanisms. Unpublished PhD Thesis, Cranfield University.

Hardy, I. A. J., Carter, A. D. and Leeds-Harrison, P. B. 2000. The occurrence of sediment in field drainage water and its significance for the transport of pesticides. *Pesticide Science*, in press.

Harrod, T. R. 1994. Runoff, soil erosion and pesticide pollution in Cornwall. In Rickson, R. J. (ed.) *Conserving Soil Resources – European Perspectives*. CAB International, Wallingford.

Haygarth, P. M., Hepworth, L. and Jarvis, S. C. 1998. Forms of phosphorus transfer in hydrological pathways from soil under grazed grassland. *European Journal of Soil Science*, **49**, 65–72.

Hodge, C. A. H., Burton, R. G. O., Corbett, W. M., Evans, R. and Scale, R. S. 1984. Soils and their use in Eastern England. *Soil Survey of England and Wales Bulletin*, **13**.

Hollis, J. H., Hallet, S. H. and Keay, C. A. 1993. The development and application of an integrated database for modelling the environmental fate of herbicides. *Proceedings of Brighton Crop Protection Conference – Weeds*. British Crop Protection Council, Farnham, Surrey, 1355–1364.

Johnson, A. C., Haria, A. H., Bhardwarj, C. L., Williams, R. J. and Walker, A. 1996. Preferential flow pathways and their capacity to transport isoproturon in a structured clay soil. *Pesticide Science*, **48**, 225–237.

Jones, R. L., Harris, G. L., Catt, J. A., Bromilow, R. H., Mason, D. J. and Arnold, D. J. 1995. Management practices for reducing movement of pesticides to surface water in cracking clay soils. In *Proceedings of Brighton Crop Protection Conference – Weeds*. British Crop Protection Council, Farnham, Surrey, 489–498.

Kwaad, F. J. P. M. 1994. A splash delivery ratio to characterise soil erosion events. In Rickson, R. J. (ed.) *Conserving Soil Resources – European Perspectives*. CAB International, Wallingford.

Main, D. S., Kirkwood, R. C. and Gettinby, G. 1995. The fate of isoproturon and diflufenican in an agricultural sandy loam soil under laboratory and field conditions including simulation modelling. In *British Crop Protection Council Monograph No. 62: Pesticide Movement to Water*. BCPC, Farnham, Surrey, 287–294.

McBride, M. B. 1994. *Environmental Chemistry of Soils*. Oxford University Press, Oxford.

Reichert, J-M. and Norton, L. D. 1994. Aggregate stability and rain-impacted sheet erosion of air-dried and prewetted clayey surface soils under intense rain. *Soil Science*, **158**(3), 159–169.

Sanders, R. M. 1997. The characterisation of drainflow sediments from agricultural soils using magnetic, radionuclide and geochemical techniques. Unpublished MSc Thesis, Coventry University.

Thompson, R. and Oldfield, F. 1986. *Environmental Magnetism*. Allen & Unwin, London.

Trafford, B. D. and Massey, W. 1975. A design philosophy for heavy soils. *Field Drainage Experimental Unit Technical Bulletin 75/5*. MAFF, London.

Walden, J., Slattery, M. C. and Burt, T. P. 1997. Use of mineral magnetic measurements to fingerprint suspended sediment sources: approaches and techniques for data analysis. *Journal of Hydrology*, **202**, 353–372.

Williams, R. J., Brooke, D. N., Clare, R. W., Matthiessen, P. and Mitchel, R. D. J. 1996. *Rosemaund Pesticide Transport Study 1987–1993*. Report No. 129. Institute of Hydrology, Wallingford.

Worrall, F., Parker, A., Rae, J. E. and Johnson, A. C. 1999. A study of suspended and colloidal matter in the leachate from lysimeters. *Journal of Environmental Quality* **28**, 595–604.

Young, R. A. and Wiersma, J. L. 1973. The role of rainfall impact in soil detachment and transport measurements. *Water Resources Research*, **9**, 1629–1636.

13 Tracing Phosphorus Movement in Agricultural Soils

R. M. DILS[1] and A. L. HEATHWAITE[2]
[1]*Environment Agency, EHS National Centre, Wallingford, UK*
[2]*Department of Geography, University of Sheffield, UK*

INTRODUCTION

The introduction of agri-environment schemes in Europe and the US has arisen from increasing concern regarding environmental degradation of freshwaters, coupled with the demand for cost-effective sustainable agriculture (Withers and Jarvis, 1998). These reforms include a commitment to reduce nutrient losses (mainly nitrogen (N) and phosphorus (P)) from agricultural land. From a farming perspective, the loss of nutrients can reduce profitability as crop and animal productivity has to be maintained by importing nutrients in the form of commercial fertilisers and feedstuffs. From an environmental perspective, the loss of plant nutrients in biologically available forms can degrade water quality, by accelerating eutrophication in sensitive water bodies. A wide range of control mechanisms exist to reduce nutrient losses from agriculture which include both source management (soil nutrient status/manure management) and transport control (land management) (Sharpley and Withers, 1994; Heathwaite and Sharpley, 1999). These mechanisms may be operated through voluntary codes of practice, regulatory measures or financial instruments. However, in order to achieve this balance between agricultural and environmental sustainability, a thorough understanding of the mechanisms transporting nutrients from agricultural land to aquatic systems is essential.

In this chapter we examine how P losses from a small agricultural catchment in England can be traced using a combination of in-situ instruments to intercept P in mobile soil water, small-scale field experiments using solutes and dyes to monitor the rate of soil water movement, and stream water quality monitoring to examine the amounts and forms of P exported from the catchment. The research focuses on P rather than N, as P is generally the nutrient controlling primary productivity in surface freshwaters (Vollweider, 1968).

NUTRIENTS IN AGRICULTURAL AND AQUATIC SYSTEMS

Phosphorus and nitrogen are key plant nutrients in both agricultural and aquatic systems. However, the over-enrichment of surface waters with bioavailable nutrients

Tracers in Geomorphology. Edited by Ian D. L. Foster. © 2000 John Wiley & Sons Ltd.

may lead to undesirable ecological effects. This process, known as eutrophication, causes the excessive growth of macrophytes and micro-organisms, which can lead to invertebrate and fish mortalities through either de-oxygenation or the production of toxins by certain algal species. Nutrients are delivered to the aquatic environment from easily identifiable point sources (e.g. sewage effluent) or from more discrete diffuse sources (e.g. runoff from agricultural land). Their relative contribution varies both spatially and temporally, in response to factors including land use and management, climate, technological advances and environmental legislation. With stringent water quality standards being set for many large point-source nutrient discharges in response to the Urban Waste Water Treatment Directive (91/271/EEC), the relative contribution of diffuse nutrient sources (dominantly agriculture) to deteriorating water quality has increased over the last decade in the UK (Foy and Withers, 1995). Consequently, there is mounting pressure on the agricultural industry to identify the nature and extent of nutrient imbalances, and to determine the mechanisms of nutrient loss, so that management strategies to control the problem can be implemented.

Nutrient Transport Mechanisms

Nitrogen and phosphorus exhibit contrasting behaviour in agricultural soils with respect to both cycling and transport mechanisms. The phosphate anion (PO_4^{-3}) is strongly bound in the soil's solid phase and the solid-solution exchange is primarily controlled by chemical processes (mainly adsorption/desorption and dissolution/precipitation). In contrast, the nitrate anion (NO_3^-) is ineffectively retained by the soil, and transformations of nitrogen compounds are largely driven by biological processes (mainly mineralisation and nitrification). Consequently, whereas nitrate is readily leached through the soil profile or lost to the atmosphere as other forms of nitrogen (e.g. through ammonia volatilisation, denitrification or nitrification), phosphate losses in edge-of-field runoff are smaller and usually associated with the solid (particulate) phase, i.e. soil erosion. Although the majority of this soil-bound P is not immediately available for biological uptake, in the long term between 10 and 90 percent is potentially bioavailable through in-stream transformations including dissolution, desorption and mineralisation (Böstrom et al., 1988). Therefore, even though most P is lost in bound or insoluble forms in quantities that are agronomically and economically insignificant, these losses can cause the eutrophication of P-sensitive waters following transformation into bioavailable forms (Sharpley and Withers, 1994).

Tracing the pathways of phosphate movement in agricultural soils is more complex than for nitrate which can be treated as a conservative solute (Ball and Trudgill, 1995). [32]P radioisotope tracers have been used on a very small scale to trace phosphate movement. Ahuja et al. (1981), for example, conducted a study of rainfall–soil–runoff P interactions by using ^{32}P as a tracer and applying simulated rainfall to generate runoff from pre-wetted soil boxes. However, due to the short half-life of [32]P (approximately 13 days), the amount of radioactive P required to trace hillslope P movement over several months would exceed present national and EC health and safety guidelines. As there is no alternative, more stable P isotope, or an element that behaves in a similar manner to P, P is routinely traced in its natural state. This is achieved by intercepting all or part of the soil water flow and channelling it into a collection device. The flow velocity and relative contribution of different surface and subsurface hydrological pathways is subsequently determined using radioactive tracers, conservative solute tracers or fluorescent dyes (Atkinson, 1978).

A brief description of the current state of knowledge regarding P transport and tracing is given below.

Particulate P Transport

Sediment-bound P has been reported to be the dominant component of the total P load lost annually in runoff from several catchments in the UK (e.g. Heathwaite, 1997; Hodun and Burt, 1997; Catt et al., 1998). High erosion rates of arable soils and the strong affinity for phosphate ions by particulate material can explain this finding. On a shorter timescale, suspended sediment (SS) and particulate P (PP) concentrations are often reported to correlate with river discharge. For example, significant relationships were determined for particulate inorganic matter and particulate inorganic P ($r^2 = 0.89$), and particulate organic matter and particulate organic P ($r^2 = 0.93$) for the Gelbæk River which drains intensively farmed land (8.5 km^2) in Denmark (Kronvang, 1990, 1992).

Despite particulate material acting as a source, carrier and sink for P and many other environmental contaminants (e.g. pesticides, pathogens and heavy metals), the measurement of SS is rarely a constituent of routine water quality monitoring. Consequently, the amount of data pertaining to fluvial sediment transport rates and loads is limited (Foster and Walling, 1994). Several techniques have therefore been developed to trace the origin of fluvial sediment material. Foster and Walling (1994) used a combination of physical and chemical properties (bulk density, particle size, mineral magnetics, radiometry (^{137}Cs) and chemical analyses) to trace the movement of sediment at the catchment scale. Their work confirmed that the main sources of sediment in streamflow include eroded soil transported in surface and subsurface pathways, eroded stream-bank material, and re-suspended bed sediment.

Soluble P Transport

Few large-scale experiments tracing the movement of water and dissolved nutrients through the soil profile under natural field conditions have been reported. This is because they tend to be expensive and/or time-consuming, requiring permanently installed field equipment to measure soil moisture content and to sample soil water for nutrients. In the UK, notable field-scale experiments tracing soluble P movement include work undertaken on the plots of the Broadbalk continuous wheat experiment at Rothamsted (Heckrath et al., 1995), and 1 ha hillslope lysimeters that form part of the Rowden Drainage Experiment, Devon (Haygarth and Jarvis, 1996). An alternative approach to soil moisture instrumentation has been the application of conservative solutes such as bromide (Br$^-$) and chloride (Cl$^-$) to trace water movement in agricultural soils (e.g. Jury et al., 1982; Butters et al., 1989; Roth et al., 1991). Bromide is usually the preferred tracer due to low natural background concentrations compared to the more ubiquitous chloride ion that is present in crustal rock, manures and fertilisers. The toxicity of Br$^-$ to living organisms, however, has limited its use in large-scale field experiments.

METHODOLOGY

In this chapter we report how the combination of stream water quality monitoring data, soil water data, and data generated from solute and dye tracing experiments can be used to trace P loss from a small agricultural catchment. At the catchment scale, the relationship between (i) suspended solids and particulate P, and (ii) dissolved solids and dissolved P fractions in storm-flow was investigated to help determine the sources and pathways of P exported from the catchment. Phosphorus fractionation in different soil water pathways (overland flow, matrix flow, macropore flow and groundwater) was determined in order to explain the P losses observed in streamflow and drainflow. Simulated rainfall experiments using Br$^-$ and dye as tracers were used to quantify the rate of soil water movement and the relative contribution of different pathways to agricultural runoff.

Site Characteristics

The Pistern Hill catchment, which drains 120 ha of mixed agricultural lowland in south Derbyshire, UK, was selected for this study (National Grid Reference SK352197). Geologically the site is dominantly glacial till, Triassic marl and Triassic sandstone, and the soils are a complex patchwork of 10 different soil series dominated by the Hodnet and Salop Series (slightly stoney sandy loam or clay loam over moderately permeable clay). All land in the catchment is managed in accordance with Ministry of Agriculture Fisheries and Food fertiliser recommendations (MAFF, 1994). Consequently, the soil has a moderate plant-available P status, average values of 25.0 and 34.1 mg Olsen P kg^{-1} soil having been determined for grassland and arable soils, respectively (ADAS Soil P index 2/3). All fields in the catchment are under-drained by permanent tile drains at 0.7 m depth to reduce the spatial extent and duration of waterlogging, thereby increasing the length of the grazing season and reducing the problem of soil poaching and compaction. For a more detailed site description, see Dils and Heathwaite (1996, 1999) and Dils (1997).

Field Monitoring

Stream discharge at the catchment outlet was continuously monitored during the study period (December 1993 to March 1996). The stream water sampling programme combined (i) weekly manual sampling to measure baseline water quality, and (ii) intense automatic sampling during and after storm events to monitor short-term fluctuations. High-magnitude, infrequent storm events were deliberately selected for intense monitoring as it is well documented that the majority of annual P loss from small rural catchments occurs during low-frequency, high-intensity storm events (Haygarth and Jarvis, 1998). The monitored storm events were diverse in character and included the first autumnal storms, snow-melt periods and a summer storm.

During storm-runoff events, automatic water samplers were triggered to sample at fixed time intervals, ranging from 30 min during the quick-flow phase to 2 h as discharge returned to baseflow. Weekly water samples were analysed for P fractions, and storm samples were analysed for P fractions and sediment concentrations.

To trace the movement of P down the hillslope and through the soil profile, a series of soil water sampling instruments were installed in a grassland field adjacent to the catchment outlet. This field was selected for intense monitoring as land management practices, soil P status and the Hydrology of Soil Types (HOST) classification (Boorman *et al.*, 1995) were representative of other grassland fields in the catchment. Six sampling nests were located at 50 m intervals within two 150 m hillslope transects (Figure 13.1). Each sampling nest consisted of an overland flow trough (5.5 m² drainage area), three macropore samplers (installed at an angle of 45° and open at 0–15, 15–30 and 30–45 cm depths) based on a design by Simmons and Baker, 1993), a porous ceramic suction cup (20 cm depth, 0.3 bar pressure), and four lined boreholes (PVC pipes open at 50, 100, 150 and 200 cm depths). Further details on equipment design are given in Dils and Heathwaite (1996) and Dils (1997). Water samples were collected from the devices on 15 occasions during the monitoring period where possible. The water table depth was measured at 25 m intervals along the two hillslope transects using lined boreholes (perforated PVC pipes, 200 cm in length). To investigate the contribution of artificial subsurface drainage to P export from the field, an automatic water sampler was positioned at the base of the hillslope to sample discharge from a tile drain outlet. The sampler was activated by a float-switch, which was positioned in the main stream channel, and set to sample drainflow simultaneously with streamflow when storm runoff was generated. As continuous discharge measurements at the drain outlet could not be made, instantaneous discharge measurements were taken manually at the time of water quality sampling. All water samples were collected from the field-site within 12 h of generation.

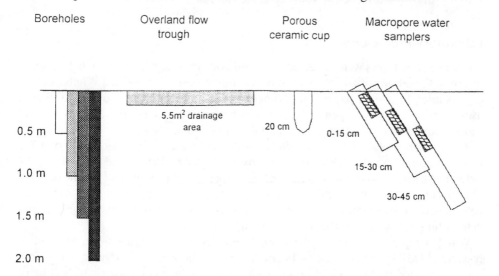

Figure 13.1 Soil water sampling equipment used in the experimental grassland field in the Pistern Hill catchment

Field Tests to Elucidate Pathways of Soil Water Movement

In the experimental grassland field, the infiltration capacity of the soil was measured using a double-ring constant head infiltrometer (Burt, 1978). Measurements were taken at the same mid-slope location in the experimental grassland field every three months during the study period and less frequently in areas where the infiltration capacity was potentially reduced by surface compaction, e.g. tractor wheelings, drinking trough.

The field bromide tracing experiment aimed to elucidate the pathways of subsurface soil water movement and provide an estimate of soil water velocity. This experiment involved the surface application of 20 g Br$^-$ (200 ml of 0.1 mg Br$^-$l^{-1} as aqueous KBr) over five 1 m^2 areas in the experimental grassland field, each containing a nest of water sampling devices. A sampling nest consisted of (i) three macropore water samplers at depths of 0–15 cm, 15–30 cm and 30–45 cm, and a suction cup sampler (20 cm, 60 bar pressure) to investigate Br$^-$ leaching losses under unsaturated conditions; and (ii) a lined borehole open at 200 cm to investigate losses to groundwater. Bromide-free water was applied at a rate of approximately 2 mm h^{-1} over a 24 h period to each of the five sampling areas. Soil water samples were collected prior to the experiment to obtain natural background measurements, and at four intervals after the Br$^-$ application (1, 4, 18 and 24 h) to monitor solute movement. Bromide tracing was complemented by a Methylene Blue dye trace study to stain soil water flow pathways and thus provide visual evidence of the operational routes. This involved the surface application of Methylene Blue dye (0.1 g l^{-1}) to a 0.5 m^2 area followed by repeated water additions to flush the dye into the soil profile. The dyed profile was exposed by digging a soil pit adjacent to the dyed area and removing vertical slices in 5 cm increments. The tracing experiments were performed in March 1995 when the antecedent soil moisture status was high (31 percent at 0–5 cm, declining to 20 percent at 100 cm) following a winter of above-average rainfall.

Laboratory Procedures

Stream water samples were collected in 600 ml polyethylene bottles in the field, and on return to the laboratory a 250 ml aliquot was filtered using a pre-dried Whatman HF/C glass microfibre filter paper (for sediment analysis), and a 10 ml aliquot was filtered using a pre-washed 0.45 μm millipore membrane filter paper (for P analysis). All water samples were stored in the dark at 4 °C prior to analysis.

The concentration of total suspended solids (SS) was determined by drying the residue retained on the HF/C glass microfibre paper at 105 °C and calculating the difference between (i) the final weight of the paper + dry residue, and (ii) the original filter paper weight (Allen, 1989). The concentration of total dissolved solids (DS) was determined by evaporating the filtrate from the suspended solids procedure and determining the weight of the residue (Allen, 1989).

Water samples were analysed for operationally defined forms of P. Molybdate reactive P (MRP), considered to be analogous to the soluble inorganic P fraction, was determined within 24 h of collection by the standard colorimetric technique of Murphy and Riley (1962). Total P (TP) and total dissolved P (TDP) were determined

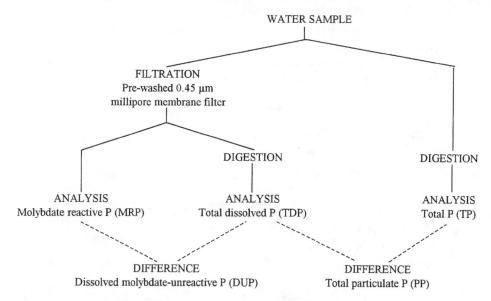

Figure 13.2 Physico-chemical fractionation of P in soil water and fresh water samples (adapted from Johnes and Heathwaite, 1992)

on unfiltered and filtered (0.45 μm) aliquots respectively using an alkaline persulphate microwave digestion technique with a detection limit of $12\,\mu g$ P l^{-1} (Johnes and Heathwaite, 1992). Particulate P (PP) and dissolved unreactive P (DUP) were determined by difference: PP = TP − TDP and DUP = TDP − MRP. The procedures employed for physico-chemical fractionation of P are summarised in Figure 13.2.

Soil water samples generated during the Br⁻ tracing experiments were collected in 100 ml polyethylene bottles in the field, and filtered on return to the laboratory using pre-dried Whatman HF/C glass microfibre paper (0.2 μm). Analysis was performed using a liquid chromatographic technique (Dionex, DX-100 Ion Chromatograph).

RESULTS AND DISCUSSION

P Losses in Drainflow and Streamflow during Storm Events

The relationship between the concentration of (i) suspended solids and PP, and (ii) dissolved solids and dissolved P fractions, was measured in streamflow and drainflow during Storm 1 (19–21 December 1995; Figure 13.3) and Storm 2 (9–14 February 1996; Figure 13.4). Storm 1 (13.0 mm rainfall) marked the transition of the groundwater system from a deep autumnal position to full winter recharge, whereas Storm 2 (15.0 mm rainfall) was a double-phased winter event preceded by high antecedent moisture conditions (27 percent at 0–7.5 cm depth), high stream discharge $(0.043\,m^3 s^{-1})$ and sub-zero temperatures.

In both streamflow and drainflow, elevated SS concentrations were recorded at the start of flow events which coincided with PP concentration peaks (Figures 13.3(a),(b)

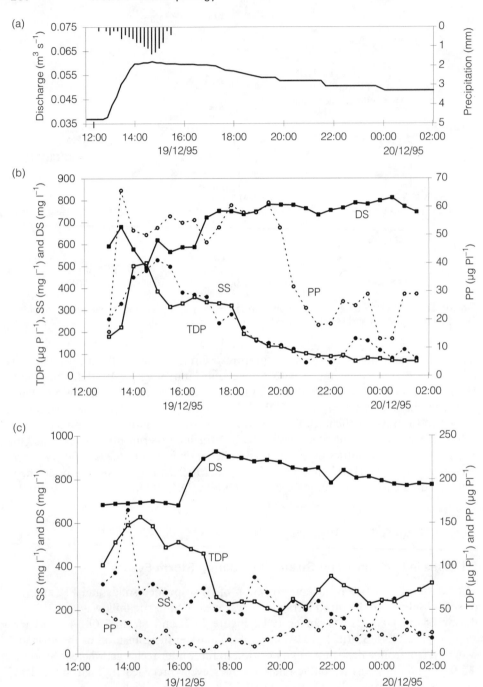

Figure 13.3 (a) Precipitation and discharge, (b) total dissolved P (TDP), particulate P (PP), suspended solids (SS) and dissolved solids (DS) response in streamflow, and (c) TDP, PP, SS and DS response in drainflow in the Pistern Hill catchment during Storm 1 (19–20 December 1995)

and 13.4(a),(b)). During Storm 2, for example, maximum streamflow values of 1180 mg SS 1^{-1} and 316 μg PP 1^{-1} coincided with peak flow on 12 February 1996. The immediacy of the quick-flow PP response in the stream initially suggested that PP originated in the stream system, i.e. re-suspension of sediment-bound P. However,

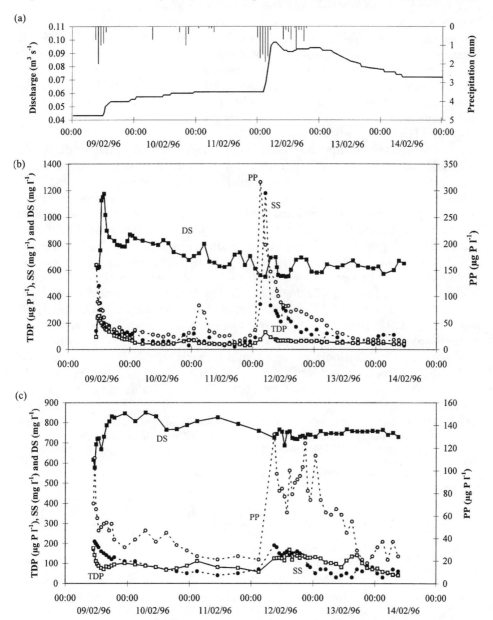

Figure 13.4 (a) Precipitation and discharge, (b) total dissolved P (TDP), particulate P (PP), suspended solids (SS) and dissolved solids (DS) response in streamflow, and (c) TDP, PP, SS and DS response in drainflow in the Pistern Hill catchment during Storm 2 (9–14 February 1996)

simultaneous monitoring of drain discharge showed that subsurface delivery of terrestrially derived particulate material was also a critical source of PP during the stream's quick-flow phase ($> 0.3\,\mathrm{g\,PP\,h^{-1}}$). Flushing of sediment-associated material was followed by a rapid decay in SS and PP concentrations during the recession limb of the quick-flow period to pre-storm levels (usually within 24 h). This response indicated that SS and PP were mobilised during the quick-flow phase when the energy of the system was high, but as rainfall ceased, discharge receded, and the readily mobile soil/sediment P store was exhausted, the ability of the system to transport particulate material decreased. Consequently, positive relationships were found between SS and PP concentrations, which were statistically significant in all cases (Table 13.1). In agricultural catchments in the UK, a similar relationship has been reported by (i) Heathwaite *et al.* (1990) and Fraser *et al.* (1999) in surface runoff; (ii) Bottcher *et al.* (1981) and Heckrath *et al.* (1995) in subsurface drainage water; and (iii) Walling and Peart (1980) and Heathwaite *et al.* (1989) in river storm-flow. The results of the Pistern Hill study and published studies confirm the importance of P attached to particulate material (soil and sediment) to the total P load exported from agricultural land.

In streamflow, DS concentrations were significantly greater than SS concentrations and consequently accounted for 75 percent and 83 percent of the total solids load for Storms 1 and 2 respectively (Figures 13.3(a) and 13.4(a)). Similarly, the DS load accounted for 69 percent and 88 percent of the drainflow total solids load in Storms 1 and 2, respectively (Figures 13.3(b) and 13.4(b)). Unlike the significant positive relationship found between SS and PP, the relationship between DS and soluble P fractions was inconsistent (Table 13.1). During Storm 2, statistically significant positive relationships were determined between the concentration of DS and MRP, and DS and TDP in streamflow, indicating that other solutes behaved similarly to the orthophosphate ion (Table 13.1). However, in Storm 1 streamflow and drainflow, significant negative relationships were found between soluble P fractions and DS (Table 13.1), in contrast to positive relationships determined between DS and other soluble ions (e.g. $Cl^-\,(p < 0.05)$, $NO_3^-\,(p < 0.01)$, $Ca^{2+}\,(p < 0.01)$ and $Fe^{3+}\,(p < 0.01)$).

Table 13.1 Correlation coefficients (r) determined between suspended solids and particulate P (PP), and dissolved solids and soluble P fractions (MRP, DUP and TDP) in streamflow and drainflow during Storm 1 (19–21 December 1995) and Storm 2 (9–14 February 1996) in the Pistern Hill catchment

| Solids | Storm | Flow | No. of samples | P fraction | | | |
				MRP	DUP	TDP	PP
Suspended solids	1	Stream	26	–	–	–	+0.67**
(SS)	1	Drain	48	–	–	–	+0.78**
	2	Stream	70	–	–	–	+0.70**
	2	Drain	54	–	–	–	+0.56**
Dissolved solids	1	Stream	26	−0.71**	−0.32	−0.81**	–
(DS)	1	Drain	48	−0.47**	−0.28*	−0.58**	–
	2	Stream	70	+0.43**	+0.14	+0.37**	–
	2	Drain	54	−0.59**	−0.55**	−0.51**	–

* $p < 0.05$, ** $p < 0.01$

The results can be explained by hydrological controls, e.g. mixing of different source waters; the low concentration of orthophosphate ions in drainflow (μgP l^{-1}) compared to the total ionic concentration (mg DS l^{-1}); and/or the dissimilar behaviour of the orthophosphate ion, as large or singly charged anions tend to form soluble salts with simple cations, e.g. NO_3^-, SO_4^{2-} and Cl^-, whereas triply charged anions form less soluble salts, e.g. PO_4^{3-} (Lindsay, 1979). These results suggest that the concentration of dissolved solids cannot be used to trace dissolved P.

Soil Water Transport in Hillslope Hydrological Pathways

The solute (Br$^-$) and Methylene Blue dye tracing experiments were used to determine the rate of subsurface water movement in grassland soils of the research catchment, and its partitioning between different hydrological pathways. The dye trace visually showed that the majority of flow was through soil micropores, but where new water intercepted old root and earthworm channels, the soil matrix was bypassed, providing evidence of macropore flow. These conduits were discrete and infrequently observed, but their dimensions (3–7 mm wide and present beyond 60 cm depth) suggested that they may provide an efficient transport route for water. From the Br$^-$ tracing experiments it was estimated that 9 percent of new water moved through the unsaturated zone via macropores (calculated as a percentage of the total amount of Br$^-$ applied intercepted by macropore samplers over the 24 h period). Transport through macropores was rapid, with Br$^-$ breakthrough at 45 cm occurring within an hour of application. Similarly, a pulse of Br$^-$ was measured in all suction cups located at 20 cm depth after an hour of rainfall, but the tortuous nature of matrix pathways meant that maximum Br$^-$ concentrations were delayed by a further 3 h (Figure 13.5).

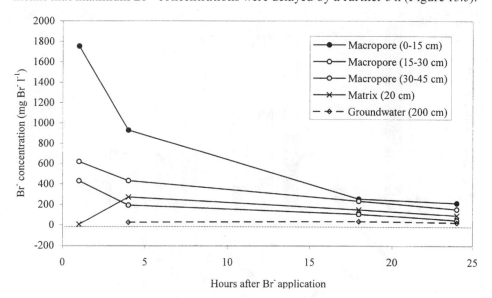

Figure 13.5 Comparison of Br$^-$ breakthrough in macropore, matrix and groundwater flow during tracing experiments performed in the experimental grassland field in the Pistern Hill catchment (mean values shown)

The slower transport of water through the soil matrix enhanced the dilution effect by increasing the contact time between the Br⁻ tracer and the newly infiltrating Br⁻-poor water. For example, peak matrix flow concentrations at 20 cm depth averaged 133 mg Br^-l^{-1} (range: 73–271 mg Br^-l^{-1}, $n = 5$) compared to 538 mg Br^-l^{-1} in macropore samplers open at 15–30 cm (range: 506–618 mg Br^-l^{-1}, $n = 3$).

The rapid transport of water through the soil profile meant that Br⁻ was detected in groundwater at 200 cm depth within 4 h of rainfall, although peak concentrations were delayed for a further 14 h. These results suggest that the macropore network was efficient at transporting soil water to groundwater or directly to the stream network. The rates of water transfer in different hydrological pathways are combined in Figure 13.5 which provides a summary of the study findings.

Phosphorus Transport in Hillslope Hydrological Pathways

The concentration and relative contribution of different P fractions transported in infiltration-excess overland flow, unsaturated subsurface flow (matrix and macropore) and groundwater was determined on 15 occasions during the study period (Figure 13.6). The aim was to obtain P signatures for surface and subsurface hydrological pathways, which in combination with the Br⁻ tracing experiment results, could explain the storm-flow P response observed in drainflow and streamflow. The P signatures exhibited significant temporal and spatial variability in response to both natural factors (e.g. climate, season and slope position) and land management practices (e.g. timing of manure applications). This variability is discussed in detail in Dils (1997) and Heathwaite and Dils (in press).

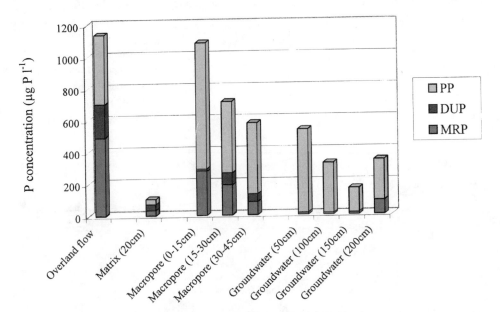

Figure 13.6   Phosphorus signatures of surface and subsurface hydrological pathways in the experimental grassland field in the Pistern Hill catchment (mean values shown)

Infiltration-excess overland flow was characterised by high P concentrations (mean: 1136 µgTP l^{-1}; range: 213–2483 µgTP l^{-1}, $n = 60$), dominated by the soluble phase (mean TDP:TP ratio of 0.62; mean MRP:TDP ratio of 0.70). This initially suggested that large quantities of potentially bioavailable P were rapidly transferred from the shallow zone of P-rich topsoil to mobile water moving overland and downslope towards the stream. However, although infiltration-excess overland flow was observed in all eight storm events, only small volumes of water were generated (4–40 ml event^{-1} m^{-2}). Infiltration capacity experiments confirmed that the rainfall intensity rarely exceeded the infiltration capacity of the soil in the grassland field. Infiltration rates measured in January, March and August ranged from 16 to 105 mm h^{-1} (mean: 65 mm h^{-1}; $n = 15$) whereas the maximum rainfall intensity measured was 4.4 mm h^{-1}. The operation of this pathway was restricted to areas where the natural infiltration capacity of the soil was reduced, primarily as a result of agricultural activities (e.g. farm gateway: 0.4 mm h^{-1}), or as a result of poaching and puddling by sheep grazing in winter.

Segregating the infiltration-excess overland flow data by hillslope position showed a statistically significant downslope increase in the mean concentration of all P fractions (Mann-Whitney U-test, $p < 0.05$), with values of 812, 1077 and 1455 µgTP l^{-1} determined for top-, mid- and base-slope positions, respectively ($n = 20$). This suggested that mobilised P was deposited at the hillslope base where it accumulated. Possibly of greater concern is the loss of this P during storm events when the water table is high and a variable source area (VSA) adjacent to the stream channel is hydrologically activated (10–20 m zone). Under these conditions, the generation of saturation-excess overland flow may transport soil P directly to the stream, or release adsorbed soil P under the reducing conditions associated with waterlogging. The influence of season and farming practices on P losses in overland flow is considered in detail in Heathwaite and Dils (in press).

High TP concentrations were measured in macropore flow throughout the monitoring period, with mean values of 1181, 717 and 578 µgTP l^{-1} determined at 0–15, 15–30 and 30–45 cm depths, respectively (range: 28–4838 µg TP l^{-1}, $n = 118$). Analysis of the data confirmed that this decrease in TP concentration with depth was statistically significant (Mann Whitney U-tests, all $p < 0.05$). At all three depths, macropore flow was dominated by PP, with the mean PP:TP ratio ranging from 0.68 at 0–15 cm to 0.75 at 30–45 cm (mean concentration: 842 µg PP l^{-1} (0–45 cm)). This may represent P-rich particulate material that has been dislodged from the ground surface and washed into well-defined cracks (Cooke, 1976), or P-rich particulate material that has been dislodged from the sides of soil aggregates and fissures by the turbid flow conditions within macropores. If the latter theory is correct, the decrease in macropore TP concentrations with depth may reflect the decrease in the amount of TP in soil with depth (320 mg TP kg^{-1} at 0–7.5 cm compared to 145 mg TP kg^{-1} at 40 cm depth). These findings provide preliminary evidence that macropores are an important transport pathway for PP in grassland soils. Despite the high PP:TP ratios measured in macropore flow, the transport of soluble P via this pathway should not be ignored. Concentrations of 10 µg l^{-1} for soluble inorganic P have been suggested as capable of promoting noxious algal growth in freshwaters (Vollenweider, 1968).

In contrast to P-rich macropore flow in the topsoil, matrix flow at 20 cm depth contained only c. 100 μg TP l^{-1}, and was dominated by soluble P fractions (mean TDP:TP ratio of 0.67). This statistically significant disparity in both the form and concentration of P transported in matrix flow compared to macropore flow (Mann-Whitney U-tests, $p < 0.05$), may result from the slower transport of soil water in tortuous pathways around soil aggregates. This may increase the contact time and degree of interaction between percolating water and soil adsorption sites, increasing the likelihood of phosphate sorption reactions in the topsoil. The difference in pore size between the macropore samplers (500 μm) and porous cup samplers (5 μm) may also contribute to the observed differences.

Phosphorus concentrations in groundwater are typically small (< 10 μg TP l^{-1}) as losses of surface-applied P are retained by P-poor subsoil with a high phosphate adsorption capacity (Lowrance *et al.*, 1984). However, high mean P concentrations were measured in groundwater in the research catchment decreasing from 500 μg TP l^{-1} at 50 cm to 200 μg TP l^{-1} at 150 cm, and increasing again to 360 μg TP l^{-1} at 200 cm. This pattern reflected soil P trends of an exponential decrease in P concentrations with depth to 150 cm and an increase below 150 cm due to the presence of P-rich marine clay bands in the Carboniferous shale parent material.

Linking P and Br⁻ Signatures

Combining the findings of the Br⁻ tracing experiment with the P signatures of hillslope hydrological pathways provided a means of tracing P movement through the soil profile. This approach was used to interpret the P response to storm events observed in tile drain discharge.

Tile drains in the research catchment were active for the majority of the year, except for 2–3 months during the summer when the soil was below field capacity. In the months when the drains were flowing, the form and flux of P in discharge was controlled by the soil moisture status in the catchment, which determined the hydrological links between P sources and the drain. Under dry, low-flow conditions (c. 0.5 l min^{-1}), TP concentrations in drainflow were low, with values ranging from 16 to 62 μg P l^{-1}, and were dominated by the soluble phase (mean TDP:TP ratio of 0.70). Based on the P signatures of subsurface hydrological pathways, the tile drains were largely fed by water percolating through the soil matrix. In contrast, drain discharge in excess of 10 l min^{-1} and large P loads (> 0.3 g TP h^{-1}) were recorded under storm-flow conditions. Figures 13.3(b) and 13.4(b) illustrate the dynamic nature of P export in drainflow during storm events, and demonstrate the importance of continuous, intense monitoring. The form of P discharged from the drain depended on the P source (e.g. topsoil, residual material within the drain, parent material, manure) and the hydrological status of the catchment which in turn determined which pathways were active. During Storm 2, for example, the initial snow-melt phase (9–11 February 1996) caused an increase in soluble P and DS concentrations as solute-rich snow-water reached the tile drains, probably via the soil matrix (Figure 13.4(b)). The small flush of sediment-associated PP is likely to represent fine soil particles which were dislodged by the freeze–thaw process and subsequently transported via macropores (biological or frost-induced) to the drain. The high PP losses measured in drainflow during the second phase of the event (12–14 February 1996) may have originated from the variable source area (saturated zone) which developed 10 m either side of the stream channel. Water-logging of soils is known to cause aggregates to expand and break-up (Brady, 1990). The PP load may also have been supplemented by ground-water sources as the water table rose and intercepted the drainage network. The gradual decline in PP concentrations over a 48 h period represented either a depletion of the P source as rainfall ceased, or the dilution of flow by large volumes of P-poor water, e.g. matrix flow.

CONCLUSIONS

The loss of the nutrient phosphorus from agricultural soils is of increasing national concern because of the undesirable ecological effects caused by the process of eutrophication in freshwaters. This loss may be as small as 5 percent of the total amount of P applied annually in the form of manures and fertilisers (Daniel et al., 1994). However, as there is no easy way of labelling P, tracing its source and movement through agricultural soils is difficult without disturbing the natural system. The total P load exported annually from agricultural catchments can be determined by flow-weighted water quality monitoring or predicted using modelling techniques such as export-coefficient modelling (Johnes, 1996). However, neither method provides any evidence of the processes that cause P to be released from the soil profile or the pathways linking the P source to the receiving water body. In terms of P management it is critical to understand both source and transport factors as all fields do not contribute equally to the P load entering a water body.

Identification and quantification of the processes and pathways transporting P through agricultural systems can be gained from small-scale, replicated experiments performed at the point, plot, hillslope and field scales. However, the key to tracing P loss from its source to the stream system is actually identifying a link between these spatial scales. In this research we used the P signature of hydrological pathways, and the high rates of PP loss associated with eroded soil and sediment, as hydrochemical methods for tracing P. We then used infiltration capacity experiments and solute and dye tracing experiments to elucidate the pathways and velocity of soil water movement.

Storm chemographs showed the importance of sediment-associated P transport, with significant positive relationships being determined between suspended solids and particulate P in streamflow and drainflow. This confirmed that the erosion of agricultural soils is a major diffuse source of PP. If this relationship is found to be statistically significant for a large number of agricultural catchments, routine monitoring of turbidity and suspended sediment could provide valuable information on losses of PP. The same positive relationship did not hold between dissolved solids and dissolved P fractions as phosphate is often a minor component of the total solids load. Identifying the route by which this P is delivered to the stream system was achieved by determining a P 'fingerprint' for surface and subsurface hillslope pathways. The high PP losses usually associated with infiltration-excess overland flow were not observed in the Pistern Hill catchment as the natural infiltration capacity of the soil was high (mean: 65 mm h^{-1}), and surface capping was restricted to small localised areas (e.g. farm gateways and feeding troughs). In this catchment, the high PP losses observed in streamflow during storm events are more likely to represent eroded soil particles (topsoil and subsoil) transported in subsurface pathways via a dense macropore network and field drains. The Br$^-$ tracing experiments showed that 9 percent of 'new' water was transported via macropores with a mean concentration of 842 μg PP l^{-1}, whereas the remaining 91 percent of infiltrating water was transported in slow matrix flow which had a mean concentration of 33 μg PP l^{-1}.

Two areas requiring further investigation emerged from this research project. Advancing our understanding of phosphorus loss from agricultural land requires an improved ability to link the process scale, the observation/measurement scale and the modelling/management scale. For example, from a management perspective it is crucial to be able to determine the effectiveness of P control measures implemented at the field scale (e.g. buffer strips) at the catchment scale, by using data collected at the soil profile and hillslope plot scales. Using a nested (integrated) catchment approach is the most likely way of progressing this issue. The second research area is improved methods for identifying the dominant sources of transported sediment within a catchment, the delivery routes and the deposition processes. This is particularly important in catchments where sediment-associated P plays a major role in governing the TP export. Mineral magnetics (Foster *et al.*, 1996) and ^{137}Cs (Owens *et al.*, 1997) are two specialised techniques that are currently being used for sediment tracing, but more routine, cheaper methods need developing.

The application of P fractionation as a technique for tracing P movement through agricultural soils may play an important role in the design of future agricultural management strategies. It potentially provides a means of linking research conducted at a range of spatial scales, particularly when used in conjunction with more traditional techniques such as solute and sediment tracing.

ACKNOWLEDGEMENT

This research was funded by the Ministry of Agriculture, Fisheries and Food, Award no. AE 8750, and completed in the Geography Department, University of Sheffield.

REFERENCES

Ahuja, L. R., Sharpley, A. N. Yamamoto, M. and Menzel, R. G. 1981. The depth of rainfall–runoff–soil interaction as determined by 32-P. *Water Resources Research*, **17**, 969–974.
Allen, S. E. 1989. *Chemical Analysis of Ecological Materials*. Blackwell Scientific, London.
Atkinson, T. C. 1978. Techniques for measuring sub-surface flow on hillslopes. In Kirby, M. J. (ed.) *Hillslope Hydrology*. John Wiley, Chichester, 73–117.
Ball, J. and Trudgill, S. T. 1995. Overview of solute modelling. In Trudgill, S. T. (ed.) *Solute Modelling in Catchment Systems*. John Wiley, Chichester, 3–56.
Boorman, D. B., Hollis, J. M. and Lilly, A. 1995. *Hydrology of Soil Types: A Hydrologically-based Classification of the Soils of the United Kingdom*. IH Report No. 26, Institute of Hydrology, Natural Environment Research Council, Wallingford.

Böstrom, B., Persson, G. and Broberg, B. 1988. Bioavailability of different phosphorus forms in freshwater systems. *Hydrobiologia*, **170**, 133–155.

Bottcher, A. B., Monke, E. J. and Huggins, L. F. 1981. Nutrient and sediment loadings from a subsurface drainage system. *Transactions of the American Society of Agricultural Engineers*, **24**, 221–226.

Brady, N. C. 1990. *The Nature and Properties of Soils* (10th edition). Collier Macmillan, New York.

Burt, T. P. 1978. Runoff processes in a small upland catchment with special reference to the role of hillslope hollows. Unpublished PhD Thesis, University of Bristol.

Butters, G. L., Jury, W. A. and Ernst, F. F. 1989. Field scale transport of bromide in an unsaturated soil. 1. Experimental methodology and results. *Water Resources Research*, **25**, 1575–1581.

Catt, J. A., Howse, K. R., Farina, R., Brockie, D., Todd, A., Chambers, B. J., Hodgkinson, R. A., Harris, G. L. and Quinton, J. N. 1998. Phosphorus losses from arable land. *Soil Use and Management*, **14**, 168–174.

Cooke, G. W. 1976. A review of the effects of agriculture on the chemical composition and quality of surface and underground waters. In *Agriculture and Water Quality*, Ministry of Agriculture, Fisheries and Food, Technical Bulletin 32, London, 5–73.

Daniel, T. C., Sharpley, A. N., Edwards D. R., Wedepohl, R. and Lemunyon, J. L. 1994. Minimising surface water eutrophication from agriculture by phosphorus management. *Journal of Soil and Water Conservation*, **49**, 30–38.

Dils, R. M. 1997. Phosphorus fractionation in hillslope hydrological pathways contributing to agricultural runoff. PhD Thesis, Department of Geography, Sheffield University.

Dils, R. M. and Heathwaite, A. L. 1996. Phosphorus fractionation in hillslope hydrological pathways contributing to agricultural runoff. In Brookes, S. and Anderson, M. G. (eds) *Advances in Hillslope Processes*. John Wiley, Chichester, 229–252.

Dils, R. M. and Heathwaite, A. L. 1999. The controversial role of tile drainage in phosphorus export from agricultural land. *Water Science and Technology*, **39**, 55–61.

Foster, I. D. L. and Walling, D. E. 1994. Using reservoir deposits to reconstruct changing sediment yields and sources in the catchment of the Old Mill reservoir, South Devon, UK, over the past 50 years. *Hydrological Sciences*, **39**, 347–368.

Foster, I. D. L., Owens, P. N. and Walling, D. E. 1996. Sediment yields and delivery in the catchments of Slapton Lower Ley, South Devon, UK. *Field Studies*, **8**, 629–661.

Foy, R. H. and Withers, P. J. A. 1995. The contribution of agricultural phosphorus to eutrophication. *Proceedings of the Fertiliser Society*, **365**.

Fraser, A. I., Harrod, T. R. and Haygarth, P. M. 1999. The effect of rainfall on soil erosion and particulate phosphorus transfer from arable soils. *Water Science and Technology*, **39**, 41–45.

Haygarth, P. M. and Jarvis, S. C. 1996. Pathways and forms of phosphorus losses from grazed grassland hillslopes. In Brookes, S. and Anderson, M. G. (eds) *Advances in Hillslope Processes*. John Wiley, Chichester, 283–294.

Haygarth, P. M. and Jarvis, S. C. 1998. Phosphorus transfer from agricultural soils described by a conceptual model. In *Proceedings of the 3rd International Conference on Diffuse Pollution*, Edinburgh, 31 August–4 September 1998, 2–5.

Heathwaite, A. L. 1997. Sources and pathways of phosphorus loss from agriculture. In Tunney, H., Carton, O. T., Brookes, P. C. and Johnston, A. E. (eds) *Phosphorus Loss from Soil to Water*. CAB International, Wallingford, 205–224.

Heathwaite, A. L. and Dils, R. M. In press. Characterising phosphorus loss in surface and subsurface hydrological pathways. *The Science of the Total Environment*.

Heathwaite, A. L. and Sharpley, A. N. 1999. Evaluating measures to control the impact of agricultural phosphorus on water quality. *Water Science and Technology*, **39**, 149–155.

Heathwaite, A. L., Burt, T. P. and Trudgill, S. T. 1989. Runoff, sediment, and solute delivery in agricultural drainage basins: a scale-dependent approach. *Regional Characterisation of Water Quality (Proceedings of the Baltimore Symposium, May 1989)*. IAHS Publication **182**, IAHS Press, Wallingford. 175–190.

Heathwaite, A. L., Burt, T. P. and Trudgill, S. T. 1990. The effect of land use on nitrogen, phosphorus and suspended sediment delivery to streams in a small catchment in southwest England. In Thornes, J. B. (ed.) *Vegetation and Erosion*. John Wiley, Chichester, 161–178.

Heckrath, G., Brookes, P. C., Poulton, P. R. and Goulding, K. W. T. 1995. Phosphorus leaching from soils containing different phosphorus concentrations in the Broadbalk experiment. *Journal of Environmental Quality*, **24**, 904–910.

Hodun, O. and Burt, T. P. 1997. Storm-event transport of phosphorus in the absence of surface runoff generation. In Tunney, H., Carton, O. T., Brookes, P. C. and Johnston, A. E. (eds) *Phosphorus Loss from Soil to Water*. CAB International, Wallingford, 377–378.

Johnes, P. J. 1996. Valuation and management of the impact of land use change on the nitrogen and phosphorus load delivered to surface waters: the export coefficient modelling approach. *Journal of Hydrology*, **183**, 323–349.

Johnes, P. J. and Heathwaite, A. L. 1992. A procedure for the simultaneous determination of total nitrogen and total phosphorus in freshwater samples using persulphate microwave digestion. *Water Research*, **26**, 1281–1287.

Jury, W. A., Stolzy, L. H. and Shouse, P. 1982. A field test of the transfer function model for predicting solute movement. *Water Resources Research*, **18**, 369–374.

Kronvang, B. 1990. Sediment-associated phosphorus transport from two intensively farmed catchment areas. In Boardman, J., Foster, I. D. L. and Dearing, J. A. (eds) *Soil Erosion on Agricultural Land*. John Wiley, Chichester, 313–330.

Kronvang, B. 1992. The export of particulate water, particulate phosphorus and dissolved phosphorus from two agricultural river basins: implications on estimating the non-point phosphorus load. *Water Research*, **26**, 1347–1358.

Lindsay, W. L. 1979. *Chemical Equilibria in Soils*. John Wiley, Chichester.

Lowrance, R., Todd, R. and Asmussen, L. 1984. Nutrient cycling in an agricultural watershed: I Phreatic movement. *Journal of Environmental Quality*, **13**, 22–27.

Ministry of Agriculture, Fisheries and Food. 1994. *Fertiliser Recommendations for Agricultural and Horticultural Crops*. Reference Book 209, HMSO, London.

Murphy, J. and Riley, J. P. 1962. A modified single solution method for the determination of phosphate in natural waters. *Analytica Chimica Acta*, **27**, 31–36.

Owens, P. N., Walling, D. E., He, Q. P., Shanahan, J. and Foster, I. D. L. 1997. The use of Caesium-137 measurements to establish a sediment budget for the Start Catchment, Devon. *Hydrological Sciences*, **42**, 405–423.

Roth, K., Jury, W. A., Flühler, H. and Attinger, W. 1991. Transport of chloride through an unsaturated field soil. *Water Resources Research*, **27**, 2533–2541.

Sharpley, A. N. and Withers, P. J. A. 1994. The environmentally-sound management of agricultural phosphorus. *Fertilizer Research*, **39**, 133–146.

Simmons, K. E. and Baker, D. E. 1993. A zero-tension sampler for the collection of soil water in macropore systems. *Journal of Environmental Quality*, **22**, 207–212.

Vollenweider, R. A. 1968. *Scientific Fundamentals of the Eutrophication of Lakes and Flowing Waters with Particular Reference to Nitrogen and Phosphorus as Factors in Eutrophication*. Publication No. DAS/SAI/68.27. Organization for Economic Cooperation and Developments, Directorate for Scientific Affairs, Paris.

Walling, D. E. and Peart, M. R. 1980. Some quality considerations in the study of human influence on sediment yields. In *Proceedings of the Helsinki Symposium, June 1980*. IAHS Publication **130**, IAHS Press, Wallingford. 293–302.

Withers, P. J. A. and Jarvis, S. C. 1998. Mitigation options for diffuse phosphorus loss to water. *Soil Use and Management*, **14**, 186–192.

Section 4

TRACING FLUVIAL SEDIMENTS

14 Uncertainty Estimation in Fingerprinting Suspended Sediment Sources

J. S. ROWAN,[1] P. GOODWILL[2] and S. W. FRANKS[3]
[1]*Department of Geography, University of Dundee, UK*
[2]*Environment Agency (Southern), Worthing, UK*
[3]*Department of Civil, Surveying and Environmental Engineering, University of Newcastle, Callaghan, Australia*

INTRODUCTION

A central issue in fluvial geomorphology is the identification of suspended sediment sources at the catchment scale. In some applications it may be important to establish the spatial location of supply sites, but more often the need is for information on the type of sources involved, i.e. differentiating between topsoil losses versus channel bank erosion. Such data are increasingly sought by catchment managers to implement control measures on erosion-sensitive areas and for the development and calibration of water quality and geomorphological models. Because the scope for direct measurements is limited, indirect estimation techniques, such as sediment source 'fingerprinting', offer the greatest potential (cf. Peart and Walling, 1986).

The fingerprinting procedure is relatively simple in principle in that the properties of suspended sediment are compared with the equivalent values of potential sources. A useful tracer should be able to differentiate between potential sources and exhibit conservative behaviour during erosion and fluvial transport (Foster and Walling, 1994). The ability of a tracer to distinguish sediment sources depends on the nature of the catchment. Geochemical properties reflect spatially variable catchment properties, such as soils and underlying geology, and so are valuable indicators of spatial provenance. In other situations, properties that are independent of lithology or soil types, such as radionuclides and mineral magnetic measurements, may be more appropriate (e.g. Yu and Oldfield, 1993; Wallbrink and Murray, 1993). Because no single tracer can reliably distinguish all potential sources it has been shown that composite signatures offer the most powerful approach to maximising source area discrimination (Walling *et al.*, 1993; Collins *et al.*, 1997a).

Fingerprinting techniques have been applied over a range of timescales spanning from the event-level (Slattery *et al.*, 1995) to extended reconstructions involving

Tracers in Geomorphology. Edited by Ian D. L. Foster. © 2000 John Wiley & Sons Ltd.

sediment sinks such as floodplains (Rowan *et al.*, 1999), reservoirs (Foster and Walling, 1994) and estuaries (Yu and Oldfield, 1989). However, while the range of fingerprinting properties and applications has grown, relatively limited attention has been paid to the quality of the statistical models developed (Lees, 1997). Walling *et al.* (1993) indicated that the main challenges for the technique were to extend the range of properties used in composite sediment signatures, to address data uncertainties in source type characterisation and to refine available multivariate mixing models. This chapter is principally concerned with the latter two issues and reports methods for appraising the discriminating power of different sediment tracers and exploration of the uncertainty inherent in the current generation of mixing models which are used to 'unmix' a suspended sediment sample to reveal its respective sources.

STUDY CATCHMENT

The fieldwork for this investigation was based in the catchment of Wyresdale Park Lake, situated 10 km south of Lancaster in north-west England (Figure 14.1). The catchment area of 1.3 km² ranges in elevation from 215 m to 70 m at the inflow to Wyresdale Park Lake. Mean annual precipitation is 1175 mm. The geology is dominated by Carboniferous (Namurian) grits, sandstones and shales overlain by till and head deposits. Soils are dominated by Belmont and Brickfield Associations (Ragg *et al.*, 1984). The former are peaty, gleyed podzols; the latter are loamey and clayey surface-water gleys. Current land use is woodland (4%), improved pasture (41%) and rough pasture (55%).

Tythe Barn Brook, the main channel system flowing into Wyresdale Park Lake, has been instrumented since 1993 with continuous 15 min interval logging of turbidity (Partech IR4Oc) and water stage (Keller pressure transducer) via an eight-bit Newlog data logger employing a V-F converter. A tipping bucket rain gauge was situated on top of Nicky Nook Fell. Bulk samples of suspended sediment were collected during flood events using a submersible pump and transported to the laboratory for decanting and filtration to recover sediment samples prior to freeze-drying (Goodwill *et al.*, 1995).

FIELD DATA COLLECTION AND TRACER SELECTION

Based on extensive field reconnaissance, five apparent sediment source groups were identified broadly conforming to land-use groups, namely rough grazing, improved pasture, roadside gullies, stream banks and valley-side colluvial sources fed by shallow mass movements. A total of 60 samples were collected as shown in Figure 14.1 involving between 10 and 20 samples per source group. All field samples were obtained using a short, plastic sampling corer to a depth of 5 cm. In order to remove bias associated with grain-size effects, only the < 63μm fraction was used (cf. Walling and Woodward, 1995; Collins *et al.*, 1997b).

This preliminary investigation used a range of tracing variables. The mineralogy of the sediment was analysed in term of major element composition, i.e. Si, Al, Fe, Ca,

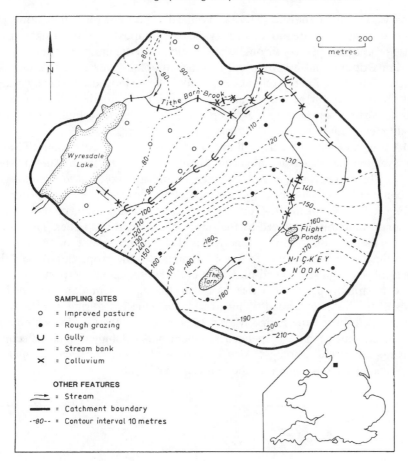

Figure 14.1   Catchment map illustrating sampling locations

Mg, Mn, Na, K, Ti and P using XRF (X-ray fluorescence spectrometry) on a Phillips PW 1400. Several elemental ratios (Fe/Mn and Fe/P) were additionally calculated. Frequency-dependent magnetic susceptibility measurements were made using a Bartington MS2 magnetic susceptibility system (χ_{LF} 0.47 kHz and χ_{HF} 4.7 kHz along with the ratio χ_{FD}), corrected for organic matter content determined via loss on ignition at 550 °C for 2 h.

SOURCE GROUP DISCRIMINATION AND SELECTION OF COMPOSITE TRACER PROPERTIES

In any formal consideration of predictive uncertainty it is clear that model performance will in part be determined by the quality of input data. To minimise over-parametrisation it is therefore necessary to define the fewest number of tracers capable of unequivocally discriminating between the previously defined source groups.

Multiple discriminant analysis (MDA), specifically the stepwise Mahalanbois procedure, offers an appropriate tool to identify the best composite sediment signature. The
stepwise procedure operates through the selection of sediment properties in order of
their relative discriminating power until all the variables have been either included in
the discriminant function or excluded as a result of being judged statistically incapable
of contributing any further significant information to the model. The best fit of the
data is associated with low within-group variability relative to high between-group
variability (cf. Hair *et al.*, 1987).

MDA was undertaken on the full initial array of 16 properties using SPSS, from
which seven were identified as statistically significant (X_{FD}, Fe/P, Fe/Mn, Ca, Mg, X_{LF}
and LOI). This subset classified all the source samples into their *a priori* defined groups
and accounted for 100 percent of the variance (cf. Collins *et al.*, 1997b). The territorial
map (Figure 14.2) shows that the source groups can be distinguished on the basis of the
discriminant functions produced by the subset of sediment properties. The plot indicates that the first discriminant function is the primary source of the differences
between the five source groups. Table 14.1 summarises the properties of the different
source groups recognised as 'end-members' in the subsequent fingerprinting analysis.

For comparative purposes Table 14.2 lists the equivalent data obtained from
analysis of the bulk suspended sediment samples collected during a series of flood
events sampled during the period March 1994–March 1995. Of obvious note is the
high SiO_2 content of both suspended sediment and catchment soils reflecting the
underlying sandstone lithology. Also of note is that the flood events spanned a
range of magnitudes as evident in the sediment yield totals.

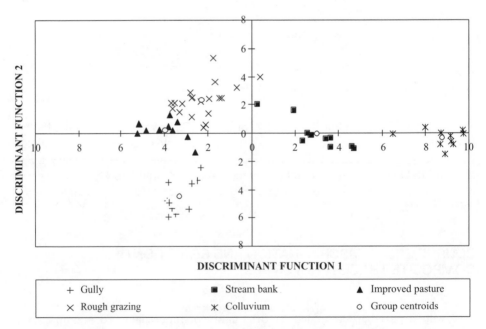

| + Gully | ■ Stream bank | ▲ Improved pasture |
| × Rough grazing | ✳ Colluvium | o Group centroids |

Figure 14.2   Territorial plot from MDA illustrating the successful discrimination of the five land-
use source groups based on a subset of the full range of sediment properties

Table 14.1   Selected tracer properties characterising Wyresdale sediment sources

Source group	LOI (%)	χ_{LF} ($\mu m^3 kg^{-1}$)	χ_{FD} (%)	Si (%)	Al (%)	Ca (%)	Mg (%)	Na (%)	K (%)	Ti (%)	Fe/Mn (−)	Fe/P (−)
Roadside gullies ($n = 10$)												
Mean	17.9	36.4	1.7	69.6	17.8	0.3	0.7	0.9	2.0	1.1	92.2	22.3
c.v. (%)	19.0	31.8	9.1	5.1	11.3	48.7	24.9	19.0	5.2	2.6	15.6	37.5
Stream bank ($n = 10$)												
Mean	13.5	10.7	3.1	67.5	18.4	0.3	1.1	0.7	2.3	1.1	112.1	37.0
c.v. (%)	14.7	14.4	2.5	5.86	9.2	14.3	23.0	14.2	9.4	4.0	28.5	37.6
Improved pasture ($n = 10$)												
Mean	21.6	22.1	2.2	71.1	17.6	0.7	0.6	0.7	1.8	1.2	70.8	10.9
c.v. (%)	10.4	12.4	7.3	2.3	4.7	15.3	12.4	21.5	4.6	4.1	19.6	20.2
Rough grazing ($n = 20$)												
Mean	27.9	18.1	2.7	79.5	13.0	0.4	0.4	0.6	1.7	1.3	36.8	5.9
c.v. (%)	28.4	15.5	11.4	5.5	23.0	42.9	38.5	19.6	6.5	9.4	48.2	63.2
Colluvial ($n = 10$)												
Mean	10.6	14.1	3.9	66.2	19.6	0.2	0.9	0.7	2.2	1.1	144.1	56.3
c.v. (%)	17.8	22.9	4.5	2.9	6.6	22.7	8.2	9.5	4.9	2.0	14.7	16.5

Table 14.2   Properties of suspended sediment for selected flood events (1994–1995)

Sample date	Sediment yield (kg)	LOI (%)	χ_{LF} ($\mu m^3 kg^{-1}$)	χ_{FD} (%)	Si (%)	Al (%)	Ca (%)	Mg (%)	Na (%)	K (%)	Ti (%)	Fe/Mn (−)	Fe/P (−)
04/03/94	1893	26.64	16.81	3.05	73.26	15.81	0.48	0.46	0.62	1.89	1.11	72.14	17.20
12/09/94	2007	24.02	17.01	2.94	75.54	14.75	0.52	0.48	0.65	1.86	1.16	69.29	20.12
02/10/94	2473	23.92	15.94	2.79	74.89	13.99	0.59	0.52	0.63	1.92	1.13	68.53	19.67
13/11/94	4794	12.59	11.51	3.48	67.85	18.29	0.21	0.99	0.67	2.19	1.09	125.86	41.68
31/01/95	11833	13.21	12.42	3.36	67.24	19.01	0.25	1.01	0.68	2.22	1.08	119.33	38.63
18/02/95	1363	25.84	15.85	2.85	76.23	15.67	0.48	0.53	0.64	1.90	1.18	61.15	18.61

TREATMENT OF UNCERTAINTY IN THE MODELLING PROCEDURE

To date, all previous attempts to estimate sediment source contributions have relied upon multivariate mixing models using constrained linear programming. Such models are over-determined if $m \geq n$ (where m is the number of tracer properties, and n is the number of distinct source groups) and thus require optimisation procedures to determine the relative contributions made by each source group. The model is represented as a series of simultaneous equations, which are solved by iteration. The robustness of the optimised solution is gauged by the 'goodness of fit' or likelihood function. In this case the likelihood measure was the explained variance (cf. Nash and Sutcliffe, 1970) which is termed the efficiency (E):

$$E = 1 - \frac{\sum_{i=1}^{m}(\hat{x}_i - x_i)^2}{\sum_{i=1}^{m}(x_i - \bar{x}_i)^2} \tag{14.1}$$

where m is the total number of sediment properties, $\bar{x}_i$ is the mean of the source group properties, and x_i and $\hat{x}_i$ are the measured and predicted values for property i (where $i = 1, 2, \cdots, m$). The linear mixing model used to calculate $\hat{x}_i$ is of the following form:

$$\hat{x}_i = \sum_{j=1}^{n}(a_{ij}b_j) \qquad \text{subject to the constraints} \qquad \begin{array}{c} \sum\limits_{j=1}^{n} b_j = 1 \\[2mm] 0 \le b_j \le 1 \end{array} \qquad (14.2)$$

where n is the number of source groups, a is the mean value of property i ($i = 1, 2 \cdots, m$) of source group j ($j = 1, 2, \cdots n$) and b is the contributory coefficient of source group j.

The validity of such optimisation procedures has, however, been questioned (cf. Beven, 1996), who highlighted problems of equifinality whereby the same likelihood value (i.e. goodness of fit) can be generated by a variety of different parameter combinations. In the case of the fingerprinting mixing models this indicates that a single optimised solution is only one of a subset of statistically equivalent solutions, yet each of these may give very different model results in terms of the contributory coefficients assigned to the component source groups.

To examine this uncertainty issue more fully a programme called SPARSSE was developed (systematic parameter space search engine). This programme systematically worked through the entire range of contributory coefficient combinations available and calculated corresponding likelihood values (efficiency). This was done by varying the contributory coefficients, b_j, of each source group from 0 to 1, using incremental b_j steps of 0.02, and mean values a, for each of seven sediment properties. In the case of the Wyresdale Park suspended sediment problem, the programme generated over 300 000 realisations, as shown in Figure 14.3. Each data point on any source group plot represents the likelihood value obtained for a given b_j value.

These plots evidence the high degree of uncertainty otherwise hidden within the model structure. In Figure 14.3 the small circles on each plot equate to the single optimised solution to the mixing model (i.e. likelihood $\equiv 0.99$). However, all the plots evidence broad flat upper surfaces (good model fits) derived from a wide range of contributory coefficients. For example, likelihood scores > 0.98 can be obtained virtually throughout the full range of $b_{\text{imp pasture}}$ (improved pasture) contributions. Greater sensitivity is exhibited by colluvial sources, i.e. when $b_{\text{colluvial}}$ lies between 0 and 0.3, the likelihood score exceeds 0.98, but beyond a value of 0.4 the likelihood score falls dramatically indicating a real limit to the contribution from this source during this event. The likelihood response surfaces vary greatly between different samples, but the important lesson is that the solution is essentially non-unique whereby the optimum model is only one of a wide range of possible and statistically 'acceptable' outcomes.

Beven and Binley (1992) previously recognised the equivalence or near-equivalence of different sets of parameters in the calibration of distributed hydrological models and formulated a statistical procedure known as GLUE (Generalised Likelihood Uncertainty Estimation). The GLUE procedure uses likelihoods as a possibilistic measure of model performance as compared to the observed behaviour of the system, and not in the restricted sense of maximum likelihood theory (Beven, 1993). The

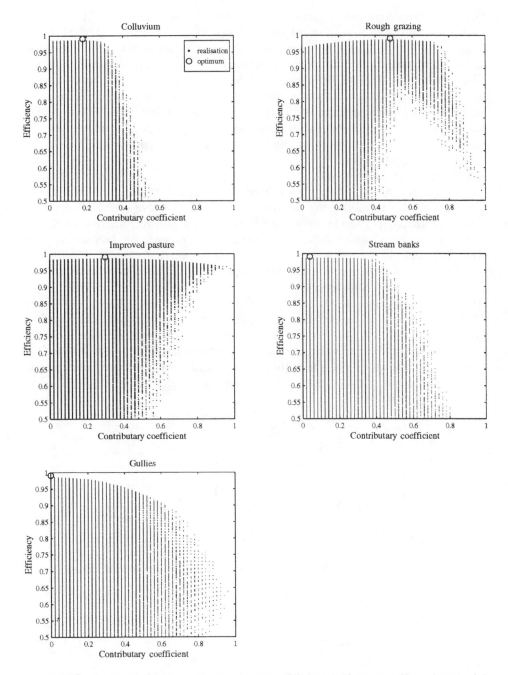

Figure 14.3 'Dotty plots' indicate likelihood scores (Efficiency) of the model based on the full range of potential source group contributions. This example shows the solution for the storm of 4 March 1994

GLUE methodology offers great potential because as a Bayesian Monte Carlo simulation approach it permits all uncertainties to be propagated through to the final model output. This can be inclusive of measurement errors, the uncertainties otherwise masked by the use of mean values in the composite signatures, and the effects of highly inter-correlated parameter sets. Ultimately the likelihoods may be updated and further constrained by direct comparison with observed data, or conditioned using *a priori* knowledge of the system, i.e. based on qualitative erosion audits. Similarly, the scheme can be used to evaluate the role of additional parameters, i.e. to incorporate weightings associated with individual tracers or particle size and organic matter enrichment effects.

The GLUE procedure has the facility to constrain the uncertainty bounds on predictions by the establishment of a threshold likelihood value, below which parameter sets are discarded as being 'non-behavioural'. The value at which this threshold is set is subjective (cf. Franks, 1998), and so to ensure rigour the threshold likelihood was here set at 0.98 which meant *c*. 15 000 model runs were still considered valid. Rescaling these likelihood values, such that they sum to 1, yields a distribution function of contributory coefficients for each respective source group (Figure 14.4). Likelihood weighted confidence intervals can then be obtained, e.g. 5th–95th percentiles, with the centroid corresponding to the modal behaviour (Beven and Binley, 1992). This procedure was applied to the six storm events sampled and the results are presented in Table 14.3. For comparative purposes the single optimised solution to each mixing problem was also included. Note that in several cases, i.e. for the flood event of 12 September 1994, the optimised solution for rough grazing and colluvial sources lies outside of the 90 percent uncertainty quantiles.

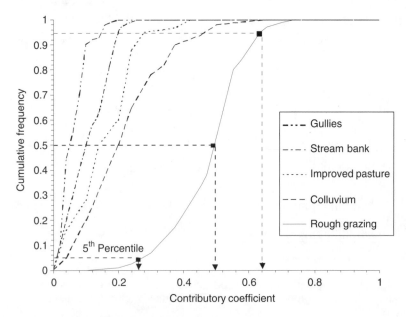

Figure 14.4 Rescaling the likelihood scores to obtain 95% quantiles for a multivariate mixing model, using the 4 March 1994 storm as an example

Table 14.3   Results of flngerprinting analysis comparing the optimised solution with the SPARSSE output

Storm event	Roadside gully		Channel banks		Improved pasture		Rough grazing		Colluvium	
	OS[a]	SPARSSE[b]	OS	SPARSSE	OS	SPARSSE	OS	SPARSSE	OS	SPARSSE
04/03/94	0	(0.00–0.14)	0.04	(0.02–0.30)	0.3	(0.03–0.48)	0.48	(0.25–0.64)	0.18	(0.02–0.22)
12/09/94	0	(0.00–0.20)	0	(0.00–0.30)	0	(0.00–0.39)	0.72	(0.32–0.67)	0.28	(0.00–0.23)
02/10/94	0	(0.00–0.17)	0.1	(0.00–0.31)	0	(0.00–0.36)	0.7	(0.41–0.69)	0.2	(0.00–0.23)
13/11/94	0	(0.00–0.23)	0.62	(0.00–0.54)	0	(0.00–0.22)	0	(0.00–0.16)	0.38	(0.35–0.77)
31/01/95	0.02	(0.00–0.28)	0.64	(0.02–0.65)	0.06	(0.00–0.25)	0	(0.00–0.19)	0.28	(0.21–0.69)
18/02/95	0	(0.00–0.09)	0.04	(0.10–0.25)	0	(0.00–0.15)	0.78	(0.65–0.78)	0.18	(0.01–0.19)

[a] OS = contributory coefficients for optimised solution.
[b] SPARSSE = 90% uncertainty quantiles.

Using the rescaled likelihood values, revised estimates of source group contributions can be produced with realistic confidence intervals. For example, for the event of 4 March 1999 the improved pasture contribution was estimated as $46 + 18 - 21$ percent (note the asymmetry of the error terms). The 90 percent quantiles ranged from ± 6 to ± 25 percent of the modal values for different source groups. Such large uncertainty boundaries have obvious implications for the usefulness of the fingerprinting scheme, particularly where these data may be important for catchment management purposes.

Figure 14.5 illustrates the results of applying the likelihood updated results to each of the flood events sampled. For the sake of clarity, modal values were plotted, but the

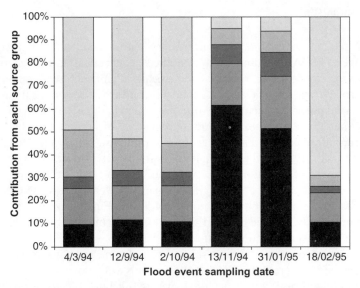

■ Colluvium ▨ Channel banks ■ Roadside gully ▢ Improved pasture ▢ Rough grazing

Figure 14.5   Illustration of inter-storm variability in sediment source contributions for selected events (1994–1995)

error bars omitted. The results evidence considerable inter-storm variability in terms of sediment sources. For four out of the six events the dominant suspended sediment source was from rough grazing, with a significant secondary component (14–25%) derived from improved pasture. This result is less surprising than might be expected considering the heavy stocking density of sheep and cattle (8 sheep ha^{-1}) and the tendency to overwinter livestock on the fell with attendant soil compaction and poaching problems.

The events of 13 November 1994 and 31 January 1995 differed greatly because they were dominated by the combination of colluvial and channel bank sources which made up 79 percent and 72 percent respectively of the total sediment load. Their relatively high sediment loads further distinguished these later events. The event of 31 January 1995 produced nearly 40 percent of the total annual load for that year and constituted one of the largest flood events since the beginning of this century (Rowan *et al.*, 2000). The contrasting behaviour of these different storms suggests that different hydrological pathways were operative. For the larger events involving stream bank and colluvial sources, it appears that activation thresholds must be exceeded before these sites make a significant contribution. The largest two events were characterised by dry antecedent conditions that enabled a build-up of sediment within the colluvial footslopes prior to the appropriate entrainment conditions being met. Both channel banks and colluvial sources are thus susceptible to grazing pressures and their relative importance for the overall sediment budget in part reflects disruption of the relative stability of the natural system due to overgrazing within the riparian zone.

CONCLUSIONS

This work set out to explore the uncertainty within conventional fingerprinting mixing models. MDA was used on a combined geochemical and magnetic susceptibility data set to define a subset of statistically significant sediment properties. However, the analysis revealed significant uncertainties exist within the mixing models which deserve explicit treatment. Likelihood weighted updates served to constrain the uncertainty but the confidence intervals remained ± 25 percent. Further work will evaluate other more powerful source group discriminators, e.g. radionuclides (Wallbrink and Murray, 1993) and additional mineral magnetic tracers (Foster and Walling, 1994). The use of qualitative observational data to further constrain the model will also be explored, i.e. by observing that during a given event no sediment moves gives the basis to exclude spurious model solutions. Finally it will be possible to investigate more fully the role of grain size and organic matter enrichment processes using Monte Carlo simulation techniques.

Control of soil erosion and diffuse sources of sediment are key elements in catchment management programmes (RCEP, 1996). Only with statistically robust predictions can the fingerprinting approach achieve its full potential to meet the applied scientific agenda of elucidating erosion dynamics and identifying erosion sensitive areas.

ACKNOWLEDGEMENTS

P.G. gratefully acknowledges the receipt of an IEBS Studentship from Lancaster University. We would also like to acknowledge the co-operation of local farmers over the duration of the project.

REFERENCES

Beven, K. J. 1993. Prophecy, reality and uncertainty in distributed hydrological modelling. *Advances in Water Resources*, **16**, 41–51.

Beven, K. J. 1996. Equifinality and uncertainty in geomorphological modelling. In Rhoads, B. L. and Thorn, C. E. (eds) *The Scientific Nature of Geomorphology*. John Wiley, Chichester, 289–313.

Beven, K. J. and Binley, A. M. 1992. The future of distributed models: calibration and predictive uncertainty. *Hydrological Processes*, **6**, 279–298.

Collins, A. L., Walling, D. E. and Leeks, G. J. L. 1997a. Use of the geochemical record preserved in floodplain deposits to reconstruct recent changes in river basin sediment sources. *Geomorphology*, **19**, 151–167.

Collins, A. L., Walling, D. E. and Leeks, G. J. L. 1997b. Source type ascription for fluvial suspended sediment based on a quantitative composite fingerprinting technique. *Catena*, **29**, 1–27.

Foster, I. D. L. and Walling, D. E. 1994. Using reservoir deposits to reconstruct changing sediment yields and sources in the catchment of the Old Mill Reservoir, South Devon, UK, over the past 50 years. *Hydrological Sciences Journal*, **39**, 347–368.

Franks, S. W. 1998. *An evaluation of single and multiple objective SVAT model conditioning schemes: parametric, predictive and extrapolative uncertainty*. Department of Civil, Surveying and Environmental Engineering Research Report No. 167.09 University of Newcastle, Australia.

Goodwill, P., Rowan, J. S. and Greco, M. 1995. Sediment routing through reservoirs, Wyresdale Park Reservoir, Lancashire, UK. *Physics and Chemistry of the Earth*, **20**, 183–190.

Hair, J. F., Anderson, R. E. and Tatham, R. L. 1987. *Multivariate Data Analysis*. Macmillan, New York.

Lees, J. A. 1997. Mineral magnetic properties of mixtures of environmental and synthetic materials: linear additivity and interaction effects. *Geophysical Journal International*, **131**, 335–346.

Nash, J. E. and Sutcliffe, J. V. 1970. River flow forecasting through conceptual models 1. A discussion of principles. *Journal of Hydrology*, **10**, 282–290.

Peart, M. R. and Walling, D. E. 1986. Fingerprinting sediment sources: the example of a small drainage basin in Devon, UK. In Hadley, R. F. (ed.) *Drainage Basin Sediment Delivery*. IAHS Publication **159**. IAHS Press, Wallingford. 41–55.

Ragg, J. M., Beard, G. R., George, H., Heaven, F. W., Hollis, J. M., Jones, R. J. A., Palmer, R. C., Reeve, M. J., Robson, J. D. and Whitfield W. A. D. 1984. *Soils and Their Use in Midland and Western England*. Soil Survey of England and Wales, Bulletin 12. Harpenden, Rothamsted.

RCEP (The Royal Commission on Environmental Pollution) 1996. *The Sustainable Use of Soil*. 19th Report of the Royal Commission on Environmental Pollution. HMSO, London.

Rowan, J. S., Price, L. E., Fawcett, C. P. and Young, P. C. 2000. Data-based mechanistic modelling as a tool in reconstructing reservoir sedimentation patterns in Wyresdale Park Lake, Lancashire, UK. *Physics and Chemistry of the Earth, in press*.

Rowan, J. S., Black, S. and Schell, C. 1999. Floodplain evolution and sediment provenance reconstructed from channel fill sequences: the upper Clyde Basin, Scotland. In Brown, A. G.

and Quine, T. A. (eds) *Fluvial Processes and Environmental Change*. John Wiley, Chichester, 223–240.

Slattery, M. C., Burt, T. P. and Walden, J. 1995. The application of mineral magnetic measurements to quantify within-storm variations in suspended sediment sources. In *Tracer Technologies for Hydrological Systems*. IAHS Publication **229**. IAHS Press, Wallingford. 143–151.

Wallbrink, P. J. and Murray, A. S. 1993. Use of fallout radionuclides as indicators of erosion processes. *Hydrological Processes*, **7**, 297–304.

Walling, D. E. and Woodward, J. C. 1992. Use of radiometric fingerprints to derive information on suspended sediment sources. In Bogen, J., Walling, D. E. and Day, T. (eds) *Erosion and Sediment Transport Monitoring Programmes in River Basins*. IAHS Publication **210**. IAHS Press, Wallingford, 153–164.

Walling, D. E. and Woodward, J. C. 1995. Tracing sources of suspended sediment in river basins: a case study of the River Culm, Devon, UK. *Marine Freshwater Research*, **46**, 327–336.

Walling, D. E., Woodward, J. C. and Nicholas, A. P. 1993. A multi-parameter approach to fingerprinting suspended sediment sources. In Bogen, J., Walling, D. E. and Day, T. (eds) *Erosion and Sediment Transport Monitoring Programmes*. IAHS Publication **215**. IAHS Press, Wallingford. 329–338.

Yu, L. and Oldfield, F. 1989. A multivariate mixing model for identifying sediment sources from magnetic measurements. *Quaternary Research*, **32**, 168–181.

Yu, L. and Oldfield, F. 1993. Quantitative sediment source ascription using magnetic measurements in a reservoir-catchment system near Nijar, S.E. Spain. *Earth Surface Processes and Landforms*, **18**, 441–454.

15 Tracing Fluvial Suspended Sediment Sources in the Catchment of the River Tweed, Scotland, using Composite Fingerprints and a Numerical Mixing Model

P. N. OWENS,[1] D. E. WALLING[1] and G. J. L. LEEKS[2]
[1]*Department of Geography, University of Exeter, UK*
[2]*Institute of Hydrology, Wallingford, UK*

INTRODUCTION

Information on the sources of the fine-grained sediment transported by rivers is needed for both geomorphological and management purposes. Such information is, for example, a key requirement in sediment budget investigations, in developing physically based distributed models of catchment sediment yield, and in interpreting sediment yield data in terms of landscape evolution. Furthermore, because there are many environmental problems linked to the presence of fine sediment in rivers, including siltation of watercourses and lakes and the transport of sediment-associated contaminants, information on fine sediment provenance is also required to underpin land management strategies and the implementation of appropriate control measures. There are several direct and indirect approaches to identifying sediment sources, and selection of the most appropriate method depends on the objective of the investigation and the nature of the study area. Over the last two decades, since the pioneering work of Wall and Wilding (1976), Oldfield *et al.* (1979), Walling *et al.* (1979) and others, the fingerprinting approach has proved to be one of the most successful and effective methods for identifying the sources of the fine-grained sediment transported by rivers (e.g. Peart and Walling, 1986, 1988; Walling and Woodward, 1992, 1995; Walling *et al.*, 1993; He and Owens, 1995; Slattery *et al.*, 1995; Walden *et al.*, 1997; Collins *et al.*, 1997a, 1998; Wallbrink *et al.*, 1998) and contemporary deposits of fine sediment in river channels (e.g. Caitcheon, 1993, 1998), on river floodplains (e.g. Owens *et al.*, 1997; Bottrill *et al.*, see Chapter 19) and in lacustrine environments (e.g. Foster and Walling, 1994).

Tracers in Geomorphology. Edited by Ian D. L. Foster. © 2000 John Wiley & Sons Ltd.

The fingerprinting approach possesses many advantages over more conventional approaches to sediment source tracing. These include its cost-effectiveness (i.e. it requires the collection and analysis of sediment and source material samples, rather than expensive long-term monitoring of sediment mobilised at a range of sites), and the potential to obtain information on both the geomorphological *type* (i.e. whether sediment is derived from the erosion of surface soil or from channel banks) and the *spatial location* (i.e. which subcatchment or geological zone) of the main sediment sources. Furthermore, and perhaps uniquely, the fingerprinting approach can also be used to determine recent historical changes in the sources of sediment deposited on river floodplains (e.g. Passmore and Macklin, 1994; Collins *et al.*, 1997b; Owens *et al.*, 1999a) and in estuaries, lakes and reservoirs (e.g. Stott, 1986; Yu and Oldfield, 1989, 1993; Foster and Walling, 1994; Foster *et al.*, 1998).

The fingerprinting approach is founded on the fact that comparison of the physical and chemical (and possibly biological) properties of the sediment with those of potential sources can be used to identify its source. This commonly involves two stages. First, a number of diagnostic properties capable of discriminating potential sediment sources are identified, either intuitively or statistically. Secondly, these fingerprint properties are subsequently used to estimate the relative importance of the various potential sources by comparing the property values for sediment samples with those for potential sources. Existing studies that have employed the fingerprinting approach to trace sediment sources have used both specific properties and composite fingerprints based on several properties. These fingerprint properties include mineral magnetic (e.g. Slattery *et al.*, 1995; Walden *et al.*, 1997) and radiometric parameters (e.g. Walling and Woodward, 1992, 1995; He and Owens, 1995; Wallbrink *et al.*, 1998). However, it is increasingly apparent that the use of composite fingerprints, incorporating several physical and chemical properties chosen from a range of different property subsets (i.e. mineral magnetic, radiometric, geochemical and organic), provide the most robust, reliable and comprehensive approach to source tracing. In order to maximise the effectiveness of the composite fingerprint, the diagnostic properties should be influenced by different environmental controls, and thus possess a degree of independence, so that in combination they afford a high degree of source discrimination. The use of multivariate mixing models (e.g. Yu and Oldfield, 1989, 1993; Walling *et al.*, 1993; He and Owens, 1995; Walling and Woodward, 1995; Collins *et al.*, 1997a, 1998), as opposed to simple qualitative comparisons of sediment and source properties (e.g. Walling and Woodward, 1992), have further improved the reliability and power of the fingerprinting approach.

This chapter reports the use of statistically verified composite fingerprints in association with a multivariate mixing model to establish the relative importance of different potential sources for the suspended sediment transported by the River Tweed in Scotland. Sediment sources have been defined in terms of two categories, representing source type (topsoil from woodland, pasture/moorland and cultivated areas, and channel bank/subsoil material) and spatial location (the four main geological/topographic zones). This investigation represents part of a larger project (cf. Walling *et al.*, 1998a,b, 1999a,b; Owens *et al.*, 1999a,b) within the NERC-funded Land–Ocean Interaction Study (LOIS) which was concerned with monitoring and modelling material fluxes from the land to the ocean within parts of the UK. Sediment sources were determined for the River Tweed at its tidal limit (at Norham), and for two of its main tributaries, the River Teviot and Ettrick Water, in order to complement other investigations undertaken within the Tweed basin by the authors within the framework of the LOIS project (e.g. Owens *et al.*, 1999b; Walling *et al.*, 1999b). These sites were also chosen because there were adjacent gauging stations for which discharge records were available; such data were an important requirement of the fingerprinting approach adopted in this investigation. Information on the sources of the suspended sediment transported by the River Tweed is also needed from a management perspective, because there is increasing concern over the detrimental effects of fine-grained sediment on salmonid habitats and the effects of recent and anticipated land-use changes within Scottish drainage basins (cf. Maitland *et al.*, 1994). To the authors' knowledge, the present study represents the first attempt to identify the sources of the fine-grained sediment transported in a large drainage basin in Scotland.

STUDY AREA AND METHODS

Study Area

Detailed information on the Tweed basin can be found in Robson *et al.* (1996) and Fox and Johnson (1997). In brief, most of the River Tweed basin is located in the Borders region of Scotland, but a small section in the south-east of the basin falls in Northumbria, England (Figure 15.1). The Tweed basin has a catchment area of $4390\,km^2$ above the Scottish Environment Protection Agency (SEPA) gauging station located at the tidal limit at Norham (National Grid Reference NT898477), while that

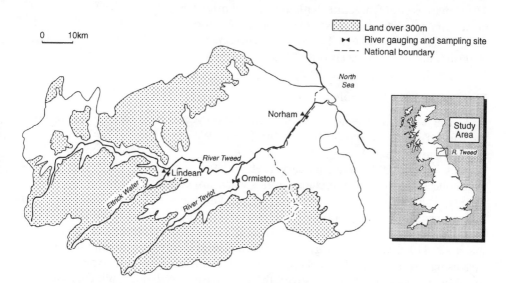

Figure 15.1 Map of the study area showing the location of the river gauging and suspended sediment sampling sites

of the River Teviot upstream of the SEPA gauging station at Ormiston (NT702280) is 1110 km^2, and that of Ettrick Water above Lindean (NT486315) is 500 km^2. For most of its length, the River Tweed is an unpolluted, gravel-bed river, and it flows approximately 160 km from its headwaters in the west, to the North Sea in the east. The mean discharge and mean annual flood (1959–1995) for the River Tweed at Norham are estimated to be 77.8 and 837 m^3 s^{-1}, respectively, while the mean discharges for the River Teviot at Ormiston (1960–1995) and Ettrick Water at Lindean (1961–1995) are 19.6 and 14.9 m^3 s^{-1}, respectively (Fox and Johnson, 1997). Suspended sediment concentrations are typically <10 mg l^{-1} under low to moderate flow conditions, but may exceed 500 mg l^{-1} during periods of high discharge.

The Tweed basin is largely rural and there are no major urban centres. Much of the basin lies above 300 m a.m.s.l. and the altitude locally exceeds 800 m in the western headwaters (Figure 15.1). In the eastern part of the basin, most of the land is flat and low-lying, except for the Cheviot Hills to the south. The underlying geology is dominated by Silurian and Ordovician greywackes, slates and shales in the west and north of the basin, and by a mixture of Devonian sandstones and Carboniferous limestones in the east (Figure 15.2(a)). There are also igneous intrusive (granite) and extrusive (basic lavas) rocks of Devonian age in the south-east of the basin in the Cheviot Hills. Glacial deposits, including boulder clay, morainic drift, and sands and gravels, overlie the bedrock in many areas (Figure 15.2(b)). These deposits are mainly confined to the lowlands, and the uplands are generally free of glacial drift deposits. The soils include well-drained brown earths (lowlands), gleys (intermediate slopes), and podzols and peats (uplands). Land use is closely related to altitude, relief and underlying geology, and varies from moorland and rough grazing in the uplands to cultivated land (mainly for cereal crops) in the lowlands (Figure 15.2(c)). Approximately 16 percent of the basin is afforested, predominantly with conifer plantations. The climate is cool temperate, with average monthly temperatures ranging from 1°C in January to 13°C in August. The mean annual precipitation (1961–1990) for the basin is 969 mm, although locally this ranges from >2000 mm in the uplands to <700 mm in the lowlands (Fox and Johnson, 1997).

The Collection of Samples

Bulk river water samples were collected from the River Tweed (NT891473) ($n = 6$), the River Teviot (NT708274) ($n = 7$) and Ettrick Water (NT485315) ($n = 4$) near to the downstream SEPA gauging stations, using a submersible pump powered by a portable generator, between January 1996 and February 1997. The water samples were collected during high discharge events when suspended sediment concentrations were typically within the range 100–500 mg l^{-1}. The sediment was recovered from the bulk water samples by continuous-flow centrifugation and the resulting sediment samples were freeze-dried prior to analysis. In order to characterise potential source materials, 119 samples (>500 g) were collected from locations throughout the study area using a stainless steel trowel. To obtain information on both the type and spatial location of the main sediment sources, the study basin was divided into the four main geological/topographic zones, namely Silurian and Ordovician (western and northern uplands), igneous (Cheviot uplands), Devonian (lowlands) and Carboniferous

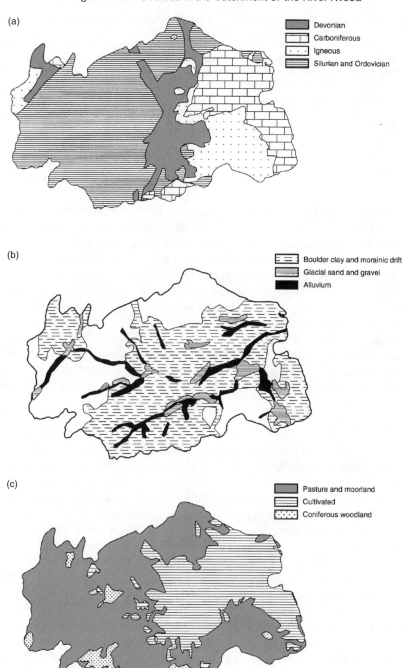

Figure 15.2 Simplified maps of the Tweed basin showing (a) solid geology, (b) Quaternary drift deposits and (c) land use (based partly on Robson *et al.*, 1996). For clarity, in (c) small areas of woodland are not shown

Table 15.1   The number of source material samples collected from each source type within each of the geological/topographic zones (see text for details)

Geological/ topographic zone	Woodland topsoil	Pasture/moorland topsoil	Cultivated topsoil	Channel bank/subsoil	Total
Silurian and Ordovician	5	16	14	22	57
Devonian	2	3	12	6	23
Carboniferous	2	4	9	6	21
Igneous	2	5	5	6	18
Total	11	28	40	40	119

(lowlands) (cf. Figure 15.2). Within each of these zones, replicate representative samples were collected from the faces of eroding channel banks, ditches and gullies (termed channel bank/subsoil material) and from the surface (*c.* top 2 cm) of woodland, pasture/moorland and cultivated areas (Table 15.1). The number of samples collected from each source type and geological/topographic zone was based, in part, on the surface area that each land use and zone occupies in the basin (cf. Figure 15.2). Thus, for example, in the case of the Silurian and Ordovician zone (which is the largest zone, occupying about half of the contributing basin; cf. Figure 15.2), 57 source material samples were collected: five from woodland areas, 16 from pasture and moorland areas, 14 from cultivated areas and 22 from channel bank locations, with each sample being collected from a different field or section of river bank. These source material samples were subsequently dried at 40°C and disaggregated prior to analysis.

Laboratory Analysis

The suspended sediment and source material samples were analysed for a range of potential diagnostic properties, selected on the basis of past experience in sediment source fingerprinting studies in UK rivers. All source materials were screened through a 63 μm sieve prior to analysis, in order to facilitate direct comparison with suspended sediment samples (additional correction procedures are described later). The concentrations of selected radionuclides (^{137}Cs, ^{226}Ra and unsupported ^{210}Pb) were determined using an EG&G Ortec HPGe well-type detector. Mineral magnetic properties (X_{LF}, X_{FD} and SIRM) were measured at Coventry University using a Bartington MS2B dual frequency sensor for the susceptibility(X) parameters and a Molspin pulse magnetiser and Molspin minispin fluxgate magnetometer for SIRM (IRM at 0.8 T). Metal concentrations (Al, Ca, Cr, Cu, Fe, K, Mg, Mn, Na, Ni, Pb, Sr and Zn) were measured using a Unicam 939 atomic absorption spectrophotometer after HCl and HNO_3 digestion (cf. Allen, 1989). Organic C and N concentrations were determined using a Carlo Erba ANA 1400 automatic nitrogen analyser. Phosphorus concentrations (inorganic, organic and total) were measured using a Pye Unicam SP6 UV/visible spectrophotometer after chemical extraction (cf. Mehta *et al.*, 1954). In order to correct for the influence of further differences in particle size composition between the <63 μm fraction of source materials and suspended sediment samples on the values for the diagnostic properties, values of specific surface area were

estimated for the samples from their particle size distributions. Particle size distributions were determined using a Coulter LS130 laser diffraction granulometer, after
standard chemical and ultrasonic pretreatment.

Statistical and Numerical Methods

A detailed description of the statistical and numerical procedures used is given in
Collins et al. (1998) and Walling et al. (1999a). In brief, a two-stage statistical
procedure was used to identify optimum sets of tracer properties for use as composite
fingerprints. First, the non-parametric Kruskal–Wallis analysis of variance H-test was
used to establish which properties exhibited significant differences between the individual potential source groups within a particular category of sources. Secondly,
multivariate discriminant function analysis was applied to the properties selected in
the first stage in order to identify the set of properties or composite fingerprint which
afforded optimum discrimination between source groups. A stepwise selection algorithm, based on the minimisation of Wilks' lambda or U statistic, was used in this
analysis. A multivariate mixing model was subsequently used to estimate the relative
contribution of the potential sediment source groups to a particular suspended sediment sample. For each of the tracer properties i in the composite fingerprint, a linear
equation is constructed which relates the concentration of property i in the suspended
sediment sample to that in the mixture representing the sum of the contributions from
the different source groups. Thus the composite fingerprint is represented by a set of
linear equations (one for each of the properties in the composite fingerprint). Instead
of solving the set of linear equations directly, the least-squares method was used, and
the proportions derived from the individual sources s were estimated by minimising
the sum of the squares of the residual R for the n tracer properties and m source
groups involved (cf. He and Owens, 1995; Collins et al., 1998; Walling et al., 1999a),
where

$$R = \sum_{i=1}^{n} \left(\frac{C_{ssi} - \left(\sum_{s=1}^{m} C_{si} P_s \right)}{C_{ssi}} \right)^2 \tag{15.1}$$

where C_{ssi} is the concentration of tracer property i in the suspended sediment sample,
C_{si} is the mean concentration of tracer property i in source group s, and P_s is the
relative contribution of source group s. The model must satisfy two linear constraints:

(a)   the contribution from each source must lie within the range 0 to 1, i.e.

$$0 \leq P_s \leq 1 \tag{15.2}$$

(b)   the sum of the contributions from all sources is 1, i.e.

$$\sum_{s=1}^{m} P_s = 1 \tag{15.3}$$

For many sediment-associated chemical elements; there is a relationship between the concentration and the specific surface area of a sediment sample (cf. Horowitz, 1991; He and Owens, 1995; He and Walling, 1996). It is, therefore, necessary to take account of differences in the particle size composition between source materials and suspended sediment (cf. Walling et al. (2000) for information on the Tweed basin). This was partly addressed by restricting analysis to the $<63\mu m$ fraction, and further correction was applied by introducing the ratio of the specific surface area of the suspended sediment sample to that of the mean value for each of the individual source groups, as follows:

$$C_{si} = C_o \left(\frac{S_{ss}}{S_s} \right) \tag{15.4}$$

where C_{si} is the particle size-corrected mean concentration of tracer property i in source group s, C_o is the original mean concentration of tracer property i in s, S_{ss} is the specific surface area of the suspended sediment sample $(m^2 g^{-1})$, and S_s is the average specific surface area of the source material samples for s $(m^2 g^{-1})$.

No corrections were introduced to take account of differences in organic matter content between source materials and suspended sediment samples (cf. Collins et al., 1997a; Walling et al., 1999a). Mean values of the particle size-corrected tracer properties for each source group were then compared with equivalent values for each individual suspended sediment sample, using the mixing model.

An assessment of the goodness-of-fit provided by the mixing model was undertaken using the procedures described in Collins et al. (1998) and Walling et al. (1999a). The mean (average for all properties within each composite fingerprint) relative errors for the mixing model calculations typically ranged between ± 7 percent and ± 14 percent, and suggest that the mixing model is able to provide an acceptable prediction of the concentrations of the fingerprint properties associated with individual suspended sediment samples.

Most of the suspended sediment load delivered by a river is usually transported during a small number of high discharge events, when sources may differ from those contributing during lower magnitude events. Because the suspended sediment samples collected from each sampling site covered a range of discharge conditions and suspended sediment concentrations, the values for the contributions of the various sources calculated for individual sampling sites have been weighted according to the suspended sediment load at the time of sampling (derived as the product of the instantaneous values of discharge and suspended sediment concentration):

$$P_{sw} = \sum_{x=1}^{n} P_{sx} \left(\frac{L_x}{L_t} \right) \tag{15.5}$$

where P_{sw} is the load-weighted relative contribution from source group s, L_x $(kg\,s^{-1})$ is the instantaneous suspended sediment load for suspended sediment sample x, L_t $(kg\,s^{-1})$ is the sum of the instantaneous loads (L_x) for all suspended sediment samples collected from that sampling site (n), and P_{sx} is the relative contribution from source group s for sediment sample x.

Values of discharge at the time of sampling were supplied by SEPA for the nearby gauging station, while suspended sediment concentrations were determined for the collected samples. The load-weighted approach provides a more realistic estimate of the proportion of the total suspended sediment load at a sampling site contributed by individual sources, than a simple average of the percentage contribution values associated with individual suspended sediment samples.

RESULTS

Establishing Composite Fingerprints

Table 15.2 presents the results of applying the Kruskal–Wallis H-test which was used to assess the ability of the tracer properties to discriminate the four different source groups representing both source type and source area. As the critical Kruskal–Wallis H value at the 95 percent confidence level for four groups (i.e. from the chi-squared distribution with three degrees of freedom) is 7.82, for those properties that exceed this critical value there is a 95 percent probability that there is a significant difference

Table 15.2 The results of using the Kruskal–Wallis test to assess the ability of each tracer property to discriminate between source types (surface soil from woodland, pasture/moorland and cultivated land, and channel bank/subsoil material) and geological/topographic zones (Carboniferous, Silurian and Ordovician, igneous, and Devonian)

Tracer property	Source type		Geological/topographic zone	
	H value[a]	P value	H value[a]	P value
^{137}Cs (mBq g^{-1})	27.93	0.001*	7.04	0.071
^{226}Ra (mBq g^{-1})	21.48	0.001*	18.52	0.001*
Unsupported ^{210}Pb (mBq g^{-1})	22.98	0.001*	10.54	0.015*
N (%)	52.42	0.001*	7.14	0.068
C (%)	56.62	0.001*	5.19	0.159
Total P (μg g^{-1})	32.79	0.001*	10.27	0.017*
Inorganic P (μg g^{-1})	22.00	0.001*	16.02	0.001*
Organic P (μg g^{-1})	39.48	0.001*	5.65	0.131
X_{LF} (μm^3 kg^{-1})	11.16	0.011*	16.20	0.001*
X_{FD} (nm^3 kg^{-1})	12.48	0.006*	5.23	0.156
SIRM (mAm^2 kg^{-1})	9.46	0.022*	30.48	0.001*
Al (μg g^{-1})	1.06	0.788	16.20	0.001*
Ca (μg g^{-1})	3.30	0.349	21.39	0.001*
Cr (μg g^{-1})	6.56	0.088	9.58	0.023*
Cu (μg g^{-1})	5.98	0.113	0.72	0.869
Fe (μg g^{-1})	10.04	0.019*	19.52	0.001*
K (μg g^{-1})	14.47	0.002*	5.46	0.142
Mg (μg g^{-1})	8.41	0.039*	1.54	0.672
Mn (μg g^{-1})	16.73	0.001*	3.80	0.285
Na (μg g^{-1})	7.51	0.058	1.09	0.778
Ni (μg g^{-1})	28.29	0.001*	8.06	0.045*
Pb (μg g^{-1})	6.56	0.088	4.76	0.191
Sr (μg g^{-1})	4.02	0.259	13.27	0.004*
Zn (μg g^{-1})	20.96	0.001*	1.85	0.603

[a] Critical H value is 7.82.
* Significant ($p < 0.05$).

between the groups. In the case of source type, 17 out of 24 tracer properties were able to differentiate between the four source groups and only seven properties failed the test. It is interesting to note that those properties which did fail this test (Al, Ca, Cr, Cu, Na, Pb and Sr) are all from the geochemical subset, indicating that many of the geochemical properties have a limited capacity to discriminate source type groups. In the case of source ascription based on the four main geological/topographic zones, 12 out of 24 tracer properties met the requirements of the Kruskal–Wallis test. Those properties which failed include examples from the radiometric (^{137}Cs), mineral magnetic (χ_{FD}), organic (C, N, organic P) and geochemical (Cu, K, Mg, Mn, Na, Pb, Zn) subsets.

Table 15.3 presents the optimum composite fingerprint as selected by multivariate discriminant function analysis for both source type and source area ascription. In the case of source type, the composite fingerprint comprises eight tracer properties, which in combination are able to classify 76 percent of the source materials into the correct source group. In the case of source area, the composite fingerprint comprising nine tracer properties is able to classify c. 76 percent of the source materials into the correct source group. Both composite fingerprints comprise tracer properties drawn from different property subsets (radionuclide, mineral magnetic, organic and geochemical), thereby confirming the need to use properties with differing environmental controls in order to obtain a composite fingerprint that affords a high degree of discrimination. Furthermore, the use of a composite fingerprint that comprises properties from several different subsets reduces any potential problems that may be associated with a particular property subset (i.e. such as property transformations during transport).

Sediment Source Ascription

The optimum composite fingerprints for distinguishing between the different source groups in each category (Table 15.3) were employed in the mixing model to estimate the relative contributions of the various source groups to the suspended sediment

Table 15.3 The results of using stepwise multivariate discriminant function analysis to identify which combination of tracer properties provides the best composite fingerprint for discriminating source materials on the basis of source type and geological/topographic zone

Source type		Geological/topographic zone	
Tracer property	Cumulative % samples classified correctly	Tracer property	Cumulative % samples classified correctly
C	53.10	Inorganic P	56.38
Total P	61.61	Fe	61.70
N	65.18	^{226}Ra	63.83
Zn	69.64	χ_{LF}	63.83
K	66.67	Total P	69.15
^{137}Cs	74.29	SIRM	70.21
χ_{LF}	76.19	Unsupported ^{210}Pb	73.40
Mn	76.19	Sr	73.40
		Ni	75.53

samples collected from the River Tweed at Norham, the River Teviot at Ormiston and Ettrick Water at Lindean, during the period January 1996 to February 1997. The results are presented in Figure 15.3.

In the case of source type (Figure 15.3(a)), application of the mixing model to the samples collected from the River Tweed indicates that most of the suspended sediment at the time of sampling was derived from the erosion of topsoil, which provided a load-weighted contribution of c. 61 percent, with channel bank/subsoil sources contributing about 39 percent. The load-weighted contribution from woodland topsoil is relatively small at about 7 percent. Although woodland (mostly coniferous) occupies about 16 percent of the basin (Scottish Office, 1994), the small proportion of sediment derived from woodland topsoil reflects the limited surface erosion and sediment yield generally associated with woodland areas (cf. Morgan, 1986). It may also reflect the distal locations of woodland relative to both the channel network and

(a) Type

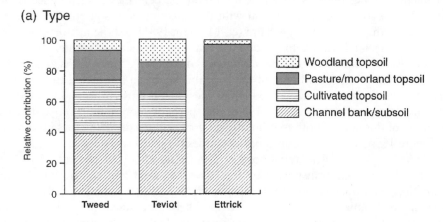

(b) Geological/topographic zone

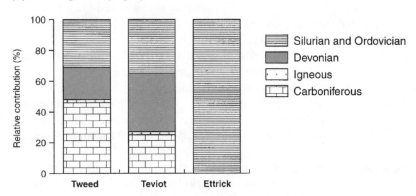

Figure 15.3 The suspended sediment load-weighted contributions of (a) topsoil from woodland, pasture/moorland and cultivated areas and channel bank/subsoil material, and (b) the four main geological/topographic zones, to the suspended sediment samples collected from the Rivers Tweed and Teviot and Ettrick Water during the period January 1996 to February 1997

the downstream sampling site at Norham (cf. Figure 15.2). Previous work by the authors in this basin (cf. Owens *et al.*, 1999b) has documented significant suspended sediment conveyance losses associated with floodplain deposition during overbank flows. For sources located in headwater areas (as in the case of most woodland), such conveyance losses would be expected to reduce the relative contribution derived from these sources to the sediment load at the downstream sampling site, whereas contributions from sources located nearer to the sampling site will be influenced less by conveyance losses. The low contribution from woodland topsoil is consistent with the results obtained for several other basins in the UK using the fingerprinting approach (e.g. Collins *et al.*, 1997a; Walling *et al.*, 1999a). The load-weighted contributions from pasture/moorland and cultivated topsoil sources are estimated to be about 20 percent and 35 percent, respectively. The contribution from pasture/moorland topsoil sources is considerably lower than might be expected based on the proportion of the catchment area occupied by this land use (about 52 percent; Scottish Office, 1994). However, this low value is consistent with the lower rates of soil erosion generally associated with pasture and moorland (cf. Morgan, 1986) and the upstream location of much of this land relative to the downstream sampling site at Norham. The load-weighted relative contribution from cultivated land (about 35 percent) is greater than the proportion of cultivated land in the contributing basin (estimated to be over 25 percent), and this value reflects the higher rates of soil erosion normally associated with cultivated land (cf. Morgan, 1986) and, perhaps more importantly, the close proximity of this land to both the channel network and the downstream sampling site (cf. Figure 15.2), and thus reduced opportunities for conveyance losses.

Channel bank/subsoil material sources contribute about 39 percent of the suspended sediment load transported by the River Tweed at Norham during the time of sampling. Because there are no signs of significant subsoil erosion or gullying in the basin, channel banks and drainage ditches are likely to represent the main sources of sediment within this source group. The relatively high contribution from channel banks reflects the well-developed banks found throughout the basin, which are frequently > 2 m in height in the middle and lower reaches of the river, the fact that the banks are often disturbed and poached by grazing animals, and the direct delivery of the eroded material into the flowing water (see Figure 15.4). This value of about 40 percent is also consistent with those obtained by other studies in the UK, which have either monitored rates of bank erosion directly (e.g. Ashbridge, 1995; Bull, 1997; Lawler *et al.*, 1999) or used the fingerprinting approach to estimate the contribution from channel banks to the annual suspended sediment load (e.g. He and Owens, 1995; Walling and Woodward, 1995; Walling *et al.*, 1999a). It is important to note that those samples collected during periods of higher discharge and higher instantaneous sediment load were characterised by a greater contribution from channel bank sources than those samples collected during periods with lesser loads. This suggests that the detachment and entrainment of channel bank material is significantly greater during high flows than during low flows. Walling *et al.* (1999a) describe a similar situation for the drainage basin of the River Ouse, in Yorkshire, UK.

The source type tracing results for the suspended sediment samples collected from the River Teviot and Ettrick Water are also presented in Figure 15.3(a). In the case of the River Teviot, the load-weighted contribution from channel bank sources is

Figure 15.4 Photograph showing bank erosion for the River Teviot upstream of Ormiston. Note the accumulation of fine-grained material at the base of the bank, which will be mobilised during the next high flow event and incorporated into the sediment load

again high at 39 percent. The contribution from woodland topsoil (15 percent) is double that estimated for the River Tweed at Norham and this reflects the increased amount of woodland in this sub-basin, particularly in the uplands (cf. Figure 15.2). The contributions from pasture/moorland and cultivated topsoil sources are similar at 21 percent and 24 percent, respectively, and these values are in line with the reduced proportion of cultivated land and increased proportion of pasture/moorland, relative to the catchment contributing to the River Tweed at Norham (cf. Figure 15.2). In the case of the samples collected from Ettrick Water, there are no major areas of cultivated land in this sub-basin, and this source was therefore not included in the mixing model. The load-weighted contribution from woodland topsoil is estimated to be 3 percent, and this low value reflects the small amount of woodland in the upstream basin. The contributions from pasture/moorland topsoil and channel bank sources are almost the same at 49 percent and 48 percent, respectively. The high value for pasture/moorland topsoil is consistent with the dominance of this land use in the basin upstream of the sampling site at Lindean. The high relative contribution from channel bank sources, when compared to the results for the Rivers Tweed and Teviot, where the contributions are < 40 pecent, may reflect the relatively low rates of erosion and sediment yield associated with pasture and moorland; rather than increased rates of bank erosion.

The mixing model results for the contributions from the four main geological/topographic zones are presented in Figure 15.3(b). In the case of the River Tweed, the largest proportion of the suspended sediment (46 percent) is derived from areas underlain by Carboniferous rocks, which occur mainly in the east of the basin. This value is higher than might be expected based on the relative proportion of the contributing catchment underlain by this rock type (about 20 percent), and probably reflects the proximity of this zone to the sampling site and the fact that it is dominated by cultivated land, which has been shown previously to be associated with proportionally higher rates of soil erosion and sediment yield. The areas underlain by Silurian and Ordovician rocks, which occur mainly in the uplands to the west and north of the basin, contribute about 31 percent of the suspended sediment. This value is lower than might be expected from consideration of the proportion of the catchment occupied by this geological/topographic zone (about 50 percent), but it is consistent with the dominance of pasture and moorland in this zone, and the relatively low contribution of this source type to sediment samples collected from the sampling site (cf. Figure 15.3(a)). Areas underlain by Devonian rocks also contribute a significant amount of sediment (21 percent) and this is in broad accordance with the relative proportion of the catchment underlain by this rock type (17 percent), its location close to the sampling site, and the importance of cultivated land within this geological/topographic zone. Although the area underlain by igneous rocks accounts for about 13 percent of the catchment, this zone only contributes about 2 percent of the sediment load sampled at Norham. When the relative sediment contributions from each geological/topographic zone are expressed as a ratio to the relative surface areas of the individual zones, the sediment contributions from the Carboniferous (2.30) and Devonian (1.24) zones can be seen to be significantly greater than those from the Silurian and Ordovician (0.62) and igneous (0.15) zones.

In the case of the River Teviot, the Devonian geological/topographic zone, which contributes 38 percent of the total sampled load, represents the most important sediment source. This partly reflects the areal importance of this rock type; which occupies about 28 percent of the contributing catchment. It also reflects the close proximity of this zone to the sampling site and the fact that much of this zone is occupied by cultivated land, which is associated with higher rates of soil erosion and sediment yield. There are also significant contributions from the Carboniferous (25 percent) and Silurian and Ordovician (35 percent) zones. In the latter case, this mainly reflects the areal dominance of this geological/topographic zone; which occupies about 64 percent of the contributing catchment. The value for the Carboniferous zone is higher than might be expected based on the proportion of the catchment that it occupies (about 5 percent) and its location relative to the sampling site. Areas underlain by igneous rocks only contribute about 2 percent of the sediment load, and this value is consistent with the limited proportion of the catchment underlain by this rock type (about 3 percent). The relative sediment contributions from each zone when expressed as a ratio to the surface area that each zone occupies in the catchment, again are higher for the Carboniferous (5.0) and Devonian (1.36) zones and lower for the Silurian and Ordovician (0.55) and igneous (0.67) zones. The catchment of Ettrick Water is only underlain by Silurian and Ordovician rocks and, although the other zones were included in the mixing model, it correctly indicated that 100 percent of the sediment was derived from this rock group. This result confirms the ability of the composite fingerprint to discriminate sediment from different spatial sources.

PERSPECTIVE

Statistically verified composite fingerprints and a multivariate mixing model have been used to identify the dominant sources of the suspended sediment transported by the River Tweed at its tidal limit, and by two of its main tributaries, the River Teviot and Ettrick Water. Sources have been classified in terms of both type (topsoil from woodland, pasture/moorland and cultivated areas, and channel bank/subsoil material) and spatial location (defined in terms of the four main geological/topographic zones). The source ascription results apply to the individual suspended sediment samples, but the values presented for the overall relative contributions of the different source groups within the two source categories have been weighted according to the instantaneous suspended sediment load at the time of sampling. It is important to use such weightings, since the relative contributions of the different sources may vary according to the magnitude of the event, but the high magnitude events will account for a major proportion of the annual sediment load.

Most of the suspended sediment transported by the study rivers is derived from topsoil sources, primarily pasture/moorland and cultivated areas, with woodland topsoil accounting for only a small proportion of the sediment load. There is, however, a substantial contribution from channel bank sources for all of the study rivers. In the case of Ettrick Water, channel banks account for nearly 50 percent of the sampled suspended sediment load. In terms of the geological/topographic zones, the suspended sediment load was predominantly derived from areas underlain by Carboniferous (River Tweed), Devonian (River Teviot) and Silurian and Ordovician (Ettrick Water) rocks. In the case of source type ascription, the results are consistent with the dominant land use in each of the contributing basins and with the relative magnitude of the erosion rates and sediment yields likely to be associated with each of the source groups. In the case of spatial location source ascription, the results are in broad agreement with the relative extent of the main geological/topographic zones in each of the contributing basins and the dominant land use in each of these zones.

The results obtained must, nevertheless, be qualified by a number of observations. First, it should be recognised that the results relate to only a limited number of suspended sediment samples collected from the three sampling sites, and that these may not be fully representative of the total sediment load at these sites. Secondly, previous studies of sediment sources in the basin of the River Ouse, in Yorkshire, undertaken by the authors as part of the LOIS project (Walling *et al.*, 1999a), have demonstrated the need to take account of the timing of sample collection relative to the hydrograph peak. The routing of sediment contributions from different parts of the catchment (and thus from different spatial sources) will cause these contributions to pass the sampling sites at different times, and the relative importance of the different spatial sources will, therefore, vary during the event. In this study, most of the suspended sediment samples were collected close to the hydrograph peak or during the falling limb, and may, therefore, have emphasised the importance of distal sources. The timing of sampling will, however, only have a limited influence on the collection of information on source type. Thirdly, the individual source groups have been represented within the mixing model by a single mean value for each tracer property. Many of the tracer properties exhibit considerable variability within the individual source groups and further research is needed to develop mixing models and optimisation procedures that can incorporate this variability. Finally, the fingerprinting approach applied in this study takes no account of possible transformations in tracer properties during transport and temporary storage between source location and downstream sampling sites. Despite these potential problems and limitations, the results presented are consistent with other evidence regarding the likely importance of potential sediment sources within the study catchments, and the fingerprinting approach can be seen to offer many advantages over existing approaches to source tracing.

ACKNOWLEDGEMENTS

Financial support for the work reported in this chapter was provided by a Special Topic research grant (GST/02/774) within the framework of the UK NERC Land–Ocean Interaction Study (LOIS) and this support is gratefully acknowledged. Thanks are due to Rachel Bronsdon, Dick Johnson (both at the Institute of Hydrology, Stirling) and Ben Waterfall for assistance with the collection of bulk suspended sediment samples, to Art Ames and Ghida Sinawi for help with the laboratory analysis, to Ian Foster and Joan Lees (both at Coventry University) for mineral magnetic analysis of suspended sediment and source material samples, to Drew McCraw (SEPA, Galashiels) for providing river discharge data, and to Terry Bacon and Barry Phillips for producing the diagrams. Comments provided by Mike Slattery and an anonymous referee have helped to improve the chapter. This chapter represents Publication No. 740 of the LOIS project.

REFERENCES

Allen, S. E. 1989. *Chemical Analysis of Ecological Materials*. Blackwell, Oxford.

Ashbridge, D. 1995. Processes of river bank erosion and their contribution to the suspended sediment load of the River Culm, Devon. In Foster, I. D. L., Gurnell, A. M. and Webb, B. W. (eds) *Sediment and Water Quality in River Catchments*. John Wiley, Chichester, 229–245.

Bull, L. J. 1997. Magnitude and variation in the contribution of bank erosion to the suspended sediment load of the River Severn, UK. *Earth Surface Processes and Landforms*, **22**, 1109–1123.

Caitcheon, G. G. 1993. Sediment source tracing using environmental magnetism: a new approach with examples from Australia. *Hydrological Processes*, **7**, 349–358.

Caitcheon, G. G. 1998. The significance of various sediment magnetic mineral fractions for tracing sediment sources in Killimicat Creek. *Catena*, **32**, 131–142.

Collins, A. L., Walling, D. E. and Leeks, G. J. L. 1997a. Source type ascription for fluvial suspended sediment based on a quantitative composite fingerprinting technique. *Catena*, **29**, 1–27.

Collins, A. L., Walling, D. E. and Leeks, G. J. L. 1997b. Use of the geochemical record preserved in floodplain deposits to reconstruct recent changes in river basin sediment sources. *Geomorphology*, **19**, 151–167.

Collins, A. L., Walling, D. E. and Leeks, G. J. L. 1998. Use of composite fingerprints to determine the provenance of the contemporary suspended sediment load transported by rivers. *Earth Surface Processes and Landforms*, **23**, 31–52.

Foster, I. D. L. and Walling, D. E. 1994. Using reservoir deposits to reconstruct changing sediment yields and sources in the catchment of the Old Mill Reservoir, South Devon, UK, over the past 50 years. *Hydrological Sciences Journal*, **39**, 347–368.

Foster, I. D. L., Lees, J. A., Owens, P. N. and Walling, D. E. 1998. Mineral magnetic characterization of sediment sources from an analysis of lake and floodplain sediments in the catchments of Old Mill Reservoir and Slapton Ley, South Devon, UK. *Earth Surface Processes and Landforms*, **23**, 685–703.

Fox, I. A. and Johnson, R. C. 1997. The hydrology of the River Tweed. *The Science of the Total Environment*, **194/195**, 163–172.

He, Q. and Owens, P. 1995. Determination of suspended sediment provenance using caesium-137, unsupported lead-210 and radium-226: a numerical mixing model approach. In Foster, I. D. L., Gurnell, A. M. and Webb, B. W. (eds) *Sediment and Water Quality in River Catchments*. John Wiley, Chichester, 207–227.

He, Q. and Walling, D. E. 1996. Interpreting particle size effects in the adsorption of ^{137}Cs and unsupported ^{210}Pb by mineral soils and sediment. *Journal of Environmental Radioactivity*, **30**, 117–137.

Horowitz, A. J. 1991. *A Primer on Sediment-trace Element Chemistry*. Lewis, Michigan.

Lawler, D. M., Grove, J., Couperthwaite, J. S. and Leeks, G. J. L. 1999. Downstream change in river bank erosion rates in the Swale–Ouse system, northern England. *Hydrological Processes*, **13**, 977–992.

Maitland, P. S., Boon, P. J. and McLusky, D. S. (eds) 1994. *The Fresh Waters of Scotland: A National Resource of International Significance*. John Wiley, Chichester.

Mehta, N. C., Legg, J. O., Goring, C. A. I. and Black, C. A. 1954. Determination of organic phosphorus in soils: I. Extraction methods. *Soil Science Society of America Journal*, **18**, 443–449.

Morgan, R. P. C. 1986. *Soil Erosion and Conservation*. Longman, UK.

Oldfield, F., Rummery, T. A., Thompson, R. and Walling, D. E. 1979. Identification of suspended sediment sources by means of mineral magnetic measurements: some preliminary results. *Water Resources Research*, **15**, 211–219.

Owens, P. N., Walling, D. E., He, Q., Shanahan, J. and Foster, I. D. L. 1997. The use of caesium-137 measurements to establish a sediment budget for the Start catchment, Devon, UK. *Hydrological Sciences Journal*, **42**, 405–423.

Owens, P. N., Walling, D. E. and Leeks, G. J. L. 1999a. The use of floodplain sediment cores to investigate recent historical changes in overbank sedimentation rates and sediment sources in the catchment of the River Ouse, Yorkshire, UK. *Catena*, **36**, 21–47.

Owens, P. N., Walling, D. E. and Leeks, G. J. L. 1999b. Deposition and storage of fine-grained sediment in the main channel system of the River Tweed, Scotland. *Earth Surface Processes and Landforms*, **24**, 1061–1076.

Passmore, D. G. and Macklin, M. G. 1994. Provenance of fine-grained alluvium and late Holocene land-use change in the Tyne basin, northern England. *Geomorphology*, **9**, 127–142.

Peart, M. R. and Walling, D. E. 1986. Fingerprinting sediment source: the example of a drainage basin in Devon, UK. In *Drainage Basin Sediment Delivery*. IAHS Publication **159**. IAHS Press, Wallingford. 41–55.

Peart, M. R. and Walling, D. E. 1988. Techniques for establishing suspended sediment sources in two drainage basins in Devon, UK: a comparative assessment. In *Sediment Budgets*. IAHS Publication **174**. IAHS Press, Wallingford. 269–279.

Robson, A. J., Neal, C., Currie, J. C., Virtue, W. A. and Ringrose, A. 1996. *The Water Quality of the Tweed and its Tributaries*. Institute of Hydrology Report No. 128. Institute of Hydrology, Wallingford.

Scottish Office 1994. *The Scottish Environment – Statistics, 1993*. Government Statistical Service, HMSO, London.

Slattery, M. C., Burt, T. P. and Walden, J. 1995. The application of mineral magnetic measurements to quantify within-storm variations in suspended sediment sources. In *Tracer Technologies for Hydrological Systems*. IAHS Publication **229**. IAHS Press, Wallingford. 143–151.

Stott, A. P. 1986. Sediment tracing in a reservoir-catchment system using a magnetic mixing model. *Physics of the Earth and Planetary Interiors*, **42**, 105–112.

Walden, J., Slattery, M. C. and Burt, T. P. 1997. Use of mineral magnetic measurements to fingerprint suspended sediment sources: approaches and techniques for data analysis. *Journal of Hydrology*, **202**, 353–372.

Wall, G. J. and Wilding, L. P. 1976. Mineralogy and related parameters of fluvial suspended sediments in northwestern Ohio. *Journal of Environmental Quality*, **5**, 168–173.

Wallbrink, P. J., Murray, A. S., Olley, J. M. and Olive, L. J. 1998. Determining sources and transit times of suspended sediment in the Murrumbidgee River, New South Wales, Australia, using ^{137}Cs and ^{210}Pb. *Water Resources Research*, **34**, 879–887.

Walling, D. E. and Woodward, J. C. 1992. Use of radiometric fingerprints to derive information on suspended sediment sources. In *Erosion and Sediment Transport Monitoring Programmes in River Basins*. IAHS Publication **210**. IAHS Press, Wallingford. 153–164.

Walling, D. E. and Woodward, J. C. 1995. Tracing sources of suspended sediment in river basins: a case study of the River Culm, Devon, UK. *Journal of Marine and Freshwater Research*, **46**, 327–336.

Walling, D. E., Peart, M. R., Oldfield, F. and Thompson, R. 1979. Suspended sediment sources identified by magnetic measurements. *Nature*, **281**, 110–113.

Walling, D. E., Woodward, J. C. and Nicholas, A. P. 1993. A multi-parameter approach to fingerprinting suspended-sediment sources. In *Tracers in Hydrology*. IAHS Publication **215**. IAHS Press, Wallingford, 329–338.

Walling, D. E., Owens, P. N. and Leeks, G. J. L. 1998a. The characteristics of overbank deposits associated with a major flood event in the catchment of the River Ouse, Yorkshire, UK. *Catena*, **32**, 309–331.

Walling, D. E., Owens, P. N. and Leeks, G. J. L. 1998b. The role of channel and floodplain storage in the suspended sediment budget of the River Ouse, Yorkshire, UK. *Geomorphology*, **22**, 225–242.

Walling, D. E., Owens, P. N. and Leeks, G. J. L. 1999a. Fingerprinting suspended sediment sources in the catchment of the River Ouse, Yorkshire, UK. *Hydrological Processes*, **13**, 955–975.

Walling, D. E., Owens, P. N. and Leeks, G. J. L. 1999b. Rates of contemporary overbank sedimentation and sediment storage on the floodplains in the main channel systems of the Yorkshire Ouse and River Tweed, UK. *Hydrological Processes*, **13**, 993–1009.

Walling, D. E., Owens, P. N., Waterfall, B. D., Leeks, G. J. L. and Wass, P. D. 2000. The particle size characteristics of fluvial suspended sediment in the Humber and Tweed basins. *The Science of the Total Environment*, **251/252**, 205–222.

Yu, L. and Oldfield, F. 1989. A multivariate mixing model for identifying sediment source from magnetic measurements. *Quaternary Research*, **32**, 168–181.

Yu, L. and Oldfield, F. 1993. Quantitative sediment source ascription using magnetic measurements in a reservoir-catchment system near Nijar, S.E. Spain. *Earth Surface Processes and Landforms*, **18**, 441–454.

16 Use of Mineral Magnetic Measurements to Fingerprint Suspended Sediment Sources: Results from a Linear Mixing Model

M. C. SLATTERY,[1] J. WALDEN[2] and T. P. BURT[3]

[1]*Department of Geology, Texas Christian University, Fort Worth, TX, USA*
[2]*School of Geography and Geology, University of St Andrews, UK*
[3]*Department of Geography, University of Durham, UK*

INTRODUCTION

Identifying the source area(s) of a stream's suspended load is a complex and difficult task. One approach (e.g. Oldfield *et al.*, 1979; Walling *et al.*, 1979; Caitcheon, 1993) has involved the use of mineral magnetic analyses to provide a compositional 'fingerprint' by which catchment sources can be compared with suspended sediments in order to investigate the likely sources of the latter. Generally, the magnetic data are analysed in one or both of two ways: qualitatively, using graphical techniques and the 'standard' interpretations given in the literature (e.g. Thompson and Oldfield, 1986; Maher, 1988; Oldfield, 1991); and statistically, using multivariate techniques that provide a more objective analysis (e.g. Oldfield and Clark, 1990; Walden *et al.*, 1992). However, in order to develop a more complete understanding of sediment delivery processes and sediment movement in fluvial environments, we need to be able to better *quantify* sediment source contributions.

Yu and Oldfield (1993) developed a method for quantifying sediment source components from magnetic measurements by using numerical models that expressed the linkage between the magnetic properties of sediments and their source components. In their study, the magnetic measurements were made on cores sampled from an infilled reservoir in south-east Spain; the source samples were within-channel sediments taken from the beds of inflowing valley streams. This approach to quantitative sediment source ascription appeared to work well, and the linear programming procedures used in Yu and Oldfield's (1993) unmixing model are relatively simple to replicate. While this approach has been applied in a limited number of case studies in other sedimentary settings (in particular, see Lees, 1994), it has not

---

Tracers in Geomorphology. Edited by Ian D. L. Foster. © 2000 John Wiley & Sons Ltd.

been extensively tested in contemporary suspended sediment systems. A notable exception is the work reported by Walling *et al.* (1993). These authors used a multi-parameter approach to fingerprinting suspended sediment sources in two small catchments in Devon, UK that included mineral magnetic measurements. Multivariate analyses were used to distinguish statistically significant sediment source groupings and an unmixing model was employed to estimate the relative contributions from these sources to the overall suspended load. Again, the general conclusion was that the technique offers an alternative approach to sediment source determination and possesses very considerable potential, particularly as a tool for sediment budget investigations.

In an earlier paper (Walden *et al.*, 1997) the authors presented a detailed review of the use of magnetic measurements for sediment source ascription in a suspended sediment system. The aim was essentially methodological, focusing specifically on criteria for modelling source–sediment linkages and techniques for data analysis. The present study leads on from this earlier work, but here our attention shifts to the environmental interpretation of the model output. Our objective was to not only determine the relative contributions from potential sources in the catchment system generally, but also to quantify how the various sediment source areas change, both within and between individual storm events.

MODEL FORMULATION

The basic form of the unmixing model used here follows Thompson (1986) and has been used in other environmental contexts (e.g. analysis of remotely sensed data; Richards, 1993). The formulation of the model used here, its limitations, and possible sources of error, are discussed fully in Walden *et al.* (1997) and therefore only a brief summary is given here. The model essentially carries out an iterative search to find the optimum combination of the source materials which, when linearly mixed, minimises the differences between the measured magnetic properties of the suspended sediment and the magnetic properties of the mathematical mixture of the sources. The optimisation process has to work within certain constraints, namely:

(1)   the proportion of each source must be lie between 0 and 1;
(2)   the sum of all source proportions should total 1.

The following notation is adopted below:

p number of magnetic parameters used in the modelling procedure;
e number of end-members (sources) used in the modelling procedure;
a_{ij} measured value of magnetic parameter j in end-member (source) i;
b_j measured value of magnetic parameter j in the natural sample being modelled;
x_i hypothetical proportion of end-member i in the natural sample where $x_i \geq 0$ and $x_i \leq 1$;
s_j simulated (modelled) value of magnetic parameter j in the simulated version of the natural sample.

$$s_j = \sum_{i=1}^{e} a_{ij} x_i$$

The minimum difference between the simulated and natural samples is found by minimising

$$\sum_{j=1}^{p} \left(\sum_{i=1}^{e} a_{ij} x_j - b_j \right)^2 = \sum_{j=1}^{p} (s_j - b_j)^2 \qquad (16.1)$$

subject to the constraint that $x_i \geq 0$ and $x_i \leq 1$ for each end-member i and $\sum_{i=1}^{e} x_i = 1$.

Each variable was weighted in the model by expressing the differences between modelled and measured values as a percentage of the modelled value (see Walden *et al.*, 1997) thus avoiding the model being dominated by variables expressed in units with different orders of magnitude, as follows:

$$\sum_{j=1}^{p} \{ [(s_j - b_j)/b_j]^* 100 \}^2 \qquad (16.2)$$

The model was run on a PC using the Solver add-in component of Microsoft's Excel^{TM} spreadsheet, which allows optimisation routines to be constructed using a SIMPLEX algorithm. Table 16.1 shows the basic structure of the model in the Excel spreadsheet. The user has to supply values to represent the properties of the source materials, the measured properties of the suspended sediment sample which has to be unmixed, and the initial starting proportions of each source from which the optimisation routine will move to find the best solution.

STUDY SITE AND METHODS

Field Setting and Sample Collection

The study catchment (National Grid Reference SP356362) is situated in the north-eastern Cotswold Hills, approximately 8 km north-east of Chipping Norton (Figure 16.1), in the south Midlands of England. The basin has a drainage area of 6.2 km^2 and its altitude ranges from 126 m.a.s.l. at its outlet to 202 m.a.s.l. on the northern divide. Slopes are gentle (around 1°) near the interfluves but steeper (> 5°) in the central part of the basin. The stream network is moderately incised into the valley floors, and bank heights are commonly less than 1 m. Two major soil types occur within the basin (Jarvis *et al.*, 1984): *brown calcareous earths* of the Aberford series (moderately stony fine loam) and *ferritic brown earths* of the Banbury series (stony fine loam). Land use is mixed arable farming with extensive autumn sowing of wheat and barley.

Field observation and mineral magnetic data, summarised later, suggested that three potential sediment source groups can be identified: Aberford series topsoils; Aberford series channel bank material; and either topsoils or channel bank material from the Banbury soil series (which could not be distinguished magnetically). A number of field samples were collected from each of these potential sources, although the highly rigorous approach to quantifying source material variation suggested by

Table 16.1   Example of the unmixing model constructed using the Solver add-in in Microsoft's Excel 5 spreadsheet. The basic structure of the spreadsheet is explained below

	A	B	C	D	E	F	G	H	I
1	**Three Source Model**								
2									
3	**Source**				**Source properties**				
4					χ	**Soft IRM**	**Mid IRM**	**Hard IRM**	**SIRM**
5	Aberford TS				10.67	141.40	395.51	64.84	601.75
6	Aberford CB				5.78	82.73	303.36	21.72	407.81
7	Banbury TS and CB				32.98	513.31	1252.00	64.21	1829.53
8									
9	**Suspended sediment sample**				**Suspended sediment properties**				
10	**No.**	**Date**	**Type**		χ	**Soft IRM**	**Mid IRM**	**Hard IRM**	**SIRM**
11	18	18/12 11:40	Storm		10.96	157.82	409.74	37.59	605.15
12									
13					**Modelled values**				
14		**Start prop.**	**Model prop.**		χ	**Soft IRM**	**Mid IRM**	**Hard IRM**	**SIRM**
15	Ab TS	0.33	0.84		11.03	150.04	420.14	59.79	629.97
16	Ab CB	0.33	0.11						
17	Ban TS & BC	0.33	0.04	**Diffs =**	0.61	5.19	2.48	3.71	3.94
18				**Diff² =**	0.38	26.92	6.13	13.78	15.52
19		**Sum =**	1.00		Sum Diffs²	62.73			

Cells	*Contents*
E5–I7	The measured parameter values used to represent the properties of the three source materials.
E11–I11	The measured parameter values for the particular suspended sediment sample that is currently being unmixed.
E15–I15	The modelled parameter values for the simulated version of the suspended sediment sample currently being unmixed. These cells are linked to E5–I7 and C15–C17 by a linearly additive formula. For example, cell E15 contains the formula = (C15*E5) + (C16*E6) + (C17*E7).
C15–C17	The modelled proportions of the three source materials. Solver is configured so that these cells are changed in order to minimise the value held in cell E19.
E17–I18	Contains the differences and squared differences between the measured and modelled parameter values. As discussed in the text, these cells can be configured in a number of ways to force the model to treat variables of different orders of magnitude with equal weight.
E19	Contains the sum of cells E18–I18. Solver is configured to minimise the value contained in this cell by changing the values held in cells C15–C17.

Lees (1994) could not be adopted here for logistical reasons. In addition, suspended sediment samples were collected using a Rock and Taylor automatic pump sampler over a period of some 12 months along with rainfall and discharge readings. Suspended sediment concentrations were generally in the region of 80–5000 mg l^{-1} for storm flows. Thus, enough sediment was extracted from the stream during storm events to enable magnetic measurements to be conducted on individual samples.

Mineral Magnetic Analysis

A series of standard mineral magnetic measurements (Thompson and Oldfield, 1986), including magnetic susceptibility (χ), frequency-dependent susceptibility (χ_{FD}), and isothermal remanence magnetisation (IRM) acquisition in magnetic fields between 20

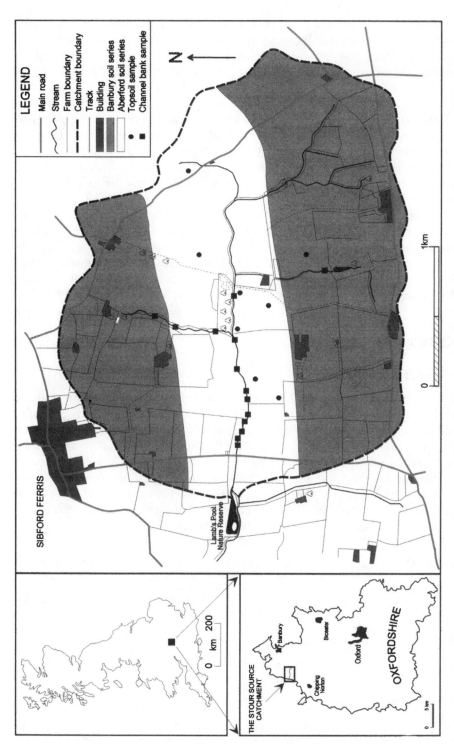

Figure 16.1 Study catchment within the Stour River system, north-west of Oxford, showing the distribution of the two major soil series and the locations of the source samples used in this work

and 1000 mT, were made upon the field samples. Given the relatively small mass available for the suspended sediment samples, calculation of χ_{FD} could not be made with any confidence and this parameter was therefore not used in the modelling procedure. Particle size analysis of the suspended sediment samples showed that material coarser than 63 µm (4 φ) was almost totally absent, suggesting that sand-grade material from the catchment was not being transported as suspended load into the stream network. Magnetic analysis on the sediment sources was therefore per-formed only on material finer than 63 µm. Analysis of size-specific magnetic prop-erties (separated using the method of Walden and Slattery, 1993) for the three sources showed that the differences in their respective particle size distributions had little influence on their magnetic properties, and therefore the bulk properties of the source samples are used here.

The mineral magnetic data were analysed qualitatively and statistically, using standard graphical and multivariate techniques, as noted earlier. The results of these analyses are presented in detail in Walden *et al.* (1997) and are therefore discussed only briefly here. The data were then subjected to the unmixing model outlined above in an attempt to quantify the proportions of each catchment source to the suspended sediment samples based upon their respective magnetic properties.

RESULTS AND DISCUSSION

Identifying Source Areas

Table 16.2 summarises the magnetic properties for the three types of catchment source materials. The Aberford topsoil and channel bank material could clearly be distin-guished on the basis of their magnetic properties. However, the Banbury topsoil and channel bank material, despite their significantly higher values, could not be distin-guished magnetically; these were therefore combined into a single potential source type. The samples from the three source materials showed statistically different magnetic properties and also plotted as separate clusters in an R- and Q-mode factor analysis (see Walden *et al.*, 1997), again suggesting that they can be distinguished on the basis of the five magnetic parameters used in the analysis.

Table 16.2 Mean magnetic data for the various sample groups (χ is expressed in units of $10^{-7} m^3 kg^{-1}$, soft IRM, mid IRM, hard IRM and SIRM are expressed in units of $10^{-5} Am^2 kg^{-1}$)

	N	χ	Soft IRM (0–20 mT)	Mid IRM (20–300 mT)	Hard IRM (300–1000 mT)	SIRM
(a) Source materials						
Aberford TS	7	10.67	141.40	395.51	64.84	601.75
Aberford CB	12	5.78	82.73	303.36	21.72	407.81
Banbury TS + CB	3	32.98	513.31	1252.00	64.21	1829.53
(b) Non-storm and storm suspended sediment samples						
Non-storm	9	6.53	110.34	338.65	34.35	483.33
Storm	24	9.35	138.1	396.12	41.31	575.53

Magnetic Measurements on the Suspended Sediment

Table 16.2 also indicates that the non-storm and storm samples exhibit generally different magnetic properties. The non-storm samples show lower concentrations of magnetic minerals and are closer in value to the Aberford channel bank samples. In contrast, the storm samples show higher concentrations of magnetic minerals and are closer in measurement to the Aberford topsoil material. This pattern makes 'environmental' sense in that topsoil contributions might increase during periods of higher flow as a result of surface runoff and overbank flow (see below). The data in Table 16.2 also suggest that the contribution to the suspended sediments from the Banbury series is much less than for the other two source types. Again, this would seem to be consistent within the field context, given the distal location of the Banbury source areas to the main channel (see Figure 16.1).

Modelling Source–Sediment Linkages

Analysis of the magnetic data indicate that the source materials and the suspended sediment samples meet the two most important criteria for successful quantitative unmixing of sediment sources, discussed initially by Lees (1994) and summarised later by Walden *et al.* (1997):

(1) The three source material types possess magnetic behaviours that can be distinguished using linearly additive parameters.
(2) The majority of the suspended sediment samples display magnetic behaviour that lies within the range of behaviours exhibited by the three source types.

Compliance with the second criteria above is particularly important in this case, because (i) it suggests that all the major source types have been identified and that no major mineralogical alteration of the suspended sediments is occurring during the weathering and transportation processes; and (ii) in numerical terms it, should be possible to 'unmix' the suspended sediments on the basis of these three source material types alone.

In total, 33 suspended sediment samples were available for unmixing, representing both low flow (non-storm) and high flow (storm) conditions. The mineral magnetic measurements and model output for these data are given in Table 16.3.

The results of the modelling process for the mean magnetic properties of both the non-storm and storm samples are consistent with the qualitative interpretation of the data presented earlier in Table 16.2 and, again, seem sensible given their environmental context. Thus, the non-storm mean is dominated by material derived from the Aberford channel banks (89 percent) but with a significant proportion of Aberford topsoil (7 percent). Most likely, this topsoil contribution is material transported during higher flow conditions, subsequently deposited and stored on the channel bed; it is then gradually re-entrained and exported during periods of lower flow (Figure 16.2). In contrast, the storm mean indicates a much greater proportion of Aberford topsoil (43 percent), with channel bank material contributing approximately 50 percent of the suspended load. The modelling procedure also confirms that the contribution to the suspended sediments from the Banbury series is significantly less than for the other two source types.

There is a clear temporal pattern to the data presented in Table 16.3, and closer examination of the model output reveals subtle changes in source component contributions during the hydrological record. For example, the first non-storm sample, a composite taken between 28 and 31 August 1992, is ascribed a 44 percent Aberford topsoil contribution by the model. Although this value significantly skews the non-storm mean, it is entirely reasonable in terms of the hydrological response of the basin.

Table 16.3 Mineral magnetic properties of all suspended sediment samples and the proportions of the three source materials ascribed by the unmixing model. The model output was produced by minimising the sum of the squared % differences between the measured and modelled versions of the five magnetic parameters (see text for explanation)

Sample	Date/time	Non-storm (NS) or storm (S)	χ	Soft IRM	Mid IRM	Hard IRM	SIRM	Modelled proportions		
								Ab TS	Ab CB	Ban CB + TS
Mean	Non-storm	$n = 9$	6.53	110.34	338.65	34.35	483.33	0.07	0.89	0.04
	Storm	$n = 24$	9.35	138.10	396.12	41.31	575.53	0.43	0.50	0.06
Mean	Storm 18/12/92	$n = 8$	10.42	153.15	397.27	38.12	588.54	0.63	0.31	0.06
	Storm 13/1/93	$n = 7$	8.36	131.57	349.55	29.80	510.92	0.39	0.58	0.04
	All samples									
1	28/8–31/8	NS	7.56	111.85	332.87	60.02	504.74	0.44	0.56	0.00
2	31/8–4/9	NS	7.18	127.54	362.97	38.01	528.52	0.02	0.90	0.08
3	4/9–7/9	NS	6.47	114.05	371.18	26.59	511.82	0.00	0.94	0.06
4	7/9–14/9	NS	4.91	91.76	327.78	13.57	433.12	0.00	0.99	0.01
5	14/9–21/9	NS	6.79	125.96	374.23	55.95	556.13	0.13	0.81	0.06
6	16/11–23/11	NS	6.67	123.44	353.40	29.17	506.02	0.00	0.93	0.07
7	26/11 11:00	NS	7.64	132.55	379.16	28.66	540.37	0.00	0.90	0.09
8	29/3–2/4	NS	5.55	76.74	252.79	27.03	356.57	0.00	1.00	0.00
9	9/4 16:45	NS	5.99	89.12	293.46	30.10	412.68	0.06	0.94	0.00
10	25/9 21:00	S	12.69	183.56	600.97	102.79	887.32	0.49	0.29	0.22
11	3/10	S	7.76	115.45	433.86	51.20	600.51	0.04	0.86	0.09
12	20/10 10:00	S	7.79	112.20	463.24	68.48	643.92	0.13	0.77	0.10
13	23/11–25/11	S	8.22	127.90	468.24	44.20	640.34	0.00	0.87	0.13
14	27/11 20:06	S	10.08	161.06	424.02	45.23	630.30	0.33	0.55	0.11
15	30/11 10:10	S	10.72	187.76	425.57	42.00	655.32	0.39	0.46	0.14
16	2/12 04:10	S	8.75	112.58	347.41	39.49	499.48	0.54	0.45	0.00
17	18/12 11:10	S	13.23	208.78	494.70	31.69	735.17	0.66	0.17	0.17
18	18/12 11:40	S	10.96	157.82	409.74	37.59	605.15	0.84	0.11	0.04
19	18/12 12:10	S	10.53	170.23	414.23	33.53	617.99	0.48	0.42	0.10
20	18/12 12:40	S	11.00	179.61	444.75	20.58	644.94	0.24	0.60	0.16
21	18/12 13:10	S	10.11	143.28	372.17	37.45	552.89	0.90	0.10	0.00
22	18/12 13:40	S	10.13	124.82	350.60	30.25	505.66	0.74	0.26	0.00
23	18/12 14:10	S	9.22	133.72	364.71	52.09	550.52	0.70	0.29	0.01
24	18/12 18:30	S	8.15	106.94	327.30	61.79	496.03	0.45	0.55	0.00
25	13/1 13:45	S	9.14	147.58	378.74	24.54	550.86	0.30	0.62	0.08
26	13/1 14:00	S	6.56	107.50	311.27	30.13	448.91	0.12	0.86	0.02
27	13/1 14:15	S	8.04	130.57	351.37	28.76	510.70	0.20	0.74	0.06
28	13/1 14:30	S	9.16	134.71	364.55	41.35	540.62	0.62	0.35	0.02
29	13/1 14:45	S	9.09	147.64	354.05	30.69	532.39	0.61	0.36	0.03
30	13/1 15:00	S	9.20	138.87	370.37	29.42	538.66	0.52	0.44	0.04
31	13/1 15:20	S	7.33	114.15	316.48	23.67	454.30	0.33	0.66	0.01
32	4/4	S	7.55	74.45	355.72	32.33	462.50	0.24	0.76	0.00
33	9/4 10:45	S	9.10	93.16	362.90	52.12	508.19	0.51	0.49	0.00

TS = topsoil; CB = channel bank.

Figure 16.2 Within-channel deposits of Aberford topsoil material. This sediment was gradually entrained and exported from the basin during periods of low flow but also 'flushed' from the system during initial stormflow generation

These four days experienced an extended period of relatively light rainfall (24.9 mm in total) with no discernible storm peak; stream discharge varied between 10.1 l s^{-1} (essentially baseflow) and 25.0 l s^{-1}. Despite this muted response, streamflow was nevertheless of sufficient velocity to export fairly significant quantities of sediment, including the within-channel deposits of Aberford topsoil (Figure 16.2). The model then ascribes a channel bank contribution of > 90 percent Aberford for the remaining eight non-storm samples, which again, with the exception of sample 5, appears hydrologically sensible given the low flow conditions. Sample 5, accumulated over a one-week period from 14 to 21 September 1992, is more difficult to explain, given the fact that discharge remained at baseflow during sampling. While we cannot offer a reasonable hydrological explanation for the increased Aberford topsoil contribution at this time, suspended sediment concentrations were noticeably elevated during this week, ranging between 25 and 50 mg l^{-1}; the topsoil contribution may therefore be the result of some anthropogenic disturbance, most likely soil moved into the channel through slope cultivation.

The 24 storm samples also produce what would appear to be sensible proportions of the three source materials given their magnetic properties and their respective flow conditions. Sample 10 was taken at 2100 hours on 25 September 1992, during the onset of a slow-moving frontal storm: 16.9 mm of rain fell in 1.5 h, yielding the highest 60-min rainfall intensity during the study period (equivalent to the two-year, 1-h rainstorm for the region). A post-storm survey revealed that the stream had in fact burst its banks during the event and had moved out into the valley floor, at places, to a distance of 30 m from the channel. Much of the vegetation along the channel was draped in sediment and there was extensive flooding and erosion of topsoil along the valley floor (Figure 16.3), providing an obvious mechanism for incorporation of

Figure 16.3 Photograph showing erosion of Aberford topsoil along the valley floor near the gauging station during the 25 September 1992 storm. The stream channel runs along the hedge-row behind the figure

greater amounts of Aberford topsoil into the suspended sediment system. Thus, the output from the unmixing model (49 percent Aberford topsoil, 29 percent Aberford channel banks, and 22 percent Banbury series) appears entirely reasonable given the hydrological context. Unfortunately, most of the recording equipment at the gauging station was damaged during this event, including the automatic pump sampler; hence we do not have a complete discharge record for the storm and just this single suspended sample obtained early on the rising limb. It is highly likely that the source component contribution from the Aberford topsoil would have been considerably larger during the period of maximum valley-floor inundation, probably > 80 percent. Data from subsequent storms confirmed this, as discussed below.

Samples 11–13 were obtained during a series of storms in October and November 1992, and the model output indicates a much smaller Aberford topsoil contribution, ranging between 0 percent and 13 percent. Although these values are similar to the mean Aberford topsoil contribution during low flow conditions (i.e. 7 percent), these three storms were all relatively small events, with peak discharges ranging between $42.9 \, l \, s^{-1}$ and $126.1 \, l \, s^{-1}$. There was no evidence of any surface runoff generation across the catchment slopes, nor was there any indication of overbank flooding and valley-floor erosion, shown earlier to be the dominant mechanism providing topsoil to the sediment load. In contrast, the next three samples (14–16) all show a shift to significantly increased Aberford topsoil contributions (33–54 percent) with a fairly consistent input from the Aberford channel banks (45–55 percent). Discharge during these storms ranged between $147.9 \, l \, s^{-1}$ (sample 14) and $227.0 \, l \, s^{-1}$ (sample 16), and suspended sediment concentrations were high, particularly during the storm on 2 December (maximum SSC $1350 \, mg \, l^{-1}$). Observations made in the field after this event revealed evidence of some valley-floor erosion, and rilling was noted on several of the catchment slopes.

The most complete set of storm data was obtained during two events, on 18 December 1992 and 13 January 1993. These storms produced the highest *measured* peak discharges (353.6 and 344.5 l s^{-1}, respectively) and, together, accounted for almost 20 percent of the annual suspended sediment yield (see Slattery and Burt, 1996). Importantly, these storms also occurred during daylight hours, and we were able to make detailed measurements of slope runoff and rill development, as well as general observations of the sediment delivery mechanisms.

The 18 December storm produced 18.7 mm of rainfall in 13 h (maximum 15-min intensity = 3.22 mm h^{-1}). Runoff was observed along compacted vehicle wheelings on several fields, and rill development was measured at a number of sites, including fields proximal to the main channel. Interestingly, despite considerable slope erosion (e.g. 42 tonnes of soil was lost from a single rill system on one field; see Slattery *et al.*, 1994), most of the sediment transported from rills was re-deposited within fields, in fans behind hedgerows, and along the valley floor. The river broke its banks at a discharge of *c.* 180 l s^{-1} and overbank flooding and erosion was the most widespread since the 25 September storm. These observations are entirely consistent with the model output data (Table 16.3) and also support the findings noted earlier regarding the increased Aberford topsoil contribution in samples 14–16. The mean contribution of the eight storm samples from 18 December is modelled as 63 percent Aberford topsoil and 37 percent Aberford channel banks. The individual samples, plotted schematically in Figure 16.4, indicate, at times, an almost exclusive Aberford topsoil contribution (e.g. samples 18 and 21), and a gradual shift in sediment source during the recession limb, with channel bank material becoming increasingly more important (e.g., sample 24 at 1830 hours is modelled as 45 percent Aberford topsoil and 55 percent Aberford channel banks). The only real anomaly appears to be sample 20, taken at 1240 hours during peak flow, which is modelled as a mixture of 24 percent, 60 percent and 16 percent Aberford topsoil, channel banks and Banbury series, respectively.

The model output data for the 13 January 1993 storm are also plotted in Figure 16.4, although here, samples were only retrieved during the first hydrograph peak due to equipment failure. This storm produced 17.7 mm of rainfall in 7 h but rainfall was more intense, generating the highest maximum 15-min rainfall intensity during the study period (31.4 mm h^{-1}). The mean of the seven storm samples is modelled as 39 percent Aberford topsoil and 58 percent Aberford channel banks, significantly different to the 18 December mean storm values. The lower overall Aberford topsoil contribution is most likely related to the fact that a series of storms had occurred just two days prior to the event (10–11 January), both large enough to result in a flushing of Aberford topsoil material from the sediment system, as noted by Slattery and Burt (1996) in an analysis of storm hysteresis loops. However, the individual samples from the 13 January storm show considerable variation in the modelled source component proportions: the first three samples (25–27) are ascribed a topsoil contribution of 12–30 percent Aberford, whereas the last four samples (28–31), taken over the storm peak, show a 33–62 percent topsoil contribution.

The results presented above suggest that, with only a few exceptions, the model output provides quantitative estimates of sediment source contributions that are consistent with the hydrological context and the changes in flow conditions through the sampling period. The absolute accuracy of these source contributions is, of course,

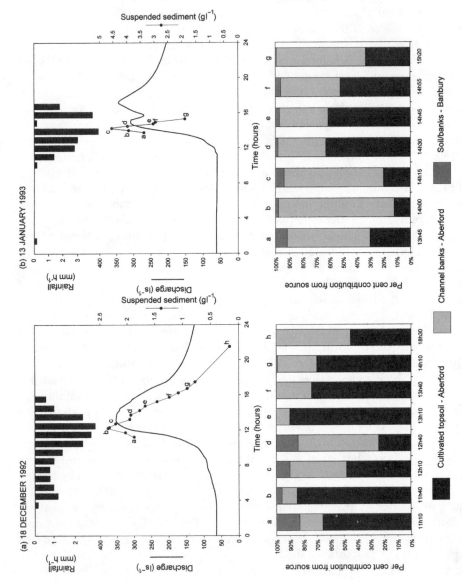

Figure 16.4 Changes in the contribution from individual sources to the suspended sediment load of the stream during two storms: (a) 18 December 1992; (b) 13 January 1993

uncertain. As indicated by Walden *et al.* (1997), qualitative and statistical analyses of these data can be used to highlight those suspended sediment samples that may prove more problematic for the model to unmix (e.g. sample 5 discussed above). A number of possible causes for the apparently less satisfactory model output for these samples can be suggested, including particular environmental circumstances (e.g. anthropogenic disturbance of the sediment system) or methodological limitations (e.g. accuracy of some magnetic parameters when measured on relatively small sample masses).

CONCLUSIONS AND DIRECTIONS FOR FUTURE RESEARCH

This study has demonstrated that sediment fingerprinting using mineral magnetic measurements is a viable approach to determining basin sediment sources. These measurements enabled us to identify the dominant sources of a stream's suspended sediment load with a degree of resolution that permitted quantification of the relative importance of different sources at different stages within storm events as well as between storms. We found that source contributions did vary under different flow conditions, with channel bank material dominating during periods of lower flow but with significantly increased contributions from topsoil during storm events. Many of the higher flow events generated overbank flow; this resulted in widespread valley-floor erosion and the incorporation of greater amounts of topsoil into the suspended sediment system.

Certainly, some caution is needed in interpretation of the quantitative output provided by the modelling process. For example, it is difficult to assess the accuracy of the quantitative estimates of source contributions as the absolute values produced by the model are dependent upon how the source materials are defined for each parameter measured (e.g. mean versus median; see Walden *et al.* (1997) for a more detailed discussion). The initial definition of the number and mean properties of the source materials is therefore both a critical component of the modelling process and, even with the most rigorous field procedures (e.g. Lees, 1994), conditional upon some element of subjective judgement on behalf of the researcher. Despite such qualifications concerning the modelling process, the general nature of the model output would seem sensible given the environmental context and is entirely consistent with field observations.

Improvements in the basic modelling procedure used here form the focus of ongoing work. In particular, this further research will address the issues of (i) the absolute accuracy of sediment source estimates produced by linear modelling, and (ii) integration of magnetic data with other compositional analyses to improve the fingerprinting of source materials and thus the model output.

REFERENCES

Caitcheon, G. G. 1993. Sediment source tracing using environmental magnetism: a new approach with examples from Australia. *Hydrological Processes*, **7**, 349–358.
Jarvis, M. G., Allen, R. H., Fordham, S. J., Hazeelden, J., Moffat, A. J. and Sturdy, R. G. 1984. *Soil Survey of England and Wales*. Bulletin 15, Harpenden.

Lees, J. A. 1994. *Modelling the magnetic properties of natural and environmental materials.* Unpublished PhD Thesis, University of Coventry, UK.

Maher, B. A. 1988. Magnetic properties of some synthetic sub-micron magnetites. *Journal of Geophysical Research,* **94**, 83–96.

Oldfield, F. 1991. Environmental magnetism: a personal perspective. *Quaternary Science Reviews,* **10**, 73–85.

Oldfield, F. and Clark, R. L. 1990. Lake sediment-based studies of soil erosion. In Boardman, J., Foster, I. D. L. and Dearing, J. A. (eds) *Soil Erosion on Agricultural Land.* John Wiley, Chichester, 201–230.

Oldfield, F., Rummery, T. A., Thompson, R. and Walling, D. E. 1979. Identification of suspended sediment sources by means of magnetic measurements. *Water Resources Research,* **15**, 211–218.

Richards, J. A. 1993. *Remote Sensing Digital Image Analysis.* Springer-Verlag, Berlin.

Slattery, M. C. and Burt, T. P. 1996. On the complexity of sediment delivery in fluvial systems. In Anderson, M. G. and Brooks, S. M. (eds) *Advances in Hillslope Processes.* John Wiley, Chichester, 635–656.

Slattery, M. C., Burt, T. P. and Boardman, J. 1994. Rill erosion along the thalweg of a hillslope hollow: a case study from the Cotswold hills, central England. *Earth Surface Processes and Landforms,* **19**, 377–385.

Thompson, R. 1986. Modelling magnetization data using SIMPLEX. *Physics of the Earth and Planetary Interiors,* **42**, 113–127.

Thompson, R. and Oldfield, F. 1986. *Environmental Magnetism.* Allen and Unwin, London.

Walden, J. and Slattery, M. C. 1993. Verification of a simple gravity technique for separation of particle size fractions suitable for mineral magnetic analysis. *Earth Surface Processes and Landforms,* **18**, 829–833.

Walden, J., Smith, J. P. and Dackombe, R. V. 1992. The use of simultaneous R- and Q-mode factor analysis as a tool for assisting interpretation of mineral magnetic data. *Mathematical Geology,* **24**, 227–247.

Walden, J., Slattery, M. C. and Burt, T. P. 1997. Use of mineral magnetic measurements to fingerprint suspended sediment sources: approaches and techniques for data analysis. *Journal of Hydrology,* **202**, 353–372.

Walling, D. E., Peart, M. R., Oldfield, F. and Thompson, R. 1979. Suspended sediment sources identified by magnetic measurements. *Nature,* **281**, 110–113.

Walling, D. E., Woodward, J. C. and Nicholas, A. P. 1993. A multi-parameter approach to fingerprinting suspended sediment sources. In Peters, N. E., Hoehn, E., Leibundgut, Ch., Tase, N. and Walling, D. E. (eds) *Tracers in Hydrology.* IAHS Publication **215**, IAHS Press, Wallingford. 328–338.

Yu, L. and Oldfield, F. 1993. Quantitative sediment source ascription using magnetic measurements in a reservoir-catchment system near Nijar, S.E. Spain. *Earth Surface Processes and Landforms,* **18**, 441–454.

17 Sediment Delivery to a Proglacial River: Mineral Magnetic, Geochemical and the Potential for Radionuclide Fingerprinting

I. D. L. FOSTER,[1] A. M. GURNELL[2] and D. E. WALLING[3]

[1]Centre for Environmental Research and Consultancy, NES (Geography), Coventry University, UK
[2]School of Geography and Environmental Sciences, University of Birmingham, UK
[3]Department of Geography, University of Exeter, UK

INTRODUCTION

In many regions of the world, meltwater from areas of permanent snow and ice cover provides the primary water resource. Meltwater draining from glacier basins is characterised by suspended sediment loads that are well above the global average (Gurnell et al., 1996) and which can present major problems for the management and use of the meltwater (Bezinge et al., 1989; Bogen, 1989). The estimation and prediction of both suspended sediment transport and meltwater discharge from glacier basins is, therefore, of major management importance. There is also a geomorphological rationale for such studies in terms of understanding the denudation systems of these areas and in drawing comparisons with other morphoclimatic zones. Geomorphological interest in the suspended sediment loads of proglacial streams centres on the very dynamic properties they confer on the sediment response and morphology of proglacial rivers when coupled with a variable discharge regime (e.g Maizels, 1983; Gurnell and Fenn, 1987), and on the degree to which this variability, combined with other factors, can be used to make inferences about subglacial processes (e.g. Gurnell, 1987; Collins, 1989; Gurnell and Warburton, 1990). Whilst there have been many independent studies of the suspended sediment output from glacier basins at a variety of timescales (e.g. Bogen, 1988; Lawler et al., 1992); of local subglacial suspended sediment transport (e.g. Stone et al., 1993; Hubbard et al., 1995); and of sediment–meltwater interactions, water pressure variations, and sediment deformation and movement at the glacier bed (e.g. Shoemaker, 1986; Clarke, 1987; Engelhardt et al., 1990), there have been few studies which have taken a whole-basin perspective on sediment sources and delivery processes.

Tracers in Geomorphology. Edited by Ian D. L. Foster. © 2000 John Wiley & Sons Ltd.

This chapter provides an outline of methods that might be used to couple patterns of suspended sediment and discharge output with controlling processes, by aiming to develop an understanding of the *relative* importance of sediment sources and delivery processes *at a glacier basin scale*. An attempt is made to demonstrate how an improved understanding can be gained of the processes controlling suspended sediment transport in proglacial rivers by coupling a knowledge of the dynamics of meltwater generation, storage and routing through glacier basins with a variety of sediment source fingerprinting techniques. Data are drawn from a number of locations, but mainly focus on a small Swiss glacier basin of 6.3 km^2 ice area where the two main types of subglacial drainage system that operate beneath temperate glaciers (distributed and conduit systems) are known to exist and to develop in a predictable manner through the melt season (Nienow *et al.*, 1998). The main rock types, in general terms, are arranged in transverse bands across the basin and produce suspended sediment which may be separated by its chemical and magnetic signatures (see below). Furthermore, the presence of extra-glacial sediment sources and the relatively slow rate of glacier movement in the context of the timescales of radionuclide fallout, offer potential to separate suspended sediment from the same lithologies but from subglacial and extra-glacial source areas on the basis of their fallout radionuclide signatures.

DISCHARGE AND SUSPENDED SEDIMENT REGIMES OF ALPINE PROGLACIAL RIVERS

The discharge of high alpine proglacial rivers is mainly generated by the melting of snow and ice. As a result of the high altitude, precipitation rarely falls as rain, and the primary energy source for snow and ice melt is incident radiation rather than advected energy. Thus the proglacial discharge regime at the annual and diurnal timescales reflects the temporal pattern of solar energy receipt, and also incorporates lag and attenuation effects which result from the varying routes and storages through which the meltwater passes as it drains to the proglacial river.

The seasonal and diurnal variability of proglacial river discharge is also reflected in the suspended sediment concentrations recorded in proglacial rivers. Indeed, the similarity in the temporal patterns of discharge and suspended sediment concentration has influenced approaches to estimating suspended sediment loads in proglacial rivers (Hodgkins, 1999). For example, suspended sediment rating curves have been estimated by linear regression analysis of the suspended sediment concentration and discharge time series. The problem with this approach is that the relationship between suspended sediment concentration and discharge in proglacial rivers is neither linear nor stable (Fenn *et al.*, 1985). Hysteresis in the relationship between suspended sediment concentration and discharge exists at a range of timescales from diurnal, through sub-seasonal and seasonal to interannual. As a result, researchers have attempted to develop the sediment rating curve methodology to incorporate other factors, including the influence of antecedent conditions (e.g. Richards, 1984) and the potential influence of englacially and subglacially routed meltwater (Gurnell and Fenn, 1984a). Furthermore, the presence of complex hysteresis in the residuals from sediment rating curves has led to the adoption of a time series approach which employs Box-Jenkins transfer functions between discharge and suspended sediment concentration (e.g. Gurnell and Fenn, 1984b; Gurnell et al., 1992). Such manipulations of the proglacial time series information produce improvements in the accuracy with which suspended sediment concentration can be estimated from discharge. However, the manipulations do not incorporate information on the many other processes that influence sediment transfer rates in proglacial streams. To achieve further understanding of the controls on suspended sediment concentration in proglacial rivers, and so to improve the accuracy with which suspended sediment loads may be estimated, it is necessary to identify how meltwater moves from its source areas to the proglacial river and to assess the degree to which it impinges upon primary and intermediate sources of sediment in transit.

Ice and snow melt occur at varying rates in different locations within the glacier basin and at different times within the ablation season. Once meltwater is produced, it can drain to the proglacial river along a variety of routes (extra-glacial, supraglacial, englacial and subglacial). It drains at different rates according to the route taken and the nature and size of the pathways through which it drains (intergranular spaces in snow pack, ice or subglacial moraine; small to very large channels, conduits and cavities; sheets of water at the snow/ice or ice/bedrock/moraine interfaces; open channels). As the meltwater drains along these pathways, it is exposed to a variety of sediment sources, and its ability to tap these sources varies according to the energy of the flow and the availability of fine sediment in the source areas. In the temperate glacial environment, the major source areas for suspended sediment are extra-glacial areas (including areas covered by snow patches and small hanging glaciers) and the subglacial environment.

Subglacial drainage below temperate glaciers may take place in major conduits or tunnel systems (Röthlisberger, 1972; Walder and Fowler, 1989); in smaller conduits which link cavities (Walder, 1986; Kamb, 1987); to a lesser extent, as a thin film or sheet between the ice and the glacier bed (Weertman, 1972); and, where permeable subglacial sediments exist, water may flow through these sediments. The proportion of meltwater carried by each of these components of the subglacial drainage system varies between glaciers, between different areas of the bed of the same glacier, and with time (Seaberg et al., 1988; Hooke, 1989; Willis et al., 1990). Differences in the stability, water pressure characteristics and seasonal dynamics of these components of the subglacial drainage system would be expected to have a major influence on the degree to which the meltwater gains access to fine sediment as it is transmitted through the glacier drainage system.

In summary, research has recognised the complexities of glacier hydrological processes, but has not proceeded to investigate specifically the influence of these processes on suspended sediment transport from glacier basins. Although many researchers have monitored the suspended sediment concentration of proglacial rivers and have made inferences about sediment source areas and routing processes from these observations, and a smaller number of researchers (e.g. Stone et al., 1993; Hubbard et al., 1995) have attempted to directly measure hydrological and suspended sediment transfer processes at the glacier bed, very few studies have previously tried to directly establish the link between sediment source areas and delivery processes operating within the entire glacier basin, establish their relative significance, and link these source areas to the suspended sediment transported in a proglacial river. Some preliminary results from an ongoing research programme are presented in an attempt to address this major gap in research through the use of sediment fingerprinting techniques coupled with hydrological analyses.

THE STUDY BASIN

The sediment source tracing methodology proposed here is based upon preliminary data provided by research undertaken within the basin of Haut Glacier d'Arolla, Valais, Switzerland (Figure 17.1). Haut Glacier d'Arolla is particularly suited to the geochemical and mineral magnetic fingerprinting techniques described below for three reasons.

First, the nature and dynamics of the subglacial drainage system have already been extensively researched (Sharp et al., 1993; Nienow et al., 1998). Both conduit and distributed drainage systems exist beneath the glacier. In simple terms, the conduit system establishes itself in late June–early July and then grows up-glacier at the expense of the distributed system until late July–early August. Although a residual distributed system remains and coexists with the major conduits, such that water moves locally between the two systems under a diurnally varying pressure gradient (Hubbard et al., 1995), nevertheless, a broad, seasonal up-glacier evolution in sub-glacial drainage occurs. Thus, both main types of subglacial drainage system that are known to exist beneath temperate glaciers occur in this basin, and they develop through the ablation season in a way that can be readily identified from field observations and from an analysis of the proglacial river discharge record (Gurnell, 1993).

Secondly, although the geology of the catchment is complex, a broad contrast can be drawn between the lower part of the glacier and the upper part of the glacier, which are underlain by different rock types (Figure 17.1). Sediment from these two zones appears to be clearly distinguishable on the basis of its chemical and magnetic signatures (see below). This up-glacier change in rock type provides an ideal basis for linking the fingerprints of sediment in the proglacial river to lithological source areas as the subglacial drainage system changes, because it is possible to assume that sediment under the upper part of the glacier represents one major lithological source whereas both sources may contribute to the soft bed under the lower part of the glacier.

Thirdly, there are meltwater and sediment inputs from snow patches and small hanging glaciers on the mountainsides above the main glacier. These are small enough to be related to single lithological sources of sediment and also contain fine sediment that will have been exposed to the deposition of fallout radionuclides. This latter property provides an approach to separating sediments delivered from beneath the main glacier from extra-glacial source areas of the same lithology.

Hydrology and Sediment Transport Characteristics of the Haut Glacier d'Arolla

Nienow et al. (1998) describe the results of dye-tracer studies which have defined the seasonal evolution of two distinct drainage systems beneath the Haut Glacier

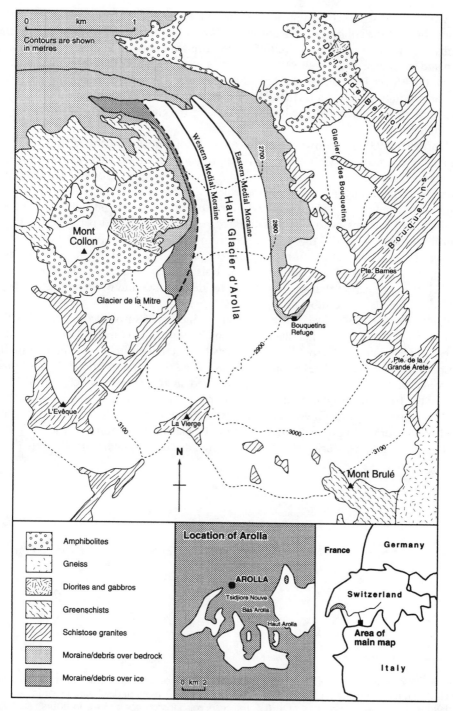

Figure 17.1 The Haut Glacier d'Arolla basin: simplified geology (white areas are areas of permanent snow and ice). (Information provided by Dr G. H. Brown)

d'Arolla – a distributed system and a conduit system. They demonstrate that the upper limit of the conduit system corresponds approximately with the position of the transient snowline. The upper limit of the distributed system is limited by the area of firn and snow generating meltwater. Nienow *et al.* (1998) show that by mid-August in 1990, the lowest 1 km of the glacier bed was largely drained by major conduits. The major conduits extended a further 2 km up-glacier, coexisting with a residual distributed system, and the remaining headward area of the glacier was largely drained by a distributed system.

Gurnell (1993) has shown how the analysis of linear elements within diurnal discharge recession curves observed in the proglacial river can be used to define the changing residence times of water within distinct hydrological reservoirs and the time periods during which each of the reservoirs is active (Figure 17.2). The two most rapidly responding (of four) linear reservoirs appear to correspond to the distributed and conduit systems, with similar estimated residence times and periods of activity during the melt season to those identified from the dye tracer experiments described above. Thus a combination of monitoring the position of the transient snowline and the upper limit of melt, coupled with an analysis of the proglacial discharge record, is sufficient to isolate the changing locations, residence times and water storage of these two components of the subglacial drainage system.

Gurnell *et al.* (1994) have shown how the production of suspended sediment by the glacier varies through distinct phases in the melt season and Gurnell (1995) associates these changes in productivity with the level of activity of the two most rapidly responding hydrological reservoirs (interpreted as the conduit and the distributed systems) and with the amount of water stored in the glacier, which can also be estimated from the proglacial discharge record. These associations can be illustrated by an analysis of the 1989 discharge record (Figure 17.3). Periods of increasing meltwater storage (more meltwater entering the glacier than can readily leave it) are associated with times during which a distributed drainage system is developing under

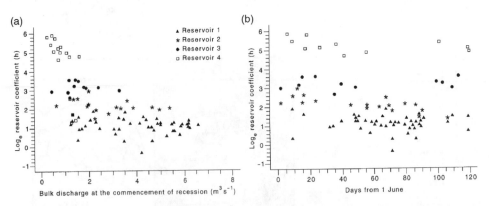

Figure 17.2 Estimated residence times for four linear hydrological reservoirs based on the analysis of diurnal discharge recession curves selected from 10 years of discharge records. (a) The relationship between residence time (i.e. the reservoir storage coefficient, in hours) and discharge at the commencement of flow recession. (b) Variation in estimates of residence time through the melt season (after Gurnell, 1995)

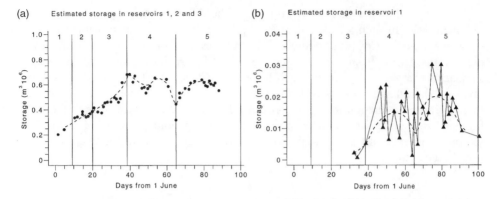

(a) Estimated storage in reservoirs 1, 2 and 3 (b) Estimated storage in reservoir 1

Figure 17.3 Estimates of meltwater storage in the Haut Glacier d'Arolla during the 1989 melt season. (a) Total meltwater storage in three linear hydrological reservoirs. (b) Meltwater storage in the most rapidly responding reservoir (the conduit system?) (after Gurnell, 1995)

the glacier, but the conduit system is not developed to any significant extent (Periods 1 to 3, Figure 17.3). These culminate in Period 3, when water storage is high and the suspended sediment concentration is relatively responsive to variations in the melt-water discharge (Table 17.1). Periods of relatively stable total storage in reservoirs 1 to 3 (Periods 4 and 5, Figure 17.3) appear to relate to phases when the area of the glacier drained by a distributed system migrates up-glacier as the conduit system extends headwards. The increasing extent of the conduit system is reflected in the higher water storage estimates for reservoir 1 during Period 5 than during Period 4 (Figure 17.3(b)). The phase of low storage in Figures 17.3(a) and (b), which occurs at the boundary between Periods 4 and 5, reflects a time of summer snowfall and flow recession from the glacier. In summary, the evidence presented in Table 17.1 and Figures 17.2 and 17.3 suggests that the nature and activity of the subglacial drainage system is probably the key to understanding the varying location of subglacial sediment source areas and the delivery of sediment from those source areas to the proglacial stream.

Sediment fingerprinting approaches should allow separation of the contribution of extra-glacial from subglacial sediment sources to test the hypothesis that the subglacial zone is the most important sediment source in glacier basins. It will also allow

Table 17.1 Simple regression relationships estimated between $\log_{10}$ suspended sediment concentration and $\log_{10}$ discharge at the best temporal match position Haut Glacier d'Arolla, 1989

Dependent variable (lag in hours)	Independent variable	Period	a	t_a	b	t_b	n	R^2
$\log_{10} S$ (0)	$\log_{10} Q$	1	1.75	6.6	0.31	3.2	137	0.072
$\log_{10} S$ (0)	$\log_{10} Q$	2	−0.15	0.6	0.86	11.7	116	0.545
$\log_{10} S$ (0)	$\log_{10} Q$	3	−3.95	28.4	1.98	47.2	506	0.815
$\log_{10} S$ (−2)	$\log_{10} Q$	4	−0.87	5.2	1.09	22.9	658	0.445
$\log_{10} S$ (0)	$\log_{10} Q$	5	−1.07	7.3	1.21	29.2	490	0.443

assessment of the changing contributions of the different lithologies to suspended sediment yield through the melt season. In particular, it should be possible to isolate subglacial contributions that contain sediment derived from lithologies in the lower part of the basin, from those which do not. Fingerprinting analyses applied to subglacial samples will provide corroborative evidence. If the hypothesis that the distributed drainage network is the major subglacial sediment source area is correct, then a gradual reduction in the proportion of subglacial sediment derived from the lithology that underlies the lower part of the glacier would be expected as the melt season progresses and the conduit system develops up-glacier at the expense of the distributed system.

SEDIMENT SOURCE TRACING

A range of diagnostic sediment properties including physical, chemical, mineral magnetic and environmental radionuclides have been used for tracing sediment sources (cf. Walling *et al.*, 1979, 1993; Walling and Kane, 1984; Peart and Walling, 1988; Foster *et al.*, 1990, 1996a, 1998; Walling and Woodward, 1992, 1995; Foster and Walling, 1994; Owens *et al.*, 1997; Walden *et al.*, 1997). Pilot studies have already been undertaken to illustrate the potential for application of these methods in the Haut Glacier d'Arolla basin.

In simple terms, the basin is dominated by two major lithological units (Figure 17.1). Furthermore, fine sediment mainly derives from either subglacial or extra-glacial sources associated with each of the two main lithological units. A combination of these factors provides four potential sources which might be quantitatively discriminated. However, sediment beneath the glacier and in old lateral moraines on the valley sides represents a variable mix of material from these two major sources as a result of transport and deposition of sediments by the glacier. Geochemical, physical and mineral magnetic fingerprinting techniques provide approaches to separating the main lithological source areas.

To illustrate the potential of such approaches, a pilot project was undertaken, where suspended sediment was sampled from two streams which are believed to drain two different lithological units. One sample site was on a proglacial stream close to the terminus of a small hanging glacier located to the east of the most downstream portion of the main glacier. The second sampling site was a tributary stream draining from the main glacier, whose catchment area had been established by dye-tracer observations (Sharp *et al.*, 1993) to be largely confined to the western headwaters of the glacier basin. Whilst a more comprehensive survey of source streams would provide a fuller characterisation of the detailed variations between sediment sources within the basin, these two sample sites provided a representation of gross differences between the two main lithological units, which hereafter will be named 'amphibolite' and 'granite', and are sufficient to illustrate the potential of a sediment fingerprinting approach. In addition, samples of bulk meltwater from the entire catchment were taken from the main proglacial stream during August 1993. Geochemical, particle size and mineral magnetic analyses were then undertaken on the < 63 μm fractions of these sediment samples.

Sediment Chemistry

Selected samples were digested using a mixture of perchloric, sulphuric and nitric acids and analysed for Na, K, Ca and Mg by atomic absorption spectrophotometry following the methods described by Foster et al. (1996b).

Discrimination between the lithological units and the relative significance of these sources in controlling the chemistry of suspended sediments in the proglacial stream are shown in Figure 17.4. The <63μm fluvial sediment derived from amphibolite is characterised by over 80 percent Mg, less than 10 percent Ca and less than 15 percent of Na+K. In contrast, fluvial sediment derived from granite areas contains roughly equal proportions of Na+K and Ca but less than 30 percent Mg. The average of eight samples collected from the proglacial stream suggests that some 85–90 percent of the <63μm sediments is derived from granite sources.

Particle Size

Particle size analysis of the < 63μm fraction of source and main stream sediment was undertaken by laser granulometry after ultrasonic dispersal following the methods described by Foster et al. (1991). Particle size distributions of suspended sediment are affected by the transport processes as well as by the particle size distribution of the sediment source and so interpretation of the controls on the particle sizes of sediment, even of this fine fraction, should be approached with caution. Nevertheless, the particle size distribution (PSD) curves (Figure 17.5) suggest that sediment derived from areas underlain by amphibolite is generally characterised by finer material at the 10th, 50th and 90th percentiles of the cumulative PSD than that derived from granite areas. Furthermore, the multimodal nature of the PSD curve for samples from the

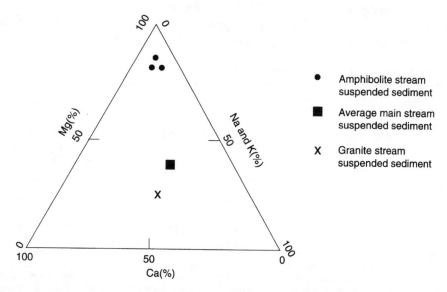

Figure 17.4 Geochemical discrimination of fluvial sediments from amphibolite and granite streams and the main stream; Haut Glacier d'Arolla, 25 August 1993

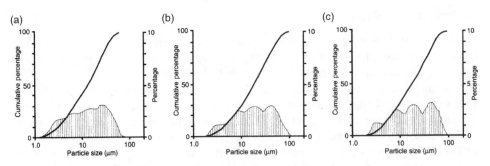

Figure 17.5 Particle size distribution curves for stream sediments from the (a) amphibolite and (b) granite streams and (c) a main stream sample; Haut Glacier d'Arolla, 25 August 1993

granite areas is less evident in the amphibolite sample although it is clearly evident in the fluvial sediments collected from the proglacial stream.

Mineral Magnetism

Measurements of low (χ_{LF}) and High (χ_{HF}) frequency magnetic susceptibility and a range of remanent magnetic properties (cf. Thompson and Oldfield, 1986; Foster *et al.*, 1998; see Table 17.2 for definitions of measured and derived magnetic signatures) were also made on the suspended sediment samples. Selected samples were used to derive IRM acquisition curves and were also measured in a Molspin Vibrating Sample Magnetometer (VSM). Good discrimination between the sources was provided by the values of χ_{LF} and HIRM (Table 17.3) and the validity of mineral magnetic finger-printing is further demonstrated by the IRM acquisition curves for the source materials and, more especially, by the VSM data of Figure 17.6.

Figure 17.7(a) shows the results of applying simple mixing models, based on average geochemical and mineral magnetic properties of potential source materials,

Table 17.2 Measured and derived magnetic measurements

Susceptibility/remanence property	Measured (M)/ derived (D)	Instrument	Units
χ_{LF}	M	Bartington MS2B	$10^{-6}\text{m}^3\text{kg}^{-1}$
χ_{HF}	M	Susceptibility meter	$10^{-6}\text{m}^3\text{kg}^{-1}$
χ_{FD}	D		$10^{-9}\text{m}^3\text{kg}^{-1}$
$\chi_{FD}\%$	D		%
$\text{IRM}_{(0.025-0.8T)}$	M	Pulse magnetiser	$\text{mAm}^2\text{kg}^{-1}$
$\text{IRM}_{(-0.1T)}$	M	Molspin fluxgate magnetometer	$\text{mAm}^2\text{kg}^{-1}$
S ratio	D	$(\text{IRM}_{(-0.1T)}/\text{IRM}_{(0.8T)})_{x-1}$	dimensionless
HIRM	D	$(\text{IRM}_{(0.8T)} \times (1 - \text{S ratio}))/2$	$\text{mAm}^2\text{kg}^{-1}$

$\chi =$ Magnetic susceptibility measured at low (LF) and high (HF) frequency. Derived measures are frequency-dependent susceptibility expressed as a concentration (FD) and as a percentage (FD%).
IRM = Isothermal remanent magnetisation acquired in a range of forward and reverse magnetic fields of varying strength (tesla).
(See Foster *et al.*, 1998 for further details.)

Table 17.3 Mineral magnetic characteristics of sources and mixtures in the main stream

Sample	χ_{LF} $(10^{-6}m^3\,kg^{-1})$	$IRM_{0.8T}$ $(mAm^2\,kg^{-1})$	$IRM_{-0.1T}$ $(mAm^2\,kg^{-1})$	S-ratio	HIRM $(mAm^2\,kg^{-1})$
Stream 1	0.132	2.227	−1.400	0.629	0.413
Stream 2	0.134	1.656	−0.880	0.532	0.388
Stream 3	0.143	2.055	−1.168	0.568	0.444
Stream 4	0.147	2.368	−1.382	0.584	0.493
Stream 5	0.135	2.001	−1.125	0.562	0.438
Stream 6	0.116	1.365	−0.762	0.558	0.302
Stream 7	0.124	1.955	−1.022	0.523	0.466
Stream 8	0.123	2.131	−1.161	0.545	0.485
Unit 1	0.537	12.345	−7.639	0.619	2.353
Unit 1	0.485	11.127	−6.606	0.594	2.260
Unit 1	0.444	9.931	−6.045	0.609	1.943
Unit 2	0.113	1.894	−1.198	0.633	0.348

Unit 1: amphibolite stream.
Unit 2: granite stream.
Stream 1–8: proglacial stream samples 25–26 August 1993.

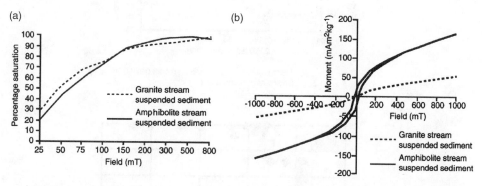

Figure 17.6 Magnetic discrimination of stream sediments draining Lithological Units 1 and 2, Haut Glacier d'Arolla, 25 August 1993. (a) IRM acquisition curves; (b) VSM hysteresis loops

to samples collected for a single day's ablation in August 1993. Although over 90 percent of the sediment appears to be derived from streams draining granite sources, there is some variability throughout the day. With high but declining total sediment concentration between 15.00 hours and 21.00 hours on 25 August, the relative importance of suspended sediment derived from the granite source decreases according to independent results obtained from a geochemical (Na) and mineral magnetic (χ_{LF}) signature. Between midnight on 25 August and 09.00 hours on 26 August, however, granite sources dominate the suspended sediment transported by the proglacial stream. For this ablation period, the highest contribution of amphibolite to the main stream sediment occurs at sediment concentrations exceeding $c.300\,mg\,l^{-1}$ and suspended sediment concentrations are significantly ($p > 0.05$) and positively correlated with stream discharge.

(a)

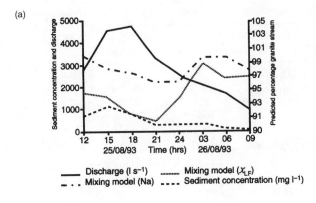

Discharge (l s⁻¹) ········ Mixing model (χ_{LF})
─ · · Mixing model (Na) ─ ─ ─ Sediment concentration (mg l⁻¹)

(b)

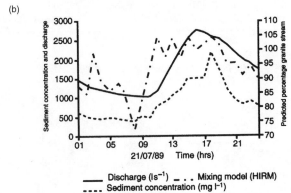

Discharge (ls⁻¹) ─ · · Mixing model (HIRM)
─ ─ ─ Sediment concentration (mg l⁻¹)

(c)

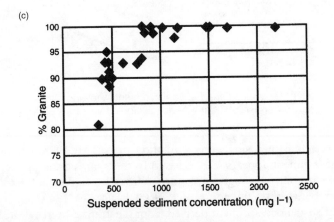

Figure 17.7 Geochemical and magnetic mixing models based on source area data from Haut Glacier d'Arolla and the chemical and mineral magnetic characteristics of proglacial stream sediments showing: (a) relative contribution of sediments to the main stream based independently on χ_{LF} and Na concentrations for samples collected on 25 August 1993; (b) relative contribution of sediments to the main stream based on HIRM measurements of filter paper residues for samples collected on 21 July 1989; (c) a three-component mixing model used to compute the proportion of fine sediment derived from granite sources plotted as a function of suspended sediment concentration for samples collected on 21 July 1989

Samples of suspended sediment have been collected from the main proglacial stream draining the Haut Glacier d'Arolla over a number of field seasons. These small volume (500 ml) samples have been field filtered and dried and the sediments retained for further analysis. Although the sample mass retained on the filter papers was inadequate for determining χ_{LF} and χ_{HF}, sufficient sensitivity was achieved for the measurement of mineral magnetic remanence properties. A magnetic mixing model was applied to samples collected during July 1989 when the sediment concentrations were considerably higher than those found in August 1993 (Figure 17.7(b)). Here, sediment concentration is again significantly correlated with stream discharge, but the contribution from the amphibolite source is predicted momentarily to exceed 25 percent before the first significant rise in sediment concentration after 09.00 hours on 21 July. In contrast to the August 1993 data, however, the data from July 1989 suggest that granite sources dominate at high sediment concentrations and discharges. These two examples suggest that the relative contribution of granite and amphibolite sources varies substantially throughout the ablation season and with diurnal variations in flow.

More sophisticated mixing models have been developed involving combinations of mineral magnetic and geochemical parameters, using a Microsoft Excel Solver solution following the procedure described by Walden et al. (1997). An example of a three-component mixing model using HIRM, SIRM and $IRM_{0.1T}$ to estimate the relative proportion derived from granite sources in the proglacial stream for 21 July 1989 (Figure 17.7(b)) is given in Figure 17.7(c). The proportional contribution of sediment from areas of granite is plotted against suspended sediment concentration and the results demonstrate that the proportion derived from granite sources increases with increasing suspended sediment concentration. At a threshold concentration of $c.800 \, \mathrm{mg} \, l^{-1}$ and above, the model predicts that the stream sediments are dominated by contributions from granite sources.

These preliminary results appear to offer a potentially successful set of methodologies based on mineral magnetic and geochemical signatures for discriminating source area contributions on the basis of lithology. They also demonstrate that the relative contribution, derived from the two potential sources, to the suspended sediment transported by the proglacial river varies in relation to suspended sediment concentration and the diurnal ablation cycle at different times of the year. The potential for source tracing is demonstrated by a simple conceptual model of sediment source linkages and fingerprint signatures presented in Figure 17.8. Here it is suggested that the relative proportion of amphibolite and granitic fluvial sediments in the proglacial stream of the Haut Glacier d'Arolla (Figure 17.8(a)) will provide characteristic chemical and mineral magnetic signatures (Figure 17.8(b)) which could be modelled using simple mixing models to identify the dominant diurnal and seasonal sediment sources to the glacial drainage network (Figure 17.8(c)). However, further research is required in order, first, to identify the particle size controls on sediment chemical and mineral magnetic signatures; secondly, to assess the variability in source characteristics; and, thirdly, to identify the degree to which the mixed lithologies in glacial deposits, which may provide contemporary sediment sources, confound attempts to accurately model the ablation limits.

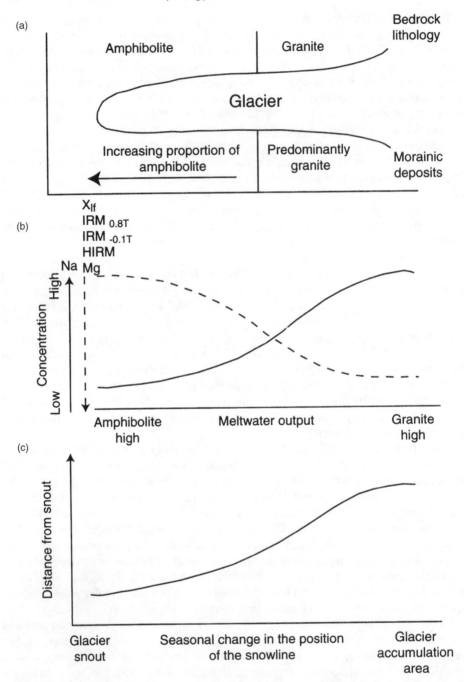

Figure 17.8 A conceptual model of chemical and mineral magnetic source discrimination for the Haut Glacier d'Arolla. (a) The spatial location of bedrock sources; (b) variations in chemical and magnetic signatures reflecting dominant spatial source; (c) seasonal change in the distance of the snowline from the proglacial stream

THE POTENTIAL FOR RADIONUCLIDES

Whereas the mineral magnetic and geochemical fingerprinting techniques described above may provide a basis for discriminating suspended sediment derived from parts of the study area underlain by different lithologies, there is a need to identify potential fingerprint properties capable of, first, discriminating sediment derived from subglacial/englacial and supraglacial/extra-glacial sources and, secondly, providing information on the possible residence times of sediment passing through the sediment delivery system from supraglacial/extra-glacial to subglacial/englacial locations before being delivered to the proglacial stream. Fallout radionuclides offer considerable potential for this purpose, since they are essentially independent of lithology and will only be associated with fine sediment derived from surface material exposed to fallout. Work undertaken in Greenland on the Mitdluagkat glacier in eastern Greenland (cf. Hasholt and Walling, 1992) has already demonstrated how caesium-137 (^{137}Cs) can provide an effective means of discriminating suspended sediment derived from glacial and proglacial sources. In that environment, the recently exposed proglacial area provides a major source of the sediment transported by the Mitdluagkat stream.

Whilst the use of ^{137}Cs has been successfully demonstrated by Hasholt and Walling (1992), two additional fallout radionuclides, namely unsupported lead-210 (^{210}Pb) and beryllium-7 (^{7}Be) would also appear to offer potential as complementary fingerprint tracers. By virtue of their different sources, fallout histories and half-lives, these additional radionuclides afford further potential for source discrimination and interpretation (cf. Wasson et al., 1987; Burch et al., 1988; Oldfield and Clark, 1990; Walling and Woodward, 1992; Wallbrink and Murray, 1993, 1994, 1996; Walling et al., 1993), although this potential has not to date been exploited in glacierised drainage basins. Caesium-137 is a human-made radionuclide, with a half-life of 30.1 years, associated primarily with the past testing of thermonuclear weapons. The temporal record of fallout reflects the history of weapons testing and the major proportion of the associated global fallout occurred during the period 1955–1970 (cf. Figure 17.9). In some parts of Europe, additional fallout was associated with the Chernobyl disaster of 1986, but information on levels of Chernobyl-derived ^{137}Cs fallout over Switzerland compiled by the Federal Commission on Radioactivity Monitoring (Commission Fédérale de Surveillance de la Radioactivité, 1987, 1989) indicates that values recorded in Valais were amongst the lowest in the country and were less than 300 mBq cm^{-2}. Data on both bomb- and Chernobyl-derived ^{137}Cs inventories compiled by the same Commission indicate that total ^{137}Cs inventories in the study area are likely to be c.400 mBq cm^{-2}, with Chernobyl inputs contributing c.30 percent of this total. The bomb-derived signal will, therefore, be dominant and Chernobyl inputs will not unduly complicate interpretation of ^{137}Cs activities.

In contrast, unsupported ^{210}Pb and ^{7}Be are natural fallout radionuclides and their annual fallout can be viewed as essentially constant through time rather than having occurred during a particular period. Lead-210 is a product of the ^{238}U decay series, with a half-life of 22.26 years. It is derived from the decay of gaseous ^{222}Rn, the daughter of ^{226}Ra. Radium-226 occurs naturally in soils and rocks and will generate ^{210}Pb which will be in equilibrium with its parent (cf. Wise, 1980). Diffusion of a small proportion of the ^{222}Rn from the soil introduces ^{210}Pb into the atmosphere and its

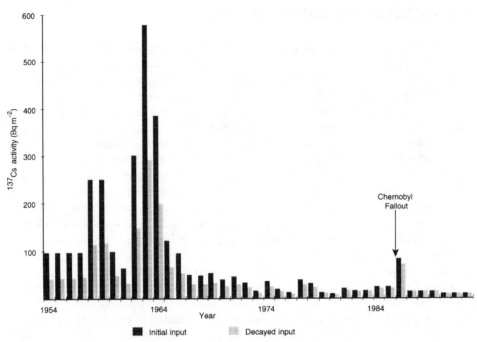

Figure 17.9 The typical temporal pattern of ^{137}Cs fallout input to sites in the northern hemisphere (based on measurements undertaken at Chilton, Oxfordshire. Data provided by the AERE Harwell Laboratory). (Decay calculated to 1992)

subsequent fallout provides an input of this radionuclide to surface soils and sediments which is not in equilibrium with its parent ^{226}Ra and is therefore termed unsupported ^{210}Pb. The amount of unsupported or atmospherically derived ^{210}Pb in a sediment sample can be calculated by measuring both ^{210}Pb and ^{226}Ra and subtracting the supported or in-situ component. Beryllium-7 is a cosmogenic radionuclide produced in the upper atmosphere by cosmic ray spallation of nitrogen and oxygen, which again arrives at the land surface as fallout (Arnold and Al-Salih, 1955; Lal and Peters, 1967). In this case the radionuclide is short-lived (half-life 53.3 days) and its behaviour in terrestrial and aquatic environments can be used to trace contemporary sediment movement.

By deriving composite fallout radionuclide fingerprints for suspended sediment and its potential sources, based on measurements of ^{137}Cs, unsupported ^{210}Pb and ^{7}Be activity, it should be possible to interpret both the dominant *source* (i.e. englacial/subglacial or supraglacial/extra-glacial) and the recent and medium-term history of the sediment (Table 17.4). Thus, for example, only sediment exposed to fallout within the past few months will contain significant levels of ^{7}Be, whereas sediment exposed for longer periods in the past will contain both ^{137}Cs and unsupported ^{210}Pb, with the levels reflecting both the timing and the duration of exposure. Because unsupported ^{210}Pb fallout has remained effectively constant through time, whereas ^{137}Cs fallout was essentially restricted to the period 1955–1970, comparison of the levels of these

Table 17.4 The potential for three radionuclides to discriminate meltwater suspended sediment sources

Radionuclide	Half-life (years)	Activity in meltwater sediments	Likely dominant source	Residence time in englacial/ subglacial locations if detected (years)
[7]Be	0.145	High	Supraglacial/ extra-glacial	< 0.5
		Low	Englacial/ subglacial	
[137]Cs	30.2	High	Supraglacial/ extra-glacial	< 45
		Low	Englacial/ subglacial	
[210]Pb (unsupported)	22.3	High	Supraglacial extra-glacial	< 70
		Low	Englacial/subglacial	

two radionuclides will provide further information on the history of the sediment involved, which will provide a basis for not only separating main-glacier subglacial sediments from supraglacial and extra-glacial sediments, but also for differentiating between extra-glacial sediment sources including exposed moraines, snow patches or small hanging glaciers.

CONCLUSIONS

This contribution has drawn upon successful attempts to use chemical and mineral magnetic signatures and the radionuclide ^{137}Cs to discriminate sediment sources in a glaciated environment and has suggested that two additional environmental radionuclides, [7]Be and ^{210}Pb, offer additional potential for discriminating extra- and englacial sediment sources. A number of specific conclusions may be drawn from the above analysis and discussion.

First, the use of mineral magnetic and geochemical signatures, coupled with multivariate mixing models, offers an effective basis for discriminating sediment sources on the basis of lithology during seasonal and diurnal ablation activity and may provide a method for testing models of glacier drainage pathways. Secondly, ^{137}Cs has been shown to be an effective tracer for differentiating fine sediment derived from extra-glacial and en- or subglacial sources. Thirdly, complementary measurements of the radionuclides ^{210}Pb and [7]Be should allow further discrimination between sediments mobilised from extra-glacial and en- or subglacial sources both recently ([7]Be) or over longer timescales (^{137}Cs and ^{210}Pb).

The development of multi-parameter fingerprinting models offers the potential to provide a powerful means of discriminating sediment sources spatially and of providing information on the timescales of sediment transfer. The data and approach presented in this chapter provide a framework that could be used to couple patterns of suspended sediment and discharge output with controlling processes by developing an understanding of the relative importance of sediment sources and delivery processes at a glacier basin scale.

Nevertheless, it is important to stress that the methodologies proposed in this chapter are aimed at a whole-catchment scale. Indeed, the preliminary results offered here reflect a very simple semi-distributed view of sediment sources based upon a crude, twofold lithological classification and a lumped hydrological perspective on the functioning of the subglacial drainage system. Even if a more complex sampling scheme was used to explore the range of sediment sources and if multi-parameter mixing models were developed, the results would still reflect considerable spatial lumping. The approach proposed in this chapter complements, both in spatial scale and in level of field effort, the detailed, process-based field investigations of subglacial environments that are currently being undertaken (e.g. Hubbard *et al.*, 1995). Whilst direct measurements of processes at the glacier bed offer higher spatial resolution and thus reveal greater mechanistic understanding, they are logistically confined to relatively small areas of the glacier bed, and process observations may only be maintained for relatively short periods of time. This chapter offers an innovative approach that can contribute to the armoury of field and laboratory techniques that may be used to develop a fuller understanding of the hydrological and sediment transfer systems of glaciers.

ACKNOWLEDGEMENTS

The authors wish to thank Joan Lees of Coventry University for making VSM measurements and the Cartographic Unit at Coventry University for drawing the diagrams. Dr Giles Brown provided the geological information for Figure 17.1.

REFERENCES

Arnold, J. R. and Ali Al-Salih, H. 1955. Beryllium-7 produced by cosmic rays. *Science*, **121**, 451–453.

Bezinge, A., Clark, M. J., Gurnell, A. M. and Warburton, J. 1989. The management of sediment transported by glacial meltwater streams and its significance for the estimation of sediment yield. *Annals of Glaciology*, **13**, 1–5.

Bogen, J. 1988. A monitoring programme of suspended sediment transport in Norway. In Bordas, M. P. and Walling, D. E. (eds) *Sediment Budgets*. International Association of Hydrological Sciences Publication **174**, 149–159.

Bogen, J. 1989. Glacial sediment production and development of hydro-electric power in glacierized areas. *Annals of Glaciology*, **13**, 6–11.

Burch, G. J., Barnes, C. J., Moore, I. D., Barling, R. and Olley, J. M. 1988. Detection and prediction of sediment sources in catchments. Use of [7]Be and [137]Cs. In *Proceedings of Hydrology and Water Resources Symposium*, ANU, Canberra, February 1988. Australian Institute of Engineers Conference Publication No. 88/1, 146–161.

Clarke, G. K. C. 1987. Subglacial till: a physical framework for its properties and processes. *Journal of Geophysical Research*, **92**, 9023–9036.

Collins, D. N. 1989. Seasonal development of subglacial drainage and suspended sediment delivery to melt waters beneath an alpine glacier. *Annals of Glaciology*, **13**, 45–50.

Commission Fédérale de Surveillance de la Radioactivité 1987. 29ème Rapport Pour les Années 1985 et 1986.

Commission Fédérale de Surveillance de la Radioactivité 1989. 30ème Rapport Pour les Années 1987 et 1988.

Englehardt, H., Humphrey, N., Kamb, B. and Fahnestock, M. 1990. Physical conditions at the base of a fast moving Antarctic ice stream. *Science*, **248**, 57–59.

Fenn, C. R., Gurnell, A. M. and Beecroft, I. 1985. An evaluation of the use of suspended sediment rating curves for the prediction of suspended sediment concentration in a pro-glacial stream. *Geografiska Annaler*, **67A**, 71–82.

Foster, I. D. L. and Walling, D. E. 1994. Sediment yields and sources in the catchment of the Old Mill Reservoir, South Devon, UK over the past 50 years. *Hydrological Sciences Journal*, **39**, 347–368.

Foster, I. D. L., Grew, R. and Dearing, J. A. 1990. Magnitude and frequency of sediment transport in agricultural catchments. In Boardman, J., Foster, I. D. L. and Dearing, J. A. (eds) *Soil Erosion on Agricultural Land*. John Wiley, Chichester, 153–171.

Foster, I. D. L., Albon, A., Bardell, K. M., Fletcher, J. L., Jardine, T. C., Mothers, R. J., Pritchard, M. A. and Turner, S. E. 1991. Coastal sedimentary deposits on the Isles of Scilly: storm-surge or tsunami deposit? *Earth Surface Processes and Landforms*, **16**, 341–356.

Foster, I. D. L., Owens, P. N. and Walling, D. E. 1996a. Sediment yields and sediment delivery in the catchments of Slapton Lower Ley, South Devon, UK. *Field Studies*, **8**, 629–661.

Foster, I. D. L., Charlesworth, S. and Proffitt, S. B. 1996b. Sediment-associated heavy metal distribution in urban fluvial and limnic systems: a case study of the River Sowe, UK. *Archiv fur Hydrobiologia*, **47**, 537–545.

Foster, I. D. L., Lees, J. A., Owens, P. N. and Walling, D. E. 1998. Mineral magnetic characterisation of sediment sources from an analysis of lake and floodplain sediments in the catchments of the Old Mill reservoir and Slapton Ley, South Devon, UK. *Earth Surface Processes and Landforms*, **23**, 685–704.

Gurnell, A. M. 1987. Suspended sediment. In Gurnell, A. M. and Clark, M. J. (eds) *Glacio-fluvial Sediment Transfer: An Alpine Perspective*. John Wiley, Chichester, 305–354.

Gurnell, A. M. 1993. How many reservoirs? An analysis of flow recessions from a glacier basin. *Journal of Glaciology*, **39**, 409–414.

Gurnell, A. M. 1995. Sediment yield from alpine glacier basins. In Foster, I. D. L., Gurnell, A. M. and Webb, B. W. (eds) *Sediment and Water Quality in River Catchments*. John Wiley, Chichester, 407–435.

Gurnell, A. M. and Fenn, C. R. 1984a. Flow separation, sediment source areas and suspended sediment transport in a pro-glacial stream. *Catena Supplement*, **5**, 109–119.

Gurnell, A. M. and Fenn, C. R. 1984b. Box-Jenkins transfer function models applied to suspended sediment concentration–discharge relationships in a proglacial stream. *Arctic and Alpine Research*, **16**, 93–106.

Gurnell, A. M. and Fenn, C. R. 1987. Proglacial channel processes. In Gurnell, A. M. and Clark, M. J. (eds) *Glacio-fluvial Sediment Transfer: An Alpine Perspective*. John Wiley, Chichester, 423–472.

Gurnell, A. M. and Warburton, J. 1990. The significance of suspended sediment pulses for estimating suspended sediment load and identifying suspended sediment sources in Alpine glacier basins. In Lang, H. and Musy, A. (eds) *Hydrology in Mountain Regions, I: Hydro-logical Measurements. The Water Cycle*. International Association of Hydrological Sciences Publication **193**, 463–470.

Gurnell, A. M., Clark, M. J. and Hill, C. T. 1992. Analysis and interpretation of patterns within and between hydroclimatological time series in an Alpine glacier basin. *Earth Surface Processes and Landforms*, **17**, 821–839.

Gurnell, A. M., Hodson, A., Clark, M. J., Bogen, J., Hagen, J. O. and Tranter, M. 1994. Water and sediment discharge from glacier basins: an arctic and alpine comparison. In Olive, L. J., Loughran, R. J. and Kesby, J. A. (eds) *Variability in Stream Erosion and Sediment Transport*. International Association of Hydrological Sciences Publication **224**, 325–334.

Gurnell, A. M., Hannah, D. M. and Lawler, D. M. 1996. Suspended sediment yield from glacier basins. In Walling, D. E. and Webb, B. W. (eds) *Erosion and Sediment Yield: Global and Regional Perspectives*. International Association of Hydrological Sciences Publication **236**, 97–104.

Hasholt, B. and Walling, D. E. 1992. Use of caesium-137 to investigate sediment sources and sediment delivery in a small glacierized mountain drainage basin in eastern Greenland. In Walling, D. E., Davies, T. R. and Hasholt, B. (eds) *Erosion, Debris Flows and Environment in Mountain Regions*. International Association of Hydrological Sciences Publication **209**, 87–100.

Hodgkins, R. 1999. Controls on suspended sediment transfer at a High Arctic glacier determined from statistical modelling. *Earth Surface Processes and Landforms*, **24**, 1–21.

Hooke, R. LeB. 1989. Englacial and subglacial hydrology: a qualitative review. *Arctic and Alpine Research*, **21**, 221–223.

Hubbard, B. P., Sharp, M. J., Willis, I. C., Nielsen, M. K. and Smart, C. C. 1995. Borehole water-level variations and the structure of the subglacial hydrological system of Haut Glacier d'Arolla, Valais, Switzerland. *Journal of Glaciology*, **41**, 572–583.

Kamb, B. 1987. Glacier surge mechanism based on linked-cavity configuration of the basal water conduit system. *Journal of Geophysical Research*, **92**(B9), 9083–9100.

Lal, D. and Peters, B. 1967. Cosmic ray produced radioactivity on the earth. *Handbook of Physics*, **46/2**, 551–612.

Lawler, D. M., Dolan, M., Tomasson, H. and Zophoniasson, S. 1992. Temporal variability of suspended sediment flux from a subarctic glacial river, southern Iceland. In Bogen, J., Walling, D. E. and Day, T. (eds) *Erosion and Sediment Transport Monitoring*. International Association of Hydrological Sciences Publication **210**, 233–243.

Maizels, J. K. 1983. Proglacial channel systems: change and thresholds for change over long, intermediate and short timescales. *Special Publication of the International Association of Sedimentologists*, **6**, 251–266.

Nienow, P., Sharp, M. J. and Willis, I. C. 1998. Seasonal changes in the morphology of the subglacial drainage system, Haut Glacier d'Arolla, Switzerland. *Earth Surface Processes and Landforms*, **23**, 825–843.

Oldfield, F. and Clark, R. L. 1990. Lake-sediment based studies of soil erosion. In Boardman, J., Foster, I. D. L. and Dearing, J. A. (eds) *Soil Erosion on Agricultural Land*. John Wiley, Chichester, 201–228.

Owens, P., Walling, D. E., He, Q., Shanahan, J. and Foster, I. D. L. 1997. The use of caesium-137 measurements to establish a sediment budget for the Start Catchment, Devon, UK. *Hydrological Sciences Journal*, **42**, 405–423.

Peart, M. R. and Walling, D. E. 1988. Techniques for establishing suspended sediment sources in two drainage basins in Devon, UK: a comparative assessment. In Bordas, M. P. and Walling, D. E. (eds) *Sediment Budgets*. International Association of Hydrological Sciences Publication **174**, 269–279.

Richards, K. 1984. Some observations on suspended sediment dynamics in Storbregrova, Jotunheim. *Earth Surface Processes and Landforms*, **9**, 101–112.

Röthlisberger, H. 1972. Water pressure in intra-and subglacial channels. *Journal of Glaciology*, **11**, 177–203.

Seaberg, S. Z., Seaberg, J. Z., Hooke, R. LeB. and Wiberg, D. W. 1988. Character of the englacial and subglacial drainage system in the lower part of the ablation area of Storglaciaren, Sweden, as revealed by dye tracer studies. *Journal of Glaciology*, **34**, 217–227.

Sharp, M., Richards, K., Willis, I., Arnold, N., Nienow, P., Lawson, W. and Tison, J. L. 1993. Geometry, bed topography and drainage system structure of the Haut Glacier d'Arolla, Switzerland. *Earth Surface Processes and Landforms*, **18**, 557–571.

Shoemaker, E. M. 1986. Subglacial hydrology for an ice sheet resting on a deformable bed. *Journal of Glaciology*, **32**, 20–30.

Stone, D. B., Clarke, G. K. C. and Blake, E. W. 1993. Subglacial measurement of turbidity and electrical conductivity. *Journal of Glaciology*, **39**, 415–420.

Thompson, R. and Oldfield, F. 1986. *Environmental Magnetism*. Allen & Unwin, London.

Walden, J., Slattery, M. C. and Burt, T. P. 1997. Use of mineral magnetic measurements to fingerprint suspended sediment sources: approaches and techniques for data analysis. *Journal of Hydrology*, **202**, 353–372.

Walder, J. S. 1986. Hydraulics of subglacial cavities. *Journal of Glaciology*, **32**, 439–445.

Walder, J. S. and Fowler, A. 1989. Channelised subglacial drainage over a deformable bed. *EOS*, **70**, 1084.

Wallbrink, P. J. and Murray, A. S. 1993. Use of fallout radionuclides as indicators of erosion processes. *Hydrological Processes*, **7**, 297–304.

Wallbrink, P. J. and Murray, A. S. 1994 Fallout of ^{7}Be in south Eastern Australia. *Journal of Environmental Radioactivity*, **25**, 213–228.

Wallbrink, P. J. and Murray, A. S. 1996. Distribution and variability of ^{7}Be in soils under different surface cover conditions and its potential for describing soil redistribution processes. *Water Resources Research*, **32**, 467–476.

Walling, D. E. and Kane, P. 1984. Suspended sediment properties and their geomorphological significance. In Burt, T. P. and Walling, D. E. (eds) *Catchment Experiments in Geomorphology*. GeoBooks, Norwich, 311–334.

Walling, D. E. and Woodward, J. C. 1992. Use of radiometric fingerprints to derive information on suspended sediment sources. In Bogen, J., Walling, D. E. and Day, T. (eds) *Erosion and Sediment Transport Monitoring Programmes in River Basins*. International Association of Hydrological Sciences Publication **210**, 153–164.

Walling, D. E. and Woodward, J. C. 1995. Tracing sources of suspended sediment in river basins: a case study of the river Culm, Devon, UK. *Marine and Freshwater Research*, **46**, 327–336.

Walling, D. E., Peart, M. R., Oldfield, F. and Thompson, R. 1979. Suspended sediment sources identified by magnetic measurements. *Nature*, **281**, 110–113.

Walling, D. E., Woodward, J. C. and Nicholas, A. P. 1993. A multi-parameter approach to fingerprinting suspended sediment sources. In *Tracers in Hydrology*. International Association of Hydrological Sciences Publication **215**, 329–338.

Wasson, R. J., Clark, R. L., Nanninga, P. M. and Waters, J. 1987. ^{210}Pb as a chronometer and tracer, Burrinjuck Reservoir, Australia. *Earth Surface Processes and Landforms*, **12**, 399–414.

Weertman, J. 1972. General theory of water flow at the base of a glacier or ice sheet. *Reviews of Geophysics and Space Physics*, **10**, 287–333.

Willis, I. C., Sharp, M. J. and Richards, K. S. 1990. Configuration of the drainage system of Midtdalsbreen, Norway, as indicated by dye-tracing experiments. *Journal of Glaciology*, **36**, 89–101.

Wise, S. M. 1980. Caesium-137 and lead-210: a review of the techniques and some applications in geomorphology. In Cullingford, R., Davidson, D. and Lewin, J. (eds) *Timescales in Geomorphology*. John Wiley, Chichester, 109–127.

18 Tracing Sediments within Urban Catchments using Heavy Metal, Mineral Magnetic and Radionuclide Signatures

S. M. CHARLESWORTH,[1] L. M. ORMEROD[2] and J. A. LEES[1]

[1]Centre for Environmental Research and Consultancy, Coventry University, UK
[2]Department of Geography and Environmental Science, University of Newcastle, Callaghan, Australia

INTRODUCTION

Tracing fine sediment movement in the urban environment has attracted little research interest in spite of the fact that urban areas are where many people live and work. The processes involved in fine sediment movement and subsequent deposition lead to impacts that will affect the majority of the population within and downstream of urban areas. However, the complexity of erosional and depositional processes makes the study of urban areas difficult. Figure 18.1 illustrates this complexity, which can be summarised into four main areas:

(1) The number and variety of sources of fine sediment within urban catchments.
(2) Subsequent modification of the sediment by, for example, authigenic or biotic mineral growth, e.g. greigite (Snowball and Thompson, 1988) or magnetotactic bacteria (Farina et al., 1990). Whilst noted in non-urban sediments, there is no reason to believe that the same processes do not occur in urban environments as well.
(3) Sporadic periods of construction work releasing pulses of sediment within the urban catchment (Hollis, 1988).
(4) Changing environmental conditions such as Eh and pH, which are affected by climatic conditions ranging from the urban heat island effect (Oke, 1988), extra nuclei for raindrop formation (Ayers et al., 1982), to problems with acid rain (Fowler et al., 1985) and eutrophication (Ryding and Rast, 1989). Changing Eh and pH conditions affect the speciation of heavy metals and hence the way in which they are transported and deposited.

Tracing fine sediment sources and movement within the urban environment seems, therefore, to be fraught with difficulties. Techniques designed to characterise sedi-

Tracers in Geomorphology. Edited by Ian D. L. Foster. © 2000 John Wiley & Sons Ltd.

ments and provide a fingerprint (Peart and Walling, 1986; Owens *et al.*, see Chapter 15) have been applied to agricultural, pastoral and other natural environments for the past 20 years (Oldfield *et al.*, 1979; Walling *et al.*, 1979), but there has been very little such application to predominantly urbanised catchments. Fingerprinting techniques, however, are considered to be far more cost-effective than field monitoring

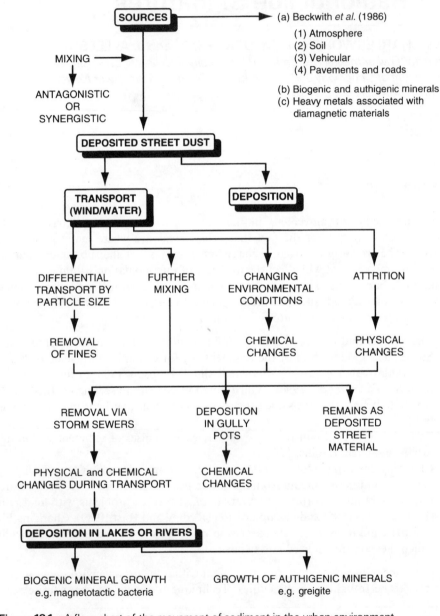

Figure 18.1   A flow chart of the movement of sediment in the urban environment

(Walling and Woodward, 1992) since field monitoring of urban areas suffers operational difficulties in terms of interference with, and loss of, equipment through vandalism.

Two case studies are presented in this chapter which use mineral magnetic characterisation and radioactive isotope techniques to trace heavy metals associated with fine urban sediments. Mineral magnetic measurements have been used in a variety of environments to characterise sediments in order to trace their origins (Thompson and Oldfield, 1986; Foster *et al.*, 1998; Lees, 1999). However, those studies that have been carried out in urban environments tend to concentrate on individual specific sources, such as fly ash and motor vehicle emissions (Hunt *et al.*, 1984). A wider remit is taken here to assess the usefulness of mineral magnetic measurements in tracing the movement of sediment through a source–transport–deposit cascade. Many studies, discussed by Charlesworth and Lees (1997), highlight a relationship between particulate-associated heavy metals and mineral magnetic characteristics. This relationship is vital if sediment tracing using mineral magnetic characterisation is to be effective. It is, therefore, further explored here within the source–transport–deposit cascade. Heavy metals themselves have been used in tracer studies, but mainly in relation to geological rather than land-use differences, or as a component of a larger suite of sediment tracers (Passmore and Macklin, 1994; Caitcheon *et al.*, 1995; Collins *et al.*, 1998). The ability of elemental analysis alone to fingerprint sources specifically within an urbanised catchment is also assessed here.

Radioactive isotopes are frequently associated with fine particulate sediments and, in particular, ^{137}Cs has been widely used both as a chronological marker, and as a sediment tracer with specific application to discriminate between topsoil and subsoil materials (Loughran *et al.*, 1986; Peart and Walling, 1986; Walling and Woodward, 1992; Caitcheon *et al.*, 1995; Wallbrink *et al.*, 1996; Collins *et al.*, 1997a,b). The two case studies presented here are based on the characterisation of fine urban sediment using particulate-associated characteristics and physical properties in sediment source tracing and sediment movement studies.

CASE STUDY 1: TRACING URBAN SEDIMENTS IN COVENTRY, UK

Figure 18.2 shows the source–transport–deposit cascade and illustrates the numbers of samples taken. Most studies of the relationship between particle size and heavy metal content tend to fractionate the sediment into arbitrary size fractions that have little relationship with the natural environment. In this study, the samples were fractionated into two particle sizes: $< 63\,\mu m$ to represent sediment accumulating in a lake basin (Förstner, 1983), and a $< 2\,mm$ bulk sample to represent the saltating–suspended load of a typical stream (Thoms, 1987).

Sampling Suite, Sampling and Analytical Methodology

Figure 18.3 shows the location of Coventry in the English Midlands and the position of the sampling points relative to the city centre. The samples were collected according to a sediment source–transport–deposit cascade as follows:

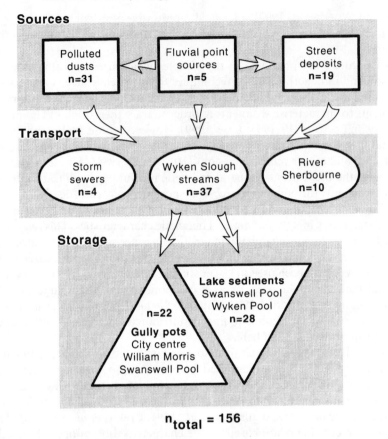

Figure 18.2   The source–transport–deposit sampling cascade

(1)   Sediment sources included a polluted dust database (Lees, 1994), fluvial point sources from an industrial estate collected by Charlesworth (1994) and street dusts from various points in the city centre. These samples were collected by hand using either a plastic scoop, e.g. for chimney fly ash, or by means of a plastic dustpan and brush in the case of dust swept from the city streets. The fluvial sediments were collected by hand-grab sampling using a plastic bag.

(2)   Sediments classified as transported were collected from storm sewers and urban streams, particularly the River Sherbourne which flows through Coventry from the north-west to the south-east and is channelised through the city centre. A second set of fluvial samples was obtained from the two streams flowing into an area known as Wyken Slough to the north-east of the city centre and on the urban periphery. These urban streams, in particular the western stream which flows through industrial and housing estates, are typically augmented by the addition of runoff from surrounding streets and roads. The sediments collected were grab samples taken by hand from surficial material in the storm sewer or stream.

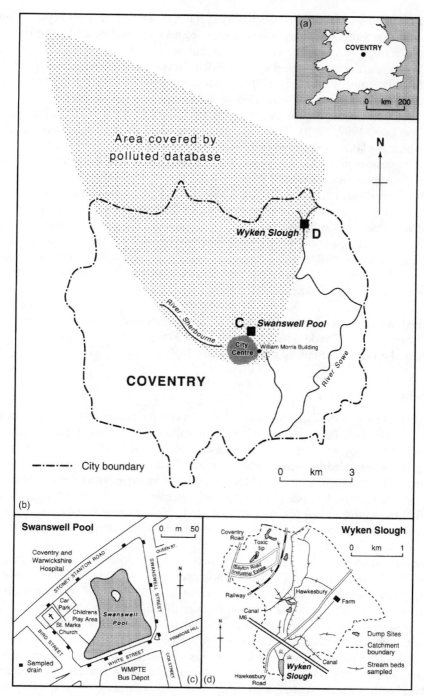

Figure 18.3 Location maps of (a) Coventry, (b) the area sampled for polluted dusts and locations of the two urban lakes, (c) Swanswell Pool gully pots, and (d) the catchment of Wyken Pool

(3)   Deposited sediment was collected from the pool associated with Wyken Slough
      and from Swanswell Pool, a small urban freshwater pond located in the centre of
      the city of Coventry. The latter site was cut off from its catchment in *c.* 1850 and
      hence has no inlets or outlets; sediment accumulating within the basin will
      therefore mostly be derived from atmospheric fallout with a relatively small
      contribution from the surrounding park and walkways. Further details of
      these catchments can be found elsewhere (Charlesworth, 1994; Charlesworth
      and Lees, 1997, 1999). These samples were taken as part of a larger study in
      which the lakes were cored on a grid system using a modified Livingstone corer
      (Charlesworth, 1994). The top 1 cm slice of the 12 cores from Swanswell Pool
      and 16 cores from Wyken Pool were used in the study as they represent the most
      recent sediment to be deposited. Deposited sediment was also collected from
      gully pots surrounding Swanswell Pool and the building housing the Geography
      Department at Coventry University. The gully pot samples were retrieved using
      a 15 × 12 cm close-mesh catch net, of the type used for tropical aquaria. This
      approach was used due to difficulties found when attempting to insert standard
      samplers through the gully pot grid and the fact that most of these grids were
      fixed into place by the road tar.

Detailed descriptions of analytical methodologies are given elsewhere (Charlesworth
and Foster, 1993; Charlesworth, 1994; Lees, 1994; Charlesworth and Lees, 1997,
1999), and a brief summary is given in Figure 18.4.

Results

The skewed frequency distribution of the fingerprint properties was demonstrated by
Charlesworth and Lees (1999) and median values are discussed throughout this
chapter since they provide a better measure of central tendency than the mean.
Table 18.1 shows median heavy metal concentrations and mineral magnetic measurements for the sediment sources and the transported and deposited sediments in both
particle size classes. Copper (Cu) is the only element to increase in concentration
through the source–transport–deposit cascade in both particle sizes; otherwise there
are very few consistent trends shown. Mineral magnetic characteristics show a similar
lack of consistent trends through the cascade.

Mineral magnetic measurements were used to characterise the sediments by comparing SIRM/χ_{LF} ratios with those of Thompson and Oldfield (1986). The resultant
scattergram (Figure 18.5) enables an assessment to be made of the concentration and
grain size of magnetic minerals. Results from both sediment particle size fractions of
the deposited urban sediments are shown in Figure 18.5, with the observed scatter in
the data reflecting differences in mineralogy and grain size. The mineralogy is dominated by ferrimagnetic multidomain grains characteristic of pollution particles
(Thompson and Oldfield, 1986). Some of the samples lie on a trend line at ratios of
1:20 and 1:10, with the scatter about these lines demonstrating the variability of the
samples and hence precluding their usefulness as tracers.

The second focus of the study was to assess the use of mineral magnetic measurements as surrogates for heavy metals analysis in the urban environment. Correlations

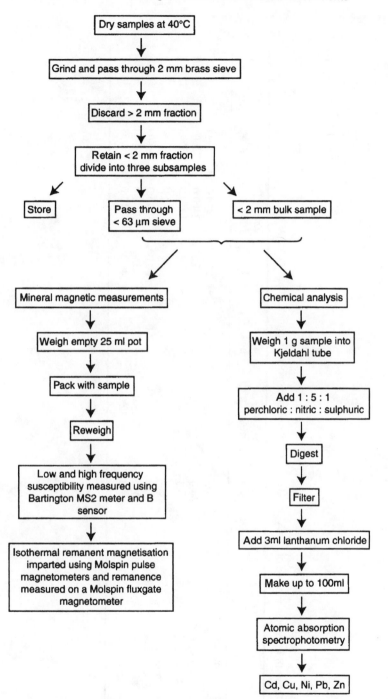

Figure 18.4 Flow chart of analytical methodologies

Table 18.1 Median values of heavy metal concentrations and mineral magnetic measurements

	X_{LF} $(10^{-6}\,m^3\,kg^{-1})$	X_{FD} $(10^{-9}\,m^3\,kg^{-1})$	SIRM $(mAm^2\,kg^{-1})$	Cd $(mg\,kg^{-1})$	Ni $(mg\,kg^{-1})$	Zn $(mg\,kg^{-1})$	Cu $(mg\,kg^{-1})$	Pb $(mg\,kg^{-1})$
< 63μm								
Sources	3.6	34.0	43.0	4.1	75.0	490.0	193.0	50.0
Transport	1.6	34.0	26.0	3.6	108.0	2400.0	260.0	94.0
Deposit	3.3	0.3	14.2	3.3	87.0	1430.0	350.0	158.0
< 2 mm								
Sources	1.8	20.0	0.45	2.1	60.0	226.0	40.0	110.0
Transport	10.3	0.28	23.0	2.3	42.0	270.0	105.0	77.0
Deposit	2.3	42.0	32.0	1.7	49.0	1017.0	195.0	107.0

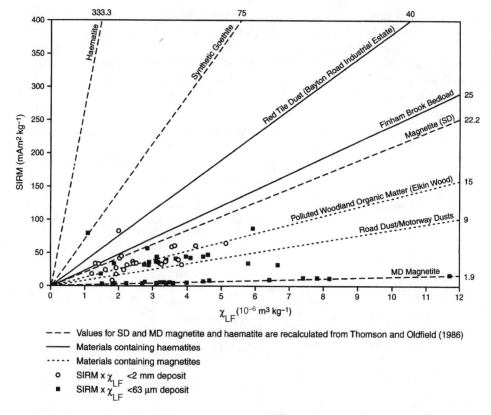

Figure 18.5 A plot of SIRM versus χ_{LF} for deposited sediment compared with published ratios for polluted materials (after Lees, 1994)

between these properties are shown in Table 18.2. The relationships shown here are inconsistent between environmental compartments and between elements, with the <2 mm transported and deposited material showing no statistically significant relationships. Mineral magnetic characteristics cannot therefore be used to trace heavy metals through the source–transport–deposit cascade. Principal component analysis further highlighted this lack of a relationship (Table 18.3), with heavy metals and mineral magnetic measurements mostly separating into components 1 and 2, the dominant component explaining up to 66 percent of the variance in the data. However, drawing envelope curves in component space using individual factor score plots (Figure 18.6) showed that the <2 mm fraction was much less variable than the < 63 μm fraction. All the envelopes overlapped showing the close relationship between samples.

Discussion

There does not appear to be a consistent relationship between heavy metals and mineral magnetic measurements in the sediments collected from the Coventry area.

The diversity of sources, and the variety of processes acting on the sediments, lead to a large variation in chemical and physical properties which makes overall characterisation difficult. Discriminating two main sources in a mixture, such as topsoil and subsoil is possible, but in the urban environment, sediments are complex mixtures of both natural and anthropogenic material. It can be argued that this study may have been undertaken at too large a scale, since the samples were taken across the

Table 18.2 Correlation found between measured parameters

	< 63 μm		< 2 mm	
	Correlated properties	Correlation coefficients	Correlated properties	Correlation coefficients
Source	SIRM: Ni	0.6181	SIRM: Cd	0.8129
$n = 34$			SIRM: Pb	0.8305
$r = 0.554$			SIRM: Cu	0.6974
$p = 0.001$			X_{LF} : Pb	0.8889
			X_{LF} : Cd	0.8961
			X_{FD} : Ni	0.8965
			X_{FD}: Cu	0.6814
Transport	X_{LF}: Cd	0.4688	none	
$n = 51$	X_{LF} : Ni	0.6990		
$r = 0.443$	X_{LF}: Pb	0.7318		
$p = 0.001$				
Deposit	SIRM: Cd	0.5630	none	
$n = 48$	SIRM: Pb	0.5207		
$r = 0.465$	X_{FD}: Cd	0.5491		
$p = 0.001$				

Table 18.3 Rotated component matrix and per cent explained variance

	Component	% explained variance		Dominant component[a]	
		< 2 mm	< 63 μm	< 2 mm	< 63 μm
Source	1	65.7	42.4	mix	hm
	2	16.3	23.9	mix	mag
Transport	1	26.8	42.0	hm	hm
	2	20.8	16.1	mag	mag
	3	18.6	13.5	hm	hm
	4	14.0	nc	X_{FD}	nc
Deposit	1	30.2	36.5	hm	mix
	2	20.9	20.6	hm	hm
	3	16.0	14.8	mag	Zn
	4	13.2	13.0	mix	X_{LF}

[a] mag = mineral magnetic characteristics; hm = heavy metals; mix = mixture of mineral magnetic characteristics and heavy metals; nc = no component.

whole of the urban area and were physically unrelated to one another. PCA plots (Figure 18.6), however, do indicate that the suites of samples have similar characteristics. Scaling down the sampling suite, i.e. to the individual catchment, may help to resolve similarities and differences between samples. However, there are very few complete catchments within an urban environment, many being truncated (e.g. Swanswell Pool) and still others having augmented catchments (e.g. Wyken Pool), so there are still difficulties associated with this approach.

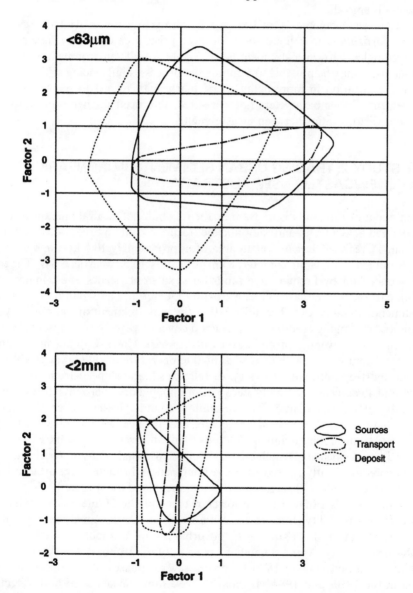

Figure 18.6 Envelope curves of factor score plots from Principal Component analysis for samples in the source–transport–deposit cascade

Categorising samples as acting as source, transported and deposited material depended on their situation on the day they were collected. Hence, gully pot sediments were deposited on the day of collection. However, should a storm ensue, then, depending on its severity, the material collected in the gully pot can become entrained and transported out of the gully pot and into the storm sewer. The same is true of material transported in rivers which can become deposited. It is arguable, therefore, whether or not it is meaningful to categorise sediments thus when their transport and deposition is episodic.

The urban environment is, at best, a frustrating location in which to carry out research, particularly in comparison with more pristine catchments where processes are less complex. However, by identifying a smaller scale environment with related components, it may be possible to examine the processes influencing sediment movement and deposition in more detail. Particularly with regard to mineral magnetic measurements, other parameters not measured here need further exploration, e.g. ARM and various ratios between measurements.

CASE STUDY 2: TRACING URBAN SEDIMENTS IN IRONBARK CREEK, NEWCASTLE, NSW, AUSTRALIA

The main aim of this study is to examine the reliability of natural and anthropogenically derived heavy metal and radionuclide tracers within an urbanised catchment.

Ironbark Creek catchment has an area of approximately 10.5 km^2 and is located 32° 52'S and 151° 38'E, approximately 8 km north-west of Newcastle, NSW. The catchment is underlain by Permian age sandstones, siltstone, shales and conglomerates forming part of the Newcastle Coal Measures. Quaternary alluvium occurs as valley fill sediments along creeks. The soils within the subcatchment are primarily yellow podzols and are highly erodible where a good cover of vegetation is absent (Matthei, 1993). The mean average precipitation for Ironbark Creek Catchment is approximately 1100 mm year^{-1}, with a mean annual temperature of 18 °C. The study area is bordered by steep hills, with peaks up to 150 m.a.s.l. and slopes varying between 10 percent and 40 percent. Originally, eucalypt-dominated dry sclerophyll forest covered most of the catchment but little of this vegetation remains. However, there are relatively large remnant stands on steeper slopes located near the headwaters of Ironbark Creek. About 16 percent of the catchment contains remnant forest, while some 65 percent is urbanised. Some open areas, including sporting fields, small areas of agricultural land and other predominantly grassed areas, are interspersed throughout the urban area and make up the remainder (19 percent) of the catchment land use (Figure 18.7).

The catchment is situated in the lower reaches of the Hunter River, which has extreme flood variability (Erskine and Saynor, 1996). It is assumed that the removal of natural vegetation for agriculture and urbanisation has increased this naturally high flood variability. Approximately 65 percent of the Ironbark Creek catchment is presently urbanised (Figure 18.7) following several phases of urbanisation, which peaked in the 1970s and 1980s. Ironbark Creek shows evidence of both direct and indirect impacts of urbanisation. It may be expected that the source of sediment stored within channels and floodplain deposits may have changed over the last

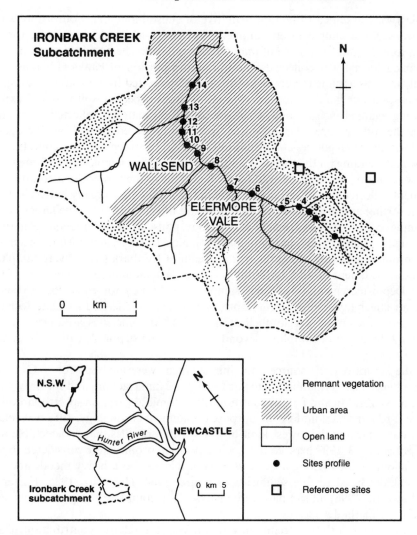

Figure 18.7   Ironbark Creek study area

few decades, reflecting both catchment and channel changes associated with urbanisation.

Method

Source samples were collected from topsoil, subsoil, channel banks and urban gutters. Upstream forest areas were used to collect both topsoil and subsoil samples as this area was considered to be relatively unaffected by anthropogenic inputs of heavy metals. Surface scrape samples (5 cm depth × 400 cm^2) (after Campbell *et al.*, 1988) were obtained from four transects perpendicular to eroded gullies located upstream of Site 1 (Figure 18.7). Samples obtained from the walls and floor of each gully (12 samples) were assumed to be representative of subsoil material. Four samples from each transect (16 samples) were sampled from undisturbed sites within the forest to obtain topsoil samples, typical of material derived from sheet and rill erosion. Channel bank material was collected from the upper portion of banks (12 samples) by trowel, and was assumed to consist of sediments deposited from upstream sources. In some cases lower bank material consisted of undisturbed subsoil material exposed following channel incision and is similar to subsoil material collected from gullies within the forested area. Urban gutter sediments were collected from lag deposits found within urban gutters and drains (30 samples) by sweeping or by collection with a trowel, depending on the thickness of the deposit. These samples were thought to be representative of sediment derived from all sources in the upslope contributing area. Sediment collected from these points were in transit and had already undergone some partial sorting and mixing by overland flow and fluvial processes. The larger number of samples collected from urban areas was in response to the heterogeneous nature of those areas. Channel and floodplain morphological features, identified from surveyed cross-profiles at 14 sites along the trunk stream of Ironbark Creek, were sampled to determine sediment source contributions over a contemporary timescale (Figure 18.7). Depth-incremental sediment samples were taken for each core (31 cores) using an auger (averaging 10 cm depth increments) or the scraper-plate technique (2 cm depth increments) (Campbell *et al.*, 1988). At some sites an auger was used to extend the depth of sampling beyond the maximum depth (26 cm) of the scraper-plate.

In the laboratory, all source and sink sediment were analysed for ^{137}Cs activity using a hyperpure germanium detector for the < 2 mm fraction, and for heavy metal (Pb, Cu, Ni, Zn, Mn and Fe) concentrations by atomic absorption spectrophotometry for the < 63 mm fraction. For the purpose of sediment chronology, 1958 was selected as the date marking the lowest limit of ^{137}Cs within profiles that is detectable today (Olley *et al.*, 1993, 1997 pers. comm.), while peaks in profiles were considered to mark peak fallout in 1964 (Longmore *et al.*, 1983). Because both heavy metals and ^{137}Cs preferentially bond to fine sediments (Tamura, 1964; Horowitz, 1985; Foster and Charlesworth, 1996) all samples were analysed for grain-size composition using the hydrometer method.

Sediment fingerprinting techniques quantifying the relative contribution of sediment sources to sediment deposited within channels and floodplains can only be applied if individual sediment sources can be sufficiently distinguished from each other (Collins *et al.*, 1997a,b). Sediments derived from subsoil material, channel banks and urban gutters could not be distinguished from each other (Table 18.4). Mann-Whitney U-tests indicated that ^{137}Cs concentrations for topsoil samples were statistically different from all the other sources at the 99.99 percent significance level or better (Table 18.5).

Results from Mann-Whitney U-tests indicated that sediment derived from urban gutters could be distinguished from all other sources, by all of the heavy metals used in this study (Pb, Cu, Ni, Zn, Mn and Fe) with highly significant results (99.99 percent) (Tables 18.6 and 18.7). Channel banks could similarly be distinguished from all other sources. However, significance levels were a little lower in most cases. Gullies and forest topsoil could be differentiated by Pb, Cu and Mn which were therefore able to

Table 18.4 ^{137}Cs concentrations and mean areal activity of catchment sources

Catchment sources	Mean ^{137}Cs concentration (mBq g^{-1})	Mean areal activity (mBq cm^{-2})
Forest topsoil, $n = 16$	6.53 ($\pm$ 0.15) SD = 0.86, CV = 13%	34.5 SD = 6.0
Subsoil material, $n = 12$	0.44 ($\pm$ 0.03) SD = 0.44, CV = 100%	1.9 SD = 2.1
Channel banks, $n = 12$	0.58 ($\pm$ 0.14) SD = 0.81, CV = 140%	n.a
Urban gutters, $n = 30$	0.38 ($\pm$ 0.14) SD = 0.43, CV = 113%	n.a.

Measurement errors are given in brackets (SD = 1 standard deviation), mean areal activity based on surface areas of 400 cm^2 for all relevent samples.

Table 18.5 Mann-Whitney U-test significance levels for differences between source pairs based on ^{137}Cs concentrations

	Gullies	Channel banks	Construction sites	Urban gutters
Forest topsoil	0.0000	0.0000	0.0000	0.0000
Gullies	–	ns	0.0089	ns
Channel banks	–	–	0.0166	ns
Construction sites	–	–	–	0.0001

ns = not significant at the 0.05 level for a two-tailed test.

Table 18.6 Means, standard deviations and CV for heavy metal data from each source

Source	Pb (mg g^{-1})	Cu (mg g^{-1})	Ni (mg g^{-1})	Zn (mg g^{-1})	Mn (mg g^{-1})	Fe (%)
Urban gutters, $n = 30$	595.8 SD = 308.7 CV = 52%	159.6 SD = 80.5 CV = 50%	24.6 SD = 8.1 CV = 33%	1631.1 SD = 1223 CV = 76%	822.1 SD = 323 CV = 39%	2.520 SD = 0.84 CV = 33%
Forest topsoil, $n = 16$	39.6 SD = 9.2 CV = 23%	28.6 SD = 5.0 CV = 17%	7.8 SD = 2.7 CV = 35%	98.9 SD = 26.4 CV = 27%	145.9 SD = 88.9 CV = 61%	0.967 SD = 0.34 CV = 35%
Gullies, $n = 12$	21.4 SD = 4.5 CV = 21%	24.8 SD = 3.1 CV = 12%	8.4 SD = 0.9 CV = 11%	101.6 SD = 16.7 CV = 16%	45.6 SD = 9.4 CV = 21%	1.122 SD = 0.2 CV = 17%
Channel banks, $n = 12$	141.5 SD = 100.0 CV = 71%	40.6 SD = 14.4 CV = 36%	11.6 SD = 4.4 CV = 38%	184.0 SD = 92.9 CV = 50%	264.3 SD = 165 CV = 63%	1.823 SD = 0.85 CV = 46%

Table 18.7 Significance levels of Mann-Whitney U-tests between sources using heavy metal concentrations.

Source pairs	Pb	Cu	Ni	Zn	Mn	Fe	Pb/Ni[a]	Pb/Fe[a]
Urban and forest topsoil	0.0000	0.0000	0.0000	0.0000	0.0000	0.0000	0.0000	0.0000
Urban and subsoil	0.0000	0.0000	0.0000	0.0000	0.0000	0.0000	0.0000	0.0000
Urban and channel banks	0.0000	0.0000	0.0000	0.0000	0.0000	0.0033	0.0016	0.0003
Forest topsoil and subsoil	0.0000	0.0168	n.s.	n.s.	0.0015	n.s.	0.0001	0.0000
Forest topsoil and channel banks	0.0032	0.0216	0.0130	0.0087	0.0347	0.0006	0.0011	0.0388
Subsoil and channel banks	0.0003	0.0027	0.0072	0.0102	0.0004	0.0073	0.0000	0.0014

Significance levels were calculated using a two-tail test. n.s. = not significant at 0.05 level.[a] Ratios able to distinguish all sources.

differentiate all sources (Table 18.7). Similar tests for heavy metal ratios revealed that Pb/Cu and Pb/Fe could also differentiate between all sources (Table 18.7).

Mixing models were applied to individual tracers to quantify the relative contribution of sediment sources to sediment mixtures found within channel or floodplain deposits. The method used in this study is similar to that described by Collins *et al.* (1998) who used an iterative approach to find the mixture of sources that minimised the error term within each sediment sink sample. Particle size adjustments were made for each source/sink comparison of individual heavy metal parameters, by using the ratio of this fraction in the sample to the average quantity found in each source respectively (Collins *et al.*, 1998). This was thought necessary because of the widely different clay fractions associated with each source, which varied from 41 percent in gully material down to 5 percent in urban gutters.

Results and Discussion

The interpretation of the results of mixing models for individual heavy metal parameters proved problematic. In many cases it was possible to obtain more than one solution to the mixing models, and topsoil amounts estimated by ^{137}Cs two-source mixing models (topsoil and subsoil) differed from topsoil amounts estimated by mixing models using individual heavy metal parameters. Furthermore, the results did not make sense from a geomorphological perspective, e.g. some samples within the forested area were estimated to contain a large proportion of sediment derived from urban areas. These results therefore demonstrate that the use of individual heavy metal tracers is not reliable.

The reliability of source/sink comparisons may be increased by using multiple tracers combined to give a composite signature (Peart and Walling, 1986; Walling *et al.*, 1993; Walling and Woodward, 1995; Collins *et al.*, 1998). Fingerprinting parameters included in models may be selected by statistical means. These include cluster analysis (Yu and Oldfield, 1989; Walling *et al.*, 1993), principal component analysis (Shankar *et al.*, 1994), and multivariate discriminant function analysis, e.g. Wilks' lambda (Collins *et al.*, 1998).

Spurious source/sink comparisons are reduced where composite signatures include variables that behave in contrasting ways within the environment (Walling *et al.*,

1993). Heavy metal parameters used in this study behave in a similar manner within the environment and consequently there is little advantage of using a composite signature in this study.

In an attempt to improve the results of individual heavy metal parameters, the percentage of topsoil estimated from ^{137}Cs for each sample was substituted into the mixing model. This had the advantage of including ^{137}Cs data into the models, possibly increasing their reliability. To integrate the results of mixing models that combined individual heavy metal parameters (Pb, Cu, Mn, Pb/Ni and Pb/Fe) and ^{137}Cs results, each heavy metal tracer was weighted according to its ability to differentiate between sources, as determined by maximum likelihood factor analysis, available in the MINITAB statistical package. This was achieved by using a combined data set of all source samples. The results of the factor analysis indicated that Pb and Pb/Ni explained the most variance; however, the other variables also explained significant amounts of variance (Table 18.8). Pb/Ni may actually provide better source/sink comparisons than Pb because of the negation of particle size effects.

The validity of any mixing model can only be truly tested by field monitoring (Collins *et al.*, 1998). However, geomorphological knowledge about the catchment and sediment movement within the catchment may be used to partially validate the results of mixing models.

Two situations were used to test the reliability of the combined mixing model results: first, subsoil sampled in the base of some valley floor cores; and secondly, samples located within upstream remnant bushland. There should be no anthropogenic input of heavy metals in either of these situations. The reliability of the mixing model results can only be verified if they can estimate the expected sources of sediment, i.e. 100 percent subsoil in the case of subsoil in the base of valley floor cores and a combination of topsoil and subsoil sources in the case of remnant bushland samples.

At least 80 percent of the undisturbed valley floor sediments was estimated as being subsoil material. Some examples of such cores are shown in Figure 18.8. Errors of up to 20 percent are present within some samples, indicated by the presence of channel bank or urban sediments. This is mainly thought to be due to averaging effects of both source and sink samples. All samples collected within remnant forest, except one sample from the channel at site 2, contained at least 80 percent topsoil and subsoil material according to the results of the combined mixing models (Figure 18.9). Again there appears to be about 20 percent error in the estimation of sediment

Table 18.8 Factor analysis results for combined sources

Variable	Factor score
Pb	0.957
Pb/Ni	0.943
Cu	0.884
Mn	0.769
Pb/Fe	0.475
% Variance	68.1

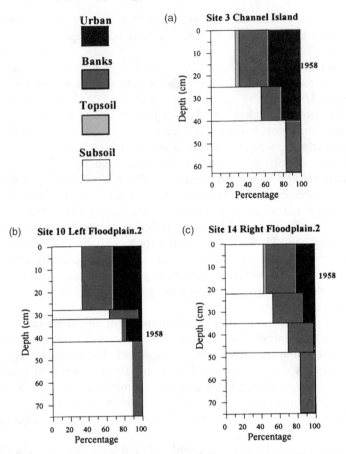

Figure 18.8   Subsoil materials in lower layers of sediment cores

sources. While most of these errors are thought to be associated with averaging effects, some anthropogenic inputs may be present, especially at site 2 because of its close proximity to the urban fringe (Figure 18.7), and therefore errors may be less than 20 percent.

It appears that the combined mixing model results used in this study can estimate source contributions, with errors generally being less than 20 percent. Therefore, an approximation of sediment sources should be possible along Ironbark Creek. Examination of results for all floodplain and channel samples revealed that there were variations in sediment source contributions between channel and floodplain cores. There also appear to be some downcore and downstream variations in sediment source contributions.

These variations could be explained by geomorphological processes. For example, generally there was less downcore variation in sediment sources for channels than floodplains, reflecting greater fluvial mixing and sorting within channels (Figures 18.9 and 18.10(d) and (e)). While subsoil materials predominated in both channel and

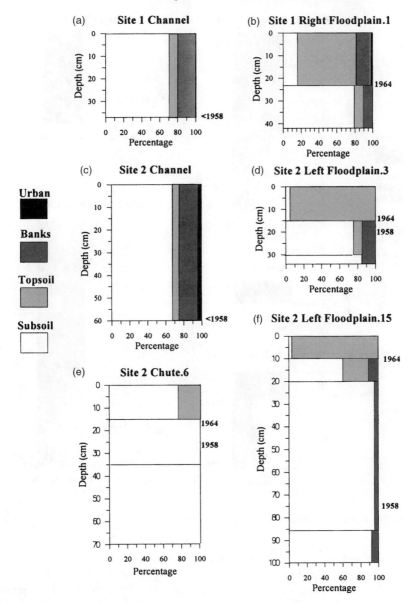

Figure 18.9 Predominance of topsoil and subsoil within forest sites

floodplain deposits, channel bank material was the second most common source within channels, suggesting a predominance of channel sources (Figures 18.8(a), 18.9(a),(c) and 18.10(b),(d)), as might be expected.

Topsoil was generally absent or present only in small amounts within channel deposits (Figures 18.8(a) and 18.10(b),(d)), except at sites 1 and 2 (Figure 18.9(a), (c)). Greater topsoil components with channel deposits at sites 1 and 2 may be

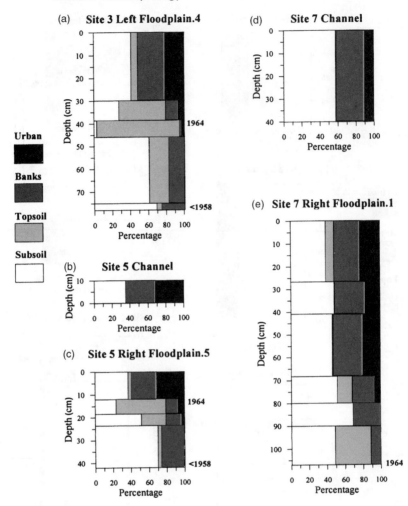

Figure 18.10   Additional channel and floodplain profiles

explained by sheet and rill erosion within the forested area. Topsoil layers occur close to the surface in floodplain cores at sites 1 and 2 (Figure 18.9(b),(d)–(f)), but occur considerably lower in cores at sites 3 to 7 (Figure 18.10(a),(c),(e)) and are generally absent below site 7 (e.g. Figure 18.8(b),(c)), indicating that the original topsoil lies below the depth of sampling. This would suggest that deposition rates are increasing in the downstream direction or, alternatively, that topsoil layers downstream of site 7 may have been eroded due the greater erosivity of floods in this reach.

In general, floodplain deposits appear to contain higher amounts of urban sediments within the upper layers. This may be associated with the expansion of urbanisation since the 1960s. Both an increase in flood magnitudes and frequencies, and greater amounts of urban sediments, could account for their deposition within floodplains in more recent years. There appears to be no obvious downstream trend concerning urban sediments, apart from a significant increase in percentage amounts below site 2, because inputs of urban sediments are related to catchment land use and drainage networks at any particular location.

There is some evidence of lateral variations in sediment sources across the flood-plain, associated with frequency and timing of inundation and particle size effects. A discussion of lateral variations is not possible here as it requires the full data set, but is discussed in Ormerod (1999).

SUMMARY

The tracing methodologies used in these two case studies have been successfully used in many different environments. However, in spite of being in different hemispheres, in different climatic regions and using different measures of physical properties, the overall conclusions of both urban studies remain broadly similar. Hence, individual heavy metal tracers proved ineffective in determining source contributions to Ironbark Creek. This suggests that heavy metal tracer parameters may only be useful when combined with other tracer parameters that behave differently within the environment, as suggested by Walling *et al.* (1993) and Collins *et al.* (1998). This may be particularly important in urban areas where the likelihood of remobilisation of heavy metals may be increased. However, the UK study highlighted the inconsistent relationship between parameters which may make the choice of tracer parameter difficult. These relationships reflect the complexity of the urban environment, perhaps suggesting that methodologies developed for more pristine environments may not be applicable here.

The UK case study took a holistic view in categorising sediments across the whole urban area into the source–transport–deposit cascade. This large scale may have exacerbated the problems encountered in discriminating between the different com-partments. On the other hand, the Australian study was undertaken at the individual catchment scale in which combined mixing models indicated that channel enlarge-ment could be the major source of sediment stored within channel and floodplain deposits. This smaller scale may therefore have been more successful in terms of establishing a more consistent relationship between sediment characteristics.

These studies have therefore highlighted the need for the development of techniques more appropriate to urban areas and for more research focused on the unique processes of such environments.

ACKNOWLEDGEMENTS

L. M. O. gratefully acknowledges the interest and suggestions of Bob Loughran, Peter Glennie for assistance in the field, and Chris Dever for laboratory assistance. Funding from the Faculty of Science and Mathematics, University of Newcastle, NSW was much appreciated in assistance to attend the BGRG Conference at Coventry, UK in September 1998. S. M. C. is grateful for suggestions made by Ian Foster for the improvement of earlier drafts of this chapter. Thanks also to the referees for con-structive comments.

REFERENCES

Ayers, G. P., Bigg, E. K., Turvey, D. E. and Manton, M. J. 1982. Urban influence on condensation nuclei over a continent. *Atmospheric Environment*, **16**(5), 951–954.

Beckwith, P. R., Ellis, J. B., Revitt, D. M. and Oldfield, F. 1986. Heavy metal and magnetic relationships for urban source sediments. *Physics of the Earth and Planetary Interiors*, **42**, 67–75.

Caitcheon, C. G., Donnelly, T., Wallbrink, P. and Murray, A. 1995. Nutrient and sediment sources in Chaffey reservoir catchment. *Australian Journal of Soil and Water Conservation*, **8**(2), 41–49.

Campbell, B. L., Loughran, R. J. and Elliott, G. L. 1988. A method for determining sediment budgets using caesium[137]. In *Sediment Budgets*. IAHS Publication **174**, IAHS Press, Wallingford. 171–179.

Charlesworth, S. M. 1994. The pollution history of two urban lakes in Coventry, UK. Unpublished PhD Thesis, Coventry University.

Charlesworth, S. M. and Foster, I. D. L. 1993. The effect of urbanisation on sedimentation: the history of two lakes in Coventry, UK. In Duck, R. W. and McManus, J. (eds) *Geomorphology and Sedimentology of Lakes and Reservoirs*. John Wiley, 15–29.

Charlesworth, S. M. and Lees, J. A. 1997. The use of mineral magnetic measurements in polluted urban lakes and deposited dusts, Coventry, UK. *The Physics and Chemistry of the Earth*, **22**(1–2), 203–206.

Charlesworth, S. M. and Lees, J. A. 1999. The distribution of heavy metals in deposited urban dusts and sediments, Coventry, UK. *Environmental Geochemistry and Health*, **475**, 1–19.

Collins, A. L., Walling, D. E. and Leeks, G. J. L. 1997a. Source type ascription for fluvial suspended sediment based on a quantitative composite fingerprinting technique. *Catena*, **29**, 1–27.

Collins, A. L., Walling, D. E. and Leeks, G. J. L. 1997b. Use of the geochemical record preserved in floodplain deposits to construct recent changes in river basin sediment sources. *Geomorphology*, **19**, 151–167.

Collins, A. L., Walling, D. E. and Leeks, G. J. L. 1998. Use of composite fingerprints to determine the provenance of the contemporary suspended sediment load transported by rivers. *Earth Surface Processes and Landforms*, **23**, 31–52.

Erskine, W. D. and Saynor, M. J. 1996. Effects of catastrophic floods on sediment yields in southeastern Australia. In *Erosion and Sediment Yield: Global and Regional Perspectives*. IAHS Publication **236**, IAHS Press, Wallingford. 381–388.

Farina, M., Esquivel, D. M. S. and Lins de Barros, H. P. S. 1990. Magnetic iron-sulphur crystals from a magnetotactic microorganism. *Nature*, **343**, 256–258.

Förstner, U. 1983. Assessment of heavy metal pollution in rivers and estuaries. In Thornton, I. (ed.) *Applied Environmental Geochemistry*. Academic Press, London, 395–423.

Foster, I. D. L. and Charlesworth, S. M. 1996. Heavy metals in the hydrologicalcycle: trends and explanation. *Hydrological Processes*, **10**, 227–226.

Foster, I. D. L., Lees, J. A., Owens, P. N. and Walling, D. E. 1998. Mineral magnetic characterisation of sediment sources from an analysis of lake and floodplain sediments in the catchments of the Old Mill Reservoir and Slapton Ley, South Devon, UK. *Earth Surface Processes and Landforms*, **23**, 685–703.

Fowler, D., Cape, J. N. and Leith, I. D. 1985. Acid inputs from the atmosphere in the United Kingdom. *Soil Use and Management*, **1**(1), 3–8.

Hollis, G. E. 1988. Rain, roads, roofs and runoff: hydrology in cities. *Geography*, **73**, 9–18.

Horowitz, A. J. 1985. A primer on trace metal–sediment chemistry. *US Geological Survey Water Supply Paper 2277*.

Hunt, A., Jones, J. and Oldfield, F. 1984. Magnetic measurements and heavy metals in atmospheric particulates of anthropogenic origin. *Science of the Total Environment*, **33**, 129–139.

Lees, J. A. 1994. Modelling the magnetic properties of natural and environmental materials. Unpublished PhD Thesis, Coventry University.

Lees, J. A. 1999. Evaluating magnetic parameters for use in source identification, classification and modelling of natural and environmental materials. In Walden, J., Oldfield, F. and Smith, J. (eds) *Environmental Magnetism: A Practical Guide*. Quaternary Research Association, Cambridge, UK, 118–139.

Longmore, M. E., O'Leary, B. M., Rose, C. W. and Chandica, A. L. 1983. Mapping soil erosion and accumulation with the fallout isotope caesium[137]. *Australian Journal of Soil Research*, **21**, 373–385.

Loughran, R. J., Campbell, B. L. and Elliott, G. J. 1986. Sediment dynamics in a partially cultivated catchment in New South Wales, Australia. *Australian Journal of Hydrology*, **83**, 285–297.

Matthei, L. 1993. *Soils and Landscapes of the Ironbark Creek Catchment*. Soil Survey Unit Miscellaneous Report 3, Hunter Catchment Management Trust and Department of Land Management.

Oke, T. R. 1988. The urban energy balance. *Progress in Physical Geography*, **12**, 471–508.

Oldfield, F., Rummery, T. A., Thompson, R. and Walling, D. E. 1979. Identification of suspended sediment sources by means of magnetic measurements: some preliminary results. *Water Resources Research*, **15**, 211–218.

Olley, J. M., Murray, A. S., Mackenzie, D. H. and Edwards, K. 1993. Identifying sediment sources in a gullied catchment using natural and anthropogenic radioactivity. *Water Resources Research*, **29**(4), 1037–1043.

Ormerod, L. M. 1999. Sedimentation rates and sediment provenance within a predominantly urbanised catchment: Ironbark Creek, New South Wales. Unpublished PhD thesis, Department of Geography and Environmental Science, University of Newcastle, NSW.

Passmore, D. G. and Macklin, M. G. 1994. Provenance of fine-grained alluvium and late Holocene land-use change in the Tyne basin, northern England. *Geomorphology*, **9**, 127–142.

Peart, M. R. and Walling, D. E. 1986. Fingerprinting sediment sources: the example of a drainage basin in Devon, UK. In *Drainage Basin Sediment Delivery*. IAHS Publication **159**, 41–55.

Ryding, S.-O. and Rast, W. (eds) 1989. *The Control of Eutrophication of Lakes and Reservoirs. Vol 1. Man and the Biosphere Series*. UNESCO, Paris.

Shankar, R., Thompson, R. and Galloway, R. B. 1994. Sediment source modelling: unmixing of artificial magnetisation and natural radioactivity measurements. *Earth and Planetary Science Letters*, **126**, 411–420.

Snowball, I. F. and Thompson, R. 1988. The occurrence of greigite in sediments from Loch Lomond. *Journal of Quaternary Science*, **3**, 121–125.

Tamura, T. 1964. Selective sorption reactions of caesium with soil minerals. *Nuclear Safety*, **5**, 262–268.

Thompson, R. and Oldfield, F. 1986. *Environmental Magnetism*. Allen and Unwin, London.

Thoms, M. C. 1987. Channel sedimentation within the urbanized River Thame, UK. *Regulated Rivers: Research and Management*, **1**, 229–246.

Wallbrink, P. J., Olley, J. M., Murray, A. S. and Olive, L. J. 1996. The contribution of subsoil to sediment yield in the Murrumbidgee River basin, New South Wales, Australia. In *Erosion and Sediment Yield: Global and Regional Perspectives*. IAHS Publication **236**, IAHS Press, Wallingford. 347–355.

Walling, D. E., Peart, M. R., Oldfield, F. and Thompson, R. 1979. Suspended sediment sources identified by magnetic measurements. *Nature*, **281**, 110–113.

Walling, D. E., Woodward, J. C. and Nicholas, A. P. 1993. A multi-parameter approach to fingerprinting suspended sediment sources. In *Tracers in Hydrology*. IAHS Publication **215**, IAHS Press, Wallingford. 329–338.

Walling, D. E. and Woodward, J. C. 1992. Use of radiometric fingerprints to derive information on suspended sediment sources. In *Erosion and Sediment Transport Monitoring Programs in River Basins*. IAHS Publication **210**, IAHS Press, Wallingford. 153–164.

Walling, D. E. and Woodward, J. C. 1995. Tracing suspended sediment sources in river basins: a case study of the River Culm, Devon, UK. *Journal of Marine and Freshwater Research*, **46**, 327–336.

Yu, L. and Oldfield, F. 1989. A multivariate mixing model for identifying sediment sources from magnetic measurements. *Quaternary Research*, **32**, 168–181.

19 Using Recent Overbank Deposits to Investigate Contemporary Sediment Sources in Larger River Basins

L. J. BOTTRILL,[1] D. E. WALLING[1] and G. J. L. LEEKS[2]
[1]Department of Geography, University of Exeter, UK
[2]Institute of Hydrology, Wallingford, UK

INTRODUCTION

Erosion and sediment yield research has attracted increasing attention in recent years, since the need for an improved understanding of the fluvial sediment system now coincides with some of the current concerns and priorities of society. Information regarding the source of suspended sediment transported by a river must be seen as an essential prerequisite in any attempt to reduce or minimise the offsite effects of erosion and sedimentation (Walling, 1990) and in the design of more effective sediment and pollution control strategies (Wolman, 1977; Vaska and Vrana, 1993). Establishing the relative importance of a range of potential sediment sources involves many practical problems, but recent developments in the use of sediment tracers or signatures to fingerprint sediment sources, and to establish their relative contribution to the suspended sediment load of a river, represent an important advance.

Sediment fingerprinting attempts to isolate property signatures for soils and other potential sediment sources in a catchment and relate these to the properties of suspended sediment (Oldfield *et al.*, 1979; Walling *et al.*, 1993). The basis of the fingerprinting approach involves two steps: first, the selection of physical or chemical properties that are diagnostic of particular sources and which can therefore clearly differentiate such sources on a statistical basis; and, secondly, comparison of the fingerprints for potential source materials with the corresponding values for sediment samples. The quest for a single diagnostic property capable of discriminating a range of sources has now been recognised to be unrealistic (Walling and Woodward, 1995) and several fingerprint properties are now commonly used to provide a means of ensuring that the results obtained are consistent and reliable. The use of composite fingerprints has the added benefit of permitting the discrimination of a greater range of sediment sources. Furthermore, multicomponent signatures are likely to prove more effective in establishing source–sediment linkages, by reducing the possibility

Tracers in Geomorphology. Edited by Ian D. L. Foster. © 2000 John Wiley & Sons Ltd.

of spurious matches which may occur with the use of individual fingerprint properties (Collins *et al.*, 1996).

To date, the fingerprinting technique for sediment source identification has been widely used to establish sediment provenance in small lake catchments, generally of less than 50 km^2 (e.g. Dearing, 1992; Foster and Walling, 1994), and in river basins of less than 300 km^2 (e.g. He and Owens, 1995; Walling and Woodward, 1995; Foster *et al.*, 1996; Walden *et al.*, 1997). These studies of smaller catchment systems have generally addressed sediment provenance in terms of source type. In small drainage basins, the requirement is often for information on the type of sediment source involved and thus, for example, whether the sediment has originated primarily from erosion of topsoil by rilling or sheetwash, from subsoil by gullying or channel scour, or from erosion of cultivated land, pasture or forest. In larger catchments there is frequently a need to consider suspended sediment origin in terms of the spatial distribution of the primary sources and to identify the relative importance of different subcatchments or areas within the larger basins. To date, considerably less attention has been directed to the application of the fingerprinting technique in establishing the spatial distribution of sediment sources in larger river basins. This chapter reports on the results of a study that utilises the fingerprinting technique to establish the spatial provenance of fluvial suspended sediment in a river basin of 10 000 km^2.

Existing studies of the spatial provenance of suspended sediment in larger drainage basins, such as those reported by Collins *et al.* (1996, 1997, 1998), have examined the sources of contemporary suspended sediment. These studies have focused upon bulk suspended sediment samples collected during flood events. Such samples have typically been collected over a short duration and therefore represent essentially instantaneous samples of the suspended sediment load at the time of sampling. To provide time-integrated and more generally representative samples of the suspended sediment transported by rivers during storm events, and therefore to provide more reliable information on the relative contributions of individual sediment sources to the overall suspended sediment load, the study reported here has explored the potential for using samples of overbank sediment collected from river floodplains to characterise both sediment output and individual subcatchment sources. The floodplains of lowland rivers commonly represent important sinks for suspended sediment transported through a river system and they are frequently characterised by extensive deposits of fine sediment resulting from the deposition of suspended sediment during overbank flood events (Walling *et al.*, 1992). Because there is a close link between in-channel suspended sediment transported during high-energy events and overbank deposits, and because overbank sediment deposits will provide a temporally integrated sample of the sediment transported during overbank flood events, the record of suspended sediment transport preserved in floodplain deposits offers considerable potential for elucidating the relative contributions of sediment sources to the sediment load transported by rivers over longer time periods than that offered by instantaneous samples. The use of overbank floodplain sediment as a proxy for suspended sediment also overcomes the many problems associated with suspended sediment sampling during high flows, which can include the forecasting of high-energy flood events which should be sampled, identification of the occurrence of high sediment concentrations within flood periods associated with the movement of the sediment wave through the sampling point, and the physical practicalities of in-channel sampling when access may be limited by floodplain inundation or safety considerations. Due to their accessibility, their advantages of providing more representative samples of the suspended sediment load transported during high discharge periods and the ease and convenience of sampling after flood events, overbank floodplain deposits offer considerable benefits as a sampling medium for investigations of contemporary sediment sources.

Previous fingerprinting studies have established the relative importance of specific sediment sources by comparing the properties of suspended sediment with the properties of soil samples collected from potential source areas (e.g. Walling and Woodward, 1992, 1995; He and Owens, 1995; Collins *et al.*, 1996, 1997, 1998). However, the use of such soil and sediment comparison in tracing studies introduces uncertainties associated with the preferential transport of certain size fractions caused by selective erosion of fines and selective deposition of the coarser fractions. It is generally accepted that particle size exerts a fundamental control on element concentrations and it is therefore undesirable to attempt to compare fingerprint properties of samples with a different particle size composition. This study adopts a new approach. By comparing the properties of sediment collected from downstream floodplains with sediment collected from upstream floodplains, the degree to which the grain-size composition of the sediment sampled at the basin outlet differs from that of the material used to characterise the potential sources is closely constrained. Assuming that material collected from the floodplains is characteristic of a mixture of sediment from upstream areas, the use of such deposits means that the geochemical characterisation of potential source areas should also be more accurate than that achieved when using soil samples.

Surface chemical reactions are extremely important in aquatic trace element–sediment interactions. Because of their large surface areas, fine-grained sediments are the main sites for the adsorption and transport of many trace elements. It is for this reason, and to ensure comparability between suspended sediment and soil sources, that many previous sediment fingerprinting studies have focused upon this fraction (e.g. Walling *et al.*, 1993; Walling and Woodward, 1995; He and Owens, 1995; Collins *et al.*, 1997). However, it is important to recognise that significant concentrations of trace elements can and do exist in coarser material. To address this fact, the fingerprinting procedures used in this study have been applied to both the <63 μm fraction and the <2 mm fraction, in order to consider whether the fingerprinting procedure employed is able to establish the source of sediment comprising a wider range of size fractions. Some examples of the use of the <2 mm fraction for sediment fingerprinting are contained in the literature, but these are primarily confined to radiometric fingerprints (e.g. Walling and Woodward, 1992).

This chapter reports two applications of the fingerprinting procedure to floodplain deposits in order to identify the sources of the <63 μm and <2 mm sediment fractions in the $10\,000\,km^2$ basin of the River Severn, UK. The first investigates the relative contributions of upstream subcatchments to recent overbank sediment collected from the floodplain at Haw Bridge which represents the non-tidal limit of the Severn Basin. The second uses the fingerprinting procedure to examine the contribution of geological sub-areas to the sediment deposited at Haw Bridge.

MATERIALS AND METHODS

Study Basin

The Severn Basin (Figure 19.1 and Table 19.1) represents the largest drainage area in England and Wales and is underlain by Ordovician (mudstones and shales), Silurian (shales and grits), Devonian (Old Red sandstone), Carboniferous (limestones and sandstones), Permo-Triassic (sandstones) and Jurassic (sandstones, mudstones and clays) strata. As a result of its diverse geology, the basin embraces a great variety of terrain types, ranging from mountainous uplands in the western headwaters, underlain by Ordovician and Silurian deposits; through rolling hills in the central areas, developed on Devonian and Carboniferous sandstones; to lowlands in the east underlain by Triassic and Jurassic deposits. Relief generally decreases from west to east and from north to south, with elevations ranging from 752 m in the north-western

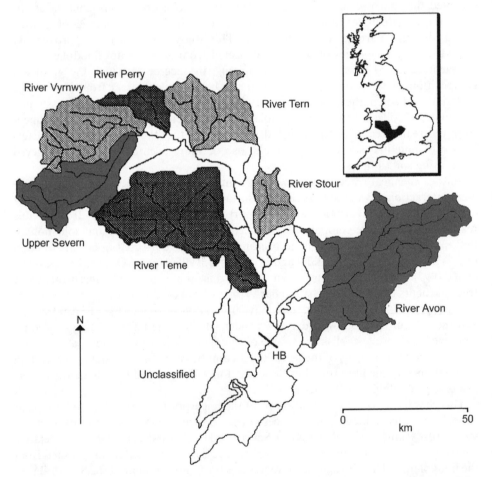

Figure 19.1 The Severn Basin showing subcatchment source areas and the location of the Haw Bridge floodplain (HB)

Table 19.1 Summary information for subcatchment sediment source areas in the Severn Basin

Catchment	Area (km²)	Geology	Precipitation (mm year⁻¹)	Relief	Land use
River Avon	2900	Jurassic clays, Triassic mudstones	600–672	Lowland, predominantly flat	Arable, grassland, urban
River Teme	1648	Silurian mudstones, Devonian sandstones	840–1200	Hilly, very little lowland	Upland pasture, arable, grassland
River Stour	373	Triassic sandstone	620–710	Lowland, predominantly flat	Arable, urban, grassland
River Tern	852	Triassic sandstone	650–750	Lowland, predominantly flat	Arable, grassland, urban
River Perry	181	Triassic sandstone	660–770	Lowland, predominantly flat	Arable, grassland, urban
River Vyrnwy	778	Ordovician and Silurian mudstones	660–2500	Mountainous uplands, wide lowland valleys	Grassland, upland pasture, moorland
Upper Severn	1033	Ordovician and Silurian mudstones	660–2400	Mountainous uplands, wide lowland valleys	Grassland, upland pasture, moorland

headwaters to <10 m in the lowlands of Warwickshire and Gloucestershire. Mean annual precipitation ranges from *c.* 2300 mm in the north-west of the catchment to <600 mm in the south-east. The land use in the western regions is dominated by pastoral farming, mixed farming assumes greater importance in the central regions, and arable farming is of major importance in the lowlands. Although a large proportion of the basin is used for farming, a number of urban areas also lie within the catchment.

Sample Collection

Collection of upstream floodplain sediment to characterise the potential sources of downstream floodplain sediment was based on representative sampling of the surface (0–2 cm) of floodplains distributed throughout the study catchment (Figure 19.2). Floodplain sediment representative of the suspended sediment output from the Severn Basin was characterised by two surface samples collected from the floodplain at Haw Bridge. The resulting sediment samples were dried and screened to 2 mm. A subsample of the <63 μm particle size fraction was also taken and the results of laboratory analysis on both size fractions were used to establish the relative contribution of potential sediment sources.

The fingerprinting technique used for source ascription is based on a comparison of the geochemical properties of sediment collected from the basin outlet with those of potential source sediments. It was therefore necessary to identify a range of properties that could be used to characterise the samples and discriminate different potential sources. The selection of these geochemical properties, which included, carbon and nitrogen, iron, manganese and aluminium, the heavy metals, chromium, copper, lead, zinc, nickel, cobalt and cadmium, the trace metals, arsenic and strontium, the cation

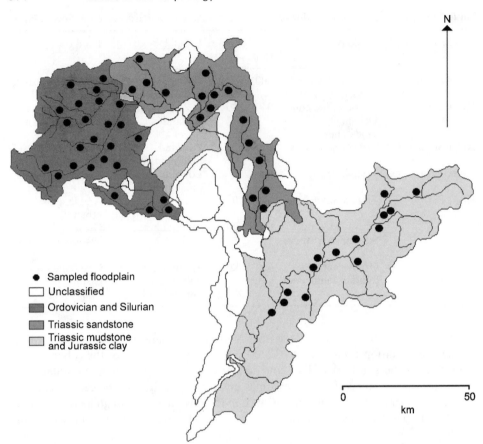

Figure 19.2   The geological source area classification for floodplain surface deposits and the location of floodplains sampled

exchange elements, potassium, magnesium, sodium and calcium, and the pyrophosphate, dithionite and oxalate extractable fractions of iron and manganese, was based on the requirement for the suite of selected properties to reflect different environmental controls, in order to maximise the discrimination between the potential sediment sources. Following nitric and hydrochloric acid extraction, total metal concentrations were determined by atomic absorption spectrophotometry (AAS). Absolute particle size composition was determined using a Malvern Mastersizer after sample pre-treatment with hydrogen peroxide to remove the organic fraction and dispersion with sodium hexametaphosphate. Carbon and nitrogen concentrations were determined by pyrolysis/thermal conductivity using a CE Instruments NA2500 elemental analyser.

FINGERPRINTING THE ORIGIN OF FLOODPLAIN SEDIMENT

To permit the provenance of sediment collected from floodplain surfaces to be examined in terms of spatial origin, the potential source areas in the Severn Basin were first classified in terms of subcatchments. The seven main tributary subcatchments in the basin, which reflect inherent variations in geology, topography, soil, land use and meteorology, were selected and brief descriptions of the characteristics of each are provided in Table 19.1. The locations of these seven subcatchment areas within the Severn Basin are shown in Figure 19.1. The geological subdivision of the Severn Basin reflected the samples available for characterising potential source areas. Given the complex nature of the geology of the Severn Basin, it was not possible to characterise sediment originating from the entire catchment using a geological zonation of source areas. No other river basin of comparable size in Britain displays such a variety of geology as that seen in the Severn Basin (Environment Agency, 1997). Eleven out of the 13 recognised periods of geological time are represented, ranging from 700 million years ago to the late Pleistocene. Given that many rock types only outcrop over small areas, or exist as linear features, it was only possible to accurately delimit three major geological zones using the floodplain sediment samples. The three geological source areas, which cover *c*. 60 percent of the study basin, are shown in Figure 19.2, which also shows the location of the floodplain sites used to characterise the different geological source areas. The chosen classification system comprises five major geological time periods and five associated dominant rock types. These comprise outcrops of Jurassic clay and Triassic mudstone in the south-east of the basin, the area of Triassic sandstone in the north-east of the catchment, and the Ordovician and Silurian mudstones and shales which outcrop in the western areas of the Severn Basin.

To establish the relative contribution of these upstream source areas to sediment deposited on the floodplain at Haw Bridge, sediment from the individual subcatchments and geological zones was characterised by replicate samples collected from 58 floodplain sites. Two samples were collected from the floodplain of the River Severn at Haw Bridge to ensure that the sediment to be sourced was representative of the floodplain as a whole. The first was taken from higher ground 40 m away from the channel and a second from lower ground closer to the channel. Significant lateral variations in sediment properties can occur over relatively small areas in overbank sediments of the same general age (Macklin *et al.*, 1994; Taylor, 1996) and, although it would have been desirable to collect more samples in order to address this variation, sample numbers were limited by the resources available to the study. However, since the area sampled represented a relatively uniform and well-defined area of significant overbank deposition, such variability was judged to be of limited importance at this site.

As indicated above, application of the fingerprinting technique commonly involves two main steps: first, the selection of a range of fingerprint properties which reflect different environmental controls and which are capable in combination of discriminating between potential sediment sources; and, secondly, the application of statistical procedures to provide an objective estimate of the relative contribution of the potential sediment sources to the suspended load of a stream or, in this case, overbank floodplain sediment collected from the basin outlet. The statistical procedures involve two main stages: first, statistical testing of the selected combination of fingerprint properties to confirm that individual sediment sources are clearly discriminated by their composite fingerprint signatures; and, secondly, application of an objective algorithm that is capable of comparing the composite fingerprint of a sediment sample with those of potential sources and providing an estimate of the relative contributions of those sources to the overbank sediment deposit.

Selection of Composite Fingerprint Signatures

It is important to test the extent to which individual sediment properties are able to discriminate the potential suspended sediment sources within the drainage basin and thus to confirm that each source area is characterised by a fingerprint that is significantly different from those of the other areas. In order to select the fingerprint signature most capable of differentiating between sediment sources it is necessary to first identify those properties that are not able to distinguish between source categories and to remove them from the subsequent analysis. Existing studies have used a number of different statistical methods for this purpose (e.g. Walling et al., 1993; Yu

Table 19.2   Results of using the Kruskal-Wallis H-test to assess the ability of individual chemical properties to discriminate between the potential sources of <63 μm and <2 mm floodplain sediment

Parameter	Subcatchment source areas		Geological source areas	
	H value <63 μm	H value <2 mm	H value <63 μm	H value <2 mm
Carbon	45.01[*]	55.81[*]	39.77[*]	40.03[*]
Nitrogen	46.18[*]	57.4[*]	40.43[*]	40.56[*]
Iron	39.33[*]	37.42[*]	10.81[*]	22.87[*]
Aluminium	38.42[*]	37.42[*]	35.63[*]	24.78[*]
Calcium	48.44[*]	50.34[*]	41.39[*]	37.66[*]
Chromium	44.99[*]	15.71[*]	24.3[*]	9.85[*]
Copper	45.22[*]	43.87[*]	24.37[*]	18.46[*]
Potassium	45.75[*]	30.16[*]	34.31[*]	24.71[*]
Magnesium	45.58[*]	25.34[*]	29.74[*]	17.12[*]
Manganese	24.58[*]	26.79[*]	8.89[*]	5.64
Sodium	32.94[*]	30.36[*]	12.48[*]	18.42[*]
Nickel	44.27[*]	32.2[*]	19.49[*]	16.75[*]
Strontium	54.79[*]	33.93[*]	41.95[*]	26.11[*]
Lead	32.38[*]	33.86[*]	6.11[*]	4.94
Zinc	39.48[*]	37.29[*]	13.77[*]	6.41[*]
Pyrophosphate iron	33.53[*]	17.5[*]	26.22[*]	1.82
Pyrophosphate manganese	16.59[*]	31.45[*]	3.43	8.11[*]
Dithionite iron	34.57[*]	25.51[*]	11.87[*]	4.17
Dithionite aluminium	18.42[*]	17.96[*]	17.03[*]	17.53[*]
Dithionite manganese	19.68[*]	20.61[*]	10.69[*]	8.36[*]
Oxalate iron	31.83[*]	23.24[*]	24.64[*]	5.83
Oxalate manganese	22.05[*]	31.83[*]	8.27[*]	8.8[*]
Cobalt	35.17[*]	36.96[*]	8.5[*]	32.08[*]
Cadmium	32.17[*]	27.18[*]	6.89[*]	12.5[*]
Arsenic	25.93[*]	40.26[*]	12.03[*]	14.69[*]
Tin	42.81[*]	33.06[*]	16.99[*]	6.15[*]
Critical value	12.59	12.59	5.99	5.99

[*] Significant at $P = 0.05$.

and Oldfield, 1993). Most recently, Collins *et al.* (1996) have proposed a two-stage verification procedure, consisting of a Kruskal–Wallis *H*-test and discriminant function analysis. This study has also adopted this approach.

The results of using the Kruskal–Wallis *H*-test to analyse the variance within the <63 μm and <2 mm source material data sets are presented in Table 19.2. Any parameters that failed at this stage to reach the critical value for the 95 percent confidence limit of 12.59 for the subcatchment classification and 5.99 for the geological zonation were removed from the analysis. Most of the parameters exceed the critical values for the 95 percent confidence level. In terms of subcatchment sources all potential fingerprint signature parameters exceeded the critical value. However, pyrophosphate extractable manganese was shown to be a poor discriminator of the <63 μm geological sources and was at this stage removed from further consideration. Five parameters, namely manganese, lead, pyrophosphate extractable iron, dithionite extractable iron and oxalate extractable iron failed to reach the critical value for the <2 mm geological source zonation and were also removed from further analysis.

The second stage of statistical selection, involving the application of multivariate discriminant function analysis to identify the property subsets that would provide composite fingerprints capable of maximising the statistical differentiation of the defined source areas, produced the results presented in Tables 19.3 and 19.4 for the

Table 19.3   Results of the stepwise discriminant function analysis of subcatchment source areas for < 63 μm floodplain sediment

Enter	Proportion of samples classified correctly (%)							
	Avon	Teme	Stour	Tern	Perry	Vyrnwy	Upper Severn	Total
Step 1 Mg	68.4	16.7	0	30	50	62.5	20	38.9
Step 2 Zn	63.2	83.3	75	30	50	62.5	40	56.9
Step 3 Fe	63.2	83.3	75	40	25	87.5	40	59.7
Step 4 Mn	68.4	100	75	40	75	75	46.7	66.7
Step 5 Py. Fe	84.2	100	75	60	100	87.5	46.7	76.4
Step 6 Py. Mn	94.7	100	75	70	75	87.5	60	81.9
Step 7 C	94.7	100	75	70	75	87.5	60	81.9

Table 19.4   Results of the stepwise discriminant function analysis of subcatchment source areas for < 2mm floodplain sediment

Enter	Proportion of samples classified correctly (%)							
	Avon	Teme	Stour	Tern	Perry	Vyrnwy	Upper Severn	Total
Step 1 N	42.8	50	33.3	16.7	50	12.5	6.7	31.1
Step 2 Sr	68.4	50	50	33.3	66.7	25	6.7	43.2
Step 3 Co	68.4	71.4	50	50	66.7	37.5	46.7	58.1
Step 4 Di. Mn	73.7	78.6	83.3	33.3	83.3	75	60	70.3
Step 5 Fe	78.9	100	83.3	33.3	83.3	75	53.3	74.3
Step 6 Pb	73.7	92.9	83.3	33.3	100	100	80	81.1
Step 7 Py. Mn	94.7	100	100	66.7	100	100	80	91.9

Table 19.5 Results of the stepwise discriminant function analysis of geological source areas for < 63 μm floodplain sediment

		Proportion of samples classified correctly (%)			
	Enter	Jurassic clay and Triassic mudstone	Triassic sandstone	Ordovician and Silurian	Total
Step 1	Sr	69.2	57.1	96.4	80
Step 2	Py. Fe	100	78.6	100	94.6
Step 3	Al	100	78.6	100	94.6
Step 4	C	100	85.7	100	96.4
Step 5	Na	100	85.7	100	96.4
Step 6	Mn	100	100	100	100
Step 7	Ca	100	100	100	100
Step 8	K	100	100	100	100

Table 19.6 Results of the stepwise discriminant function analysis of geological source areas for < 2 mm floodplain sediment

		Proportion of samples classified correctly (%)			
	Enter	Jurassic clay and Triassic mudstone	Triassic sandstone	Ordovician and Silurian	Total
Step 1	N	84.6	56.3	96.2	81.8
Step 2	Sr	69.2	68.8	100	83.6
Step 3	Al	69.2	100	100	92.7
Step 4	K	84.6	100	100	96.4
Step 5	As	84.6	100	100	96.4
Step 6	Fe	92.3	100	100	98.2
Step 7	Di. Mn	92.3	100	100	98.2
Step 8	Mg	92.3	100	100	98.2

subcatchment sources and Tables 19.5 and 19.6 for the geological sources, for the <63μm and <2mm size fractions respectively. The tables show the steps taken in each analysis and the resulting percentage of samples that were correctly apportioned to their source group. A composite signature consisting of seven parameters, namely magnesium, zinc, iron, manganese, pyrophosphate extractable iron, pyrophosphate extractable manganese and carbon, was identified as that most capable of distinguishing <63 μm sediment from the individual subcatchment sources. This seven-parameter signature was capable of correctly apportioning 81.9 percent of the overall set of samples to their source groups. The discrimination afforded by the selected signature is shown graphically in Figure 19.3. To discriminate between <2 mm sediment representative of individual subcatchments, the analysis also identified a seven-parameter signature, including nitrogen, strontium, cobalt, dithionite extractable manganese, iron, lead and pyrophosphate extractable manganese, that was capable of assigning 91.9 percent of samples to their correct subcatchment origin. The discrimination afforded by the signature for <2 mm sediment is shown in Figure 19.4. The

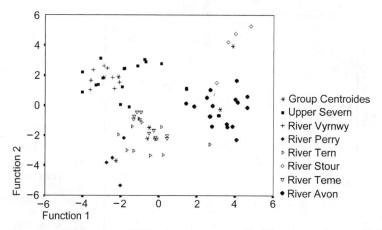

Figure 19.3 Canonical discriminant functions constructed from the first and second discriminant functions calculated for subcatchment source areas of <63 μm floodplain sediment

discriminant function analysis used to identify a composite signature for geological source areas of <63 μm sediment selected eight parameters, namely strontium, pyrophosphate extractable iron, aluminium, carbon, sodium, manganese, calcium and potassium, which in combination were capable of correctly apportioning 100 percent of the overall set of samples to their source group. The analysis applied to the geological source areas of <2 mm sediment also produced an eight-parameter signature that was capable of correctly assigning 98.2 percent of the <2 mm floodplain samples to their correct geological origins. The degree of discrimination afforded by the two geological source area signatures is demonstrated in Figures 19.5 and 19.6.

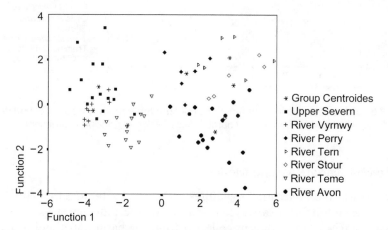

Figure 19.4 Canonical discriminant functions constructed from the first and second discriminant functions calculated for subcatchment source areas of <2 mm floodplain sediment

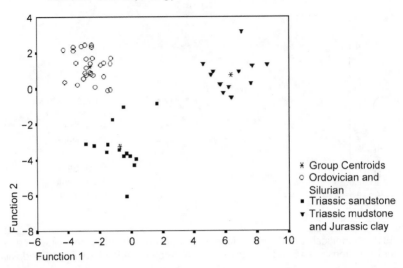

Figure 19.5 Canonical discriminant functions constructed from the first and second discriminant functions calculated for geological source areas of $< 63\mu m$ floodplain sediment

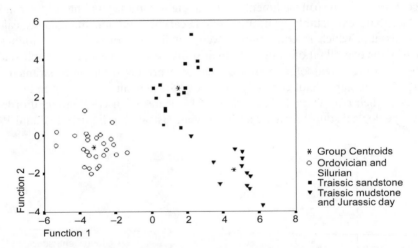

Figure 19.6 Canonical discriminant functions constructed from the first and second discriminant functions calculated for geological source areas of < 2 mm floodplain sediment

Use of a Multivariate Mixing Model

In order to use the fingerprint signatures identified by the discriminant function analysis to establish the relative contribution of the different subcatchment and geological sources to the sediment collected from the basin outlet and to maximise the accuracy of such source apportionment, a multivariate mixing model is required. The mixing model optimising algorithm proposed by Walling *et al.* (1993) was used for this purpose.

This mixing model seeks to satisfy the following set of linear constraints:

(1)   The source type contributions must all sum to unity, i.e.

$$\sum_{s=1}^{s} P_s = 1$$

(2)   The n source type contributions must all be non-negative, i.e.

$$0 \leq P_s \leq 1$$

(3)   The error term must be reduced to its minimal value, i.e.

$$E = \sum_{t=1}^{T} \left\{ \left[\left| B_t - \left(\sum_{s=1}^{S} V_{st} P_S Z_S \right) \right| \right] / B_t \right\}$$

where B_t is the concentration of tracer parameter (t) in overbank floodplain sediment obtained from the catchment outlet, P_s is the percentage contribution from each source area, V_{st} is the mean concentration of tracer parameter (t) for source area (s) and Z_s a particle size correction factor for source area (s).

The particle size of suspended sediment exerts a fundamental control over its mineralogy and geochemistry and is hence the most important factor influencing trace element concentrations. For this reason, particle size weighting factors based upon the ratio of the mean specific surface area of the catchment outlet sediment samples to the mean specific surface area of the sediment samples from each of the sources was included in the mixing models (Table 19.7). The error term was minimised by utilising the optimisation program available within the Quatro Pro database package. Initially, the source contributions P_s are set equal and the error term calculated. New errors are then calculated for a series of changes in the values of P_s and the values that minimise the error term are selected as a solution and provide the percentage source area contributions.

Table 19.7   Mean particle size correction factors for the source areas of $<63\,\mu m$ and $<2\,mm$ sediment

Source area	Correction factor	
	$<63\,\mu m$	$<2\,mm$
Subcatchment source area		
River Avon	0.722	1.068
River Teme	1.308	1.455
River Stour	0.797	1.615
River Tern	0.922	1.597
River Perry	1.036	2.063
River Vyrnwy	1.057	1.299
Upper Severn	0.993	1.272
Geological source area		
Jurassic clay/Triassic mudstone	0.666	1.115
Triassic sandstone	0.786	1.157
Ordovician/Silurian mudstone and shale	0.926	1.299

Table 19.8   Mean percentage errors between observed and predicted parameter concentrations for Haw Bridge floodplain sediment

	% Error	
Source type	$<63\,\mu m$	$<2\,mm$
Subcatchment source	19.96	20.92
Geological source	31.13	25.62

Given the four selected fingerprint signatures and particle size correction factors, the optimisation algorithm of Walling *et al.* (1993) was applied to the two Haw Bridge floodplain sediment samples to establish sediment provenance in terms of both sub-catchment and geological sources. The level of accuracy associated with the resultant model predictions is shown in Table 19.8, which presents the mean percentage error between the observed parameter concentrations in the floodplain sediment samples from Haw Bridge and the predicted concentrations based upon the optimised percentage contribution from each of the defined sources provided by the results of the mixing model. The errors produced by the application of the mixing model to the identification of the provenance of the floodplain deposits are comparable with those given by Collins *et al.* (1996, 1997, 1998) who examined the sources of contemporary suspended sediment in the Upper Severn Basin. The larger errors associated with the application of the geological signature to both size fractions may be attributable to the limited coverage of the catchment provided by the three geological zones, i.e. the mixing model still apportions sediment originating in the 40 percent of the basin not underlain by Jurassic clay and Triassic mudstone, Triassic sandstone or Ordovician and Silurian rocks to one of the three source areas.

RESULTS AND DISCUSSION

Figure 19.7 presents the results of the application of the fingerprinting technique to the identification of the relative contributions of the seven subcatchment source areas to the sediment collected from the Haw Bridge floodplain, for both the $<63\,\mu m$ and $<2\,mm$ fractions. Examination of the results based on $<63\,\mu m$ sediment indicates that the sources of deposited material remain relatively constant across the floodplain at Haw Bridge since both samples produce similar estimates in terms of the relative contributions of individual subcatchment sources. The results indicate that the River Avon, River Teme, River Vyrnwy and Upper Severn catchments are the most important sources of $<63\,\mu m$ sediment. The Rivers Teme, Vyrnwy and Upper Severn, which contribute 24 percent, 18 percent and 26 percent respectively, drain the western uplands of the Severn Basin, which are steeply sloping and receive higher levels of precipitation than the eastern areas and therefore could be expected to represent important sediment sources. The Avon, which contributes 27 percent, is a lowland river draining the lower-energy environment of the south-western Severn Basin. Its

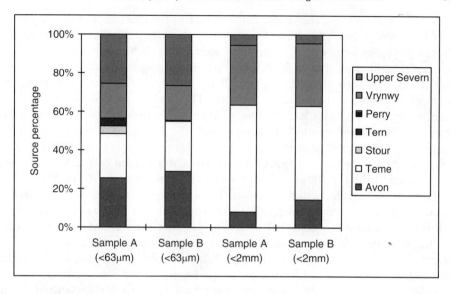

Figure 19.7 Subcatchment sources of Haw Bridge floodplain sediment established using the fingerprinting technique

relative importance as a $< 63\,\mu$m sediment source is attributable to two factors: first, the underlying Jurassic mudstone and clay bedrock, and the existence of unconsolidated drift deposits which provide readily eroded fine material; and, secondly, the high levels of arable land use in the catchment. The smaller catchments of the Rivers Perry, Stour and Tern, which drain the central areas of the Severn catchment, are shown to be insignificant sources of $<63\,\mu$m sediment.

In considering the sources of $<2\,$mm sediment, the relatively smaller central catchments of the Rivers Stour, Tern and Perry are again shown to be insignificant sediment sources. No sediment contribution is assigned to either of the three catchments, which is perhaps surprising given that all three catchments drain sandstones and are therefore likely to generate coarser sediment. Most $<2\,$mm sediment is ascribed to the catchments of the Vyrnwy and Teme (31 percent and 50 percent respectively). The River Vyrnwy is a more important source of $<2\,\mu$m sediment than of $<63\,\mu$m sediment. This is largely because the Upper Severn is ascribed a lower sediment contribution. The change in the relative importance of the northwestern upland source areas may be attributable to the coarser composition of the sediment that originates from the Vyrnwy catchment and also conveyance losses within the Upper Severn as a result of deposition of coarse material in the extensive floodplain systems of the Vale of Powys. The relatively higher contribution of the River Teme catchment to the $<2\,$mm sediment fraction at Haw Bridge may also reflect the particle size characteristics of the transported sediment. The catchment of the River Teme is dominated by Devonian sandstones and the resulting suspended sediment has a larger mean particle size than any of the other three major contributing catchments. This means that it is likely to provide a more significant contribution to the coarser sediment fraction.

The results of applying the geological signature to the identification of the sources of <63 μm and <2 mm sediment are shown in Figure 19.8. The Ordovician and Silurian geological source area contributes approximately 87 percent of the <63 μm sediment deposited at Haw Bridge. The Triassic sandstone source, which includes most of the catchments of the Rivers Tern, Perry and Stour, remains insignificant with 0 percent of sediment attributed to this area. The lowland south-west of the basin, underlain by Jurassic clays and Triassic mudstones, contributes between 11 and 14 percent of the <63 μm sediment. Examination of the geological source of the <2 mm sediment fraction deposited at Haw Bridge gives further support to the findings of the subcatchment source investigations, in that the central region of the Severn Basin, characterised by the Triassic sandstone geology and drained by the Rivers Stour, Tern and Perry, is not a major sediment source. The importance of the western uplands to sediment generation in the Severn catchment is again shown by the dominance of the Ordovician and Silurian geological source in the provenance of the <2 mm fraction. The Jurassic clay and Triassic mudstone source area contributes approximately 6 percent of the <2 mm sediment fraction. A high degree of consistency is evident between the results obtained for the geological origins of the two size fractions. For example, the Ordovician and Silurian source area accounts for 86 percent of the <63 μm fraction of sample A compared to 96 percent for the <2 mm fraction and for 89 percent of the <63 μm fraction and 90 percent of the <2 mm fraction for sample B. As with the results of the subcatchment apportioning of source, it is clear that the fine sediment derived from the Jurassic clay and Triassic mudstone source

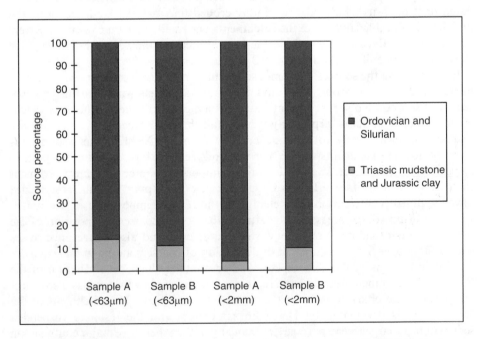

Figure 19.8 Geological sources of Haw Bridge floodplain sediment established using the fingerprinting technique

area results in a reduction of the importance of this source when examining the larger size fraction, since the coarser sediment of the western tributaries becomes more dominant.

The inconsistencies in the estimates of the relative sediment contributions from essentially similar areas, defined in terms of subcatchment and geological source, can be attributed to the fact that the geological classification of source areas employed in this study is, first, based upon the concept of dominant upstream geology and, secondly, does not represent all of the potential sediment source areas in the Severn Basin. For example, the ascription of 27 percent of the <63 μm sediment transported by the River Severn at its outlet to the subcatchment of the River Avon is significantly higher than the ascription of 13 percent to the equivalent Jurassic clay and Triassic mudstone source area by the geological approach. However, the Avon subcatchment is not underlain solely by Jurassic clay and Triassic mudstone. Its geology also includes Triassic sandstone and significant areas of rocks not included in the basic geological zonation of sediment sources, including Carboniferous sandstone. Sediment derived from Triassic or Carboniferous sandstone should therefore be assigned to the River Avon source area. Overall, however, although there are differences in the estimated contributions, similar spatial patterns are demonstrated by the results based on both the subcatchment and geological classifications of spatial sources. Both signatures indicate that the major sediment source area in the Severn Basin is the upland west of the catchment. This area is represented by the Ordovician and Silurian geological source area and the catchments of the Upper Severn, Vyrnwy and Teme.

PERSPECTIVE

This contribution has successfully employed multicomponent fingerprint signatures and a multivariate mixing model to elucidate the sources of overbank floodplain sediment deposited during approximately the last 5–10 years on the River Severn floodplain at the catchment outlet. The results of applying the fingerprinting technique to the Haw Bridge floodplain deposits have revealed, first, a pattern of homogenous deposition across the floodplain with close agreement between two samples for both subcatchment and geological source area provenance. Secondly, consideration of the source area of the two size fractions of the sediment indicates that there are only minor differences in source area contributions between the fractions. These differences in the sources of <63 μm and <2 mm sediment have been attributed to the particle size composition of the sediment from different areas.

The methodology, results and discussion presented clearly demonstrate the potential for using recent floodplain deposits in sediment source fingerprinting studies, both in terms of providing information on sediment source contributions and the potential for characterising sediment sources using a small number of samples. The use of samples collected from the surface of floodplains enables the contributions of sediment sources to be estimated more reliably and the relative importance of individual sources to be assessed over a longer period than is possible when considering the provenance of contemporary suspended sediment samples. Such samples represent essentially instantaneous samples of the suspended sediment load transported during storm events and may not be representative of the overall sediment load of the river due to temporal variations in the arrival of sediment from different source areas at the catchment outlet.

REFERENCES

Collins, A. L., Walling, D. E. and Leeks, G. J. L. 1996. Composite fingerprinting of the spatial source of fluvial suspended sediment: a case study of the Exe and Severn river basins, UK. *Geomorphologie; Relief Processus Environment*, **1996 (2)**, 41–54.

Collins, A. L., Walling, D. E. and Leeks, G. J. L. 1997. Fingerprinting the origin of fluvial suspended sediment in larger river basins: combining assessment of spatial provenance and source type. *Geografiska Annaler*, **79 (A)**, 239–254.

Collins, A. L., Walling, D. E. and Leeks, G. J. L. 1998. Use of composite fingerprints to determine the provenance of the contemporary suspended sediment load transported by rivers. *Earth Surface Processes and Landforms*, **23**, 31–52.

Dearing, J. A. 1992. Sediment yields and sources in a Welsh upland lake-catchment during the past 800 years. *Earth Surface Processes and Landforms*, **17**, 1–22.

Environment Agency 1997. *River Severn – Middle Reaches: Catchment Management Plan Consultation Report, July 1997*. Environment Agency, Tewkesbury.

Foster, I. D. L. and Walling, D. E. 1994. Using reservoir deposits to reconstruct sediment yields and sources in the catchment of the Old Mill Reservoir, South Devon, UK over the past 50 years. *Hydrological Sciences* Journal, **39(4)**, 347–368.

Foster, I. D. L., Owens, P. N. and Walling, D. E. 1996. Sediment yields and sediment delivery in the catchments of Slapton Lower Ley, South Devon, UK. *Field Studies*, **8**, 629–661.

He, Q. and Owens, P. N. 1995. Determination of suspended sediment provenance using caesium-137, unsupported lead-210 and radium-226: a numerical mixing model approach. In: Foster, I. D. L., Gurnell, A. M. and Webb, B. W. (eds) *Sediment and Water Quality in River Catchments*. John Wiley, Chichester, 207–227.

Macklin, M. G., Ridgway, J., Passmore, D. G. and Rumsby, B. T. 1994. The use of overbank sediment for geochemical mapping and contamination assessment: results from selected English and Welsh floodplains. *Applied Geochemistry*, **9**, 689–700.

Oldfield, F., Rummery, T. A., Thompson, R. and Walling, D. E. 1979. Identification of suspended sediment sources by means of magnetic measurements: some preliminary results, *Water Resources Research*, **15**, 211–219.

Taylor, M. P. 1996. The variability of heavy metals in floodplain sediments: a case study from mid-Wales. *Catena*, **28**, 71–87.

Vaska, J. and Vrana, K. 1993. Modelling of runoff and erosion from a small watershed. In: Banasik, K. and Zbikowski, A. (Eds) *Runoff and Sediment yield modelling*. Proceedings of the International Symposium, RSY-93, Warsaw, Poland, September 14–16, pp101–106. Warsaw Agricultural University Press, Warsaw.

Walden, J., Slattery, M. C. and Burt, T. P. 1997. Use of mineral magnetic measurements to fingerprint suspended sediment sources: approaches and techniques for data analysis. *Journal of Hydrology*, **202**, 353–372.

Walling, D. E. 1990. Linking the field to the river: sediment delivery from agricultural land. In Boardman, J. and Foster, I. D. L. (eds) *Soil Erosion on Agricultural Land*. John Wiley, Chichester, 129–152.

Walling, D. E. and Woodward, J. C. 1992. Use of radiometric fingerprints to derive information on suspended sediment sources. In *Erosion and Sediment Monitoring Programmes in River Basins*. IAHS Publication **210**, IAHS Press, Wallingford. Wallingford, UK, 153–164.

Walling, D. E. and Woodward, J. C. 1995. Tracing sources of suspended sediment in river basins: a case study of the River Culm, Devon, UK. *Marine and Freshwater Research*, **46**, 327–336.

Walling, D. E., Quine, T. A. and He, Q. 1992. Investigating contemporary rates of floodplain sedimentation. In Carling, P. A. and Petts, G. E. (eds) *Lowland Floodplain Rivers: Geomorphological Perspectives*. John Wiley, Chichester, 165–184.

Walling, D. E., Woodward, J. C. and Nicholas, A. P. 1993. A multi-parameter approach to fingerprinting suspended sediment-sources. In *Tracers in Hydrology*, IAHS Publication **215**, IAHS Press, Wallingford. Wallingford, UK, 329–337.

Wolman, M. G. 1977. Changing needs and opportunities in the sediment field. *Water Resources Research*, **13**, 50–54.

Yu, L. and Oldfield, F. 1993. Quantitative sediment source ascription using magnetic measurements in a reservoir-catchment system near Nijar, S.E. Spain. *Earth Surface Processes and Landforms*, **18**, 441–454.

20   Clast Travel Distances and Abrasion Rates in Two Coarse Upland Channels Determined using Magnetically Tagged Bedload

T. STOTT and A. SAWYER

Liverpool John Moores University, Liverpool, UK

INTRODUCTION

Downstream fining and rounding of coarse river sediments is well documented (McPherson, 1971; Mills, 1979; Knighton, 1980; Huddart, 1994; Ferguson *et al.*, 1996) and these changes in sediment characteristics are commonly explained in terms of selective entrainment, transport and deposition (hydraulic sorting), or the physical modification of clasts by mechanical abrasion. Such abrasion processes have been studied in the past using abrasion tanks (Kuenen, 1956; Bradley, 1970; Bradley *et al.*, 1972) and tumbling barrels (Daubree, 1879; Wentworth, 1919; Marshall, 1927; Bigelow, 1984). Under the laboratory conditions used by these workers, clast weight losses per unit distance were consistently lower than downstream reductions in weight loss derived from field sampling (Bradley, 1970; Schumm and Stevens, 1973; Adams, 1978, 1979). The observed differences between laboratory-simulated and field-measured reduction rates have been attributed to either weathering (Bradley, 1970), hydraulic sorting processes (Mackin, 1963) or the theory of 'abrasion in place' (Schumm and Stevens, 1973) whereby clasts may be abraded by vibration within the bed without net downstream movement. Laboratory simulation of this vibration yielded promising results (Schumm and Stevens, 1973), though the authors did not report significant abrasion occurring in their experiments. Brewer *et al.* (1992) used 25 mm rock cubes, shown using extensive laboratory simulation of abrasion processes to exhibit similar weight loss patterns and rates to natural channel material (Brewer, 1991), to monitor abrasion. The cuboid tracers, which were sawn from large pieces of parent bedrock, were bolted or tethered with string to the channel bedrock. The standard size of the cubes eliminated differences in surface area between tracers, and allowed easy identification of impact marks or loss of corners and edges. Weight losses sustained by 39 tracers over a six-week period in a coarse upland

Tracers in Geomorphology. Edited by Ian D. L. Foster. © 2000 John Wiley & Sons Ltd.

channel indicated the potential of 'sandblasting' as an additional 'abrasion in place' process.

AIMS

As part of a broader study investigating coarse sediment transport rates of upland streams using tracers and bedload traps, the aims of this study were (i) to assess abrasion rates of natural clasts using magnetically tagged bedload, and (ii) to compare abrasion rates of clasts in a channel flowing in a mature forested upland catchment with those in a channel flowing in a semi-natural upland grassland catchment.

By introducing 385 magnetically tagged clasts into two natural channels and tracing their movements for over two years, we believe that this chapter quantifies abrasion rates for individual test clasts in the natural field situation for the first time.

STUDY SITES

The study reaches were located within the Institute of Hydrology's Plynlimon Experimental Catchments (Newson, 1976; Kirby *et al.*, 1991) which have been operational since 1968 and contain intensive hydrological, water quality and sediment monitoring

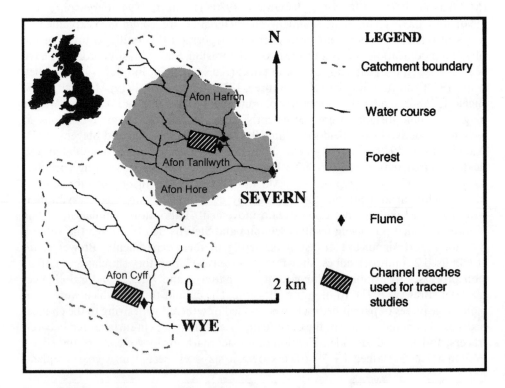

Figure 20.1   Location map to show study reaches

networks (Leeks and Roberts, 1987; Leeks, 1992; Stott and Marks, 1998). The stream channels are influenced by both the local bedrock geology (Ordovician and Silurian mudstones and shales) and glacially derived gravels, cobbles and boulders. Much of the bedload is made up from this material and reworked glacio-fluvial sediment of similar composition.

The climate is temperate with a mean annual precipitation of 2449 mm. Figure 20.1 shows the location of the catchments with the study reaches indicated. The catchment areas of the Tanllwyth (Severn) and Cyff (Wye) are 0.89 and 3.10 km^2, respectively, and both channels are at a height of approximately 350 m AOD. The streams have flashy hydrographs and over the 28 month study period from February 1995 to July 1997 maximum discharges were 2.32 m^{3}s^{-1} and 6.36 m^{3}s^{-1} with means of 0.05 m^{3}s^{-1} and 0.16 m^{3}s^{-1} in the Tanllwyth and Cyff channels, respectively. In the Tanllwyth there were an estimated 37 competent flood events when discharge exceeded 0.5 m^{3}s^{-1} and flow exceeded 1 m^{3}s^{-1} for 29 h; 0.5 m^{3}s^{-1} for 190 h; and 0.3 m^{3}s^{-1} for 488 h. In the Cyff there were an estimated 85 competent flood events when discharge exceeded 0.5 m^{3}s^{-1} and flow exceeded 1 m^{3}s^{-1} for 402 h; 0.5 m^{3}s^{-1} for 1296 h; and 0.3 m^{3}s^{-1} for 2614 h. The study reaches were both 400 m long and average gradients were 5.4 percent and 4.5 percent for the Tanllwyth and Cyff, respectively.

METHODS

Tracer clasts were manufactured by drilling a hole and implanting a small magnet and label, fixed by epoxy resin. Only less resistant clasts of Silurian mudstone, shale and greywacke were soft enough to be drilled easily so this meant that glacially derived clasts of other geological origins were not included in the sample. However, since these glacially derived clasts make up less than 5 percent of the bed material their exclusion from the tracer sample does not bias the sample significantly. A total of 385 tracers were prepared, clasts being sampled to represent the natural size distribution in the channels as closely as possible. Figure 20.2 shows the size distribution (percent by weight) of tracers and natural bed material for the Tanllwyth. The bed D_{50} values were 22.3 and 44.4 mm for the Tanllwyth and Cyff, respectively (when surface and subsurface bed material is combined). The size distribution of Cyff tracers mirrored closely those for the Tanllwyth. Clast shapes in the channels, and replicated in the sample of magnetically tagged tracers, were dominated by discs (48 percent) and blades (42 percent), with rods (8 percent) and spheres (2 percent) relatively less common. Tracers were weighted, and a, b and c axes measured, before they were first introduced into the channels in February 1995. Groups of 10 tracers were placed in an equally spaced line across the channel with no special attempt made to seed them. Tracer locations were surveyed on 10 occasions between February 1995 and July 1997. The majority of tracers became buried after 1–2 months in the channel and on locating them they were dug out of the bed, identified and replaced into the hole from which they had been excavated. From these surveys total distance travelled by individual clasts was computed and clast dimensions and masses were re-measured again in July 1997 as clasts were retrieved from the channel. At this point, 228 clasts were available for analysis, the remainder of the sample not being recovered.

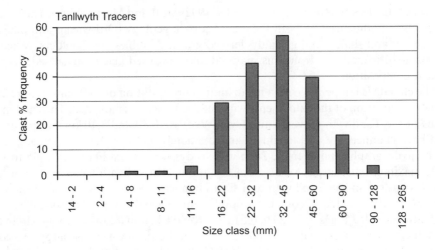

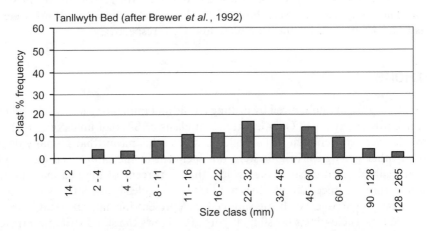

Figure 20.2   Size distributions of tracers used in this study and natural bed material in the Afon Tanllwyth, Plynlimon. (Bed D_{50} values are 22.3 and 44.4 mm for Tanllwyth and Cyff channels respectively)

RESULTS

Clast Abrasion Rates (Weight Losses)

Table 20.1 shows summary statistics for all Tanllwyth and Cyff tracers. Though tracers in the forested Tanllwyth channel appear to show higher weight losses than those in the grassland Cyff channel, the difference is not statistically significant when compared using the t-test. The mean mass of all tracers introduced to the channels was 78.51 g, which represents an actual mean weight reduction of 2.74 g per clast per year.

Table 20.1 also shows the variations in both weight loss and travel distance for clasts in the four Zingg shape classes. Rod-shaped clasts showed the greatest weight

Table 20.1 Summary statistics for tracer abrasion (% weight loss) and travel distances for Tanllwyth and Cyff channels

	n	Annual mean % weight loss	Max. % weight loss	Mean travel distance (m day^{-1})	Max. travel distance (m day^{-1})
Cyff tracers	114	2.8 ± 0.2	42.2	0.07 ± 0.03	0.43
Tanllwyth tracers	114	4.2 ± 0.2	57.2	0.14 ± 0.03	0.87
All tracers	228	3.5 ± 0.1	57.1	0.11 ± 0.02	0.87
Blades	95	2.8 ± 0.2	24.5	0.10 ± 0.03	0.73
Discs	110	3.8 ± 0.2	42.2	0.11 ± 0.03	0.50
Spheres[a]	4	*1.2 ± 0.4*	*2.5*	*0.18 ± 0.15*	*0.39*
Rods	19	5.4 ± 0.7	57.2	0.18 ± 0.09	0.87

[a] The low number of spheres in the sample means that estimates of weight loss and travel distances are unreliable. *Unreliable estimates in italics.*

losses (5.4 ± 0.7 percent) and the difference between the weight loss of rods and blades is significant (p <0.05) as shown by the *t*-test. The greater weight losses of rod-shaped clasts may be accounted for by their higher travel rates. The low number of spheres in the sample means that estimates of weight loss and travel distances are unreliable.

Figure 20.3 shows mean annual weight losses for tracers by their size class. When all tracers are combined, the trend appears to show greatest weight losses in the 11–16 mm

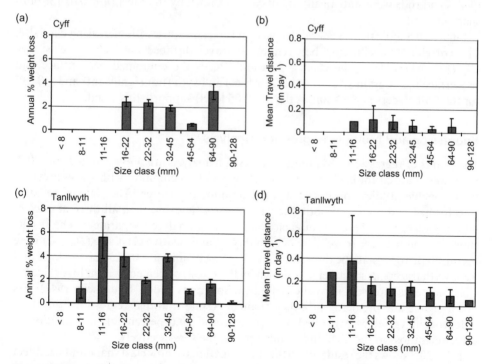

Figure 20.3 Analysis by size: weight loss of tracers and distance travelled for Tanllwyth and Cyff tracers

and 16–22 mm size classes, with a decrease in weight loss in the larger size classes. However, the small sample size in the 11–16 mm class ($n = 4$) means estimates are unreliable. The only statistically significant differences between mean weight loss and clast size class were found in the coarser size classes where the weight loss in the 32–45 mm class was significantly greater than in the 45–64 mm class ($p < 0.01$) and weight loss in the 45–64 mm class was significantly lower than in the 64–90 mm class ($p < 0.05$).

Travel Distances

Tracer travel distances in the forested Tanllwyth channel were double those in the grassland Cyff and this difference is statistically significant at the $p < 0.001$ level as shown by the t-test. The correlation coefficients between per cent weight loss and mean travel distance of 0.32 for the Cyff tracers and 0.23 for the Tanllwyth are significant at the $p < 0.01$ and $p < 0.05$ levels respectively.

Table 20.1 shows travel distances for the four shape classes. Rods and spheres showed the greatest travel rates, with means of 0.18 ± 0.09 m day^{-1} and 0.18 ± 0.15 m day^{-1}, respectively, though the small sample size for spheres means the estimate carries large error bars and is unreliable. Discs and blades had lower mean travel distances (0.11 ± 0.03 m day^{-1} and 0.10 ± 0.03 m day^{-1}, respectively) and the differences between the mean travel rates of discs and rods and between blades and rods were statistically significant as shown by the t-test ($p < 0.01$ for both shape classes).

Figure 20.3 indicates a general decrease in travel distance with increasing clast size. The correlation coefficients between mean travel distance and relative clast size (D_i/D_{50}), where D_i is the clast b axis and D_{50} is the combined bed surface and subsurface mean grain size, of -0.17 for the Tanllwyth (not significant) and -0.217 for the Cyff tracers (significant at the $p < 0.05$ level) supports this trend.

Regression and Multiple Regression Analysis

The results of regression and multiple regression analysis are presented in Table 20.2. Data were plotted for each channel separately and, in common with many previous tracer pebble studies, produced a large amount of scatter. The data were $\log_{10}$ transformed and regressed both one variable at a time and then all three at once in multiple regression. Data from each channel were analysed separately. The results show that both tracer travel distance (m day^{-1}) and relative clast size (D_i/D_{50}) were useful predictors (statistically significant at $p < 0.01$ and $p < 0.025$ for the Cyff and Tanllwyth tracers, respectively, as shown by the F test) of percent weight loss of clasts. Clast shape, as represented by the Krumbein Sphericity Index (Krumbein, 1941), was not a significant predictor in either channel. Using multiple regression, the three independent variables explained 42.5 percent and 30.6 percent of the variation in clast weight loss in the Cyff and Tanllwyth channels, respectively.

Table 20.3 shows the results of further investigations of how clast shape and size affect tracer travel distance. It can be seen that the proportion of variation in travel distance explained by sphericity and relative grain size (D_i/D_{50}) is maximised to 30.9 percent

Table 20.2   Regression and multiple regression analyses: factors affecting % weight loss (% WL) of tracers

	Multiple R	v_1	v_2	F	Significance
CYFF					
Regression					
1 Log_{10} % WL vs Log_{10} distance	31.6	1	77	8.45	$p < 0.01$
2 Log_{10} % WL vs Log_{10} sphericity	14.5	1	77	1.64	n.s.
3 Log_{10} % WL vs $Log_{10} D_i/D_{50}$	33.2	1	77	9.46	$p < 0.01$
Multiple regression					
4 Log_{10} % WL vs Log_{10} distance, Log_{10} sphericity, $Log_{10} D_i/D_{50}$	42.5	3	77	5.43	$p < 0.01$
TANLLWYTH					
Regression					
5 Log_{10} % WL vs Log_{10} distance	22.9	1	105	5.79	$p < 0.025$
6 Log_{10} % WL vs Log_{10} sphericity	10.1	1	105	1.08	n.s.
7 Log_{10} % WL vs $Log_{10} D_i/D_{50}$	21.9	1	105	5.28	$p < 0.025$
Multiple regression					
8 Log_{10} % WL vs Log_{10} distance, Log_{10} sphericity, $Log_{10} D_i/D_{50}$	30.6	3	105	3.50	$p < 0.025$

Note: v_1 and v_2 are upper and lower d.f. respectively.

Table 20.3   Regression analysis results of factors affecting tracer travel distance

Regression	R	v_1	v_2	F	Significance
CYFF					
1 Distance vs sphericity	2.6	1	118	0.08	n.s.
2 Distance vs D_i/D_{50}	27.1	1	118	9.26	$p < 0.01$
3 Distance vs (sphericity/D_i/D_{50})	30.9	1	118	12.34	$p < 0.001$
TANLLWYTH					
4 Distance vs sphericity	21.5	1	164	7.91	$p < 0.01$
5 Distance vs D_i/D_{50}	24.6	1	164	10.58	$p < 0.01$
6 Distance vs (sphericity/D_i/D_{50})	32.3	1	164	18.98	$p < 0.001$

Note: v_1 and v_2 are upper and lower d.f. respectively.

and 32.3 percent in the Cyff and Tanllwyth, respectively, by dividing sphericity by relative size to remove the effect of size. This suggests that clast shape (sphericity) does have a significant effect on travel distance after allowing for size ($p < 0.001$).

DISCUSSION

The results suggest that the difference in weight loss between tracers in the forested Tanllwyth and the grassland Cyff channels may be accounted for by the greater mean travel distances. There is no doubt that bed sediment is much more mobile in the Tanllwyth channel, where travel distances are double those in the Cyff ($p < 0.001$).

This may be attributed to the mature forest land use. It is likely that drainage ditches excavated in the 1930s when the plantation forest was established, continue to supply sediment to the main channel. Indeed, the greater production and transport of bedload in the Tanllwyth channel is discussed by Moore and Newson (1986) who report bedload yields from the Tanllwyth being six times higher than from the Cyff, based on 10 years of records. They concluded that bedload yields from the forested Tanllwyth are more predictable, which reflects the greater importance of supply-limited conditions on the grassland (unditched) Cyff catchment. The lower mean travel rates reported in this chapter for the Cyff concur with these findings.

Correlations between variables are weak but some are, nevertheless, statistically significant. The correlation between size and distance travelled is significant in the Cyff ($p < 0.05$) only giving some evidence for size selective transport in this channel where bedload seems to be supply-limited. The extent and nature of this is the subject of further analysis (Sawyer, 1999).

In terms of shape clasts, rods and spheres have the highest mean travel distances, but while rods have the highest mean weight losses, spheres have the lowest. This may be accounted for by the fact that their shapes make them more suited to rolling and, once in motion, they are less likely to be deposited due to imbrication and bed armouring. Spheres are least likely to break while in transport and predictably their mean weight loss is the lowest of the four shape classes. Spheres are, however, very uncommon in these channels and the extremely small sample size renders these inferences at best suggestive, and at worst unreliable. Rods, in contrast, seem to be more prone to breakage. Indeed, the maximum weight loss for all tracers was 57.2 percent and this clast was a rod. The next highest weight loss for rods is 13.0 percent which implies that breakage of clasts is relatively unusual and that corner rounding and edge chipping are more likely to account for the weight losses observed. However, one limitation of the approach taken is the inability to document and quantify the different weight loss mechanisms that may be operating. These mechanisms may include abrasion of clasts against bedrock and other sediment in movement, sandblasting by the sand component of the suspended or saltating load, edge/corner chipping or even splitting as illustrated above. In the light of these data, it is not possible to determine whether abrasion processes operating in each channel are different or operate in different ways. Further work based on the techniques used by Brewer *et al.* (1992) will be required to answer these questions.

Regression and multiple regression have been partially successful in explaining percent weight losses in terms of travel distance, particle shape (Krumbein's Sphericity Index) and relative clast size (D_i/D_{50}). Multiple regression was able to explain 42.5 percent and 30.6 percent of the variation in tracer percent weight loss in the Cyff and Tanllwyth, respectively, suggesting that other factors such as the precise geological composition of the clasts are almost certainly important in predicting abrasion rates. In regression analysis the least useful predictor was sphericity which was not statistically significant in either channel, whereas travel distance and relative size (D_i/D_{50}) were both statistically significant ($p < 0.01$ in Cyff; $p < 0.025$ in Tanllwyth).

In contrast to the 'abrasion in place' processes (bedload over-passing and sandblasting) monitored in the Tanllwyth channel by Brewer *et al.* (1992) and distinguished from the 'vibratory' processes proposed by Schumm and Stevens (1973), the findings of this study suggest that weight losses of natural tracers are likely to be more dependent on clast size and travel distance, with clast shape being of secondary importance. What still remains to be determined by future field observations is the proportion of time clasts spend undergoing 'abrasion in place' processes (bedload over-passing, sandblasting and vibratory processes) compared to abrasion resulting from attrition (corner rounding, edge chipping and splitting) as clasts bounce or saltate during transport. Active (radio) tracer techniques as used by Schmidt and Ergenzinger (1992) during flood events may offer another means of isolating and determining the relative importance of these two sets of abrasion processes. However, techniques that rely on drilling clasts to insert magnets (as in this study), iron rods (Reid *et al.*, 1984) or radio tags inevitably will affect the strength of clasts and may render them more prone to breakage or splitting.

These findings may help to explain observed trends in downstream fining and bed sediment character of alluvial channels and the production of fines from in-channel abrasion processes. They add significantly to a relatively small data set from which natural in-channel abrasion rates may be assessed. Consideration should be given to including them as a component in models which attempt to predict downstream changes in shape and size as well as the production of fines from abrasion processes.

CONCLUSIONS

(1)   Mean annual weight loss of natural tracers studied in two coarse upland channels was 3.5 ± 0.2 percent with weight losses for tracers in the forested Tanllwyth channel (4.2 ± 0.2 percent) being higher than for tracers in the grassland Cyff channel (2.8 ± 0.2 percent). Mean travel distance for tracers in the forested Tanllwyth channel (0.14 ± 0.03 m day^{-1}) was double that for the Cyff (0.07 ± 0.03 m day^{-1}) with the difference significant at $p < 0.001$ level.

(2)   Clast shapes in the channels, and replicated in the sample of magnetically tagged tracers, were dominated by discs (48 percent) and blades (42 percent), with rods (8 percent) and spheres (2 percent) relatively less common. In this study rods showed both the highest weight losses (5.4 ± 0.7 percent of their mass per year) and the highest travel distances (0.81 ± 0.09 m day^{-1}) though the small number of spheres in our sample means that this finding is tentative and will have not any general applicability.

(3)   Regression and multiple regression have been partially successful in explaining percent weight losses in terms of travel distance, particle shape (Krumbein's Sphericity Index) and relative clast size (D_i/D_{50}). Multiple regression was able to explain 42.5 percent and 30.6 percent of the variation in tracer percent weight loss in the Cyff and Tanllwyth, respectively. The least useful predictor was sphericity which was not statistically significant in either channel, whereas travel distance and relative size (D_i/D_{50}) were both statistically significant ($p < 0.01$ in Cyff; $p < 0.025$ in Tanllwyth). The proportion of variation in travel distance explained by sphericity and relative grain size (D_i/D_{50}) is maximised to 30.9 percent and 32.3 percent in the Cyff and Tanllwyth, respectively, by dividing sphericity by relative size to remove the effect of size. This suggests that clast

shape (sphericity) does have a significant effect on travel distance after allowing for size ($p < 0.001$).

ACKNOWLEDGEMENTS

The authors wish to acknowledge financial support from Liverpool John Moores University provided in the form of a University Research Studentship to A.S. and an equipment grant. Staff at the Institute of Hydrology's Plynlimon Offices provided field support and flow data. Professors Rob Ferguson and Dave Huddart kindly made helpful comments on a earlier draft of the manuscript and comments from Dr Jeff Warburton greatly improved the final manuscript.

REFERENCES

Adams, J. 1978. Data for New Zealand pebble abrasion studies. *New Zealand Journal of Science*, **21**, 607–610.

Adams, J. 1979. Gravel size analysis from photographs. *Journal of the Hydraulics Division, Proceedings of the American Society of Civil Engineers*, **105**, 1247–1255.

Bigelow, G. E. 1984. Simulation of pebble abrasion on coastal benches by transgressive waves. *Earth Surface Processes and Landforms*, **9**, 383–390.

Bradley, W. C. 1970. Effect of weathering on abrasion of granitic gravel, Colorado River (Texas). *Geological Society of America Bulletin*, **81**, 61–80.

Bradley, W. C., Fahnestock, R. K. and Rowekamp, E. T. 1972. Coarse sediment transport by flood flows on Knick River, Alaska. *Geological Society of America Bulletin*, **83**, 1261–1284.

Brewer, P. A. 1991. Sediment reduction processes in natural rivers. Unpublished PhD Thesis, University College of Wales, Aberystwyth.

Brewer, P. A., Leeks, G. J. L. and Lewin, J. 1992. Direct measurement of in-channel abrasion processes. In Bogen, J., Walling, D. E. and Day, T. (eds) *Erosion and Sediment Transport Monitoring in River Basins*. Proceedings of the Oslo Symposium, 24–28 August 1992. International Association of Hydrological Sciences Publication **210**, IAHS Press, Wallingford. 21–29.

Daubree, A. 1879. *Etudes Synthetiques de Geologie Experimentale*, 2 volumes. Dunod, Paris.

Ferguson, R., Hoey, T., Wathen, S. and Werritty, A. 1996. Field evidence for rapid downstream fining of gravels through selective transport. *Geology*, **24**, 179–182.

Huddart, D. 1994. Rock-type controls on downstream changes in clast parameters in sandur systems in Southeast Iceland. *Journal of Sedimentary Research*, **64**(2), 215–225.

Kirby, C., Newson, M. D. and Gilman, K. 1991. *Plynlimon Research: The First Two Decades*. Institute of Hydrology Report **109**, Wallingford.

Knighton, A. D. 1980 Longitudinal changes in size and sorting of streambed material in four English rivers. *Geological Society of America Bulletin*, **91**(1), 55–62.

Krumbein, W. 1941. Measurement and geological significance of shape and roundness of sedimentary particles. *Journal of Sedimentary Petrology*, **11**, 64–72.

Kuenen, P. H. H. 1956. Experimental abrasion of pebbles: 2. Rolling by current. *Journal of Geology*, **64**, 336–368.

Leeks, G. J. L. 1992. Impact of plantation forestry on sediment transport processes. In Billi, P., Hey, R. D., Thorne, C. R. and Tacconi, P. (eds) *Dynamics of Gravel Bed Rivers*. John Wiley, Chichester, 651–670.

Leeks, G. J. L. and Roberts, G. 1987. The effects of forestry on upland streams with special reference to water quality and sediment transport. In Good, J. E. G. and Institute of

Terrestrial Ecology (eds) *Environmental Aspects of Plantation Forestry in Wales, No. 22*, Institute of Terrestrial Ecology, Merlewood Research Station, Grange-over-Sands, 9–24.

Mackin, J. H. 1963. Rational and empirical methods of investigation in Geology. *Benchmark Papers in Geology*, **13**, 135–163.

Marshall, P. 1927. The wearing of beach gravels. *Transactions and Proceedings of New Zealand Institute*, **58**, 507–532.

McPherson, H. J. 1971. Downstream changes in sediment characteristics in a high energy mountain stream channel. *Arctic and Alpine Research*, **3**, 65–79.

Mills, H. H. 1979. Downstream fining of pebbles – a quantitative review. *Journal of Sedimentary Petrology*, **49** (1), 295–302.

Moore, R. J. and Newson, M. D. 1986. Production, storage and output of coarse upland sediments: natural and artificial influences as revealed by research catchment studies. *Journal of the Geological Society*, **143**, 1–6.

Newson, M. D. 1976. The physiography, deposits and vegetation of the Plynlimon experimental catchments. Institute of Hydrology Report **30**, Wallingford.

Reid, I., Brayshaw, A. C. and Frostick, L. E. 1984. An electromagnetic device for automatic detection of bedload motion and its field applications. *Sedimentology*, **31**, 269–276.

Sawyer, A. 1999. Forestry effects on sediment storage, routing and abrasion in upland streams. Unpublished PhD Thesis, Liverpool John Moores University.

Schmidt, K-H. and Ergenzinger, P. 1992. Bedload entrainment, travel lengths, step lengths, rest periods – studied with passive (iron, magnetic) and active (radio) tracer techniques. *Earth Surface Processes and Landforms*, **17**, 147–165.

Schumm, S. A. and Stevens, M. A. 1973. Abrasion in place: a mechanism for rounding and size reduction of coarse sediments in rivers. *Geology*, **1**, 37–40.

Stott, T. A. and Marks, S. 1998. Bank erosion and suspended sediment dynamics: responses to timber harvesting in mid Wales, UK. In *Proceedings of the International Symposium on Comprehensive Watershed Management (ISWM-'98)*, 7–10 September 1998, Beijing, China. International Research Training Centre on Erosion and Sedimentation, Beijing, 213–220.

Wentworth, C. K. 1919. A laboratory and field study of cobble abrasion. *Journal of Geology*, **27**, 507–522.

21 Influence of Bed Material Shape on Sediment Transport in Gravel-Bed Rivers: A Field Experiment

J. WARBURTON and T. DEMIR
Department of Geography, Science Laboratories, University of Durham, UK

INTRODUCTION

As a result of improved theoretical comprehension and carefully conducted field and laboratory experiments (Wiberg and Smith, 1987; Ashworth and Ferguson, 1989; Carling *et al.*, 1992) understanding of bedload transport in upland gravel-bed rivers and mountain streams has advanced considerably in recent years. Although many studies have examined bedload transport in relation to grain size, the importance of grain shape has not been examined in the same detail (Komar and Li, 1986; Carling *et al.*, 1992; Schmidt and Gintz, 1995). This is somewhat surprising given the importance attached to particle shape in early studies of fluvial sediments (Wentworth, 1922; Krumbein, 1941).

The importance of grain shape for sediment transport in coarse-bed river systems can be viewed in several contexts which relate to both the static and dynamic behaviour of sedimentary particles. First, shape influences the entrainment of individual particles. Particle shape also affects the mode of transport (e.g. sliding versus pivoting); and along with the geometry of the particle, the surface–pocket relationship (Komar and Li, 1986). Secondly, shape is important in controlling interparticle interactions which produce microforms on the surface of river beds (Reid *et al.*, 1992; Tribe and Church, 1999). Thirdly, the *en masse* behaviour of similar-shaped particles determines the bulk properties of a river bed, e.g. imbricate structures occur in rivers with a dominance of platy or disc-shaped particles. Surface roughness is also influenced by particle shape although there is no simple relationship between these parameters and the grain size of the gravel bed (Kirchner *et al.*, 1990; Carling *et al.*, 1992).

The general aim of this project is to examine these effects using a combination of field and laboratory experiments. The specific experiment reported here is designed to quantify the selective transport of different shapes of coarse river gravel and determine their spatial sorting in an upland gravel-bed stream system. Preliminary results

Tracers in Geomorphology. Edited by Ian D. L. Foster. © 2000 John Wiley & Sons Ltd.

are presented for three reaches and are also aggregated for the channel system as a whole.

EXPERIMENTAL SITE

In order to investigate the effect of shape on the transport of coarse river gravel a field experiment was set up at the Moor House National Nature Reserve, North Pennines, England (Figure 21.1). This site proved ideal because of the combination of good

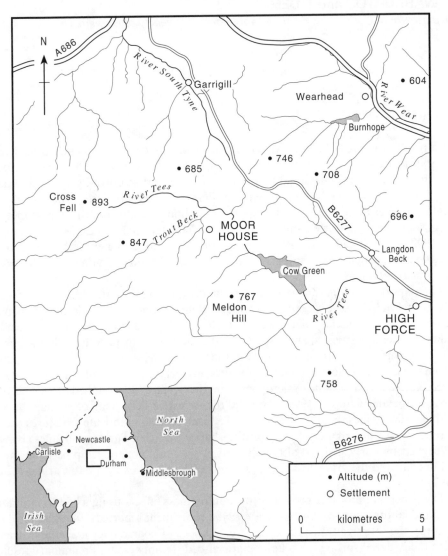

Figure 21.1 Map showing the location of the Moor House National Nature Reserve in the North Pennines and the location of the two study rivers

Table 21.1 Details of bed material tracers and site characteristics at the three reaches. Observations of the proportion of tracers moved were made during a survey undertaken eight months after the tracers were introduced into the channel

Site	Colour of tracer	Bed material size		No. of tracers	Tracers		
		D_{50}	D_{90}		No movement	Moved	Missing
		(mm)					
Lower Tees	Orange	50	100	900	470	332	98
Trout Beck	Yellow	69	120	900	627	238	35
Upper Tees	White	51	98	900	850	50	0

access, the flashy nature of the river regime and the presence of an existing research infrastructure/data archive on the hydrology and geomorphology of the site.

The field experiment was carried out on three experimental reaches: two on the River Tees (lower and upper sites), and one on Trout Beck (Table 21.1). The Trout Beck joins the River Tees between the upper and lower sites (Figure 21.1). Discharge is gauged on the Trout Beck approximately 400 m upstream of the confluence with the River Tees. The site consists of a compound Crump weir which was established in 1971. However, a full flow record is not available due to operating difficulties in the 1980s. Since 1991 the site has been run and maintained by the National Rivers Authority (the Environment Agency from 1 April 1996) and discharge data are recorded at 15 min intervals. A second gauging site was established on the upper Tees just upstream of the experimental reach. This consisted of a Campbell CR10X data logger connected to a Druck pressure transducer. Stage is recorded every 15 min in line with the Environment Agency procedures used at Trout Beck. The river at the gauging site is relatively straight and for the majority of the stream cross-section it flows over bedrock. Results based on a preliminary rating show the site is functioning well and a similar flow regime is recorded to that at Trout Beck. Trout Beck has higher peak discharges than the upper Tees. All three reaches have similar slopes and bed material sizes (Table 21.1). Slope is highest at Trout Beck and lowest at the Upper Tees.

Gauging the flow at these two sites and summing their values provided a means of estimating the discharge at the lower Tees site below the confluence. Preliminary stream gauging at all three sites, during a period of relatively steady flow, suggests the sum of the two upstream discharges provides a good estimate of the discharge below the confluence. This design around the confluence was deliberate in order that bedload transport could be observed on three rivers in close proximity and also that the gauging of discharge of the two tributaries would provide an estimate of discharge at the lower Tees site.

THE TRACING EXPERIMENT

A total of 900 tracers were prepared for each of the three sites (Table 21.1). The size range of the tracers was selected as being between 32 and approximately 256 mm. This range represented the coarsest two-thirds of the grain-size distribution. Selecting the coarsest end of the grain-size distribution is significant because this is where size and

Table 21.2   Comparison of shape distributions between the natural
bed material and the tracer material at the Upper Tees experimental site

Bed material %	Sphere	Blade	Rod	Disc
Natural	15	19	13	53
Tracer	25	25	25	25

shape sorting are usually most pronounced. Three size classes were defined: 32 to
<64 mm, 64 to <128 mm and >128 mm. In each of the smaller size classes 400 tracers
were prepared. In the larger size class, because of their lower mobility, only 100 tracers
were used. Within each size class equal proportions of different-shaped particles were
included. Careful screening of the particles on the basis of the Zingg (1935) shape
classification produced four distinct shape classes: discs, spheres, blades and rods.
Particles plotting close to the shape boundaries were not used in the experiment. The
tracers are similar in terms of lithology and density and match the coarse size fraction
of the natural bed material (Table 21.1). Tracers differ in that they have a wider range
of particle shapes (with equal numbers of spheres, rods, discs and blades) and greater
roundness. The natural bed material shows a greater proportion of disc-shaped
particles (Table 21.2).

Each of the tracer particles was first drilled and a small ferromagnetic magnet
placed inside. The magnet was sealed in place using a silicon gel, then the whole
particle was painted with masonry paint. Each particle was then labelled with an
identification number. All tracers were measured (a, b and c axes) in order to provide
an independent means of identification if the paint labels were erased. The great
advantage of magnetic tracers is that they can be relocated even if buried, covered
in algae, abraded of the tracer paint or hidden in murky water (Schick et al., 1988).

The tracers were placed randomly on the channel bed in a 3 m wide strip, 2–3 m in
from the banks. Sites were visited weekly to gauge movement and, where noticeable
movement had occurred, the sites were resurveyed as soon as discharge conditions
permitted. Tracer positions were mapped with reference to a series of monumented
sections set out along the banks adjacent to the experimental reaches. The position of
individual particles was surveyed using two methods: laser tape survey and total
station. Movements <2 m from the starting strip were not included in the analysis.
The aim of the experiment was not to reproduce the exact field conditions of the natural
bed material, but to see how different particle shapes behaved in a fluvial setting.

INITIAL RESULTS

At the outset it was decided that it would not be possible in the initial experiments to
visit the site after every flood. Therefore, bedload movement events could not be
related to individual floods. However, because the main interest is the relative movement between different shape classes this was not seen as a major drawback.

Figure 21.2 shows the combined transport data for the three reaches. The boxplots
show that there is a general decrease in distance transported with size which is

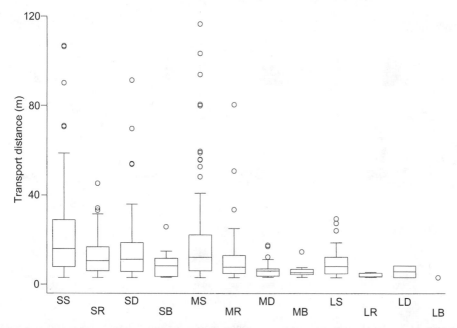

Figure 21.2 Boxplots comparing the distribution of transport distances for the different particle size and shape categories. Data for all three experimental reaches are pooled. Outliers greater than 1.5 times the interquartile range are shown as circles. Tracer particles are described by their size (S, small; M, medium; L, large) and shape (S, sphere; B, blade; R, rod; D, disc)

consistent in all shape classes. Spheres are transported the greatest distance along with small numbers of discs and rods. These results demonstrate size-selective transport and somewhat weaker shape-selective transport. Differentiating between weight, size and shape reveals some consistent patterns in transport distances (Figure 21.3). Generally, all shapes show a decrease in the distance and frequency of transport as size increases. Spheres are transported most frequently and over the greatest distances whilst blades show the least movement. Although there are some between-site differences, similar patterns of weight and transport distance occur for all three reaches and the channel system as a whole (Figure 21.4). Combining the data for the three reaches is justified because of the similarity in reach characteristics, their close proximity in the fluvial system and the general nature of the problem studied. Spheres (S) are consistently transported the greatest distances together with smaller numbers of discs (D) and rods (R). Transport is greatest at the Trout Beck and lower Tees reaches. There has been little movement at the upper Tees possibly due to the local bed topography at the head of the study reach.

   Figure 21.5 shows the result of the first surface survey 23 days after the tracers were first introduced into the Trout Beck. Most of the movement occurred in one flood event in the week prior to the survey. Peak discharge during this period was $13.7 \, m^3 s^{-1}$ on 6 December 1997. Flows at this time were generally bankfull with some local overbank flow. Results are plotted in terms of shape of the tracer particles and size. Only particles that have moved more than 2 m beyond the start line are

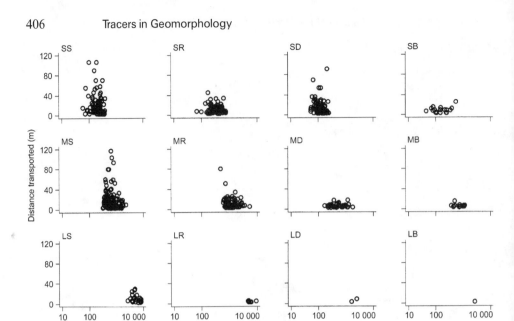

Figure 21.3 Weight versus transport distance graphs for the different shape and size categories. Data from all three experimental reaches are pooled and particles are described by their size (S, small; M, medium; L, large) and shape (S, sphere; B, blade; R, rod; D, disc)

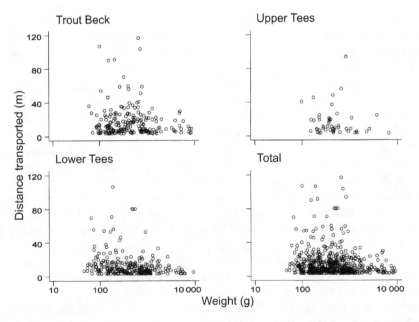

Figure 21.4 Particle weight and transport distance data shown for the three experimental reaches and pooled for all reaches (Total).

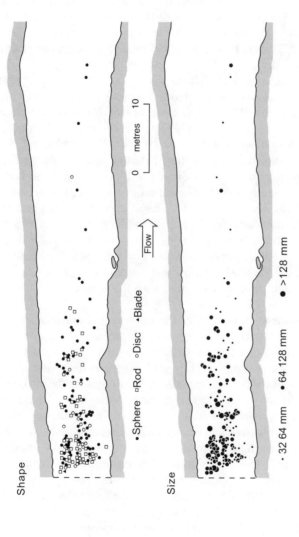

Figure 21.5 Maps of in-channel bedload sorting at Trout Beck experimental reach, 23 days after the start of the experiment. Data are plotted with respect to shape and size classes

plotted. This corresponds to approximately 17 percent of the original 900 particles. Looking at size first, it is clear that there is preferential movement of the finer size classes. Some large particles moved, but the majority of the transport is confined to the first 10 m downstream. Of the two smaller size classes, equal numbers appear to have moved, and both classes show a similar distribution in the channel. The general pattern of movement clearly shows a decay in the frequency of movement with distance down channel. Tracers that moved the greatest distances are located along the line of the channel thalweg.

The same data plotted in terms of shape show that sphere-shaped particles are transported by far the greatest distance. Rods also move preferentially and moved up to 35 m from the start line. Discs show lower travel distances and blades hardly appear to have moved at all. This is also reflected in the number of tracers in each shape class that have moved: spheres 75, rods 48, discs 25 and blades 4. Subsequent surveys of the site have shown similar patterns, though some of the smaller spheres and rods had moved further downstream. Similar results are seen at the Tees sites, but distances moved are generally lower. These findings are consistent with those of Schmidt and Gintz (1995) who carried out a similar experiment in a steep mountain stream using artificially moulded particle shapes. Similar experiments by Schmidt and Ergenzinger (1992) also concur with these results; however, the particle shapes do not correspond exactly with those reported here. For example, the 'ball' shapes of Schmidt and Ergenzinger show elongation and flattening (a axis, 122 mm; b axis, 95 mm; c axis, 62 mm) and cannot be equated with the spheres used in these experiments.

Because tracer roundness differed from the natural bed material, a tilting table experiment was designed to clarify the mechanistic behaviour of different particle shapes on beds of varying gravel roughness. The measured friction angle (Kirchner *et al.*, 1990), or pivoting angle (Komar and Li, 1986), expresses the resistance to removal of the grain by the flow. Results of an experiment investigating the effects of particle form and roundness on four differing roughnesses demonstrate several important effects. First, a slightly lower friction angle occurs with increasing roundness. However, this effect is not as great as the increase in friction angle associated with larger (coarser) roughness. Also particle shape and the orientation of non-equidimensional particles (e.g. rods) is much more significant in determining the friction angle than clast roundness. The implication of this is that although roundness may influence entrainment thresholds, particle size remains the more important factor. Therefore, although tracer roundness differs from the natural bed material roundness, this is thought to be a secondary influence on sediment transport dynamics at the study site.

CONCLUSION AND DISCUSSION

Based on these initial results it can be generally concluded that although there is size selectivity in transport there is also a clear pattern of shape-selective transport. Sphere-shaped particles are transported the longest distances and in the greatest numbers. Rods and discs also move preferentially, but blades hardly move at all.

Particle form and orientation are important in determining movement but roundness appears to be a secondary factor. However, it should be noted that the shapes used in these experiments were deliberately selected; in many natural settings variability in particle form is much less pronounced and patterns of downstream changes in particle shapes are not always apparent (Huddart, 1994).

An important limitation of the present experimental design relates to the division of the tracers into arbitrary shape classes. Natural bed material generally shows a continuous range of size and shape, and by categorising bed particles a lot of information on the precise size and shape of each tracer is lost. Furthermore, by creating shape categories the scope for statistical analysis is reduced. In particular, evaluating the relative importance of size over shape sorting cannot be easily undertaken. An alternative approach would be to do a multiple regression analysis of distance travelled on grain size and a sphericity index.

Further analysis of both field and laboratory results will enable the relative magnitude of these effects to be evaluated in a more detailed quantitative manner once additional field data become available. In particular, as time progresses and the tracers become better mixed with the natural bed material, subject to higher flow events and buried, it will be interesting to see if the shape sorting evident at the surface in the early surveys is still apparent over greater transport distances. Whether these patterns persist as particles undergo further movement and become better mixed with the bed material remains to be seen. Because very few tracers were found buried in the bed material (Table 21.1), a more comprehensive survey of the three sites will be undertaken in the summer months when flows are low and access to the stream channel is easier. In this way, a picture of the three-dimensional distribution of the tracers will be built up. In addition, the transport distance data need to be evaluated in terms of within-reach variability of sediment size and bed roughness because smaller tracers may not be able to exert a shape influence over a very coarse bed. A detailed survey of the three reaches has been undertaken to estimate these effects. Future analysis will examine relative sediment size effects (D_i/D_{50}) and their influence on sediment transport dynamics.

The design based around the confluence is useful because it provides observations on three sites in close proximity and maximises resources available for stream gauging and discharge estimation. Finally, the use of magnetic tracers provides a very useful field method for evaluating the transport of coarse bed material in active upland river systems.

ACKNOWLEDGEMENTS

This project has been jointly funded by a Turkish Government Scholarship (T.D.) and the University of Durham (J.W.). The co-operation of John Adamson (ITE), English Nature and the Environment Agency (for providing Trout Beck discharge data) is gratefully acknowledged. Technical assistance has been provided by the Department of Geography, University of Durham. Reviews by Rob Ferguson and Tim Stott were gratefully received.

REFERENCES

Ashworth, P. J. and Ferguson, R. I. 1989. Size selective entrainment of bedload in gravel bed streams. *Water Resources Research*, **25**, 627–634.

Carling, P. A., Glaister, M. S. and Kelsey, A. 1992. Effect of bed roughness, particle shape and orientation on initial motion criteria. In Hey, R. D., Billi, P., Thorne, C. R. and Tacconi, P. (eds) *Dynamics of Gravel-Bed Rivers*. John Wiley, Chichester, 23–38.

Huddart, D. 1994. Rock-type controls on downstream changes in clast parameters in sandur systems in Southeast Iceland. *Journal of Sedimentary Research*, **64**(2), 215–225.

Kirchner, J. W., Dietrich, W. D., Iseya, F. and Ikeda, H. 1990. The variability of critical shear stress, friction angle and grain protrusion in water-worked sediments. *Sedimentology*, **37**, 647–672.

Komar, P. D. and Li, Z. 1986. Pivoting analysis of the selective entrainment of sediment by shape and size with application to the gravel threshold. *Sedimentology*, **33**, 425–436.

Krumbein, W. C. 1941. Measurements and geological significance of shape and roundness of sedimentary particles. *Journal of Sedimentary Petrology*, **11**, 64–72.

Reid, I., Frostick, L. E. and Brayshaw, A. C. 1992. Microform roughness elements and the selective entrainment and entrapment of particles in gravel-bed rivers. In Hey, R. D., Billi, P., Thorne, C. R. and Tacconi, P. (eds) *Dynamics of Gravel-Bed Rivers*. John Wiley, Chichester, 253–275.

Tribe, S. and Church, M. 1999. Simulations of cobble structure on a gravel streambed. *Water Resources Research*, **35**(1), 311–318.

Schick, A. P., Hassan, M. A. and Lekach, J. 1988. A vertical exchange model for coarse bedload movement – numerical considerations. *Catena Supplement*, **10**, 73–83.

Schmidt, K.-H. and Ergenzinger, P. 1992. Bedload entrainment, travel lengths, step lengths, rest periods – studied with passive (iron, magnetic) and active (radio) tracer techniques. *Earth Surface Processes and Landforms*, **17**, 147–165.

Schmidt, K.-H. and Gintz, D. 1995. Results of bedload tracer experiments in a mountain river. In Hickin, E. J. (ed.) *River Geomorphology*. John Wiley, Chichester, 37–54.

Wentworth, C. K. 1922. A field study of the shapes of river pebbles. *US Geological Survey Bulletin*, **730**, 103–114.

Wiberg, P. L. and Smith, J. D. 1987. Calculations of critical shear stress for motion of uniform and heterogeneous sediments. *Water Resources Research*, **23**, 1417–1480.

Zingg, T. H. 1935. Beitrag zur Schötter analyse. *Schweizerische Mineralogische und Petrographische Mitteilungen*, **15**, 39–140.

Section 5

TRACERS FOR COASTAL TRANSPORT STUDIES

22 Coastal Shingle Tracing: A Case Study using the (Electronic Tracer System) (ETS)

M. W. E. LEE,[1] M. J. BRAY,[2] M. WORKMAN,[1] M. B. COLLINS[1] and D. POPE[3]

[1]School of Ocean and Earth Science, University of Southampton, Southampton Oceanography Centre, UK.
[2]Department of Geography, University of Portsmouth, UK
[3]Department of Civil Engineering, University of Brighton, UK

INTRODUCTION

In the UK alone, over 900 km of coastline is protected by shingle or mixed (sand and shingle) beaches (Fuller and Randall, 1988). A knowledge of how these features behave is thus essential and to gain this coastal engineers are required to make field measurements of littoral shingle transport. Tracers represent one of the three main techniques available to undertake such measurements; the others being the monitoring of topographic change adjacent to cross-shore barriers (e.g. Russell, 1960) and trapping (e.g. Chadwick, 1989).

Tracers can be utilised in a number of different ways to measure longshore transport rate (Madsen, 1989). Within the present study, the 'spatial integration method' was identified as the most appropriate method to adopt, due to it having been used successfully on both sand (e.g. Kraus et al., 1982; Ciavola et al., 1997) and shingle (e.g. Wright, 1982; Nicholls, 1985) beaches in the past. Its concept is to monitor, in space and time, the behaviour of a cloud of tracers. In practice, this involves injecting tracers along a cross-shore profile and periodically sampling the bed to identify the position of the tracer centroid (the centre of mass or centre of volume). This is then used to give the velocity of longshore movement of the tracer, from which a longshore drift rate is calculated (equation (22.1)).

$$Q = \bar{U} \cdot m \cdot n \qquad (22.1)$$

where Q is the longshore drift rate (m^3 s^{-1}), $\bar{U}$ is the average longshore transport velocity (m s^{-1}), m is the average width of the mobile beach (m), and n is the average thickness of the moving sediment layer (m).

Modern tracer studies not only aim to measure the transport rate at a particular site but also to relate the measured rate to the environmental conditions prevailing. The

Tracers in Geomorphology. Edited by Ian D. L. Foster. © 2000 John Wiley & Sons Ltd.

reason for doing this is to try to provide a means of predicting transport rate at other sites under similar conditions. The most simple predictive relationship available assumes that transport rate is linearly related to the longshore component of wave power (Komar, 1998) (equation (22.2)).

$$Il = K \cdot Pl \tag{22.2}$$

where Il is the immersed weight longshore transport rate (N s^{-1}), Pl is the longshore component of wave power (Watts m^{-1}), and K is a dimensionless coefficient of proportionality. It can be seen from equation (22.2) that transport rate is not expressed as a 'drift rate' (Q) in the model but, instead, as an 'immersed weight longshore transport rate' (Il). The use of Il instead of Q in the relationship has two main advantages: it enables values for K to be dimensionless; and values of Il take into account the density of the sediment undergoing transport. Equation (22.3) shows the relationship between Q and Il.

$$Il = (\rho_s - \rho) \cdot g \cdot a' \cdot Q \tag{22.3}$$

where Il is the immersed weight longshore transport rate (N s^{-1}), ρ_s is the density of the transported sediment (kg m^{-3}), ρ is the density of water (kg m^{-3}) and a' is a porosity factor (equal to $1-n$, where n is the porosity of the sediment).

This chapter presents details of the field and laboratory techniques recently used to make measurements of Il on a shingle/mixed beach in the UK. The measurements made are then used to suggest values for the coefficient of proportionality (K) in equation (22.2). The tracer results are also used to undertake a preliminary study of some of the processes occurring during sediment transport and interpretations are developed for the findings made. It should be noted that in addition to the new data presented within the present study, some data from Bray et al. (1996) have been re-analysed following their investigation of tracing techniques.

To help place the current study into context, a critical review of previous tracer studies, in terms of their contributions to knowledge of shingle beach geomorphology, is presented prior to details of the present experiment being described.

PREVIOUS STUDIES

Three distinct types of study can be identified within the literature, i.e. those relating to (i) identification of pathways and directions of transport; (ii) detection of pebble sorting and grading mechanisms, and (iii) volumetric measurements of beach transport.

Transport Pathways

Qualitative knowledge of the existence, character and frequency of littoral transport may be developed from the results of tracer studies (Table 22.1). For example, several studies have suggested that there is only very limited exchange of shingle between beaches and the offshore bed (Steers and Smith, 1956; Kidson et al., 1958; Kidson and Carr, 1959) and that transport is normally limited to shallow water depths in the

Table 22.1 Summary of tracer studies concerned with the identification of transport pathways

Author	Location	Landform	Tracer method	Timescale	Contribution
Kidson et al. (1958) Kidson and Carr (1959)	Orfordness, Suffolk, UK	Shingle spit, seabed and inlet	Radioactive tagging	Several studies of up to 12 months	Transport crosses inlet; limited transport offshore; drift reversals
Steers and Smith (1956)	Scolt Head Island, Norfolk, UK	Seabed	Radioactive tagging	5 weeks	Slow onshore feed to beach
Crickmore et al. (1972)	Shoreham, W. Sussex, UK	Seabed	Radioactive tagging	19 months	Slow onshore movement within 12 m contour; negligible movement seaward of 18 m contour
Nicholls (1985)	Hurst Spit, Hampshire, UK	Shingle spit	Aluminium pebbles	Two studies of 1–2 months	Drift along spit; losses to seaward via tidal channel
Cooper (1996)	Elmer, W. Sussex, UK	Offshore breakwater	Aluminium and fluorescent	Several studies over 4 weeks	Transport through scheme

absence of strong tidal currents (Crickmore et al., 1972). This has been confirmed by more recent studies where very high tracer recoveries have indicated that shingle (unlike sand) is retained on the beach even during storm conditions (Bray, 1996; Bray et al., 1996). Such behaviour helps to explain the fringing nature of shingle beaches, their propensity for onshore movement and their general resilience to storms and rising sea levels (Carter and Orford, 1984, 1993; Orford et al., 1995). Shingle, therefore, comprises an efficient material for coastal defence and because transport is primarily by drifting along beaches its movements can be controlled or disrupted by groynes and offshore breakwaters. The manner in which structures interfere with transport has also been studied using tracers. For example, Jolliffe (1979) detected transport in a single preferred direction across harbour breakwaters, while Cooper (1996) demonstrated that transport was possible around salients and between bays within an offshore breakwater scheme. In the vicinity of inlets, transport of shingle from beaches to the nearshore by ebb tidal currents has, however, been identified (Nicholls, 1985). Transport has also been demonstrated to occur across an ebb tidal delta, allowing shingle drift to bypass an inlet (the River Orwell, Suffolk) and return to the beach further down-drift (Kidson et al., 1958; Kidson and Carr, 1959). Such results have obvious implications for confirming output paths in sediment budget studies (Nicholls and Webber, 1987) and for defining the nature of transport cell and sub-cell boundaries (Bray et al., 1995; Leafe et al., 1998).

Sorting and Grading

Many shingle beaches exhibit size and/or shape sorting of their sediments, and these trends are often attributed to progressive abrasion (in the case of down-drift fining), or the selective transport of differently sized and shaped clasts. Tracers have provided a direct means of testing the latter explanation, and several studies have reported the existence of relationships (Table 22.2). Several experiments on different beaches in west Dorset have suggested that preferential transport of larger pebbles, towards

Table 22.2 Summary of key tracer studies concerned with differential transport and sorting

Methods	Authors	Contributions	Comments
Tracing techniques reliant on visual searches (e.g. use of painted pebbles)	Jolliffe (1964)	Found larger pebbles moved faster than small ones	Tracers were frequently unrepresentative of indigenous material
	Carr (1971, 1974)	C-axis length was best correlated with transport distance	
	Gleason *et al.* (1975)	Max transport distance was always associated with small material	
	Caldwell (1983)	Best correlations found to be with c-axis length	
Tracing techniques capable of remote detection (aluminium tracers)	Wright (1982)	No significant correlations between size or shape parameters and transport distance	Tracers more representative than earlier studies. Individual particles tracked but usually over short periods
	Nicholls (1985)	C-axis was most significant dimension but sometimes axis ratios/indices were equally or better correlated	
	Bray (1996)	Position on beach exerted stronger influence than size or shape	

zones of greater wave exposure, may explain highly developed longshore grading patterns (Carr, 1971; Bray, 1996, 1997). However, other results are inconsistent; relationships are relatively weak and different parameters explain transport on different occasions as wave conditions vary. Improved methods, providing high tracer recoveries and tracking of individual pebbles, have yet to clarify these uncertainties. The sorting processes appear complex and longshore beach grading may develop incrementally, as a cumulative result of differential transport operating over extended periods; this is not easily studied using tracers.

Shingle Drift Volumes

Relatively few estimates of drift have been derived using tracers due to the stringent requirements involving recovery of high proportions of injected tracers. Experiments utilising aluminium (Wright, 1982; Nicholls, 1985; Bray, 1996) and electronic tracers (Bray *et al.*, 1996) have enabled calculations of drift for periods of a single tide to several weeks to be made. These studies have revealed that transport is related to wave energy flux, with frequent reversals of direction and 'bursts' of rapid drift involving the disturbance of significant thicknesses of sediment during storms. Shingle beaches, therefore, appear to be more dynamic than would be supposed from the study of their morphology alone. It is difficult to validate tracer-derived drift volumes as the time periods covered are much shorter than those typically used to infer drift from morphological changes. However, work is currently in progress to apply intensive

GPS measurements of morphological change to derive independent drift estimates with which to validate the results of simultaneous tracer studies (Van Wellen *et al.*, 1999).

On the basis of relationships established between measured transport and coincident longshore wave energy flux, annual net drift can be estimated from wave climate data. Such estimates are key elements within sediment budget studies where an approximate validation may be afforded by their capacity to balance the other components of the budget. In west Dorset, tracer-derived drift estimates were applied to understand the transport of gravels released by cliff erosion towards zones of storage in pocket beaches (Bray, 1996, 1997). The predictive relationships so far developed using tracer data are relatively inconsistent. Although this may result from methodological inconsistencies, even studies adopting improved techniques and standardised procedures reveal a range of transport efficiencies. Site-specific variations may result from differences in grain size and sorting, beach slope, tidal range and wave exposure, such that predictive models have to be calibrated individually to the sites at which they are to be applied.

Synopsis

Tracers, deployed in a consistent and rigorous manner, form an important technique with which to investigate the behaviour of shingle beaches. They are especially valuable in providing insights into the processes of morphological change and in providing quantitative estimates of transport over short periods for use in the development and validation of predictive numerical and physical models. There is the potential to apply tracing techniques alongside other methods of shingle beach investigation which focus upon morphological inference at a meso-scale spatial and temporal level (Carter and Orford, 1993; Orford *et al.*, 1995).

It, nevertheless, remains uncertain how widely the results of the studies discussed may be applied beyond their study sites. Although many of the general concepts introduced do appear to hold true elsewhere, it should be appreciated that most of the experiments were of limited duration (Table 22.1) and could not include the full range of conditions likely to occur. For these reasons it is advisable that additional information, e.g. sedimentological and morphological measurements (Taggart and Schwartz, 1988; Van Wellen *et al.*, 1999), should also be analysed to provide validation and evidence of behaviour at scales beyond those which can be tackled directly using tracers. Future work should be directed towards (i) applying the techniques outlined in this chapter more widely to elucidate the factors associated with inter-site variation, and (ii) continuing to improve the resolution and precision of the techniques themselves.

STUDY SITE

The chosen study site was the shingle/mixed beach at Shoreham-by-Sea, West Sussex (Figure 22.1). The linear, uninterrupted and un-engineered nature of a considerable length of the beach (~1.5 km, Figure 22.1) was a major factor in its selection as the

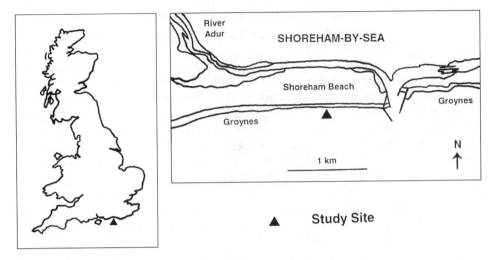

Figure 22.1   Location of the study site

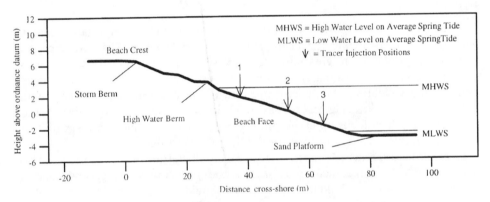

Figure 22.2   Cross-shore profile showing the major morphological features at the study site

study site. Also important was the existence of a sand platform, at the toe of the beach, which becomes exposed on extreme spring tides (Figure 22.2) as, in the experience of the authors, offshore transport and subsequent loss of tracers seldom occurs where such features exist. Seaward of the sand platform, the bed slopes gently giving way to an uncomplicated bathymetry (Hydrographer of the Navy, 1974). The beach itself has a steeply sloping face, with a series of modest high water berms and a major storm berm at its crest (Figure 22.2); features that are typical of shingle beaches both in the UK and elsewhere (Carter and Orford, 1993). It can be seen from Figure 22.2 that the site is dominated by a large tidal range (5.7 m on Mean Springs). These large tides often combine with storm waves which approach the site from the SW and S (48 percent of offshore waves approach from between S and W: Gifford Associated Consultants, 1997). The predicted maximum annual wave height offshore is 4.0 m and the 100 year extreme is estimated to exceed 5.9 m (Hague, 1992). The tidal currents in the area are weak and vary between $0.1\,m\,s^{-1}$ and $0.7\,m\,s^{-1}$ (Gifford Associated Consultants, 1997). The net annual drift along the beach is estimated to be between $15\,000\,m^3$ and $20\,000\,m^3$, on average, with the material moving from W to E until it is intercepted by the harbour breakwater (Figure 22.1) (Scott Fitzpatrick Wilson, 1996).

METHODS

Tracer Technique

The criteria that a tracer technique must satisfy to be effective are well known (Madsen, 1989). These criteria form a good basis on which to evaluate tracing systems. In particular, they highlight the importance of recovering a high proportion of the injected tracers and not systematically excluding any one part of the tracer population upon recovery (e.g. deeply buried tracers). The electronic tracer system (ETS) was developed to overcome the inability of all pre-existing shingle tracer systems to detect deeply buried tracers following storms (Nicholls, 1985). The new system provides more effective detection by transmitting electromagnetic signals (of an optimum frequency) through the sediment. The transmitter is actually in the form of a small (23 mm diameter), battery-powered circuit which is encapsulated within a loaded resin to produce a tracer with the size and shape characteristics of an actual (flint within the present study) pebble.

The circuit transmits a coded signal, allowing each pebble to be detected and identified to a depth of 1 m using a specialised receiver at the beach surface. Details of the development of the prototype system are given by Workman et al. (1994) and Voulgaris et al. (1999). To offer a baseline against which to compare the electronic system, the most effective pre-existing tracer system was also used within this study, i.e. aluminium tracers (Wright et al., 1978). Details of this comparison are given by Bray et al. (1996) and aluminium tracer data are used where appropriate to supplement the electronic pebble data in the present contribution.

During the experiment six different tracer 'forms' (a form being a particular size and shape of pebble) were deployed. These forms were the same for both the electronic and aluminium pebbles and each was an exact replica of a natural beach pebble. The characteristics of each of the forms are presented in Table 22.3.

To assess how representative of the natural material these tracers were in terms of grain size it was necessary to sample the indigenous material. This involved taking a large number (55) of (15–20 kg) sediment samples from sites across the beach profile during the deployment. Although the individual samples were smaller than was necessary to define the grain-size distribution precisely (Gale and Hoare, 1992), the sampling strategy was thought to be the best possible considering the likely existence of temporal variability and the limited resources available. Sieving of the samples showed the mean grain size for the beach to be 16.34 mm (SD 9.37 mm), with the tracers always representing between 9.0 percent and 28.1 percent of the indigenous material's grain-size distribution during the study.

The representativeness of the tracer, in terms of grain size, is often used as a measure of the quality of a tracer study. In a similar way, the number of tracers

Table 22.3 Characteristics of the tracer 'forms' used within the study

Form identity code	Axis lengths (mm)			Roundness	Sphericity	OP index	Flatness	Shape (Zingg)
	a	b	c					
LR	57	49	44	0.31	0.88	1.49	1.21	Sphere
MR	61	34	30	0.44	0.76	7.54	1.58	Rod
SR	44	34	31	0.86	0.86	3.82	1.26	Sphere
LA	70	55	35	0.37	0.68	−1.43	1.79	Disc
MA	67	38	25	0.2	0.64	5.1	2.1	Blade
SA	58	45	24	0.17	0.6	−2.84	2.15	Disc

Table 22.4 Number of tracers injected and recovery rate achieved for each search

Search identity code[a]	Duration (no. of tides)	No. of tracers injected	Tracer recovery rate (%)
1a	3	60	80 (cumulative)
2a	1	139	86
2b	2	49	71
3a	1	84	99
3b	2	84	76
3c	2	84	84
3d	6	84	70
4a	2	108	78
5a	1	54	91
6a	1	54	98

[a] The numeric part of the code distinguishes between different tracer injections and the alphabetical part between different searches.

injected and the proportion of these recovered are also used to make this sort of assessment (Bray *et al.*, 1996). These data are shown in Table 22.4.

It should be noted that all of the tracer searches were carried out over a single low tide, except for Search 1a which followed a very high energy event and required four low tides to complete. The data presented in Table 22.4 compare very well with those of other studies. For example, Bray (1996) injected between 58 and 134 tracers and achieved daily recovery rates of 13–25 percent in high energy conditions, with these values rising to a maximum of 99 percent in low energy conditions.

Field Procedure

Throughout the study, the injection of the tracers was carried out in a way that aimed to satisfy the following criteria:

- to allow the effect of cross-shore injection position on transport distance and burial depth to be determined;
- to allow direct comparison of the transport distances and burial depths of the different tracer forms;

- to avoid having redundant tracers, i.e. avoid introducing tracers at a depth where no movement occurs (this would exclude tracers from the analyses leading to a reduction in sample size).

The tracers were injected at the sediment surface, at three different cross-shore positions (Figure 22.2). To allow comparability of these injection sites, equal numbers of each tracer form were injected at each site. The total number of tracers injected on each occasion depended on the wave energy level: the greater the energy, the greater the likely movement of the tracers and hence the larger the number used. However, poor recovery rates were achieved with the aluminium tracers under high energy conditions hence data collected using this system have been excluded from some calculations.

Tracer searches were conducted by systematically scanning the beach surface at low tide. Both the number of searches associated with each injection and the length of time (i.e. number of tides) between searches varied (Table 22.4). There were two reasons for varying the length of time between searches. The first reason was to allow a particular wave energy to be ascribed to each transport measurement; this could only be done if wave energy was approximately constant between searches hence the length of the studies was adjusted to achieve this. The second reason was that for the tracers to be as representative of the indigenous material as possible they needed to be well mixed with the natural material. The amount of transport and mixing of the tracer depended on the wave energy, hence the length of time needed for mixing varied with wave energy and the study lengths were adjusted to take this into consideration also.

Having searched the beach surface and identified the locations at which signals were being received it was necessary to determine how deeply the tracers were buried, which was achieved by digging each one up. Following excavation, the x and y coordinates of each tracer were measured with a total station, the z coordinate was measured using a tape, and the identity was taken from the engraved identification number on each particle.

To make best use of the tracer data (i.e. for purposes such as the calibration of equation (22.2)) it was necessary to make simultaneous measurements of wave characteristics (height, period, wavelength, angle of approach and water depth, all at the breakpoint). A star array of resistance wave poles, known as the Inshore Wave Climate Monitor (IWCM), was installed for this purpose (Chadwick et al., 1995). However, the very high energy conditions experienced during the first experiment caused the system considerable damage, resulting in the loss of directional measurements and intermittent wave height measurements from that time onwards. This meant that wave approach angle had to be measured with a prismatic compass for the remainder of the study (one measurement was always made at high water and others were taken when time allowed or when direction changes were apparent). Visual estimates of wave height were made against graduated poles whenever wave angle was being measured, and these were used within the analysis when IWCM measurements were not available.

In much the same way that it was necessary to make wave measurements for use alongside the tracer data in equation (22.2), it was necessary to know the width of the beach over which transport was occurring for use in equation (22.1). To achieve this, the distance between the shingle toe and the high water berm was measured, at low tide, using a total station.

As already mentioned, sediment samples were collected in order to assess how representative the tracer was of the indigenous material. Each sample was taken down to a depth equal to the thickness of the moving sediment layer (as indicated by the tracers) and was dried prior to sieving at half phi intervals.

Data Analysis

The raw data provided by the field studies took the following form:

- tracer injection positions (x, y and z);
- recovery positions for each tracer (x, y and z) on each search;
- beach profile measurements;
- IWCM (and visual) wave height, angle and period measurements;
- grain-size analysis data.

The processing of these data into a form suitable for use in equations (22.1)–(22.3) is described in the following sections. No additional processing was required to then use the data to investigate the processes of sediment transport.

Calculation of Longshore Transport Rate

Equations (22.1) and (22.3) are the relationships used within the present study for the calculation of longshore transport rate. Both equations are simple in form, although variability exists in the way that certain parameters within the relationships may be derived. Consistent methods are needed if estimates of K are to be meaningful and be compared with those of other studies. Thus, it is necessary to set out the manner in which the input variables for each of the two equations were calculated.

Each tracer form was given equal computational weight within the calculations of longshore transport rate, as was each of the three tracer injection beds. Thus, it was necessary to calculate a transport rate for every combination of tracer form and injection position (18 in total). This meant determining the mean transport velocity and the mean transport layer thickness for each combination. Fractional transport rate calculations of this type have been recommended by Madsen (1989).

(i) Average Longshore Transport Velocity (U) (equation (22.1)) This parameter was calculated by dividing the mean longshore travel distance of the tracers (having a particular combination form and injection position) in transport by their average duration of coverage by the tide. Here, tracers in transport are defined as those that have moved 0.3 m, or more, in either the x or y orientations.

The mean longshore travel distance was calculated as the difference between the average pre-tide and average post-tide positions of the moving tracers. This mean distance was then converted to a mean velocity ($m\,s^{-1}$) by dividing it by the mean duration of coverage, the latter being the product of the following:

- the number of tides the tracer was in transport for;
- the number of hours for which the tracer was covered by water during each tide (6); and
- the number of seconds per hour (3600).

(ii) The average width of the mobile layer (m) (equation (22.1))   The following assumptions were made when calculating this parameter:

- whenever the tracer was submerged it was mobile;
- the tracer was present across the full width of the active beach;
- during each tidal cycle the tide rose linearly to its maximum level and then fell linearly back to its low water level;
- at low tide the water was at the toe of the shingle beach and at high tide it created a high water berm.

Having made these assumptions the average active beach width was calculated as half of the distance from the toe of the shingle up to the high water berm formed at the swash limit.

(iii) The average thickness of the moving sediment layer (n) (equation (22.1))   The depth to which the tracer material (having a particular combination of form and injection position) was mixed within the bed was used to evaluate this parameter. Bray (1996) used the boundary between non-moving tracers and moving tracers to identify the base of the moving sediment layer and hence calculate its thickness. However, within the present investigation it was often the case that few tracers became immobile. This was due to a number of factors, including the experiments being shorter in duration than Bray's (1996) and tracers being injected at the sediment surface to minimise redundancy. In such circumstances, Bray (1996) defined the base of the moving layer as the level above which 95 percent of the moving tracer existed. This definition was used here.

It should be noted that this parameter is often regarded as the most uncertain in the calculation of transport rates using tracers (Komar, 1998). This is because the thickness of the moving sediment layer may vary during the course of a tracer deployment and yet none of the methods commonly used to quantify the thickness are capable of recording such variability. Equipment has recently been deployed on UK shingle beaches (Van Wellen *et al.*, 1997) with the aim of identifying changes in mixing layer thickness. The results of these studies are currently undergoing analysis (Lee *et al.*, In submission.

(iv) Equation (22.3) parameters   Other than for Q, values for all of the parameters within equation (22.3) are either well established or they appear in previous studies. For completeness, the values used are presented below.

- *Density of the transported sediment*(ρ_s). Since the sediment on the beach is almost entirely flint, chert and sand, the density of the material has been taken as that of quartz, i.e. $2650 \, \text{kg m}^{-3}$.

- *Density of seawater* (ρ). This was taken as $1030\,\mathrm{kg\,m^{-3}}$ (after Chadwick, 1989).
- *Acceleration due to gravity* (g). This was taken as $9.81\ \mathrm{m\ s^{-2}}$.
- *Porosity factor* (a'). This was taken as 0.68 (after Chadwick, 1989).

Calculation of Pl (equation (22.2))

Variability has occurred, in the past, in the way that the longshore component of wave power (*Pl*) has been calculated. To help overcome this problem, Bodge and Kraus (1991) presented some conventions for the calculation (e.g. in relation to the choice of equation for the calculation of wave group celerity). These conventions were adopted during the present study.

RESULTS AND DISCUSSION

The longshore transport rates (*Il* and *Q*) derived from the tracer data and the corresponding values for the longshore component of wave power (*Pl*) and the coefficient of proportionality (*K*) (equation (22.2)) are presented in Table 22.5. The *Pl* values show that the data cover a wide range of energy conditions; this is essential if a reliable assessment of sediment transport at any particular site is to be made (Schoones and Theron, 1993). However, the values of *K* presented in the table (between 0.06 and 0.54) are more notable than those of *Pl*, the values being higher than has been reported by previous investigators. Bray *et al.* (1996) (using some of the same data as have been analysed here) found values between 0.02 and 0.36, but prior to their study, the highest value of *K* appearing in the literature was 0.044 (Bray, 1996). The explanation suggested by Bray *et al.* (1996) for the high *K* values was that *K* increases with increasing wave energy (their measurements being amongst the first for high energy conditions). However, both the analysis of the data by Bray *et al.* (1996) and that presented here show high values of *K* occurring at comparatively low *Pl*

Table 22.5 Tracer-derived longshore transport rates and corresponding coefficients of proportionality (see equation (22.2))

Search identity code[a]	Q (m^3 h^{-1}) (−ve indicates transport to E)	Il (N s^{-1})	Pl (W m^{-1})	K
1a	198	594.4	2203.7	0.27
2a	12.9	38.6	71	0.54
2b	−4.1	12.2	221.5	0.06
3a	−0.8	2.4	26.4	0.09
3b	1.5	4.4	58.9	0.07
3c	1.6	4.7	71.9	0.07
3d	0.8	2.4	14.1	0.17
4a	−0.3	0.9	5	0.18
5a	−81.7	245.4	1151.7	0.21
6a	−4.3	12.9	136.2	0.09

[a] The numeric part of the code distinguishes between different tracer injections and the alphabetical part between different searches.

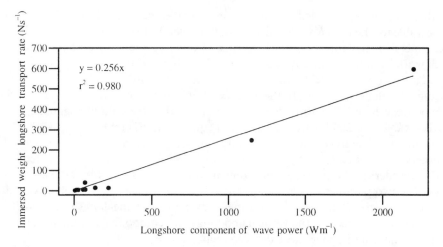

Figure 22.3 Tracer-derived immersed weight longshore transport rate plotted against long-shore component of wave power

values (recovery $I_2 + 1$ of Bray *et al.* (1996) and recoveries 2a, 3d and 4a presented here). A more clear picture of the trends is achieved by plotting *Il* against *Pl*, as in Figure 22.3.

Although the data in Figure 22.3 are scattered, they shed doubt on the suggestion that *Pl* is responsible for the elevated levels of *K*; instead they indicate a linear relationship between *Pl* and *Il*. It is also clear from the plot that the data set is relatively small and that the form of the relationship depends heavily upon just two data points. For this reason any findings based on the data presented here must be treated with caution. Further work is needed to increase the number of measurements in high energy conditions.

A possible alternative explanation for the high *K* values is that the tracer was not adequately incorporated into the natural material during the studies in question, i.e. it remained near the sediment surface where it would be transported rapidly. Very little work has been done on the subject of tracer mixing/incorporation and no guidelines exist for the length of time that tracers should remain in the bed before transport measurements are made. Of the studies that have been conducted, those of Carr (1971) and Caldwell (1981) are perhaps the most informative. Both studies reached a similar conclusion, i.e. that under certain wave conditions, tracers which were uncharacteristic of the natural material were 'rejected' from it and as a result were found in high concentrations at the sediment surface. As a consequence of this lack of information, some preliminary tests were carried out to assess the degree of tracer mixing during the study. The two types of test conducted were as follows:

- checking the depths to which tracers were mixed upon each recovery (in order to assess whether mixing depths were uncharacteristically shallow when high *K* values were recorded); and,
- searching for trends between the duration of a study and the depth of mixing of the tracer (to determine whether longer studies had greater mixing depths).

Neither type of test provided any indication that poor incorporation of the tracer was responsible for the high K values identified. However, both tests are very basic and it is recommended that further studies be carried out into the nature of tracer mixing. Such work is, in fact, under way (Lee *et al.*, In Submission).

Acquisition of more field data may well be needed in order to identify the cause of the high K values presented. In the absence of firm evidence against a linear relationship between Il and Pl a straight line has been fitted to the data in Figure 22.3. This yields a value of 0.256 for the value of K in equation (22.2) ($r^2 = 0.98$).

Having used the tracer data to calculate longshore transport rates in a similar way to others (e.g. Nicholls, 1985; Van Wellen *et al.*, 1999), the analyses were extended to begin considering the processes of sediment transport responsible for the observed rates. The two parameters within equation (22.1) for which the tracers provide data are the thickness of the moving sediment layer (n) and the longshore velocity of movement ($\bar{U}$), hence it is the vertical and longshore components of the tracer movement that have undergone more detailed analysis here.

Bray (1996) investigated the vertical mixing of shingle tracers during studies along the Dorset coast (UK) and similar studies have been carried out on sand beaches (Sherman *et al.*, 1994). The duration of studies on sand beaches is usually less than that on shingle beaches (the latter having a maximum temporal resolution of one tide) but in both environments a linear relationship between mixing depth and breaking wave height has been identified (Sherman *et al.*, 1994; Bray, 1996). To investigate whether the data collected here also show this type of relationship, the maximum depth to which the tracer was mixed during each period of transport (defined using the 95 percent method of Bray, 1996) has been plotted against significant wave height (Hs) (Figure 22.4). Despite the scatter of the data, a linear relationship, of the form $y = 0.184x + 0.0254$, fits the plotted points reasonably well ($r^2 = 0.78$). This relationship is remarkably similar to that given by Bray (1996) ($y = 0.182x + 0.016$) for the mixed beach at Charmouth (Dorset, UK). The similarity between the data presented

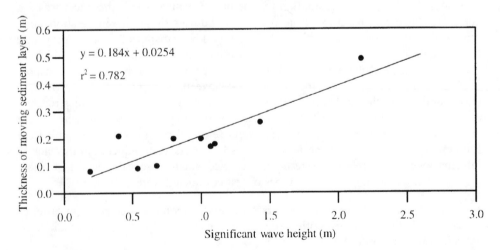

Figure 22.4 Thickness of the moving sediment layer plotted against significant wave height at breaking

here and those of Bray (1996) is encouraging since it suggests that the mixing depth relationships being presented in this type of study may not be too site-specific and hence may be of use for the prediction of mixing depths at other sites. The results of field campaigns in 1996 and 1997 are presently being analysed to evaluate the consistency of the predictive relationship.

Having investigated the relationship between the tracer mixing depth and the wave height at breaking, another aspect of the vertical movement of the tracers considered to be worthy of study was how a tracer's form (size and shape) and injection position affect its mixing depth. This type of investigation is important because if it could be shown that form and injection position affect a tracer's behaviour, then to be able to represent indigenous material properly during a tracer study it would be essential to use tracers that encompassed the full range of forms present on the beach and to inject these at a wide range of positions on the beach profile. The importance of the longshore travel distance of tracers (as well as mixing depth) in the calculation of transport rates (equation (22.1)) means that the effect of form and injection position on this parameter is equally important to study as their effect on mixing depth.

The preliminary stage of these investigations was to inspect the data visually. To best achieve this plots similar to those in Figure 22.5 were constructed. The values for each of the 18 bars in each plot are mean values, calculated from all of the tracers recovered in the particular category that the bar represents. One of each type of plot was constructed for each of the 10 tracer recoveries referred to in Table 22.4. Qualitative (i.e. visual) assessment of the data suggested that there were no consistent trends in the way the six tracer forms behaved relative to each other or the way the three injection beds behaved relative to each other. In order to carry out a more quantitative assessment, statistical procedures were used; the test applied was a two-way analysis of variance (two-way ANOVA). This test is designed for situations where two variables (e.g. form and injection position), each having a number of categories (e.g. the six types of tracer form shown in Figure 22.5), may exert an influence on a third variable (e.g. longshore travel distance or mixing depth). Using the investigation of the effect of form and injection position on longshore travel distance as an example, three stages of the test can be identified. The first stage considers whether the forms show the same behaviour (in terms of longshore transport distance) in each of the three injection beds; if they do then the second and third parts of the test become relevant. These latter parts assess whether all six forms behave in the same way and whether all three injection beds behave in the same way. The assumptions of the test are that the values for each of the combinations of injection bed and form are a random sample from a normal population and that in the populations the variances of all groups are equal. However, Norusis (1991) states that the test gives good results even if the normality and variance assumptions do not quite hold. It is also important to have a number of measurements within each of the 18 categories since the test involves calculating 'within-group variances'.

Having taken these requirements into consideration, it was clear that the test could only be applied to those data where both large numbers of tracers had been injected and high recovery rates had been achieved. It was found that the data collected during Searches 2a, 3a and 4a met this requirement. Prior to applying the test, it was also necessary to investigate whether the equal variance assumption was valid (despite its apparent flexibility). To achieve this, the Levene test (Norusis, 1991) was applied to the longshore movement and recovery depth data from Searches 2a, 3a and 4a in turn. The results given by the test are presented in Table 22.6.

It can be seen from the table that the data only appear to meet the equal variance assumption in two out of six cases at the 95 percent level and three out of six cases at

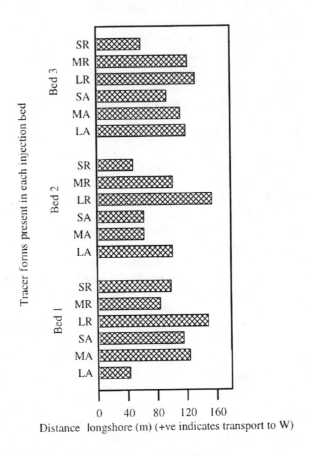

Distance longshore (m) (+ve indicates transport to W)

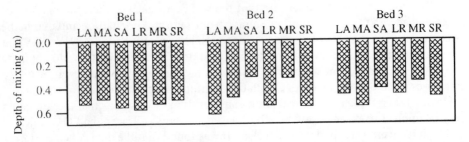

Figure 22.5 Examples of plots prepared for the visual inspection of transport distance and mixing depth data

Table 22.6   Results of the Levene test for equality of variance, when applied to the data from Searches 2a, 3a and 4a

Null hypothesis	Significance (two-tailed)	Reject at 95% level	Reject at 99% level
No differences between category variances for Search 2a longshore distance data	0	Yes	Yes
No differences between category variances for Search 3a longshore distance data	0	Yes	Yes
No differences between category variances for Search 4a longshore distance data	0.035	No	No
No differences between category variances for Search 2a recovery depth data	0.006	Yes	No
No differences between category variances for Search 3a recovery depth data	0.2	No	No
No differences between category variances for Search 4a recovery depth data	0	Yes	Yes

Table 22.7   Results of the Levene test for equality of variance, when applied to the logarithmically transformed data from Searches 2a, 3a and 4a

Null hypothesis	Significance (two-tailed)	Reject at 95% level	Reject at 99% level
No differences between category variances for Search 2a log longshore distance data	0.015	Yes	No
No differences between category variances for Search 3a log longshore distance data	0.012	Yes	No
No differences between category variances for Search 4a log longshore distance data	0.183	No	No
No differences between category variances for Search 2a log recovery depth data	0.368	No	No
No differences between category variances for Search 3a log recovery depth data	0.252	No	No
No differences between category variances for Search 4a log recovery depth data	0	Yes	Yes

the 99 percent level. In the light of these findings it was felt inappropriate to apply the two-way ANOVA to the data directly; instead the data were transformed logarithmically with the aim of trying to improve the extent to which they met the variance assumption. Table 22.7 shows the results obtained following the application of the Levene test to the transformed data.

It can be seen from the table that the transformed data fulfil the ANOVA's requirements, with respect to equal variance, in five out of six cases at the 99 percent level and three out of six cases at the 95 percent level. The two-way ANOVA was thus applied to the transformed data (Table 22.8).

Despite having applied the ANOVA to the transformed data, the findings presented in Table 22.8 are valid in terms of the original variables (i.e. the findings relate to longshore distance and recovery depth as well as to log distance longshore and log

Table 22.8 Two-way ANOVA test (Norusis, 1991) results

Null hypothesis	Search 4a (very low wave energy, $H_S = 0.18$ m)[a]		Search 3a (low wave energy, $H_S = 0.54$ m)[a]		Search 2a (moderate wave energy, $H_S = 0.80$ m)[a]	
	Significance	Reject (95% level)	Significance	Reject (95% level)	Significance	Reject (95% level)
Distance of tracer movement longshore						
No difference in log transport distance for the three injection sites	0.007	Yes	0.185	No	0	Yes
No difference in log transport distance for the six tracer forms	0.254	No	0.167	No	0.092	No
No difference in the behaviour of the tracer forms between the injection sites in terms of log transport distance	0.717	No	0.538	No	0.992	No
Depth of tracer recovery						
No difference in log burial depth for the three injection sites	0.021	Yes	0.302	No	0.684	No
No difference in log burial depth for the six tracer forms	0.421	No	0.321	No	0.08	No
No difference in the behaviour of the tracer forms between the injection sites in terms of log burial depth	0.205	No	0.691	No	0.248	No

[a] H_S here represents significant wave height at breaking.

recovery depth) (Challenor, P., pers. comm.). It can be seen from the table that on no occasion could a link be found between distance travelled longshore and tracer form (at the 95 percent level) whilst on two out of three occasions a link could be found between injection position and travel distance. This is in good agreement with previous studies which have often found links between particular aspects of tracer form and transport distance to be intermittent (Carr, 1971; Gleason *et al.*, 1975; Caldwell, 1983; Bray, 1996) and in some cases have found no links at all (e.g. Wright, 1982). Where links have been found between transport distance and form, these have been best developed in high wave energy conditions in a number of studies (Jolliffe, 1964; Bray, 1996; Cooper *et al.*, 1996). The absence of a high wave energy data set to which the ANOVA could be applied, within the work presented here, may thus explain the failure to find any link between tracer form and longshore travel distance. The finding that tracer injection bed did have an influence on transport distance, whereas form did not, is in good agreement with the work of Bray (1996) who found that cross-shore position was usually a stronger influence on transport distance than tracer form.

Few studies have investigated the existence, or otherwise, of vertical sorting of tracers. Both Wright (1982) and Nicholls (1985) had limited ability to investigate this phenomenon due to the short detection range ($\sim$25 cm) of the tracer technique they used. Neither worker identified any vertical sorting due to tracer form or injection position. In contrast, Bray (1996) was able to identify such relationships and he concluded that a tracer's cross-shore position was a more important influence upon its burial depth than tracer form. Again, the findings presented here agree well with those of Bray (1996).

Although a great deal of care was taken during the collection and analysis of the data presented here, the study is not without its limitations. In particular, the following points should be considered:

(1) The tracer population only represented a small proportion of the indigenous sediment, the finer material being particularly poorly represented. This fact may have had an influence upon the magnitude of the transport rates measured using the tracers. However, as already described, the tracer appears to have been reasonably well mixed (from preliminary analysis). This, in turn, suggests that unrealistic results were not a consequence of poor tracer incorporation. Also in support of the view that the tracer-derived transport rates are realistic is the fact that no differential movement according to tracer form was found during the study (and little consistent evidence for its existence has been found by other investigators), hence the failure to represent the smaller sediment may not have been important. However, aspects of this argument are circular, since the failure to identify differential movement within this study may have been due to the tracer forms used not being sufficiently distinct, i.e. not covering a wide enough range of shapes and sizes. Such an explanation for the failure of the tracer studies to identify differential transport according to form is, however, in conflict with the lack of any (visually) detectable longshore size or shape grading at the site.

(2) A comparatively small number of transport rate measurements were made (10), hence the data obtained and the findings presented here are only representative of a limited range of hydrodynamic conditions. The data set to which the

two-way ANOVA was applied was still smaller and hence the results of these analyses are even more limited in terms of the conditions they represent.

(3) It can be seen from Table 22.7 that in the case of the Search 4 recovery depth data, the equal variance assumption of the two-way ANOVA does not hold (despite having transformed the data). Thus the results of this particular run of the ANOVA should be treated with caution.

CONCLUSIONS

This chapter reveals how tracers are being used to provide insights into the processes of transport and sorting on shingle beaches. The following points represent the main findings of the Shoreham study:

(1) Analysis of the transport process during the study suggested that on no occasion did the tracer forms manufactured show differential movement in either a longshore or a vertical direction. The analysis did, however, indicate that movement in both orientations was intermittently dependent upon the injection position of the tracer. These findings support the conviction that rigorous experimental design (especially with respect to tracer injection position) is essential if tracer results are to accurately represent the behaviour of the indigenous material.

(2) Values between 0.06 and 0.54 are presented for K in the relationship $Il = K \cdot Pl$. These values are higher than any previously identified in studies of this type.

(3) Since they have only been found at the Shoreham site (to date), the high K values must be regarded as site-specific and as such they support the re-calibration of predictive drift equations for each site to which they are applied.

(4) The high K values do not appear to be linked to either high energy conditions (as suggested by Bray et al., 1996) or, from preliminary analyses, to poor mixing of the injected tracer population. The variability of K during the study may indicate that some hydrodynamic and/or sedimentological factor, which is not included within the relationship $Il = K \cdot Pl$, is an important influence on the longshore drift at the site.

(5) Vertical mixing depths, measured during the study, appear to have been linearly related to significant wave height. The relationship identified ($y = 0.184x + 0.0254$) is very similar to that established for the mixed beach at Charmouth, Dorset, UK (Bray, 1996). Such similarity suggests that prediction of mixing depths may be possible at other sites.

These findings suggest that if reliable tracing techniques are applied within a rigorous experimental framework then they offer a viable alternative to the use of topographic measurements for the quantification of longshore transport rate. Especially advantageous is the capability to measure during high energy storm conditions when beaches evolve most rapidly. High energy measurements of this sort are currently few in number, and the number of sites at which measurements have been made is limited. In order to gain a more representative view of longshore shingle transport in the future, it is important that these two limitations of the data set currently available are addressed. As with other techniques, the considerable advances that have recently been made in the measurement of longshore transport rate using tracers are encouraging. However, these advances do not diminish the value of simultaneously using a number of measurement techniques to provide validation for each other.

ACKNOWLEDGEMENTS

The authors would like to acknowledge the support of MAFF during the development of the Electronic Tracer System (ETS) and thank the following for their invaluable and varied contributions: T.T. Coates, J. Smith, P. Boyce, Dr T. Mason, Dr N. Cooper, Dr S. Wallbridge, B. Duane, Dr D. King, P. Challenor, Dr J. Morfett and Prof. J. Hooke.

REFERENCES

Bodge, K. R. and Kraus, N. C. 1991. Critical examination of longshore transport rate magnitude. In Kraus, N. C., Gingerich, K. J. and Kriebel, D. L. (eds) *Proceedings of Coastal Sediments '91*, Seattle. ASCE, New York, 139–155.

Bray, M. J. 1996. Beach budget analysis and shingle transport dynamics: West Dorset. Unpublished PhD Thesis, Department of Geography, London School of Economics, University of London.

Bray, M. J. 1997. Episodic shingle supply and the modified development of Chesil Beach, England. *Journal of Coastal Research*, **13**, 1035–1049.

Bray, M. J., Carter, D. J. and Hooke, J. M. 1995. Littoral cell definition and budgets for central southern England. *Journal of Coastal Research*, **11**, 381–400.

Bray, M. J. Workman, M., Smith, J. and Pope, D. 1996. Field measurements of shingle transport using electronic tracers. In *Proceedings of the 31st MAFF Conference of River and Coastal Engineers*. Ministry of Agriculture, Fisheries and Food, London, 10.4.1–10.4.13.

Caldwell, N. E. 1981. Relationship between tracers and background beach material. *Journal of Sedimentary Petrology*, **51**, 1163–1168.

Caldwell, N. E. 1983. Using tracers to assess size and shape sorting of pebbles. *Proceedings of the Geologists' Association*, **94**, 86–90.

Carr, A. P. 1971. Experiments on longshore transport and sorting of pebbles: Chesil Beach, England. *Journal of Sedimentary Petrology*, **41**, 1084–1104.

Carr, A. P. 1974. Differential movement of coarse sediment particles. In *Proceedings of the 14th Conference on Coastal Engineering*, Copenhagen. ASCE, New York, 851–870.

Carter, R. W. G. and Orford, J. D. 1984. Coarse clastic barrier beaches: a discussion of the distinctive dynamic and morphosedimentary characteristics. *Marine Geology*, **60**, 377–389.

Carter, R. W. G. and Orford, J. D. 1993. The morphodynamics of coarse clastic beaches and barriers: a short and long term perspective. *Journal of Coastal Research: Special Issue No. 15*, 158–179.

Chadwick, A. J. 1989. Field measurements and numerical model verification of coastal shingle transport. In Palmer, M. H. (ed.) *Advances in Water Modelling and Measurement*, BHRA, 381–401.

Chadwick, A. J., Pope, D., Borges, J. and Illic, S. 1995. Shoreline directional wave spectra part 1. An investigation of spectral and directional analysis techniques. *Proceedings of the Institute of Civil Engineers, Water, Maritime and Energy*, **112**, 198–208.

Ciavola, P., Taborda, R., Ferreira, O. and Dias, J. 1997. Field measurements of sand transport and control processes on a steep meso-tidal beach in Portugal. *Journal of Coastal Research*, **13**, 1119–1129.

Cooper, N. J. 1996. Evaluation of the impacts of shoreline management at contrasting sites in Southern England. Unpublished PhD Thesis, Department of Geography, University of Portsmouth.

Cooper, N., Bray, M. and King, D. 1996. Field measurements of fine shingle transport. Presented at Tidal '96, Symposium for Practising Engineers, University of Brighton, 12–13 November 1996.

Crickmore, M. J., Waters, C. B. and Price, W. A. 1972. The measurement of offshore shingle movement. In *Proceedings of the 13th Conference on Coastal Engineering*, Vancouver. ASCE, New York, 1005–1025.

Fuller, R. M. and Randall, R. E. 1988. The Oxford Shingles, Suffolk, UK, classic conflicts in coastline management. *Biological Conservation*, **46**, 95–114.

Gale, S. J. and Hoare, P. G. 1992. Bulk sampling of coarse clastic sediments for particle size analysis. *Earth Surface Processes and Landforms*, **17**, 729–733.

Gifford Associated Consultants 1997. *South Downs Shoreline Management Plan: Selsey Bill to Beachy Head*. Report to South Downs Coastal Group.

Gleason, R., Blackley, M. W. L. and Carr, A. P. 1975. Beach stability and particle size distribution, Start Bay. *Journal of the Geological Society of London*, **131**, 83–101.

Hague, R. C. 1992. *UK South Coast Shingle Study: Joint Probability Assessment*. HR Wallingford Ltd. Report No. SR 315.

Hydrographer of the Navy 1974. *Navigational Chart: Selsey Bill to Beachy Head*. NC/UK-HD/1652.

Jolliffe, I. P. 1964. An experiment designed to compare the relative rates of movement of different sizes of beach pebbles. *Proceedings of the Geologists' Association*, **75**, 67–86.

Jolliffe, I. P. 1979. *West Bay and the Chesil Bank, Dorset: Coastal Regime Conditions, Resource Use and the Possible Environmental Impact of Mining Activities on Coastal Erosion and Flooding*. Department of Geography, Bedford College, University of London. Report to West Dorset District Council and Dorset County Council.

Kidson, C. and Carr, A. P. 1959. Movement of shingle over the sea bed close inshore. *Geographical Journal*, **125**, 380–389.

Kidson, C., Carr, A. P. and Smith, D. B. 1958. Further experiments using radioactive methods to detect movement of shingle over the sea bed and alongshore. *Geographical Journal*, **124**, 210–218.

Komar, P. D. 1998. *Beach Processes and Sedimentation* (2nd edition). Prentice Hall, New Jersey.

Kraus, N. C., Isobe, M., Igarashi, H., Sasaki, T. O. and Horikawa, K. 1982. Field experiments on longshore transport in the surf zone. In Edge, W. L. (ed.) *Proceedings of the 18th Conference on Coastal Engineering*, Cape Town, South Africa. ASCE, New York, 969–988.

Leafe, R., Pethick, J. and Townend, I. 1998. Realizing the benefits of shoreline management. *Geographical Journal*, **164**, 282–290.

Lee, M. W. E, Stapleton, K. and Bray, M. in submission. Measurement of shingle transport. In Coates, T. T. (ed.) *Collaborative Shingle Beach Transport Project, Final Technical Report*. HR Report, HR Wallingford Ltd.

Madsen, O. S. 1989. Tracer theory. In Seymour, R. (ed.) *Nearshore Sediment Transport*. Plenum Press, New York, 103–114.

Nicholls, R. J. 1985. The stability of shingle beaches in the eastern half of Christchurch Bay. Unpublished PhD thesis, Department of Civil Engineering. University of Southampton.

Nicholls, R. J. and Webber, N. B. 1987. The past, present and future evolution of Hurst Castle Spit, Hampshire. *Progress in Oceanography*, **18**, 119–137.

Norusis, M. J. 1991. *The SPSS Guide to Data Analysis for SPSS/PC+* (2nd edition). SPSS, Chicago.

Orford, J. D., Carter, R. W. G., McKenna, J. and Jennings, S. C. 1995. The relationship between the rate of mesoscale sea-level rise and the rate of retreat of swash-aligned gravel-dominated barriers. *Marine Geology*, **124**, 177–186.

Russell, R. C. H. 1960. The use of fluorescent tracers for the measurement of littoral drift. In Johnson, J. W. (ed.) *Proceedings of the 7th International Conference on Coastal Engineering*, The Hague, Netherlands. Council on Wave Research, Richmond, CA 418–444.

Schoonees, J. S. and Theron, A. K. 1993. Review of the field data base for longshore sediment transport. *Coastal Engineering*, **19**, 1–25.

Scott Fitzpatrick Wilson Ltd. 1996. *Shoreham and Lancing Sea Defence Strategy Plan*. Final Report to National Rivers Authority, Southern Region.

Sherman, D. J., Nordstrom, K., Jackson, N. and Allen, J. R. 1994. Sediment mixing-depths on a low-energy reflective beach. *Journal of Coastal Research*, **10**, 297–305.

Steers, J. A. and Smith, D. B. 1956. Detection of movement of pebbles on the sea floor by radioactive methods. *Geographical Journal*, **122**, 343–345.

Taggart, B. E. and Schwartz, M. L. 1988. Net shore-drift direction determination: a systematic approach. *Journal of Shoreline Management*, **3**, 285–309.

Van Wellen, E., Chadwick, A. J., Bird, P. A. D., Bray, M., Lee, M. W. E. and Morfett, J. 1997. Coastal sediment transport on shingle beaches. In Thornton, E. B. (ed.) *Proceedings of Coastal Dynamics '97*, Plymouth, UK. ASCE, Reston, VA, 38–47.

Van Wellen, E., Chadwick, A. J., Lee, M. W. E., Baily, B. and Morfett, J. 1999. Evaluation of longshore sediment transport models on coarse grained beaches using field data: a preliminary investigation. In Edge, W. L. (ed.) *Proceedings of the 26th International Conference on Coastal Engineering*, Vol. 3, Copenhagen. ASCE, Reston, VA, 2640–2653.

Voulgaris, G., Workman, M. and Collins, M. B. 1999. Measurement techniques of shingle transport in the nearshore zone. *Journal of Coastal Research*, **15**(4), 1030–1039.

Workman, M., Smith, J., Boyce, P., Collins, M. B. and Coates, T. T. 1994. *Development of the Electronic Pebble System*. HR Report SR 405, HR Wallingford Ltd.

Wright, P. 1982. Aspects of the coastal dynamics of Poole and Christchurch Bays. Unpublished PhD Thesis, Department of Civil Engineering, University of Southampton.

Wright, P., Cross, J. S. and Webber, N. B. 1978. Aluminium pebbles: a new type of tracer for flint and chert pebble beaches. *Marine Geology*, **27**, M9–M17.

23 Tracing Beach Sand Provenance and Transport using Foraminifera: Preliminary Examples from North-west Europe and South-east Australia

S. K. HASLETT,[1] E. A. BRYANT[2] and R. H. F. CURR[1]

[1]Quaternary Research Unit, School of Geography and Development Studies, Bath Spa University College, UK
[2]School of Geosciences, University of Wollongong, Australia

INTRODUCTION

Foraminifera are marine Sarcodine Protozoa that possess tests (shells) that are preservable in the fossil record. These tests may either be constructed using organically cemented detritus (agglutinating or arenaceous forms), or secreted using calcium carbonate (calcareous forms). Their ecology embraces planktonic and benthonic modes, although planktonic forms generally inhabit the open ocean and seldom live in coastal waters in any abundance, while benthonic foraminifera exist on substrates from abyssal plains to high intertidal areas. There are many species of foraminifera that are niche-specific, making them ideal for palaeoenvironmental analysis (Boersma, 1978; Brasier, 1980; Murray, 1991; Culver, 1993).

In coastal studies, foraminifera have been employed in a number of investigations, as indicators of Quaternary sea-level change (Scott and Medioli, 1978, 1986; Gehrels, 1994; Haslett et al., 1998a,b), for establishing coastal palaeoenvironments and sedimentary biofacies (Murray and Hawkins, 1976; Martin and Liddell, 1989; Kotler et al., 1992; Boomer and Godwin, 1993; Haslett, 1997a,b), as sediment transport indicators in tidal (Brasier, 1981; Thomas and Schafer, 1982; Wang and Murray, 1983; Michie, 1987; Murray, 1987; Gao and Collins, 1992, 1995; Cole et al., 1995), wave-dominated (Moore, 1957; Jones, 1958; Loose, 1970; Pizat, 1970; Blanc-Vernet, 1974; Blanc-Vernet et al., 1979; Seibold and Seibold, 1981; Murray et al., 1982; Sneh and Friedman, 1984; Vénec-Peyré and Le Calvez, 1986; Snyder et al., 1990; Davaud and Septfontaine, 1995) and aeolian environments (Glennie, 1970; Kameswara Rao et al., 1989), and as monitors of coastal environmental pollution (Alve and Nagy, 1986; Alve, 1990, 1991a,b, 1995a,b, 1996; Sharifi et al., 1993; Bernhard and Alve, 1996; van

Geen, 1999). The value of foraminifera as sediment transport indicators in tidal environments has been realised (Murray, 1987), and is being developed (e.g. Gao and Collins, 1995). Although the same is also true to an extent for some wave-dominated environments, the study of foraminiferal assemblages on sand-grade beaches has been neglected.

Foraminifera seldom live on beaches (Murray, 1973) and their occurrence in these environments is due to post-mortem transport, therefore identified species with known ecologies can act as sediment provenance tracers, and depending on their source area, can indicate transport processes. The aim of this study is to explore hypothetical beach foraminifera assemblages emplaced under various wave hydro-dynamic conditions (fairweather, storm and tsunami), in order to assess the potential of foraminifera as sediment provenance and transport tracers in sand-enriched beach environments. New data on samples collected from wave-dominated environments at a number of north-west European and south-east Australian beach sites are then used to evaluate the models.

MODELLED FORAMINIFERA ASSEMBLAGES

Three different wave hydrodynamic conditions are modelled in relation to foramini-fera transport and deposition on sand-grade beaches. Foraminifera are simplified into two groups representing planktonic and benthonic models of life. Coastal zone morphology is based upon Reading and Collinson's (1996) definitions:

(a)   A *beach* occurs between the landward limit of swash action and Mean Low Water (MLW). Additional beach morphological subdivisions include the *back-shore* which lies between the landward limit of swash action and Mean High Water (MHW), the *foreshore* between MHW and MLW, and the *beach face* which is the slope (*c.*<16°) seaward of a berm (Pethick, 1984) where swash and backwash are active (the beach is usually flat on the landward side of a berm (*syn.* backshore) where occasional swash (mainly under storm or high tidal conditions) may occur, but backwash is eliminated by percolation of the swash into the beach sediment).

(b)   The *shoreface* occurs between MLW and the mean fairweather wave base.

(c)   The *offshore-transition zone* occurs between mean fairweather and storm wave bases.

(d)   The *offshore zone* occurs seaward of the mean storm wave base.

The following models of hydrodynamics and sediment transport processes operat-ing at the coast provide a basis upon which exploratory studies and tests can be performed.

Fairweather Wave Conditions

According to Stokes' wave theory, where wave orbital motion is not closed, mass transport velocities and direction vary throughout the water column down to wave base, with a strong onshore flow at the bed and an offshore return flow at mid depths (Pethick, 1984). Under these conditions, it is expected that onshore movement of foraminifera would be predominantly by bedload transport of larger benthonic species that lived on the shoreface (Figure 23.1). Smaller species suspended by bottom turbulence may be carried onshore, but if suspended higher in the water column are more likely to be entrained into the offshore return flow. Surface waters under fair-weather conditions also flow onshore and may hold small planktonic foraminifera in suspension. Smaller suspended benthonic and planktonic specimens on reaching the beach face may not be deposited if energy levels are too high, but would be retained in suspension by backwash activity and transported alongshore to low energy environments, such as estuaries. Under fairweather conditions, a foraminiferal beach assemblage comprising mainly larger benthonic tests may be recognised.

Storm Wave Conditions

The hydrodynamics of storm wave conditions are different from fairweather waves. The principal difference derives from onshore winds which blow across the sea surface, setting up onshore surface currents (Allen, 1982). This onshore flow is balanced by offshore bottom currents. This results in a net offshore transport of bedload sediment, so that finer sediment deposited in the offshore-transition zone during fairweather conditions is overlain by coarser sediment derived from the beach and shoreface (Johnson and Baldwin, 1996). Therefore, during storm conditions, it is expected that benthonic species will be transported offshore (Figure 23.1), except

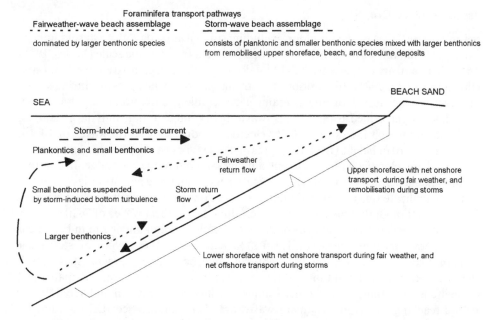

Figure 23.1   Hydrodynamic conditions and foraminifera transport pathways in the inner shelf environment

for the upper shoreface and swash zone where transport onto the beach may occur (Reading and Collinson, 1996). This offshore transport may account for the presence of nearshore foraminifera in offshore deep water > 100 m (e.g. Pizat, 1970; Blanc-Vernet, 1974; Blanc-Vernet et al., 1979), although they may also mark the position of former shorelines (Murray, 1979a). Small benthonic specimens held in suspension may be entrained into onshore flowing surface currents and delivered to the beach. Murray (1987) states that in the Celtic Sea waves associated with Force 10 south-westerly gales can suspend sediment, including foraminifera, down to depths of 180 m, which he argues may become entrained in the tidal currents of the severely macrotidal Severn Estuary.

In addition to small benthonic species, planktonic species living in the surface waters may also be transported onshore during storms. Indeed, Murray (1976) suggests that planktonic foraminifera are almost entirely allochthonous in continental shelf waters and sediment, and that the size and abundance of specimens decreases with distance from their open ocean habitat, as they settle out landward across the shelf. Thus, only very small planktonic specimens (c. 140 μm) are found in estuarine environments (Murray and Hawkins, 1976; Murray, 1980). Based on Murray's (1976) model, it may be argued that under storm conditions the size and/or abundance of planktonic foraminifera delivered to a given coastline may be expected to increase.

Under storm conditions, a foraminiferal beach assemblage may be characterised by planktonic and smaller benthonic foraminifera specimens. However, such assemblages can be skewed by specimens reworked from pre-storm upper shoreface and beach sediments.

Tsunami Wave Conditions

Seismic disturbance of the sea-floor, including submarine slides, and asteroid impacts can create tsunami waves which may propagate on a pan-oceanic scale (Myles, 1985). At sea, tsunami are of low wave height but long wavelength, compared to storm waves (Iida and Iwaski, 1981). In addition, the strong onshore winds present during storms are likely to be absent during tsunami events, unless both occur simultaneously. Calculations performed using Stokes' wave theory indicate that only large tsunami 5 m high can match the bottom drift velocities produced by storm waves of 10–15 s period and 7–10 m height (Bryant, in press). These types of storm waves can produce onshore drift velocities sufficient to entrain fine sand out to the shelf edge. In contrast, earthquake-generated tsunami, with typical open-ocean wave heights of less than 1 m, can only replicate this type of sediment transport on the inner shelf. In 20 m depth of water, even though the shoreline may be eroding, large storm waves of 7–10 m height can produce onshore bottom drift of 2.0 m s^{-1} while tsunami under 2 m in height can only generate current speeds of 1.0 m s^{-1}. While both these types of storm and tsunami wave types have similar heights and hydrodynamic characteristics outside the surf zone, they differ in that storm waves will inevitably break and dissipate much of their energy within the surf zone. Tsunami, being long waves, are unlikely to break before reaching the shore. Tsunami waves are also characterised by long duration times for onshore flow near the bed. These can exceed that generated by a single storm wave by a factor of 40–90 times. Finally, tsunami and storms differ in that tsunami wave trains generally consist of no more than around 12 large waves, whereas storms can generate waves for periods of hours if not days. Hence transport of foraminifera under tsunami waves is not sustained. Upon flooding the shore, tsunami effectively transport sediment landward through turbulent suspension and translation; but some of this material may return seaward under stronger and more prolonged backwash.

Resultant foraminifera assemblages are expected to comprise species (both planktonic and benthonic) that originally occurred within the surf zone or were previously deposited on the shore. Also, because of the relatively rapid transport and deposition, size sorting may not occur and assemblages of mixed test sizes may result, although this is dependent on an initial unsorted source. This assemblage may not be very different from storm-emplaced assemblages, and is unlikely to be characteristic. The main difference between emplacement of foraminifera by storm waves and by tsunami occurs in the run-up. Tsunami waves can generate run-up heights 30 times greater than their wave height and can sweep several kilometres inland. This penetration inland can only be duplicated on flat coastlines by storms if they are accompanied by a significant storm surge. Storm waves are incapable of flinging debris beyond cliff tops whereas tsunami can override complete headlands up to 130 m high (Bryant *et al.*, 1997).

BEACH FORAMINIFERA ASSEMBLAGES FROM NORTH-WEST EUROPE AND SOUTH-EAST AUSTRALIA

To explore the validity of the theoretical models discussed above, a number of beach sediment samples were collected from a number of wave-dominated coastlines in north-west Europe (Figure 23.2; Bird and Schwartz, 1985) and south-east Australia (Figure 23.3; Short and Wright, 1981, 1984; Short and Hesp, 1982; Bird and Schwartz, 1985). Samples of 500 g were collected from the beach face and air dried. No sieving of the sediment was undertaken, but foraminifera were concentrated using a flotation technique, by immersing the sediment in sodium polytungstate (cf. Savage, 1988), stirring, and decanting the floated fraction onto filter paper. Foraminifera float due to the presence of air in chambers of the dried tests, and so can be recovered from the floated fraction by picking, using a 000 sable paintbrush. The non-floated fraction was retained and examined for any specimens that failed to float. European species were identified with reference to Murray (1979b), and the Australian species with reference to Albani (1979) and Yassini and Jones (1988, 1995).

The foraminiferal assemblages are given in Tables 23.1 and 23.2. Generally, foraminifera abundance in the samples is relatively low with between 10 and 193 tests recovered. The preservational condition of most tests is generally good. Most identified species are capable of either epiphytic, epifaunal or infaunal life in relatively shallow water on the inner shelf, although some (*Elphidium williamsoni* and *Trochammina inflata*) only live on tidal flats and saltmarshes. These intertidal species only occurred together at Barneville, and indicate that longshore currents are actively transporting sediment to the beach from nearby tidal embayments, such as Baie du Mont St Michel (Larsonneur, 1989). Test size is generally large (*c.* 250–500 μm), and small (*c.* 140 μm) planktonic foraminifera were only encountered at Gerroa, which is

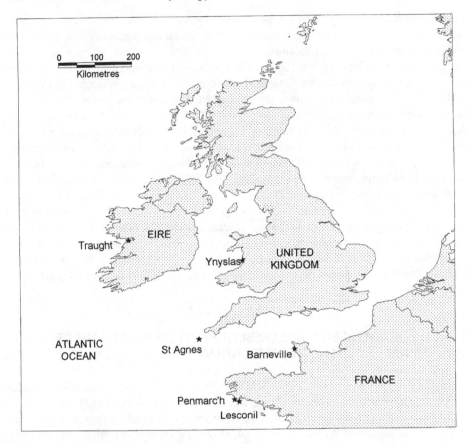

Figure 23.2   European beach sand sample locations

located at the downdrift end of Seven Mile Beach (Wright, 1970), from Shoalhaven in the south to Gerroa in the north, and implies onshore and alongshore transport of tests. All assemblages, according to the theoretical models introduced above, are compatible with transport and deposition during fairweather wave conditions, when large benthonic tests are derived from shallow inner shelf waters.

Detrended Correspondence Analysis

The conspicuous variance in species composition is explored here using Detrended Correspondence Analysis (DCA), a useful ordination technique for recent and fossil assemblages (Davies, 1998), which can be performed on low abundance (>10) raw counts (rather than percentages). All counting groups are included in the DCA. Species ordinations are plotted separately for European and Australian samples against the first two ordination axes in Figures 23.4 and 23.5. These describe the relationship between species composition and the most important controlling environmental factors. Clusters of points are interpreted as assemblage groups, which have

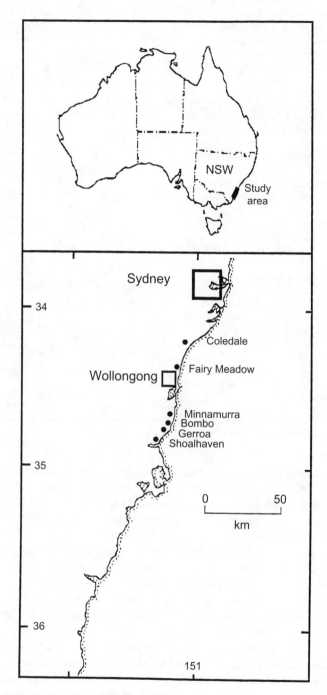

Figure 23.3 Australian beach sand sample locations

Table 23.1 Raw foraminiferal assemblage counts for European beach sand samples, with author attributions and sample collection dates. Environmental information given for each species indicates its life habitat

Species	Barneville	Lesconil	Penmarc'h	St. Agnes	Traught	Ynyslas	Environment
1 *Ammonia beccarii* var. *aberdoveyensis* Haynes	0	0	0	0	0	2	Inner shelf
2 *Ammonia beccarii* var. *batavus* Hofker	21	0	1	0	6	9	Inner shelf
3 *Brizalina variabilis* (Williamson)	0	0	0	1	0	0	Inner shelf
4 *Cibicides lobatulus* (Walker & Jacob)	0	138	2	177	21	0	Inner shelf
5 *Elphidium macellum* (Fichtel & Moll)	9	11	4	9	69	20	Inner shelf
6 *Elphidium williamsoni* Haynes	2	0	0	0	1	0	Intertidal
7 *Elphidium* sp. indet.	1	0	0	0	0	0	
8 *Glabratella millettii* (Wright)	0	0	0	0	0	1	Inner shelf
9 *Lamarckina haliotidea* (Heron-Allen & Earland)	2	0	0	0	0	0	Inner shelf
10 *Massilina secans* (d'Orbigny)	0	0	0	0	0	2	Inner shelf
11 *Planorbulina mediterranensis* d'Orbigny	0	4	0	0	0	0	Inner shelf
12 *Quinqueloculina seminulum* (Linne)	2	16	6	4	1	54	Inner shelf
13 *Rosalina williamsoni* (Chapman & Parr)	0	0	0	2	0	0	Inner shelf
14 *Trochammina inflata* (Montagu)	1	0	0	0	0	0	Intertidal saltmarsh
15 Unidentified specimens	2	0	0	0	0	1	
TOTAL	40	169	13	193	98	89	
Collection date	Aug. 95	Sep. 97	Sep. 96	Sep. 96	Aug. 98	Mar. 94	

Environmental information from Murray (1991).

similar environmental constraints. Species falling outside these clusters are not considered further. With reference to the dominant species these assemblage groups can be cross-referenced with faunal associations of Murray (1991) to aid interpretation. Table 23.3 summarises this information and indicates that in both European and Australian samples two prominent assemblage groups may be recognised. Group 1 in each case consists of species (although not the same characteristic species) that inhabit phytal and/or sandy substrates, which generally reflect non-turbid, clear water environments. Group 2 is characterised by *Ammonia beccarii* in both European and Australian samples, which tolerates muddy substrates, and therefore more turbid environments. An examination of the occurrence of this species in the samples corroborates this interpretation. In Europe, *Ammonia beccarii* is most abundant at Barneville, which is situated on a macrotidal coastline, close to the muddy Baie du Mont St Michel (Larsonneur, 1989), and in Australia it only occurs in samples from Seven Mile Beach (at Shoalhaven Heads and Gerroa), which is proximal to the outlet of the Shoalhaven River, which has a high fine sediment discharge into the inner shelf region (Wright, 1970). Therefore, it appears that species composition in beach sand assemblages is determined by the substrate conditions in the shoreface and

Table 23.2   Raw foraminiferal assemblage counts for Australian beach sand samples, with author attributions and sample collection dates. Environmental information given for each species indicates environments in which tests have been recovered, whether alive and *in situ*, or dead and derived

Species	Coledale	Fairy Meadow	Minnamurra	Bombo	Gerroa	Shoalhaven Heads	Environment
1 *Ammonia beccarii* (Linne, 1767)	0	0	0	0	2	5	Lagoonal to estuarine
2 *Anomalina nonionoides* (Parr, 1932)	4	1	0	4	1	2	Lagoonal and intertidal[a]
3 *Bulimina* spp.	0	0	0	1	0	2	Intertidal to middle shelf
4 *Cibicides refulgens* (de Montfort, 1808)	0	0	0	16	7	12	Intertidal to inner shelf
5 *Cibicidoides floridanus* (Cushman, 1918)	3	0	2	2	2	5	Intertidal to continental slope
6 *Cornuspira foliaceus* (Philippi, 1844)	1	0	0	1	0	0	Estuarine and inner shelf
7 *Elphidium advenum* (Cushman, 1922)	0	0	0	0	1	0	Lagoonal to inner shelf
8 *Elphidium crispum* (Linne, 1758) group	3	2	2	11	12	30	Intertidal to inner shelf
9 *Glabratella australensis* (Heron-Allen & Earland, 1932)	2	0	0	0	3	3	Intertidal to sheltered embayments
10 Miliolid group[b]	4	1	2	21	14	10	Intertidal to inner shelf
11 *Parrelina* spp.	2	4	0	1	1	14	Intertidal to inner shelf
12 Planktonic foraminifera	0	0	0	0	2	0	Stenohaline (open ocean)
13 *Rosalina australis* (Parr, 1932)	2	1	2	4	1	4	Tidal channels and inner shelf[a]
14 *Textlaria candeiana* (d'Orbigny, 1839)	0	0	0	4	0	0	Estuarine to middle shelf
15 *Trochulina dimidiata* (Jones & Parker, 1862) group	29	12	2	56	17	20	Intertidal to inner shelf
TOTAL	50	21	10	121	63	107	

All samples collected during April 1998.
[a] Yassini and Jones (1988). All other environmental information from Yassini and Jones (1995).
[b] Miliolid group includes all species belonging to the Soborder Miliolina.

offshore-transition zone. This is potentially valuable in (palaeo)environmental studies where species composition in (palaeo)beach sand could indicate particle size in the source area regardless of the collected samples' particle size.

DISCUSSION

If the proposed models are accepted, the foraminifera results from the beach sand samples all indicate transport from the inner shelf region and deposition under fair-weather wave conditions. Indeed, only very few non-inner shelf species were encountered in the present study of beach sand samples. This is not surprising as most samples were collected from beach faces at the end of the summer. However, other

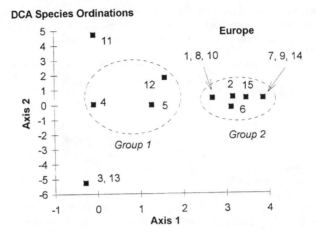

Figure 23.4 Results of Detrended Correspondence Analysis of European beach sand fora-miniferal assemblages (Axes 1 and 2). Assemblage Group 1 represents species indicative of a non-turbid source environment, whilst Assemblage Group 2 represents a turbid source environment. Species numbers are those given in Table 23.1

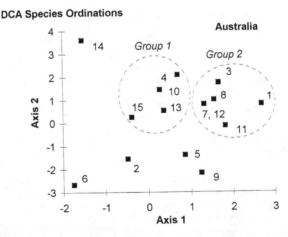

Figure 23.5 Results of Detrended Correspondence Analysis of Australian beach sand fora-miniferal assemblages (Axes 1 and 2). Assemblage Group 1 represents species indicative of a non-turbid source environment, whilst Assemblage Group 2 represents a turbid source environment. Species numbers are those given in Table 23.2

factors not addressed in detail by the models need evaluation. Taphonomic processes may play a significant role in determining resultant foraminiferal beach assemblages. The models focus on the delivery of new tests to the beach environment; however, tests that become integrated components of beach sediment may also be reworked to produce cumulative assemblages, although physical damage, abrasion and dissolution should increase with residency time. Also, temporal changes in wave energy may produce transient assemblages. For example, an assemblage emplaced under

Table 23.3 Summary of assemblage groups derived from Detrended Correspondence Analysis for both European and Australian beach sand samples. A comparison is made with faunal associations of Murray (1991). Assemblage group 1 in both cases represents a non-turbid source environment, whilst assemblage group 2 represents a turbid source

Species ordination group	Characteristic species	Additional species	Murray's association	Shelf substrate type	Depth
North-west Europe					
Group 1	*Cibicides lobatulus*	None	*C. lobatulus*	Seaweed	0–900 m
	Elphidium crispum	None	*E. crispum*	Seaweed	0–25 m
	Quinqueloculina seminula	None	*Q. seminula*	Sand	20–120 m
Group 2	*Ammonia beccarii*	8 others	*A. beccarii*	Muddy sand	0–60 m
Eastern Australia					
Group 1	*Trochulina dimidiata*	Miliolid group *Cibicides refulgens* *Rosalina australis*	*T. dimidiata*	Shelly coarse sand	5–7 m
Group 2	*Ammonia beccarii*	*Elphidium crispum* *Parrelina* spp.	*A. beccarii*	Mud, sand	0–25 m

fairweather conditions may be reworked during a storm, and small tests transported as bedload under fairweather conditions may become suspended under storm wave energy conditions and transported alongshore. These taphonomic effects are difficult to incorporate into the models at present and further research is required; in particular, high temporal resolution sampling at individual sites, and less of an *ad hoc* approach than was adopted in this preliminary study. The same is true regarding the role of tidal currents.

It is possible that the sample location along a beach profile is also important, as the active beach face may, for example, always yield an assemblage indicative of fairweather conditions as described by the models, as smaller specimens may not be deposited here due to prohibitively high energy conditions. They may be deposited, however, from swash on the landward side of a berm during high tides in fairweather conditions, or by berm-overtopping storm waves. Samples from the landward side of the berm may therefore yield very different assemblages and could be biased toward assemblages attributable to storm wave conditions or tsunamis, which are more likely to overtop berms. This is obviously an area upon which further studies will be based.

The only comparable work to the present study is that of Davaud and Septfontaine (1995) who investigated transported foraminifera in a barrier and lagoon sedimentary environment on the Tunisian coast in the Mediterranean. They encountered foraminifera in barrier (beach) and lagoon sediments which were epiphytes that lived on seagrass leaves, similar to DCA Group 1 here (Table 23.3). However, shoreface sediments lacked foraminifera. The seagrasses form subtidal meadows in the offshore-transition zone between water depths of 4 and 35 m, therefore the onshore transport of this assemblage is considered to occur during storm conditions when the wave base is lowered to within the depth range of the meadows. They also argue that the lack of foraminifera in the shoreface suggests that fairweather waves do not actively transport tests onshore. While the present study suggests that if storm waves do introduce epiphytic tests onto the shoreface they may be rapidly transported to the beach by fairweather waves in the immediate post-storm period, this is not tenable for their lagoon assemblages, which must have been introduced by storm waves over-topping the barrier. Davaud and Septfontaine (1995) also suggest that foraminifera preservation could reflect transport processes, in that bedload-transported specimens would be more abraded than those transported in suspension.

The use of foraminifera as sediment provenance and transport tracers may not be the best technique to employ when studying contemporary coasts, as more sophistic-ated tracing devices provide more fully quantitative results (e.g. Lee *et al.*, see Chapter 22). However, foraminifera have great potential in palaeoenvironmental studies. This study demonstrates that species composition can be used to distinguish between onshore transport from a shelf source and alongshore transport from an intertidal source (which could allow current development to be established through time, given adequate stratigraphic material). Species composition is also a discriminator of the sedimentological characteristics of the shelf source area, whether turbid or non-turbid, which could be extremely useful in monitoring changing sedimentological regimes through time. The relationship of foraminifera to hydrodynamic conditions requires further investigation as certain parts of a beach profile will only yield evidence of a particular hydrodynamic state. For example, and according to the model predictions, sediments from active beach faces could yield an assemblage indicative of fairweather conditions, whereas sediment landward of berms could yield assemblages emplaced by overtopping storm or tsunami waves.

CONCLUSION

Foraminifera have a demonstrated value in coastal sediment tracer studies, and this study addresses their potential for sand-grade beaches. The models that have been introduced here are simplistic, but have facilitated this preliminary investigation, which has provided encouraging results. The models predict that deposition under fairweather wave conditions is characterised by an assemblage of large benthonic foraminifera tests, and that an assemblage of small planktonic and benthonic tests, mixed with reworked large benthonic specimens, is characteric of storm wave and, probably, tsunami deposition. Samples collected from European and Australian beach faces comprise almost exclusively large benthonic tests, indicating fairweather deposition. However, variation in species composition appears to reflect substrate conditions at source, and can also be used to distinguish between onshore and alongshore derived specimens. The need for further investigations is apparent, espe-cially the study of possible spatial and temporal assemblage variations along a beach profile, the effects of taphonomic processes, and the role of tidal currents super-imposed on wave transport. The use of foraminifera may not be cost-effective in studying sediment provenance and transport to contemporary beaches, but its poten-tial application to palaeoenvironmental studies is considerable, as it provides a means for establishing shelf palaeoenvironments and coastal palaeohydrodynamics, includ-ing temporal changes in onshore and alongshore current activity.

ACKNOWLEDGEMENTS

This study has been partially funded by Bath Spa University College. Australian samples were collected whilst S.K.H. was a visiting researcher in the School of Geosciences, University of Wollongong, and he thanks colleagues there for their hospitality. S.K.H. would also like to thank Samantha Burchell, Maya Burchell Haslett and Elinor Burchell Haslett for assistance in the field and to Fiona Strawbridge for drafting Figure 23.2.

REFERENCES

Albani, A. D. 1979. *Recent Shallow Water Foraminifera from New South Wales*. Australian Marine Science Association Handbook No. 3, Sydney.

Allen, J. R. L. 1982. *Sedimentary Structures: Their Character and Physical Basis*. Developments in Sedimentology, volumes 30A & B. Elsevier, Amsterdam.

Alve, E. 1990. Variations in estuarine foraminiferal biofacies with diminishing oxygen conditions in Drammensfjord, SE Norway. In Hemleben, C. (ed.) *Paleoecology, Biostratigraphy, Paleoceanography and Taxonomy of Agglutinated Foraminifera*. Kluwer, Amsterdam, 661–694.

Alve, E. 1991a. Foraminifera, climate change, and pollution: a study of late Holocene sediments in Drammensfjord, southeast Norway. *The Holocene*, **1**, 243–261.

Alve, E. 1991b. Benthic foraminifera in sediment cores reflecting heavy metal pollution in Sørfjord, western Norway. *Journal of Foraminiferal Research*, **21**, 1–19.

Alve, E. 1995a. Benthic foraminiferal responses to estuarine pollution: a review. *Journal of Foraminiferal Research*, **25**, 190–203.

Alve, E. 1995b. Benthic foraminiferal distribution and recolonization of formerly anoxic environments in Drammensfjord, southern Norway. *Marine Micropaleontology*, **25**, 169–186.

Alve, E. 1996. Benthic foraminiferal evidence of environmental change in the Skagerrak over the past six decades. *Norges Geologiske Undersøkelse Bulletin*, **430**, 85–93.

Alve, E. and Nagy, J. 1986. Estuarine foraminiferal distribution in Sandebukta, a branch of the Oslo Fjord. *Journal of Foraminiferal Research*, **16**, 261–284.

Bernhard, J. M. and Alve, E. 1996. Survival, ATP pool, and ultrastructural characterization of benthic foraminifera from Drammensfjord (Norway): response to anoxia. *Marine Micropaleontology*, **28**, 5–17.

Bird, E. C. F. and Schwartz, M. L. (eds) 1985. *The Worlds Coastline*. Van Norstrand Reinhold, New York.

Blanc-Vernet, L. 1974. Microfaune de quelques dragages et carottages effectués devant les côtes de Tunisie (Golfe de Gabés) et de Libye (Tripolitaine). *Géologie Méditeranéenne*, **1**, 9–26.

Blanc-Vernet, L., Clairefond, P. and Orsolini, P. 1979. La mer pélagienne: les foraminifères. *Géologie Méditeranéenne*, **61**, 171–209.

Boersma, A. 1978. Foraminifera. In Haq, B. U. and Boersma, A. (eds) *Marine Micropaleontology*. Elsevier, New York, 19–77.

Boomer, I. and Godwin, M. 1993. Palaeoenvironmental reconstruction in the Breydon Formation, Holocene of East Anglia. *Journal of Micropalaeontology*, **12**, 35–46.

Brasier, M. D. 1980. *Microfossils*. Unwin Hyman, London.

Brasier, M. D. 1981. Microfossil transport in the tidal Humber Basin. In Neale, J. W. and Brasier, M. D. (eds) *Microfossils from Recent and Fossil Shelf Seas*. Ellis Horwood, Chichester, 314–322.

Bryant, E. A. in press. *Tsunami: The Underrated Hazard*. Cambridge University Press.

Bryant, E. A., Young, R. W., Price, D. M., Wheeler, D. and Pease, M. I. 1997. The impact of tsunami on the coastline of Jervis Bay, southeastern Australia. *Physical Geography*, **18**, 441–460.

Cole, A. R., Harris, P. T. and Keene, J. B. 1995. Foraminifers as facies indicators in a tropical, macrotidal environment: Torres Strait–Fly River delta, Papua New Guinea. *Special Publication of the International Association of Sedimentologists*, **24**, 213–223.

Culver, S. J. 1993. Foraminifera. In Lipps, J. H. (ed.) *Fossil Prokaryotes and Protists*. Blackwells, Boston, 203–247.

Davaud, E. and Septfontaine, M. 1995. Post-mortem onshore transportation of epiphytic foraminifera: recent example from the Tunisian coastline. *Journal of Sedimentary Research*, **A65**, 136–142.

Davies, P. 1998. Numerical analysis of subfossil wet-ground molluscan taxocenes from overbank alluvium at Kingsmead Bridge, Wiltshire. *Journal of Archaeological Science*, **25**, 39–52.

Gao, S. and Collins, M. 1992. Modelling exchange of natural trace sediments between an estuary and adjacent continental shelf. *Journal of Sedimentary Petrology*, **62**, 35–40.

Gao, S. and Collins, M. 1995. Net sand transport direction in a tidal inlet, using foraminiferal tests as natural tracers. *Estuarine, Coastal and Shelf Science*, **40**, 681–697.

Gehrels, W. R. 1994. Determining relative sea-level change from salt-marsh foraminifera and plant zones on the coast of Maine, USA. *Journal of Coastal Research*, **10**, 990–1009.

Glennie, K. W. 1970. *Desert Sedimentary Environments: Developments in Sedimentology 14*. Elsevier, Amsterdam.

Haslett, S. K. 1997a. Late Quaternary foraminiferal biozonation and its implications for the relative sea-level history of Gruinart Flats, Islay. In Dawson, A. G. and Dawson, S. (eds) *The Quaternary of Islay and Jura*. Quaternary Research Association, Cambridge, 99–104.

Haslett, S. K. 1997b. An Ipswichian foraminiferal assemblage from the Gwent Levels (Severn Estuary, UK). *Journal of Micropalaeontology*, **16**, 136.

Haslett, S. K., Davies, P., Curr, R. H. F., Davies, C. F. C., Kennington, K., King, C. P. and Margetts, A. J. 1998a. Evaluating Late Holocene relative sea-level change in the Somerset Levels, southwest Britain. *The Holocene*, **8**, 197–207.

Haslett, S. K., Davies, P. and Strawbridge, F. 1998b. Reconstructing Holocene sea-level change in the Severn Estuary and Somerset Levels: the foraminifera connection. *Archaeology in the Severn Estuary*, **8** (for 1997), 29–40.

Iida, K. and Iwasaki, T. 1981. *Tsunamis: Their Science and Engineering*. Reidel, Dordrecht.

Johnson, H. D. and Baldwin, C. T. 1996. Shallow clastic seas. In Reading, H. G. (ed.) *Sedimentary Environments: Processes, Facies and Stratigraphy* (3rd Edition). Blackwell Science, Oxford, 232–280.

Jones, D. J. 1958. Displacement of microfossils. *Journal of Sedimentary Petrology*, **28**, 453–467.

Kameswara Rao, K., Wasson, R. J. and Krishnan Kutty, M. 1989. Foraminifera from Late Quaternary dune sands of the Thar Desert, India. *Palaios*, **4**, 168–180.

Kotler, F., Martin, R. E. and Liddell, W. D. 1992. Experimental analysis of abrasion and dissolution resistance of modern reef-dwelling foraminifera: implications for the preservation of biogenic carbonate. *Palaios*, **7**, 244–276.

Larsonneur, C. 1989. La Baie du Mont-Saint-Michel: un modele de sedimentation en zone tempérée. *Bulletin de la Institut de Geologie du Bassin d'Aquitaine*, **46**, 5–73.

Loose, T. L. 1970. Turbulent transport of benthonic foraminifera. *Contributions from the Cushman Foundation for Foraminiferal Research*, **21**, 164–166.

Martin, R. E. and Liddell, W. D. 1989. Relation of counting methods to taphonomic gradients and biofacies zonation of foraminiferal sediment assemblages. *Marine Micropaleontology*, **15**, 67–89.

Michie, M. G. 1987. Distribution of Foraminifera in a macrotidal tropical estuary: Port Darwin, Northern Territory of Australia. *Australian Journal of Marine and Freshwater Research*, **38**, 249–259.

Moore, W. E. 1957. Ecology of Recent foraminifera in northern Florida Keys. *American Association of Petroleum Geologists Bulletin*, **41**, 727–741.

Murray, J. W. 1973. *Distribution and Ecology of Living Benthic Foraminiferids*. Heinemann, London.

Murray, J. W. 1976. A method of determining proximity of marginal seas to an ocean. *Marine Geology*, **23**, 102–119.

Murray, J. W. 1979a. Recent benthic foraminiferids of the Celtic Sea. *Journal of Foraminiferal Research*, **9**, 193–209.

Murray, J. W. 1979b. *Synopses of the British Fauna (New Series), No. 16: British Nearshore Foraminiferids*. Academic Press, London.

Murray, J. W. 1980. The foraminifera of the Exe Estuary. *Devon Association for the Advancement of Science* (Special Volume), **2**, 89–115.

Murray, J. W. 1987. Biogenic indicators of suspended sediment transport in marginal marine environments: quantitative examples from SW Britain. *Journal of the Geological Society, London*, **144**, 127–133.

Murray, J. W. 1991. *Ecology and Palaeoecology and Benthic Foraminifera*. Longman, London.

Murray, J. W. and Hawkins, A. B. 1976. Sediment transport in the Severn Estuary during the past 8000–9000 years. *Journal of the Geological Society, London*, **132**, 385–398.

Murray, J. W., Sturrock, S. and Weston, J. 1982. Suspended load transport of foraminiferal tests in a tide- and wave-swept sea. *Journal of Foraminiferal Research*, **12**, 51–65.

Myles, D. 1985. *The Great Waves*. McGraw-Hill, New York.

Pethick, J. 1984. *An Introduction to Coastal Geomorphology*. Edward Arnold, London.

Pizat, C. 1970. Hydrodynamisme et sédimentation dans le golfe de Gabès (Tunisie). *Tethys*, **21**, 267–296.

Reading, H. G. and Collinson, J. D. 1996. Clastic coasts. In Reading, H. G. (ed.) *Sedimentary Environments: Processes, Facies and Stratigraphy* (3rd Edition). Blackwell Science, Oxford, 154–231.

Savage, N. M. 1988. The use of sodium polytungstate for conodont separations. *Journal of Micropalaeontology*, **7**, 39–40.

Scott, D. B. and Medioli, F. S. 1978. Vertical zonations of marsh foraminifera as accurate indicators of former sea-levels. *Nature*, **272**, 528–531.

Scott, D. B. and Medioli, F. S. 1986. Foraminifera as sea-level indicators. In van de Plassche, O. (ed.) *Sea-level Research: A Manual for the Collection and Evaluation of Data*. Geo-Books, Norwich, 435–456.

Seibold, I. and Seibold, E. 1981. Offshore and lagoonal benthic foraminifera near Cochin (southwest India) – distribution, transport, ecological aspects. *Neues Jahrbuch für Geologie und Paläontologie, Abhandlung*, **162**, 1–56.

Sharifi, A. R., Croudace, I. W. and Austin, R. L. 1993. Benthic foraminiferids as pollution indicators in Southampton Water, southern England, UK. *Journal of Micropalaeontology*, **10**, 109–114.

Short, A. D. and Hesp, P. A. 1982. Wave, beach and dune interactions in southeastern Australia. *Marine Geology*, **48**, 259–284.

Short, A. D. and Wright, L. D. 1981. Beach systems of the Sydney region. *Australian Geographer*, **15**, 8–16.

Short, A. D. and Wright, L. D. 1984. Morphodynamics of high energy beaches: an Australian perspective. In Thom, B. G. (ed.) *Coastal Geomorphology in Australia*. Academic Press, Sydney, 43–68.

Sneh, A. and Friedman, G. M. 1984. Spit complexes along the eastern coast of the Gulf of Suez. *Sedimentary Geology*, **39**, 211–226.

Snyder, S. W., Hale, W. R. and Kontrovitz, M. 1990. Assessment of postmortem transportation of modern benthic foraminifera of the Washington continental shelf. *Micropaleontology*, **36**, 259–282.

Thomas, F. C. and Schafer, C. T. 1982. Distribution and transport of some common foraminiferal species in the Minas Basin, eastern Canada. *Journal of Foraminiferal Research*, **50**, 229–265.

Van Geen, A. 1999. A record of estuarine water contamination from the Cd content of foraminiferal tests in San Francisco Bay, California. *Marine Chemistry*, **64**, 57–69.

Vénec-Peyré, M. T. and Le Calvez, Y. 1986. Foraminifères benthiques et phénomènes de transfert: importance des études comparatives de la biocénose et de la thanatocénose. *Bulletin de la Museum d'Histoire Naturelle, Paris*, **8**, 171–184.

Wang, P. and Murray, J. W. 1983. The use of foraminifera as indicators of tidal effects in estuarine deposits. *Marine Geology*, **51**, 239–250.

Wright, L. D. 1970. The influence of sediment availability on patterns of beach ridge development in the vicinity of the Shoalhaven River Delta, NSW. *Australian Geographer*, **11**, 336–348.

Yassini, I. and Jones, B. G. 1988. Estuarine foraminiferal communities in Lake Illawarra, NSW. *Proceedings of the Linnean Society of New South Wales*, **110**, 229–266.

Yassini, I. and Jones, B. G. 1995. *Foraminiferida and Ostracoda from Estuarine and Shelf Environments on the Southeastern Coast of Australia*. University of Wollongong Press, Wollongong.

Section 6

TRACERS IN PALAEOENVIRONMENTAL INVESTIGATIONS

24 Holocene Sediment Erosion in Britain as Calculated from Lake-basin Studies

D. N. BARLOW and R. THOMPSON

Department of Geology and Geophysics, The University of Edinburgh, UK

INTRODUCTION

Myers (1993) estimated that, globally, approximately 75 billion tonnes of soil are eroded annually, the majority of which come from the world's croplands. Thus, each decade the global soil budget is being depleted by *c.* 7 percent (Walling, 1988). According to Pimental (1976), in the past two centuries the US has lost one-third of its topsoil. As a consequence roughly 80 percent of the world's agricultural land is deemed to be suffering from 'moderate to severe erosion' and a further 10 percent from 'slight to moderate erosion' (Speth, 1994). The variation in rates of erosion across the world is huge. Fournier (1960) reported that with average erosion rates of 1000–$2000\,t\ km^{-2}\ year^{-1}$, losses in Asia, Africa and South America are greatest; losses are lowest in the US and Europe where the average yield is *c.* 0–$600\,t\ km^{-2}\ year^{-1}$. These losses from predominantly agricultural areas contrast with those associated with undisturbed forests which range from only 0.4 to $5\,t\ km^{-2}\ year^{-1}$ (Bennett, 1939). A number of authors have attempted to calculate the financial implications of erosion in terms of both on- and off-site costs. Brown (1948) estimated that the impacts of sediment erosion downstream in the US cost in the region of $175 million annually, and Walling (1988) translates this into a 1988 value of *c.* $1000 million. Pimentel *et al.* (1995) suggested that, whilst the resulting decline in soil fertility in the US costs approximately $27 billion, the off-site environmental impact equates to an additional $17 billion (1992 dollars) a year. Thus he suggested that, in the US, the annual cost of sediment erosion resulting from agriculture is in the region of $44 billion per year, equivalent to about $100 per hectare of pastureland and cropland.

In Britain, quantitative estimates of erosion rates remain poor, particularly from an historical perspective. As reported by Moore and Newson (1986), long records of erosion are unusual in Britain. Consequently relatively little information on long-term erosion rates in British catchments is available. Work has tended to consider the relatively recent time period, particularly the past two centuries. This British work covers many important changes, including variations in erosion associated with shifts

Tracers in Geomorphology. Edited by Ian D. L. Foster. © 2000 John Wiley & Sons Ltd.

in agricultural and forestry practices, and urbanisation. However, other fundamental changes concerning historical land-use patterns and other human activities have been largely undocumented.

Here we use a catchment-based approach to study erosion. Sediment yield and flux estimates have been obtained from either reservoir re-survey data or lake sediment multi-core studies. Multiplying the sediment yield by a sediment-delivery ratio enables an erosion rate to be obtained. However, such experimental approaches are both time-consuming and expensive to undertake. It is thus desirable to predict sediment flux using a simple model. At present, sediment erosion models range from (i) simple relationships between sediment yield and a single physical catchment characteristic, e.g. the catchment to lake ratio (Dearing and Foster, 1993); through (ii) empirical equations relating the rate of erosion to a range of physical characteristics, e.g. the Universal Soil Loss Equation (USLE) of Wischmeier and Smith (1978); to (iii) highly detailed and complex models. At the moment all have their disadvantages. We discuss the development of simple regression models which link catchment characteristics to sediment flux within British lake catchments.

MODELS OF SEDIMENT EROSION

Linking Sediment Accumulation to Sediment Yield

As a starting point, the equation

$$Y = M/A \tag{24.1}$$

can be employed to determine sediment yield in lake catchments. Here Y is the sediment yield (t km^{-2} year^{-1}), M is the mass of material deposited in the basin annually (t year^{-1}) and A is the catchment area (km^2). However, as Walling (1983, 1988) has emphasised, sediment yield determined from lake sediment-based studies does not take account of the deposition of material during transport, *en route* from source to sink, whether it be in river channel or overland within the catchment. Sediment yield is therefore a function not only of the rate of soil loss but also of the efficiency with which it is delivered (Jackson *et al.*, 1986). Thus in order to relate sediment yield to erosion the sediment-delivery ratio, D, is an essential factor that must be considered. Haan *et al.* (1994) define the sediment-delivery ratio as

$$D = G/(Y \cdot A) \tag{24.2}$$

where G is the gross erosion occurring in the catchment per year (t year^{-1}). Sediment yield in lake catchments therefore becomes

$$Y = (M \cdot D)/A \tag{24.3}$$

There are a variety of difficulties in selecting a sediment-delivery ratio, D, for a given catchment as there are a range of factors that can influence it. Indeed Haan *et al.* (1994: 293) state that, 'It should be pointed out that the degree of understanding of sediment-delivery ratios is probably less than any other area of sedimentation.' Nevertheless a number of researchers have attempted to quantify the significance of the various processes involved. Vanoni (1975) suggests that in basins larger than $1\,km^2$, often less than 25 percent of the material eroded reaches a given point downstream, whilst theoretical work undertaken by Trimble (1981) suggests that, in fact, sediment delivery may fall to a mere 6 percent. The American Society of Civil Engineering (ASCE, 1975) have adopted the empirical relationship

$$D = 0.36A^{-0.2} \qquad (24.4)$$

between delivery ratio, D, and drainage basin area, A. Sediment delivery is seen to vary from more than 90 percent in some small catchments to less than 10 percent in the largest catchments. The general decrease in sediment-delivery ratio with catchment size is often attributed to a 'headwater' effect. Small lake or reservoir catchments tend to lie in the upper reaches of river systems where slopes tend to be steeper and erosion tends to predominate over deposition. Larger lake and reservoir catchments, in contrast, tend to lie in the lower reaches of river systems, with gentler slopes and more extensive floodplains which provide more scope for sediment retention.

Relationships between Sediment Yield and Catchment Characteristics

On a global scale, links between sediment yield and a number of physical parameters, such as catchment area and relief, have been investigated for various regions of the world. On the basis of discharge and sediment data for 60 large catchments, Strakhov (1967) produced a map illustrating the global pattern of erosion. He found that in large basins variations in suspended sediment yield of between $1\,m^3km^{-2}year^{-1}$ and $4000\,m^3km^{-2}year^{-1}$ and dissolved sediment yield of between $1\,m^3km^{-2}year^{-1}$ and $450\,m^3km^{-2}year^{-1}$ can be accounted for by physiography, soil type, vegetation cover and climate. Strakhov identifies two particular zones of erosion. First, a temperate moist belt in the northern hemisphere is broadly bounded to the south by the annual $+10\,°C$ isotherm. This zone is characterised by an annual precipitation of between 150 and 600 mm. It has low erosion rates, typically less than $10\,t\,km^{-2}year^{-1}$. His second zone includes parts of North America, South America, Africa and South East Asia. It corresponds to the area between the $+10\,°C$ isotherm in the northern hemisphere and the $+10\,°C$ isotherm in the southern hemisphere. His second zone is characterised by an average annual precipitation of between 1200 and 1300 mm. Here erosion is high, typically between 50 and $100\,t\,km^{-2}year^{-1}$, though rising to values in excess of $1000\,t\,km^{-2}year^{-1}$ in the Indus, Ganges and Brahmaputra basins. Britain, along with most of Europe, falls into Strakhov's low erosion zone.

Links between spatial scale and yield have been studied by many authors. An important early study is that of Brune (1950) who investigated sediment loads for a range of drainage basins in the Sangamon River Watershed, Illinois. He too noted that average rates of sediment production decreased with increasing drainage area. Following on from Brune's work, Flaxman and Hobba (1955) surveyed sedimentation in 38 stockponds in the Columbia River Basin. They observed that drainage basin area was one of the five main factors accounting for 80 percent of the variation in sediment accumulation in their stockponds. Langbein and Schumm (1958) employed American gauging-station data for 94 catchments, and reservoir sedimentation data for 163 catchments, to study the relationship between precipitation and erosion. They found that sediment yields reached a peak at the transition zone between desert shrub and grassland conditions. Much lower yields were characteristic of both particularly dry regions and more humid regions. Langbein and Schumm (1958) suggested that the low sediment yields in very dry regions could be explained by the low runoff resulting from precipitation levels of less than 300 mm year^{-1}. Schumm (1963) also noted the effect of the relief ratio (maximum basin relief/length) on sediment yield. He found that an exponential increase in annual sediment yield was caused by the relief ratio in drainage basins of area 2.6 km^2 and greater.

Amongst others, Dearing and Foster (1993) have postulated links between sediment yield and the ratio of catchment area to lake area. They plotted the relationship between catchment to lake ratio and sediment yield for 20 studies of erosion in different environments in the world (Figure 24.1) and proposed that the data can be divided into two groups. One group represents sites with recent maximum sediment yield under cultivation/moorland while the second group illustrates maximum sediment yields under forest. Both groups of sites display a decrease in sediment yield as catchment to lake ratio increases. Dearing and Foster (1993) proposed that the negative correlation could be explained by two factors: first, the increase in storage as catchment area increases, and secondly, the erosion pathways between slopes, channels and the lake increased in importance at a slower rate than catchment area. They went on to suggest that for sites where the catchment to lake ratio is less than 10, sediment is more likely to originate from slope or surface processes than from channel

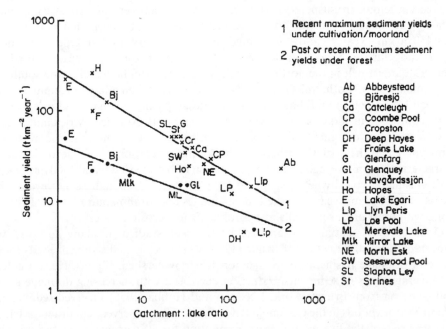

Figure 24.1 The relationship between catchment to lake ratio and sediment yield for 20 studies (from Dearing and Foster, 1993). Dearing and Foster (1993) have divided the data into two groups. The upper set comprises yield estimates obtained from catchments that are cultivated or are moorland. The lower set comprises sediment yield estimates from forested catchments

banks. In contrast, they proposed that in larger catchments, where the catchment to lake ratio is greater than 10, a channel network is supported and thus the significance of channels as a sediment source increases.

Modelling Sediment Yields and Processes

Following on from the work of Flaxman and Hobba (1955), Langbein and Schumm (1958) and Schumm (1963), more complex models of sediment yields and processes have been developed. Traditionally such models have tended to be based on empirical equations, though more recently much attention has been focused on what Foster (1990) terms 'process-based technology'. One empirical model is that of Fournier (1960). Using data from 78 drainage basins, Fournier derived the following equation:

$$\log Qs = 2.65 \log p^2/P + 0.46 (\log H)(\tan S) - 1.56 \qquad (24.5)$$

where Qs is mean annual sediment yield $(g\,m^{-2})$, p is the highest mean monthly precipitation (mm), P is mean annual precipitation (mm), H is mean catchment altitude (m), and S is the mean basin slope (degrees).

The Universal Soil Loss Equation

The most widely employed empirically based model is the Universal Soil Loss Equation (USLE). This well-known model was developed by Wischmeier and Smith (1978) from a database consisting of more than 10 000 plot-years of data. Plots studied ranged in length from 11 to 189 m and spanned a variety of soils, slope steepness, vegetation and climate in eastern North America. The model relates mean annual soil loss to rainfall erosivity, soil erodibility, slope length, slope steepness, a crop management factor and finally an erosion control practice factor. The USLE was designed for estimating inter-rill and rill erosion over time on small field plots. The equation was not designed to estimate soil loss for specific storm events. A number of authors have expressed concerns regarding the application of the USLE to larger areas (Meade, 1982). As noted by Foster (1982), the USLE is not designed to predict gully or channel bank erosion or to account for the deposition of material on hill slopes or channels, and hence assumes a sediment delivery ratio of one. Thus catchment studies which incorporate the USLE to estimate erosion need to add a sediment delivery term, D. Other limitations that have been identified in the USLE include the narrow database upon which it was built, i.e. American agricultural sites, along with theoretical problems, e.g. the lack of interaction terms.

As a result of the limitations posed by empirical models, efforts have been made towards the development of models that are better suited to predicting the distribution of sediment loss and runoff spatially on an individual storm basis as well as estimating total soil loss. Further improvements in erosion modelling are more likely to arise from models that incorporate key hydrological and erosion processes rather than from small developments based on the USLE. However, as noted by Rose et al. (1988, cited in Dickinson et al., 1990), contemporary understanding of the processes surrounding the transport and detachment of soil remains inadequate and hence hampers efforts to obtain reliable input data and to validate models. The development of physically based models is still therefore at an early stage. Indeed Morgan (1995) reports that, in practical terms, estimates of erosion obtained from empirical models are often more reliable than those based on physical processes.

The Universal Soil Loss Equation combines catchment characteristics to estimate mean annual soil loss:

$$E = R.K.LS.C.P \qquad (24.6)$$

where E is the mean annual soil loss in tonnes per hectare (t ha^{-1}), R is the rainfall erosion factor, K is the soil-erodibility factor, LS is the slope factor, C is the crop-management factor, and P is the erosion-control factor.

METHODOLOGY AND RESULTS

Sediment Flux in British Catchments

In order to investigate links between catchment/land-use characteristics and sediment flux within British catchments, a database has been compiled for 30 sites. At each of these sites sediment-yield data are available from lake, or reservoir, sediment studies (Barlow, 1998). Mean sediment yields over a minimum time period of 50 years are available for each of these sites. In addition, 11 primary catchment and land-use characteristics, plus seven derived characteristics, have now been determined (Barlow,

Table 24.1 Parameters employed in regression analysis and their potential influence on sediment flux

Parameter	Relationship to sediment flux
Mean annual precipitation	Soil loss closely related to rainfall through (i) detaching power when raindrops strike surface; (ii) rainfall contribution to runoff
Maximum mean monthly precipitation	Employed to determine p^2/P
p^2/P	Indicates concentration of rainfall in one month, measure of rainfall intensity
Lake perimeter	Significance of lake bank erosion
Catchment area	Area of potential erosion
Log catchment area	Area of potential erosion. The log accounts for the effect of storage in larger catchments which results in a reduction in sediment yield with increasing area
Lake area	Area of sediment deposition
Catchment area: lake area ratio	Frequently plotted against sediment yield in the literature
River lengths	Indicate significance of river bank erosion
Lake altitude	Influence on rainfall and vegetation
Mean catchment altitude	Influence on rainfall and vegetation. Perhaps related to slope gradient/catchment area?
Soil erodibility	Resistance of soil to (i) detachment and (ii) transport
Vegetation	Soil protection offered by vegetation cover
Slope gradient	Velocity of surface runoff
Length of slope	Volume of surface runoff
USLE sediment yield	Surface erosion, assuming no sediment storage
LS	USLE combined slope length/gradient factor
R	USLE rainfall erosivity factor

Table 24.2 Land-use and catchment characteristics determined for 30 catchments

Lake	National grid reference	Mean annual precipitation (mm)	Max. mean monthly precipitation (mm)	Lake perimeter (km)	Catchment area (km²)	Lake area (km²)	Stream length (km)	Slopes (%)	Lake altitude (m)	Average catchment altitude (m)	Catchment soil erosion susceptibility	Vegetation
North Esk Reservoir	NT155582	1077	116	1.8	7	0.10	16.0	14.7	340	462	2.12	0.010
Semer Water	SD918874	1375	161	2.4	43.6	0.26	96.6	7.5	248	532	27.05	0.012
Gormire	SE 505833	825	82	1.0	0.3	0.07	0.0	22.5	160	228	0.09	0.005
Glenfarg	NO16110	969	108	4.0	23.5	0.41	11.6	9.0	497	572	3.46	0.563
Loe Pool	SW648250	1032	125	6.0	55	0.44	47.6	4.1	5	99	19.86	0.394
R. Loch of Glenhead	NX450805	2360	267	1.2	1	0.125	1.6	19.6	300	370	0.22	0.010
Loch Valley	NX445817	2360	267	4.0	1.86	0.501	1.6	18.1	330	385	0.34	0.010
Loch Enoch	NX445851	2360	267	4.8	1.86	0.500	1.8	12.1	500	545	0.44	0.010
Merevale	SP300970	639	63	1.6	1.95	0.065	4.4	5.2	110	150	0.82	0.007
Llyn Geirionydd	SH605763	2555	329	2.8	3.90	0.26	6.0	10.4	190	479	1.71	0.007
Llyn Goddionduon	SH753586	2555	329	1.2	0.25	0.062	0.4	17.1	244	367	0.09	0.005
Seeswood	SP327905	639	63	1.6	2.21	0.067	4.0	2.0	125	145	0.93	0.379
Old Mill Reservoir	SX850522	1090	131	0.4	1.58	0.019	2.4	22.1	45	160	0.62	0.387
Kelly Reservoir	NS223685	1767	200	1.0	3.40	0.054	12.0	5.2	200	262	0.81	0.010
Llyn Peris	SH570620	2330	310	6.4	38	0.500	70.0	22.0	100	594	11.51	0.010
Lambieltham	NO502134	738	72	0.5	2.29	0.012	3.6	1.5	102	127	0.52	0.550
Harperleas	NO212053	949	93	1.8	3.44	0.162	6.8	9.8	259	360	0.52	0.010
Drumain	NO223043	949	93	0.4	1.53	0.020	2.6	4.4	231	278	0.24	0.010
Cullaloe	NT188875	796	82	3.6	4.13	0.162	4.0	6.8	89	144	0.85	0.552
Hornsea Mere	TA190447	652	66	2.8	16.70	1.200	12.0	2.4	0	13	8.61	0.384
Broomhead	SK260960	980	104	3.6	21.96	0.485	65.6	10.3	180	419	12.24	0.011
Chew	SE040020	1604	168	2.2	2.92	0.30	16.4	5.2	490	522	2.620	0.012
Deanhead	SE040415	1357	151	1.2	2.00	0.068	10.0	11.0	305	410	1.681	0.012
Gorple Upper	SE920315	1478	164	1.6	3.80	0.219	4.0	9.5	350	411	3.581	0.010
Gorpley	SE910230	1512	166	1.2	2.80	0.072	9.2	16.5	260	354	2.264	0.011
Ingbirchworth	SE215060	1006	109	1.6	7.72	0.217	4.4	5.3	260	308	3.301	0.017
Kinder	SK055883	1175	119	2.6	8.95	0.300	23.6	18.1	280	517	5.969	0.012
Mixenden	SE060290	1087	116	0.4	0.77	0.092	0.4	6.3	260	311	0.407	0.012
Snailsden	SE135040	1542	177	1.2	0.84	0.040	4.0	5.4	420	452	0.733	0.010
Widdop	SD930330	1326	144	2.4	8.90	0.039	4.4	15.2	320	408	7.399	0.010

1998). Table 24.1 summarises these 18 parameters and their potential influence on sediment flux. As shown in Table 24.2, the 30 catchments are characterised by a very broad range of land uses, soil types, altitudes, stream lengths and lake and catchment areas and thus are taken to constitute a representative cross-section of British sites.

Multi-core Studies of Sediment Yield

All 30 sites used in our study have been subjected to multiple-core studies or reservoir surveys. They provide records of sediment yield over long time periods, generally over at least the past 100 years and often over thousands of years. Using average sediment yields over centennial time periods eliminates the short-term flux variability encountered in stream-monitoring estimates of sediment flux. At Loe Pool (O'Sullivan *et al.*, 1982) exceptionally high sediment yields associated with intensive mining activity in the catchment in the period 1860–1938 are reported. For this one site the sediment yield used is for the shorter period 1938–1981, when agriculture was the dominant catchment activity.

Of the 30 sediment-yield estimates, 14 had been determined from reservoir re-surveys and 16 from multiple lake-sediment cores. The procedures used in the calculation of sediment yield at lake and reservoir sites are set out in Table 24.3. For lake or reservoir sites where direct measurements of carbonate or biogenic silica content were not available, an average value, obtained from measurements made at other sites, has been applied to calculate the inorganic sediment flux. The flux and yield estimates from the multi-core studies for the 30 catchments are tabulated in Table 24.4. The mean yield is $45\,t\,km^{-2}\,year^{-1}$, with a range of $1.8–260\,t\,km^{-2}\,year^{-1}$.

Catchment Characteristics

Eleven main characteristics have been determined for each of the 30 catchments. These are catchment area, river length, catchment slope, altitude (lake and catchment),

Table 24.3 Twelve-step procedure for determining sediment yield using multi-core methods

Step	Procedure
1	Collect multiple cores
2	Correlate cores
3	Determine dry weights and dry densities
4	Establish a chronology
5	Determine the mean dry mass accumulation rate
6	Multiply (5) by the area of active sedimentation
7	Divide the total mass of material by the number of years in each time period to give a combined influx of allochthonous and authochthonous material
8	Determine the average organic content
9	Determine the carbonate content
10	Determine the biogenic silica (diatom) component
11	Subtract (8), (9) and (10) from the bulk influx. The result is the influx of minerogenic material per year
12	Convert the influx into yield by dividing by the catchment area

soil type (susceptibility to erosion), land use/vegetation, precipitation (mean and maximum), lake perimeter and lake area. Morgan (1995) gives a very comprehensive discussion of such catchment and land-use characteristics and describes alternative approaches to their estimation, while Barlow (1998) sets out in detail the methods used here to determine each of the 12 catchment characteristics.

Table 24.4   Sediment flux and catchment yields within the 30 catchments studied

Lake	Source of flux data[a]	Lake sediment yield (t km^{-2} year^{-1})	Lake sediment flux (t year^{-1})	USLE slope length factor	USLE sediment erosion (t km^{-2} year^{-1})	USLE erosion × catchment area (t year^{-1})	Sediment delivery ratio (SDR)	USLE erosion × SDR (t km^{-2} year^{-1})	USLE sediment flux (t year^{-1})
North Esk Reservoir	A	23.4	163.8	3.23	99.9	689	0.24	24.4	168.2
Semer Water	B	13.8	730.0	4.85	373.3	16182	0.17	63.1	2734.7
Gormire	B	44.6	8.7	2.14	94.6	27	0.44	42	11.8
Glenfarg	C	31.3	735.6	3.00	1348.2	31130	0.19	257.5	5945.9
Loe Pool	D	12.0	660.0	4.67	758.0	41358	0.16	122.8	6699.9
R. Loch of Glenhead	E	33.3	31.7	2.44	139.1	115	0.36	50.6	41.8
Loch Valley	E	66.1	122.7	2.19	118.4	161	0.32	37.6	51.0
Loch Enoch	E	89.4	166.3	2.42	117.4	160	0.32	37.3	50.8
Merevale	F	8.5	16.5	3.08	79.1	149	0.32	24.9	47.0
Llyn Geirionydd	G	12.5	48.7	3.10	137.1	499	0.27	37.6	136.7
Llyn Goddion duon	H	29.5	7.4	2.02	110.5	21	0.48	52.5	9.9
Seeswood	I	11.2	24.8	3.26	1071	2295	0.31	328.8	704.6
Old Mill Reservoir	J	69.0	109.0	2.61	1498	2338	0.33	492.8	769.3
Kelly Reservoir	K	36.9	125.5	3.41	71.8	240	0.28	20.3	67.8
Llyn Peris	L	10.6	402.8	4.12	250.9	9407	0.17	43.7	1636.8
Lambieltham	M	1.8	4.1	3.31	2721.5	6200	0.31	830.1	1890.9
Harperleas	M	11.5	39.6	3.17	42.0	138	0.28	11.8	38.6
Drumain	M	3.3	5.0	3.16	34.7	52	0.33	11.5	17.3
Cullaloe	M	26.2	108.2	2.82	2361.9	9372	0.27	640.1	2539.8
Hornsea Mere	B	42.0	770.0	2.61	4035.8	60537	0.21	827.3	12410.1
Broomhead	N	31.8	698.3	4.46	279.2	5996	0.19	54.2	1163.2
Chew	N	78.5	229.2	2.84	294.8	772	0.29	85.8	224.7
Deanhead	N	33.7	67.4	3.14	352.4	681	0.31	110.3	213.1
Gorple Upper	N	27.6	104.9	2.91	313.3	1122	0.28	86.5	309.7
Gorpley	N	129.1	361.5	2.66	355.1	969	0.29	104.0	283.8
Ingbirchworth	N	79.8	616.1	3.08	206.1	1546	0.24	49.3	369.5
Kinder	N	50.9	455.6	3.46	392.6	3396	0.23	91.1	787.9
Mixenden	N	9.5	7.3	2.68	180.3	122	0.38	68.3	46.3
Snailsden	N	260.2	218.6	2.83	242.1	194	0.37	90.3	72.3
Widdop	N	81.1	721.8	2.78	328.7	2912	0.23	76.6	678.5

[a] Source of sediment yield/flux estimates: A, Lovell *et al.* (1973); B, Barlow (1998); C, McManus and Duck (1985); D, O'Sullivan *et al.* (1982); E, Flower *et al.* (1987); F, Foster *et al.* (1985); G, Snowball and Thompson (1992) and Dearing (1992); H, Bloemendal (1982); I, Foster *et al.* (1986); J, Foster and Walling (1994); K, Ledger *et al.* (1980); L, Dearing *et al.* (1981); M, Duck and McManus (1987); N, Butcher *et al.* (1993).

MODELLING YIELD AND FLUX IN BRITISH CATCHMENTS

Finding a relationship between sediment deposition in a lake and catchment erosion is not straightforward. Empirical relationships found by earlier workers between (i) catchment area and lake deposition, or between (ii) sediment yield and the ratio catchment area: lake area are not entirely satisfactory. Hence a more quantitative approach is sought, employing a statistical approach that uses catchment characteristics to improve, or modify in some way, the empirical relationships of earlier workers such as Brune (1950), Fournier (1960) and Dearing and Foster (1993).

USLE and sediment flux in British catchments

The results of our USLE calculations for the British catchments are set out in the final column of Table 24.4 for a standard slope, 22 m long. The average estimated soil loss at our 30 sites is $310 \, t \, km^{-2} \, year^{-1}$. Figure 24.2 illustrates predicted sediment flux for each of the 30 catchments studied using (i) the Universal Soil Loss Equation alone and (ii) the Universal Soil Loss Equation estimate multiplied by a sediment-delivery ratio derived from the ASCE (1975) empirical relationship of equation (24.4). Figure 24.2 also compares sediment fluxes predicted using the USLE with our flux estimates based on the multi-core studies.

The USLE estimates of sediment flux are, with two exceptions (Snailsden and Loch Enoch), considerably greater than the multi-core flux estimates. The differences

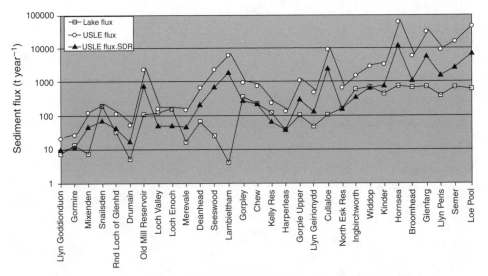

Figure 24.2 A comparison of sediment flux to 30 British lakes/reservoirs using (i) the USLE alone, (ii) the USLE multiplied by the US Soil Conservation Service sediment delivery ratio, and (iii) the lake/reservoir estimates of sediment flux. The lake sediment flux estimates are generally lower than those obtained using the USLE multiplied by the US Soil Conservation Service sediment delivery ratio. The lake sediment fluxes have been calculated on the assumption of 100 percent lake/reservoir trap efficiency

between the USLE and multi-core estimates of sediment flux can be largely attributed to the effect of sediment storage in catchments. Modifying the USLE sediment flux estimates by a sediment-delivery term makes the USLE flux estimates more comparable with those determined from the multi-core studies. Nevertheless at some sites, particularly the seven largest catchments in the database, there is considerable disagreement between the yield estimates obtained from lake sediments and those predicted using the Universal Soil Loss Equation. The discrepancy between USLE predicted sediment flux and lake sediment flux at Lambieltham is particularly marked. Duck and McManus (1987) suggest that the low sediment yield from the Lambieltham catchment results from reservoir management practices. A bypass channel has prevented water and sediment reaching the reservoir.

Regression Models of Sediment Flux

In an attempt to predict sediment flux into British lakes and reservoirs more accurately regression techniques have been employed to construct simple empirical models relating sediment flux to catchment and land-use characteristics. Table 24.5 lists the correlation coefficients between flux and yield with the 12 catchment characteristics. As would be expected, flux and catchment area have a significant positive correlation. Figure 24.3 demonstrates this relationship between sediment flux and catchment. However, the relationship

$$\text{Flux} = 18.5(\text{Catchment area}) \qquad (24.7)$$

is rather weak, having an R^2 of only 0.52 and so is only a poor model of flux. In equation (24.7) the coefficient is the average yield, namely 18.5 t km^{-2} year $^{-1}$. In Table 24.5 we can also see that yield correlates weakly with altitude (both lake and catchment). Yield is also seen to be inversely correlated with catchment area for our 30 catchments, as found by Dearing and Foster (1993) and also by many earlier studies that have reported decreases in sediment yield with increasing catchment area.

In order to try to improve the flux model, stepwise regression analysis of all 12 land-use and catchment characteristics, plus the four parameters derived from them for inclusion in the Universal Soil Loss Equation, has been performed. Stepwise regression analysis uses the F-statistic to determine whether any particular variable should be included in the equation. By adopting the usual F-value of 4, this variable selection form of regression analysis generated the following equation:

Table 24.5 Correlation coefficients between sediment flux and the 12 catchment and land-use characteristics of the 30 catchments studied

	Catchment characteristics											
	1	2	3	4	5	6	7	8	9	10	11	12
Sediment flux	−0.19	−0.16	0.50	0.71	0.73	0.57	0.58	−0.09	−0.01	0.17	0.16	0.20
Sediment yield	0.17	0.13	−0.10	−0.24	−0.17	−0.07	−0.21	0.09	0.42	0.21	−0.20	0.30

1 Mean annual ppt; 2 Max. mean monthly; 3 Lake perimeter; 4 Catchment area; 5 Log catchment area; 6 Lake area; 7 Stream length; 8 Slope steepness; 9 Lake altitude; 10 Catchment altitude; 11 Vegetation; 12 Soil erodibility

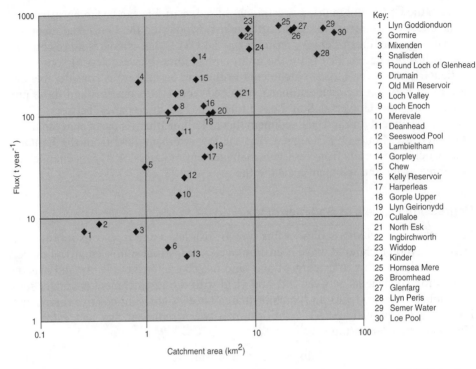

Figure 24.3 The relationship between sediment flux and catchment area for 30 British sites. An increase in sediment flux with increasing catchment area can be observed. However, the relationship has an R^2 value of only 52 percent and is thus relatively weak

$$\text{Flux} = -54.6 + 11.7\ \text{Catchment area} + 347\ \text{Lake area} + 264\ \text{Soil erodibility} \quad (24.8)$$

The relationship has an improved R^2 of 0.64 and a correlation coefficient of 0.8. In a further attempt to improve on the simple relationship between catchment area and sediment flux of equation (24.5) (Figure 24.3), and to account for the progressive increase in sediment storage as the catchment area increases, we have regressed the log of catchment area and sediment flux. The relationship between the log of catchment area and sediment flux (Table 24.6) is significantly stronger than that observed between catchment area and sediment flux, with an R^2 of 0.66 and a correlation coefficient of 0.81. It takes the following form:

$$\text{Flux} = 42.2 + 378\log(\text{Catchment area}) \quad (24.9)$$

Table 24.6 Summary of the R^2 values and correlation coefficients obtained using various combinations of catchment characteristics to determine sediment flux

R^2	Correlation coefficient	Variables
0.75	0.87	Log catchment area, soil erodibility factor, USLE erosion rate
0.66	0.81	Log catchment area
0.64	0.8	Catchment area, lake area, soil erodibility factor
0.52	0.71	Catchment area

Indeed when the log of catchment area is added to the stepwise regression analysis, the log of catchment area is the only variable to be selected.

If the F-value is reduced to 3 then fewer parameters are removed from the full regression model during variable selection and the following regression equation is produced:

$$\text{Flux} = 67.0 + 298 \log(\text{Catchment area}) + 243 \,(\text{Soil erodibility}) \\ + \, 0.0057 \,(\text{USLE erosion rate}) \tag{24.10}$$

This relationship has an R^2 of 0.75 and a correlation coefficient of 0.87 (Table 24.6). However, it must be remembered that such a low F-value can lead to over-fitting.

Table 24.6 summarises the R^2 and correlation coefficients that result from employing various combinations of catchment characteristics to determine sediment flux. All the correlation coefficients and relationships of Table 24.6 are highly significant with p-values below 0.01. In selecting the most appropriate of these competing regression equations to estimate sediment fluxes, a balance between a strong correlation and a simple empirical model should be sought. With a larger data set the formal technique of cross-validation could be used to assess the number of variables to include in the model. The simple regression relationship employing the log of catchment area alone is seen to provide a reasonable account of sediment flux.

Power-law Relationships and Flux

An improvement on using the log of catchment area could be the use of power-law relationships of the type used in equation (24.4), e.g. flux = yield × area^n. The simplest power-law relationship found for the British sites is plotted in Figure 24.4. More involved power-law relationships were explored to try to improve on the fit of Figure 24.4. However, the results were very similar to those of the regression work. The power-law models consistently selected catchment area as the main predictor, with slope as an additional parameter. Once again, neither climatic factors nor the USLE yield estimates were found to be significant variables.

In summary, simple regression models involving catchment area, lake area and possibly soil erodibility can explain up to 66 percent of the variance of the flux at the 30 British sites analysed. By far the most dominant of these variables is catchment area. This parameter alone explains over 50 percent of the variance. Following Brune (1950), Boyce (1975), Walling (1983, 1988) and many others, we attribute the strong relationship between catchment area and flux to the role of sediment delivery in modulating sediment fluxes within catchments.

DISCUSSION

The Universal Soil Loss Equation

The Universal Soil Loss Equation (USLE) is the most widely used soil-erosion model available, and remains one of the simplest to use. However, this study has illustrated that estimates of sediment erosion determined for British catchments using the USLE

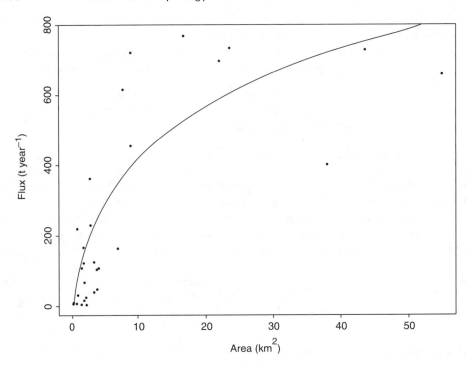

Figure 24.4 Power-law relationship between sediment flux and catchment area for the 30 British sites of Figure 24.3

tend to be considerably greater than those from the multi-core studies. The USLE only estimates surface-sheet erosion; it does not include gully or channel erosion and consequently it might be expected to underestimate erosion. One of the major limitations of the USLE is that it was designed for small plot studies rather than lake catchments. Consequently the slope-length factor was not designed to accommodate the downhill slope-lengths observed in catchments. Indeed Hickey *et al.* (1994) state that 'the largest problem in using the, USLE' has been the calculation of the cumulative downhill slope-length factor.

In other parts of the world a variety of studies employing the USLE to estimate soil erosion have similarly found that the USLE-determined erosion rates are higher than those obtained using other techniques. For example, Busacca *et al.* (1993) compared estimates of erosion in an agricultural watershed in Idaho, USA, using the Revised Universal Soil Loss Equation (RUSLE) with estimates determined using ^{137}Cs as a sediment tracer. They found that the RUSLE significantly overestimated erosion. Similarly, Harden (1993), working on an agricultural drainage basin in Andean Ecuador, noted that upland soil-erosion estimates determined using the USLE were consistently higher than estimates extrapolated from rainfall-stimulation experiments. Kusumandari and Mitchell (1997) compared rates of erosion determined using the USLE with those determined using the Agricultural Non-Point Source Pollution (AGNPS) model in a forested basin in West Java, Indonesia. The rate of erosion determined using the AGNPS model was found to be about half that predicted by the USLE. Taken together, all these results suggest that rates of erosion predicted using the USLE in UK catchments may be too high. However, from our compilations it is difficult to ascertain whether such high USLE flux estimates result from (i) overestimates of sediment erosion obtained using the USLE, or from (ii) underestimates of the sediment-delivery ratio.

A Sediment-Delivery Model

Sediment delivery remains an extremely complex and limiting factor in relating lake-sediment fluxes to erosion rates in catchments. Whilst lake-sediment flux estimates are an ideal way of determining the mass of material reaching a given point, insufficient data on rates of erosion in British catchments prevent the determination of more accurate estimates of sediment delivery. Consequently any attempts to develop our understanding of the factors that influence the delivery ratio, and quantify the importance of different factors, are limited. Simple sediment-delivery models, which can be more readily and easily applied to catchments, are very desirable. Such models enable the identification of catchments where further, more detailed, studies may be warranted in order to test hypotheses relating, for example, sediment delivery to slope-lengths or gradients. Our models indicate that the log of catchment area is more strongly related to sediment flux than catchment area alone. The soil type within a catchment also has a significant impact on its tendency to erode on Holocene time-scales.

Sediment-Delivery Ratios in the Larger Catchments

The quantitative nature of our physically based models of Table 24.6 can be used to highlight an important point that has not been elaborated in previous studies. This concerns the larger British catchments such as that of Semer Water. Our models can be used to estimate theoretical volumes of material stored in sediment sinks by transforming them into simple mass-balance relationships (using equations (24.3) and (24.4)). At Semer Water, for example, by combining our estimates of sediment-delivery ratio with the volume of sediment in the lake we estimate the volume of sediment stored within the catchment to be about 50 million cubic metres. This sediment volume is equivalent to a mean sediment thickness over the whole of the catchment of 1.2 m. However, parts of the Semer Water catchment are characterised by slopes of steep gradient and although there are some areas where sediment accumulation may occur, it seems highly improbable that these are sufficient to result in a mean sediment thickness of >1 m over the entire catchment. Thus, either significant quantities of sediment are being lost through the lake outflow, or the sediment-delivery ratios underestimate the proportion of sediment entering the lake. Similarly, at Gormire, we estimate from our models that over one million cubic metres of sediment should remain in the catchment. However, at Gormire, steep slopes drain almost exclusively straight into the lake and thus there is again virtually no scope for sediment storage. Furthermore there is no outflow and so no scope for sediment loss. Thus we are left with the paradoxical situation that while the flux of sediments in

British catchments, and the variation of sediment-delivery ratio with catchment area, are in excellent agreement with other northern hemisphere studies, the volumes of sediment stored in the larger catchments appear to be too low to account for 'missing' sediment.

CONCLUSIONS

(1) In the United Kingdom monitored records of sediment yield covering time periods of more than a few years are rare.

(2) Sediment-flux estimates, based on multi-core studies, have here been assembled for 30 British lake catchments.

(3) The British sediment fluxes are similar to those found for other places in the temperate zone.

(4) Universal Soil Loss Equation estimates of sediment yield for the British catchments are higher than those of the multi-core studies.

(5) A strong relationship ($R^2 = 0.75$) has been found between, on the one hand, sediment flux and, on the other, catchment area and soil-erosion susceptibility for the British catchments.

(6) Multi-core studies of sediment accumulation in lakes/reservoirs confirm the view that small catchments (< 1 km^2) provide the best estimates of sediment yield (i.e. soil loss) because sediment-delivery ratios are close to one. Hence they provide a lower bound on sediment flux in Britain.

(7) The apparently lower sediment yields of the larger catchments (> 10 km^2) can be reconciled with those of the smaller catchments by appealing to the relationship between catchment area and sediment-delivery ratio found in many parts of the world.

(8) In mass-flux terms, a major imbalance is found, with millions of cubic metres of sediment apparently missing from large upland catchments.

(9) Sediment-delivery ratio remains one of the most poorly understood and poorly quantified concepts in studies of sediment erosion.

ACKNOWLEDGEMENTS

DNB was supported by a tied NERC studentship. The database development and modelling work developed out of the NERC-funded HULAP (HUmber LAkes Project) coordinated and led by F. Oldfield as part of LOEPS. The fieldwork at Semer Water and Gormire formed an integral part of HULAP.

REFERENCES

ASCE (American Society of Civil Engineering) 1975. *Sedimentation Engineering*. Manuals and Reports on Engineering Practices No. 54, American Society of Civil Engineering, New York.
Barlow, D. 1998. A lake sediment study of particulate flux in the Humber catchment. Unpublished PhD Thesis, University of Edinburgh.

Bennett, H. H. 1939. *Soil Conservation*. McGraw-Hill, New York.

Bloemendal, J. 1982. The quantification of rates of total sediment influx to Llyn Goddionduon, Gwynedd. Unpublished PhD Thesis, University of Liverpool.

Boyce, R. C. 1975. *Sediment Routing with Sediment Delivery Ratios, in Present and Prospective Technology for Predicting Sediment Yields and Sources*. USDA Agriculture Research Service, Publication ARS-S40, 61–65.

Brown, C. B. 1948. Perspectives on sedimentation-purpose of conference. In *Proceedings of the 1st Federal Interagency Sedimentation Conference*. US Bureau of Reclamation, Washington, D.C.

Brune, G. M. 1950. The dynamic concept of sediment sources. *Transactions of the American Geophysical Union*, **31**(4), 587–594.

Busacca, A. J., Cook, C. A. and Mulla, D. J. 1993. Comparing landscape-scale estimation of soil-erosion in the Palouse using Cs-137 and RUSLE. *Journal of Soil and Water Conservation*, **48**(4), 361–367.

Butcher, D. P., Labadz, J. C., Potter, A. W. R. and White, P. 1993. Reservoir sedimentation rates in the southern Pennine Region, UK. In McManus, J. and Duck, R. W. (eds) *Geomorphology and Sedimentology of Lakes and Reservoirs*. John Wiley, Chichester, 73–92.

Dearing, J. A. 1992. Sediment yields and sources in a Welsh upland lake-catchment during the past 800 years. *Earth Surface Processes and Landforms*, **17**, 1–22.

Dearing, J. A. and Foster, I. D. L. 1993. Lake sediments and geomorphological processes: some thoughts. In McManus, J. and Duck, R. W. (eds) *Geomorphology and Sedimentology of Lakes and Reservoirs*. John Wiley, Chichester, 5–14.

Dearing, J. A., Elner, J. K. and Happey-Wood, C. M. 1981. Recent sediment flux and erosional processes in a Welsh upland lake-catchment based on magnetic susceptibility measurements. *Quaternary Research*, **16**, 356–372.

Dickinson, W. T., Wall, G. L. and Rudra, R. P. 1990. Model building for predicting and managing soil erosion and transport. In Boardman, J., Foster, I. D. L. and Dearing, J. A. (eds) *Soil Erosion on Agricultural Land*. John Wiley, Chichester, 415–428.

Duck, R. W. and McManus, J. 1987. Sediment yields in lowland Scotland derived from reservoir surveys. *Transactions of the Royal Society, Edinburgh; Earth Sciences*, **78**, 369–377.

Flaxman, E. M. and Hobba, R. L. 1955. Some factors affecting rates of sedimentation in the Columbia River basin. *Transactions of the American Geophysical Union*, **38**(2), 293–303.

Flower, R. J., Battarbee, R. W. and Appleby, P. G. 1987. The recent palaeolimnology of six acid lakes in Galloway, south-west Scotland. Diatom analysis, pH trends, and the role of afforestation. *Journal of Ecology*, **75**, 797–824.

Foster, G. R. 1982. Modeling the erosion process. In Hann, C. T., Johnson, H. P. and Brakensiek, D. L. (eds) *Hydrologic Modeling of Small Watersheds*. American Society of Agricultural Engineers, St Joseph, Michigan, 297–382.

Foster, G. R. 1990. Process-Based Modelling of Soil Erosion by water on Agricultural Land. In Boardman, J., Foster, I. D. L. and Dearing, J. A. (eds) *Soil Erosion on Agricultural Land*. John Wiley, Chichester, 429–445.

Foster, I. D. L. and Walling, D. E. 1994. Using reservoir deposits to reconstruct changing sediment yields and sources in the catchment of the Old Mill Reservoir, South Devon, UK, over the past 50 years. *Hydrological Sciences – Journal des Sciences Hydrologiques*, **39**(4), 347–368.

Foster, I. D. L., Dearing, J. A., Simpson, A. D. and Appleby, P. G. 1985. Lake catchment based studies of erosion and denudation in the Merevale catchment, Warwickshire, UK. *Earth Surface Processes and Landforms*, **10**, 45–68.

Foster, I. D. L., Dearing, J. A. and Appleby, P. G. (1986). Historical trends in catchment sediment yields: a case study in reconstruction from lake-sediment records in Warwickshire, UK. *Hydrological Sciences – Journal des Sciences Hydrologiques*, **31**(3), 427–443.

Fournier, F. 1960. *Climat et Érosion: La Relation Entre l'Érosion du Sol par l'Eau et les Précipitations Atmosphériques*. Presses Universitaires France, Paris.

Haan, C. T., Barfield, B. J. and Hayes, J. C. 1994. *Design Hydrology and Sedimentology for Small Catchments*. Academic Press, London.

Harden, C. P. 1993. Upland erosion and sediment yield in a large Andean drainage basin. *Physical Geography*, **14**(3), 254–271.

Hickey, R., Smith, A. and Jankowski, P. 1994. Slope length calculations from a DEM within ARC/INFO grid. *Computers, Environment and Urban Systems*, **18**(5), 365–380.

Jackson, W. L., Gebharelt, K. and Van Haveven, B. P. 1986. Use of the Modified Universal Soil Loss Equation for average annual sediment yield estimates on small rangeland drainage basins. In *Drainage basin sediment delivery*, (Proceedings of the International Symposium on Sediment Delivery, Albevquerque, New Mexico, August 1986). IAHS Publication **159**, IAHS Press, Wallingford. 413–422.

Kusumandari, A. and Mitchell, B. 1997. Soil erosion and sediment yield in forest and agroforestry areas in West Java, Indonesia. *Journal of Soil and Water Conservation*, **52**(5), 376–380.

Langbein, W. B. and Schumm, S. A. 1958. Yield of sediment in relation to mean annual precipitation. *Transactions of the American Geophysical Union*, **39**, 1076–1084.

Ledger, D. C., Lovell, J. P. B. and Cuttle, S. P. 1980. Rate of sedimentation in Kelly Reservoir, Strathclyde. *Scottish Journal of Geology*, **16**, 281–285.

Lovell, J. P. B., Ledger, D. C., Davies, I. M. and Tipper, J. C. 1973. Rate of sedimentation in the North Esk Reservoir, Midlothian. *Scottish Journal of Geology*, **9**(1), 57–61.

McManus, J. and Duck, R. W. 1985. Sediment yield estimated from reservoir siltation in the Ochil Hills, Scotland. *Earth Surface Processes and Landforms*, **10**, 193–200.

Meade, R. H. 1982. Sources, sinks and storage of river sediment in the Atlantic Drainage of the United States. *Journal of Geology*, **90**, 235–252.

Moore, R. J. and Newson, M. D. 1986. Production, storage and output of coarse upland sediments: natural and artificial influences as revealed by research catchment studies. *Journal of the Geological Society, London*, **143**, 921–926.

Morgan, R. P. C. 1995. *Soil Erosion and Conservation* (2nd edition). Longman Scientific and Technical, Harlow.

Myers, N. 1993. *Gaia: An Atlas of Planet Management*. Anchor and Doubleday, New York.

O'Sullivan, P. E., Coard, M. A. and Pickering, D. A. 1982. The use of laminated sediments in the estimation and calibration of erosion rates. In *Recent Developments in the Explanation and Prediction of Erosion and Sediment Yield* (Proceedings of the Exeter Symposium, July 1982). IAHS Publication **137**, IAHS Press, Wallingford. 385–396.

Pimental, D. 1976. Land degradation: effects on food and energy resources. *Science*, **194**, 149–155.

Pimentel, D., Harvey, C., Resosudarmo, P., Sinclair, K., Kurz, D., McNair, M., Crist, S., Shpritz, L., Fitton, L., Saffouri, R. and Blair, R. 1995. Environmental and economic costs of soil erosion and conservation benefits. *Science*, **267**, 1117–1123.

Schumm, S. A. 1963. *The Disparity between Present Rates of Denudation and Orogeny*. US Geological Survey Professional Paper 454-H, 1–13.

Snowball, I. and Thompson, R. 1992. A mineral magnetic study of Holocene sediment yields and deposition patterns in the Llyn Geirionydd catchment, north Wales. *The Holocene*, **2**(3), 238–248.

Speth, J. G. 1994. Towards an effective and operational international convention on desertification Internation Negotiating Committee, International Convention on Desertification, United Nations, New York. United Nations, New York.

Strakhov, N. M. 1967. *Principles of Lithogenesis*, Vol. 1. Consultants Bureau, New York.

Trimble, S. W. 1981. Changes in sediment storage in the Coon Creek Basin, Driftless area, Wisconsin, 1853–1975. *Science*, **214**, 181–183.

Vanoni, V. A. (ed.) 1975. *Sediment Engineering*. American Society of Civil Engineering Manuals and Reports on Engineering Practices No. 54. ASCE, New York.

Walling, D. E. 1983. The sediment delivery problem. *Journal of Hydrology*, **65**, 209–237.

Walling, D. E. 1988. Erosion and sediment yield research – some recent perspectives. *Journal of Hydrology*, **100**, 113–141.

Wischmeier, W. H. and Smith, D. D. 1978. *Predicting Rainfall Erosion Losses*. USDA Agricultural Research Service Handbook 537.

25 Sediment Fingerprinting as a Tool for Interpreting Long-term River Activity: The Voidomatis Basin, North-west Greece

R. H. B. HAMLIN,[1] J. C. WOODWARD,[1] S. BLACK[2] and M. G. MACKLIN[3]
[1]School of Geography, University of Leeds, UK
[2]PRIS, University of Reading, UK
[3]Institute of Geography and Earth Sciences, University of Wales, Aberystwyth, UK

INTRODUCTION

Information on sediment provenance is a fundamental requirement for many geological and geomorphological investigations. Provenance data have been used to calibrate models of tectonic uplift and displacement, to investigate the formation of sedimentary sequences and for large-scale palaeogeographic reconstructions (Haughton *et al.*, 1991). Within process geomorphology provenance information has proved to be of considerable value in elucidating catchment sediment dynamics, providing information on temporal and spatial variations in sediment sources (e.g. Yu and Oldfield, 1989; Walling and Woodward, 1995; Collins *et al.*, 1998). These studies have focused mainly on either tracing the source of contemporary suspended sediment (e.g. Walling *et al.*, 1979; Walling and Woodward, 1992, 1995) or on fine-grained sediments deposited in lake basins and estuaries (e.g. Yu and Oldfield, 1989, 1993; Dearing, 1992; Hutchinson, 1995). They have demonstrated that the source of sediment from catchment areas of contrasting land use and/or lithology can be established (e.g. Walling and Woodward, 1995).

Many provenance studies have employed qualitative or semi-quantitative methods to identify sediment source areas or types (e.g. Wood, 1978; Walling *et al.*, 1979; Oldfield *et al.*, 1985; Woodward *et al.*, 1992). More recently, however, techniques for obtaining more robust, quantitative constraints of fine sediment source have been developed. These 'quantitative fingerprinting' approaches (Peart and Walling, 1986) assume that the physical and chemical properties of the sediment of interest directly reflect the relative contributions from catchment source areas. Certain physical and chemical properties are first selected to differentiate the source area materials (soils, sediments, bedrock, etc.), and then a multivariate mixing model is used to determine the relative contribution of each individual source type (e.g. Walling *et al.*, 1993; Yu

Tracers in Geomorphology. Edited by Ian D. L. Foster. © 2000 John Wiley & Sons Ltd.

and Oldfield, 1989). A wide variety of approaches have been employed to establish sediment provenance, including geochemistry (Passmore and Macklin, 1994; Collins *et al.*, 1997a,b, 1998), mineral magnetics (Walling *et al.*, 1979; Yu and Oldfield, 1989, 1993; Dearing, 1992; Hutchinson, 1995; Foster *et al.*, 1998), radionuclide concentrations (Walling and Woodward, 1992; Olley *et al.*, 1993; Hutchinson, 1995), SEM analysis (de Boer and Crosby, 1995), mineralogy (Wood, 1978; Woodward *et al.*, 1992), sediment colour (Grimshaw and Lewin, 1980) and particle-size distributions (Kurashige and Fusejima, 1997).

Whilst there has been some work on sourcing Holocene floodplain deposits (Woodward *et al.*, 1992; Passmore and Macklin, 1994; Collins *et al.*, 1997b; Schell *et al.*, see Chapter 26), the potential for *quantitative* provenancing in elucidating long-term changes (10^2–10^5 years) in fine sediment sources remains largely undocumented. Over these timescales, the flux and provenance of fine-grained fluvial sediment may be influenced by environmental controls such as climate variations, vegetation dynamics, land-use change or tectonic activity. Thus data on changes in Late Pleistocene and Holocene sediment provenance may provide valuable information on long-term river behaviour and catchment dynamics. This study applies a quantitative fingerprinting technique to Pleistocene and Holocene fluvial deposits in the Voidomatis basin, a steepland catchment in north-west Greece. Results are presented for both the chronology and provenance of the fine-grained matrix within coarse alluvial fills, and fine-grained palaeoflood slackwater deposits (SWDs). The chapter builds on previous work in the area (Woodward *et al.*, 1992) to explore the variation in sediment sources during and since the Late Pleistocene, and how this may be related to long-term environmental change.

STUDY AREA: PHYSICAL CHARACTERISTICS AND PREVIOUS RESEARCH

The Voidomatis River basin (384 km^2) drains the western side of the Pindus Mountains in Epirus, north-west Greece (Figure 25.1). The river is steep (average gradient 0.016), with elevations within the catchment ranging from over 2400 m along the watershed to *c.* 390 m on the Konitsa basin (Lewin *et al.*, 1991). River incision driven by long-term tectonic uplift has resulted in the development of deep bedrock gorges (Figures 25.1 and 25.2). The largest of these is the Upper Vikos Gorge which reaches depths of almost 1000 m (Bailey *et al.*, 1997). Immediately downstream the Voidomatis flows through the Lower Vikos Gorge and out into the Konitsa basin (Figure 25.1). In the Lower Vikos Gorge the Voidomatis is a 10–20 m wide, meandering gravel-bed river, with extensive bars and prominent riffles.

Rainfall is concentrated from October to May. Intense and prolonged storms are common, and daily rainfall totals of over 115 mm have been recorded. Mean annual precipitation ranges from 1079 mm over the Lower Vikos Gorge to 1702 mm in the particularly mountainous Tsepelovon area (Figure 25.1).

Alluvial fills in the lower part of the basin display two distinctive types of lithofacies. First, high units 8–15 m above the modern channel are quite frequently preserved, especially in wider gorge sections where there has been less reworking.

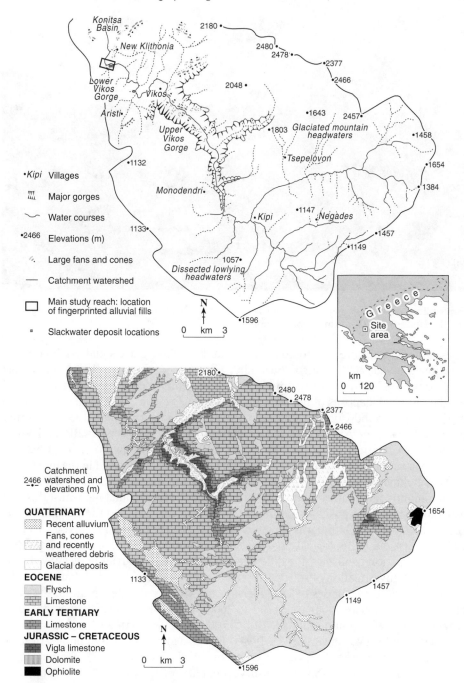

Figure 25.1   Maps of the Voidomatis catchment showing physical characteristics, study sites and geology (after IGME, 1968, 1970). (*Terra rossa* occurs as a thin soil on top of some of the exposed limestone areas. Dolomite-enriched flysch is present in the extreme east of the catchment)

Figure 25.2   The Voidomatis catchment with the Lower Vikos Gorge evident in the foreground and the Upper Vikos Gorge in the background

These alluvial units are composed of well-rounded coarse gravels and small boulders in a fine-grained sandy matrix. Secondly, in near-channel areas lower units exist approximately 2–4 m above the modern channel, composed of coarse sandy gravel overlain by thick deposits of bedded sands and silts. Previous work in the area (Lewin *et al.*, 1991) has subdivided the chronology of the older, coarse-grained sediments into the Late Pleistocene Aristi (*c.* 28 200 (±7000) to 24 300 (±2600) years BP) and Vikos (*c.* 24 300 (±2600) to 19 600 (±3000) years BP) units (Woodward *et al.*, 1992). In addition, a very old (>150 000 years BP) Kipi unit was also identified in the upper part of the catchment (Lewin *et al.*, 1991), but this is not discussed here. The younger, predominantly fine-grained sediments were termed the Klithi unit, and whilst this unit was shown to be composite in nature, its deposition was dated to the last 1000 years (Lewin *et al.*, 1991). Dating was achieved by means of ESR, TL and ^{14}C (Lewin *et al.*, 1991), whilst clast lithological analysis and semi-quantitative sourcing of the fine (<63 μm) fraction confirmed the differences between these alluvial units (Woodward *et al.*, 1992). Investigation of soil development on these units identified that there were at least two different phases of Late Pleistocene Aristi-type sedimentation, the more recent at *c.* 28 200 (±7000) to 24 300 (±2600) years BP, and an earlier phase of sedimentation with a minimum age of *c.* 85 000 years, estimated by means of a soil weathering index (Woodward *et al.*, 1994). The palaeoenvironmental implications of these results have been discussed in detail by Macklin *et al.* (1997). We have now extended that work by mapping and surveying parts of the valley floor in detail, and by applying recent developments in radiometric dating and fine sediment fingerprinting to the alluvial deposits of the Voidomatis basin.

In addition to the major valley-floor alluvial units, fine-grained palaeoflood slack-water deposits (SWDs) have been identified at three locations in the catchment

(Figure 25.1) (Lewin *et al.*, 1991). These sequences present an opportunity to invest-igate the occurrence and significance of high-magnitude floods over the Late Pleis-tocene, which has rarely been explored in the Mediterranean region. It is well known that extreme flood events can play an important role in valley-floor development (e.g. Baker, 1977; Gupta, 1983; Coxon *et al.*, 1996), especially in high threshold environ-ments. SWDs are typically fine-grained sediment sequences ($D_{50} < 2$ mm), resulting from the rapid deposition of suspended sediment during large floods in sheltered areas, where flow velocities suddenly decrease and reverse eddying predominates (Baker *et al.*, 1983). Much of the previous work on such sedimentary features has been in the USA (Kochel and Baker, 1982, 1988; Ely and Baker, 1985; O'Connor *et al.*, 1994; Ely, 1997) and Australia (Baker and Pickup, 1987; Gillieson *et al.*, 1991; Wohl, 1992; Saynor and Erskine, 1993), with only very recent work providing detailed analysis of SWDs in the Mediterranean region (cf. Benito *et al.*, 1998; Greenbaum *et al.*, 1998). The stable channel cross-sections and large river stage increases during floods in the gorges of the Voidomatis basin are conducive to the long-term preserva-tion of these deposits. They have been identified in areas of ineffective flow during large floods, such as tributary mouths, areas of channel expansion and in the floor of rockshelters and caves, preserved at 9–11 m above the channel bed.

FIELD AND LABORATORY METHODS

Study Sites and Dating Techniques

Geomorphological mapping and surveying techniques were employed to identify the number, extent and height of river terraces. A reach encompassing the southern corner of the Konitsa basin and the mouth of the Lower Vikos Gorge was selected for detailed study (Figure 25.1) as it contained extensive, well-preserved terrace surfaces with good exposures in river cut sections. It was common to find the coarse-grained Late Pleistocene alluvial fills cemented by calcite as a result of the carbonate-rich character of the surrounding bedrock. This accumulation of second-ary calcite has firmly cemented clasts of all sizes, normally within the upper 1–1.5 m of the sediments. The calcite formation was dated by means of uranium-series disequi-libria (^{230}Th/^{234}U method) and provides a minimum age for the alluvial deposits. Uranium-series radionuclides were measured by high-resolution alpha spectrometry and ICP-MS at Lancaster University. Uranium and thorium isotopes were separated on ion-exchange resins and electrodeposited onto stainless steel planchets (see Black *et al.*, 1997; Kuzucuoglu *et al.*, 1998). Corrections were made for decay of excess ^{234}U and detrital ^{230}Th, on the assumption that these were present at precipitation of the calcite deposits. A correction for the detrital component was made from isochron plots after successive total dissolutions were performed, following the methods of Bischoff and Fitzpatrick (1991). In all cases the slopes of the isochrons are best determined by a method of least-squares fitting which takes account of the errors in both variables (after York, 1969).

Three main sites with SWDs have so far been identified within the catchment. These are the Boila and Old Klithonia Bridge sites at the bottom of Lower Vikos Gorge, and the Tributary site which is located in a left bank tributary in the middle reaches of the Lower Vikos Gorge, just upstream of the Spiliotissa Monastery (Figure 25.1). Dating of the Boila sediments was achieved by AMS ^{14}C analysis of charcoal, whilst a vein of cemented sands in the Tributary site SWDs yielded a uranium-series age.

Field Sampling

For the fingerprinting procedure, 52 source samples were collected from all major geological formations within the catchment (Figure 25.1). Target samples were then collected from all the deposits of interest. Two samples of fine sediment were taken from each alluvial unit. Material was collected from different parts of naturally exposed sections in an attempt to account for any heterogeneity in sediment properties. In the case of the cemented alluvial units, care was taken to ensure that these were collected from the lower, unconsolidated parts of the sediments which were free from alteration by secondary calcite and pedogenic weathering (Woodward et al., 1992, 1994). The SWDs were logged in detail, and samples were carefully collected from each sedimentary layer. Samples of contemporary fluvial fines were collected from channel margin locations at eight sites along the main channel.

Determination of Fine Sediment Properties

A combination of geochemical and magnetic analyses was chosen to quantify the differences between source materials, as they can provide reliable quantitative information for a number of independent parameters. The use of two contrasting types of sediment property, which reflect different environmental controls, is likely to yield more reliable fingerprinting results (Walling et al., 1993).

For geochemical analyses, the < 2 mm fraction of the source materials and target sediments was used. The trace elements Ba, Cu, Cr, La, Nd, Ni, Pb, Rb, Sr, V, Y, Zn and Zr were determined by X-ray fluorescence (XRF) using a ARL 9400 spectrometer. Both low-frequency specific magnetic susceptibility (χ) and frequency-dependent magnetic susceptibility ($\chi_{FD}\%$) were determined for the <1 mm sample fraction using a standard Bartington MS2 magnetic susceptibility meter. Magnetic susceptibility is largely a function of the concentration of ferrimagnetic minerals, particularly magnetite (Fe_3O_4) (Thompson and Oldfield, 1986), although it is also sensitive to changes in magnetic grain size. Frequency-dependent magnetic susceptibility, however, is largely a function of the concentration of ultrafine ($<0.03\ \mu m$) superparamagnetic grains within the sample (Dearing, 1994).

ALLUVIAL STRATIGRAPHY AND CHRONOLOGY

Phases of Aggradation and Incision

Geomorphological mapping and survey work in the southern part of the Konitsa basin (main study reach, Figure 25.1) has identified a series of river terraces and palaeochannels. The broad valley floor at this point has facilitated good preservation of alluvial sediments. Five high (15.5–8.5 m) coarse-grained alluvial units, and three lower (4.5–2.0 m) predominantly fine-grained units were identified (Figure 25.3). The coarse-grained units are equivalent to the previously described Aristi-type sediments (Lewin *et al.*, 1991). All but one contained an exposed cemented upper layer from which samples were taken for uranium-series dating. These yielded Late Pleistocene ages for these sediments, culminating in a depositional phase that ended at *c.* 25 000±2000 years BP (Figure 25.3). These dates provide a minimum age for the deposits, with the calcite cement assumed to have formed soon after the depositional

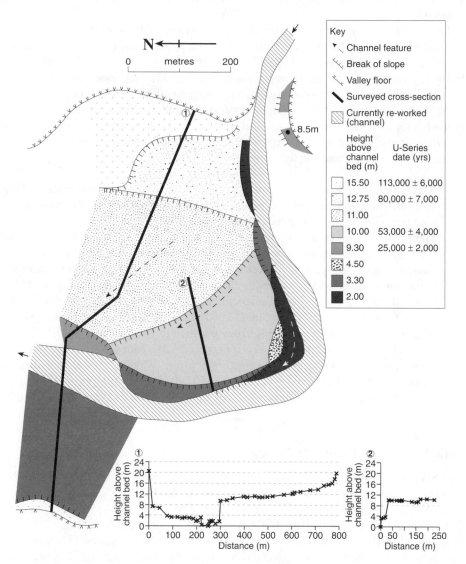

Figure 25.3   Geomorphological map of the Konitsa basin study reach (location shown on Figure 25.1)

phase of the detrital material. This is consistent with the observed U–Th isochrons which show very good correlations ($r^2 > 0.98$) indicating that the cement probably formed during a short time period. These features therefore provide clear evidence for five phases of aggradation and incision over the Late Pleistocene, when the channel bed was between 8 and 15 m higher than at present. The lower units are Late Holocene alluvial fills, and represent the multi-phase Klithi unit sediments described by Lewin *et al.* (1991) and Woodward *et al.* (1994). Radiocarbon-dated charcoal from a preserved hearth buried within a unit of this type in the Lower Vikos Gorge yielded an age of AD 1420 to 1650 (Beta-109186). This hearth was overlain by >1.5 m of fine-grained overbank sediments.

These dates can be compared with climate proxy records. Three records have been selected for this study (see Figure 25.4): (i) the long pollen record from Lake Ioaninna (Tzedakis, 1994; Tzedakis *et al.*, 1997), only 30 km to the south of the Voidomatis basin; (ii) an oxygen isotope record from the central Mediterranean Sea (Paterne *et al.*, 1986); and (iii) the high-resolution GRIP oxygen isotope record (GRIP Members, 1993; Thouveny *et al.*, 1994). The dating results from the Konitsa basin reach are plotted with 2 sigma uncertainties and are shown with dates from alluvial units elsewhere in the Lower Vikos Gorge which exhibit notable similarities (Figure 25.4). These results demonstrate the following:

(1) A major phase of Late Pleistocene aggradation occurred at approximately 25 000 years BP, which on all palaeoclimate records is associated with the pronounced cold period of the Last Glacial Maximum (LGM) (Figure 25.4).

(2) The earlier phases of aggradation took place around 55 000, 78 000 and 113 000 years BP. Greater uncertainty with older dates can make correlation to environmental phases less precise, especially in view of the fluctuating and complex nature of Late Pleistocene climate change. However, it is likely that at 55 000 years BP the area was also experiencing cold conditions (early Oxygen Isotope Stage (OIS) 3), whilst 78 000 and 113 000 years BP seem to have been periods of transitional climate, changing from warmer to cooler periods (OIS 5a to 4 and OIS 5e to 5d, respectively) (Figure 25.4). Mapping in the Konitsa basin identified another phase of sedimentation, which although not directly dated, can be bracketed to between *c*. 80 000 and 53 000 years BP (Figure 25.3). All of these dated units represent major periods of Aristi-type aggradation.

Periods of High-Magnitude Flooding

The SWDs are laminations of sand and silt (e.g. Figures 25.5 and 25.6), typical of flood deposition in high-level slackwater zones. At the Boila rockshelter (Kotjaboupoulou *et al.*, 1997), a series of flood events occurred between 13 960 ± 260 (Beta-109162) and 14 310 ± 200 (Beta-109187) years BP (2 sigma uncertainties), overtopping the lip of the cave and depositing the flood sediments (Figures 25.5 and 25.6). This was an unstable period of climatic change encompassing warming after the LGM (Figure 25.4). A uranium-series date on the Tributary site SWDs of 21 250 ± 2500 years BP indicates the occurrence of large floods during the cold conditions of

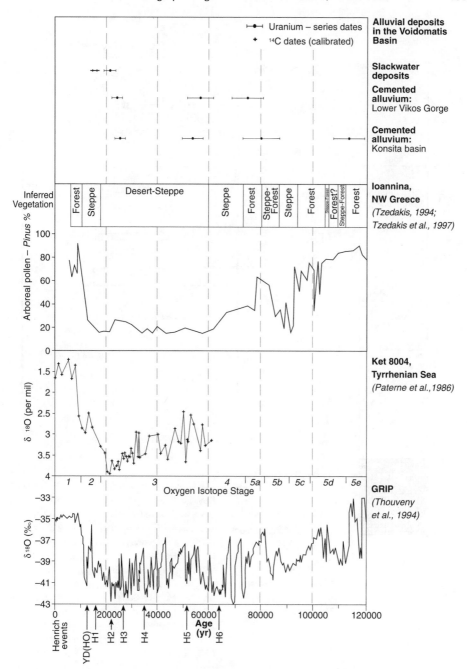

Figure 25.4   Climate proxy records and dating results from alluvial deposits in the Voidomatis basin. All results are shown in calendar years, with 2 sigma uncertainties. Radiocarbon dates, when too old for dendrochronological calibration, have been calibrated according to the results of Bard *et al.* (1990, 1992)

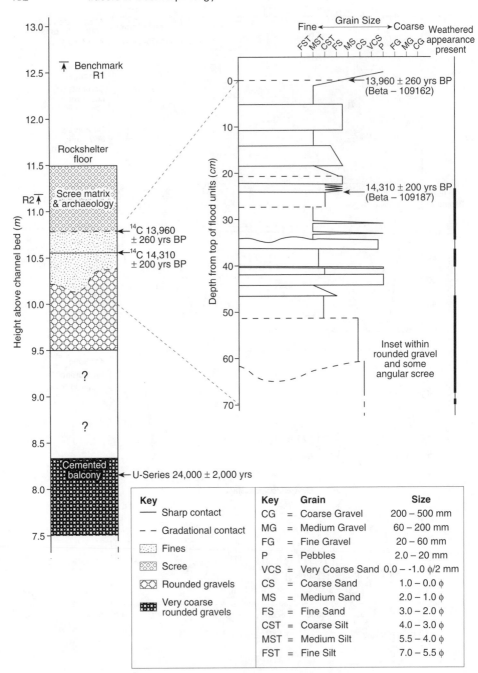

Figure 25.5 Sedimentary log of the Boila rockshelter sequence, showing the detailed stratigraphy of the palaeoflood slackwater deposits. (Conventional ^{14}C ages are shown with 2 sigma uncertainties)

Figure 25.6   The upper metre of deposits excavated in the Boila rockshelter. The laminated slackwater deposits are clearly visible at the base of the section beneath the angular scree sediments

the LGM (Figure 25.4). Unfortunately, we have been unable to directly date the other SWD at the Old Klithonia Bridge site due to the absence of any preserved organic material for ^{14}C or cemented horizons for uranium-series. However, previous dating results (see Macklin *et al.*, 1997: 359) and its stratigraphic position in relation to a dated sequence indicate that these sediments were deposited after a phase of LGM aggradation (*c.* 25 000 ± 2000 years BP), but probably no later than the flood sediments at Boila (*c.* 14 000 years BP).

In summary, uranium-series and ^{14}C dating techniques have allowed the timing of Late Pleistocene aggradation and high-magnitude flooding within the Voidomatis basin to be established. However, as we have noted, information on fine sediment provenance has the potential to enhance our understanding of long-term fluvial activity and catchment sediment dynamics. We now describe how quantitative sediment fingerprinting can be applied as a tool to elucidate temporal changes in fluvial sediment source, and link changing catchment characteristics to the genesis of these deposits.

FINGERPRINTING PROCEDURE

The methodology, which builds on previous fingerprinting approaches (e.g. Yu and Oldfield, 1989, 1993; Walling *et al.*, 1993; Collins *et al.*, 1997a, 1998), has three key stages:

Stage 1: Theoretical Framework

There are several key assumptions associated with our approach:

(1) Variations in the properties of the fine sediment transported through the fluvial system reflect changing catchment characteristics and associated changes in sediment supply.
(2) Such catchment changes may be brought about by the influence of various environmental controls (e.g. climate and vegetation change, tectonics and land-use change) and their variation through time.

Two main potential source areas of fine sediment have been identified in the Voidomatis basin (Woodward *et al.*, 1992). First, in the east of the catchment is a high, mountainous area dominated by Jurassic to Eocene limestone. This area was glaciated during the Pleistocene and extensive moraines are present in the headwater area of Tsepelovon (Figure 25.1) (Woodward *et al.*, 1995; Smith *et al.*, 1998). Secondly, to the south is a low-lying Late Eocene to Miocene flysch basin (beds of sandstones intercalated with softer fissile siltstones) (Figure 25.1) (Bailey *et al.*, 1997). This material is highly erodible and currently exhibits a dissected semi-badland topography. Major variations in sediment source during both the Pleistocene and Holocene are therefore likely to be marked by changes in the proportion of flysch- or limestone-derived material in the fluvial sedimentary record.

Stage 2: Selection of a Suite of Variables which Together Differentiate Unequivocally all the Potential Sediment Sources

It is important to select an appropriate combination of sediment properties to effectively differentiate between all potential sediment sources. A range of statistical procedures are therefore employed in three steps.

(i) Non-parametric Kruskal–Wallis H-test to Determine which Parameters Successfully Differentiate the Source Groups

Using geological maps, air photographs, field data and XRD (X-ray diffraction) information, seven distinct geological source groups were identified (Table 25.1). In addition, alluvium is also assumed to be a potential source, thus allowing for the possible incorporation of reworked sediments. Alluvium source samples were only taken from Late Pleistocene deposits, as the more recent alluvial sediments would not have been present when most of the target samples were deposited (*c.* 113 000–14 000 years BP). The lithologies which most commonly outcrop in the catchment, namely limestone and flysch, demanded a larger number of source samples to ensure that they were adequately represented (Table 25.1). The collection of only a single dolomite sample was due to difficulty in differentiating it from limestone in the field, and its generally limited outcrop in the catchment (Figure 25.1).

All analytical data were initially made dimensionless, dividing the values for each parameter by the maximum recorded value. This was done to normalise the data,

Table 25.1   List of geological source groups used in the fingerprinting analysis

Geological source group	No. of samples
Limestone	13
Till	4
Flysch	18
Dolomite-enriched flysch	2
Dolomite	1
Ophiolite	3
Terra rossa	4
Reworked alluvium	7

thereby ensuring that each parameter exerted an equal influence in the fingerprinting calculations (Verrucchi and Minisale, 1995). A non-parametric test was used as the data were found to have neither uniform distributions nor equal variances, thus making parametric tests unsuitable. Kruskal–Wallis H-tests were employed to determine those elements which significantly differentiate between these eight source groups.

With the exception of $X_{FD}\%$, all parameters produce values for H_{calc} greater that H_{crit} at the 99.9 percent significance level (Table 25.2). This means that for all parameters apart from $X_{FD}\%$ there is a 99.9 percent probability that the differences between the mean parameter values for each source group are not attributable to random variation. This exceptionally high level of significance is due to the distinct differences between the geological source groups, which will encourage accurate

Table 25.2   Results of Kruskal-Wallis *H*-tests on the different parameters used to differentiate geological sources. Critical values are also shown. (A calculated value greater than the critical value indicates that the null hypothesis can be rejected, and the parameter does statistically significantly differentiate source groups)

Parameter	*H*-value
Barium (Ba)	45.18
Chromium (Cr)	46.65
Copper (Cu)	38.22
Lanthanum (La)	39.29
Neodymium (Nd)	38.82
Nickel (Ni)	46.67
Lead (Pb)	45.46
Rubidium (Rb)	44.98
Strontium (Sr)	44.96
Vanadium (V)	43.61
Yttrium (Y)	47.55
Zinc (Zn)	45.11
Zirconium (Zr)	47.78
Mass-specific magnetic susceptibility (X)	43.02
% frequency-dependent magnetic susceptibility ($X_{FD}\%$)	14.30

$H_{crit} = 14.07$ (95%), 24.32 (99.9%).

fingerprinting results. The H-value for $\chi_{FD}\%$ is less than the 99.9 percent significance level, as it is not so successful at differentiating the sources, due to the low concentration of ultrafine superparamagnetic grains in these samples. Frequency-dependent magnetic susceptibility is therefore rejected at this point. In contrast, low-frequency magnetic susceptibility provided effective discrimination and has been shown to be the most reliable mineral magnetic parameter in experimental evaluations of quantitative fingerprinting procedures (Lees, 1997). All the geochemical (XRF) parameters and the low-frequency magnetic susceptibility pass the test and are used in the next stage.

(ii) Assess Parameters for Reproducibility and Long-term Stability in Geological Sediments

To produce accurate results it is important that the geochemical and magnetic data are reliable and reproducible. To assess this, repeat analyses of the same sample aliquot were carried out on a random 10 percent selection from the total sample set. The replicate data were divided by their mean, and a weighting (1 minus the standard deviation) was calculated (Table 25.3). This provides an assessment of the reproducibility of analytical measurements (Collins *et al.*, 1997a). Table 25.3 shows that all elements show very good reproducibility with the exception of La, and in particular Nd. It is well known that under certain conditions, some trace elements can be prone to alteration in sediments, often due to diagenetic effects (e.g. Farmer and Lovell, 1984). However, we have selected a combination of trace elements for this study that would be expected to be stable in this environment over the timescale of interest. Nevertheless, both La and Nd are Light Rare Earth Elements (LREE), and known to sometimes be prone to long-term instability in sediments. This could be a problem if their primary concentrations in old alluvial sediments have been altered. For reasons of both poor reproducibility and potential long-term instability, La and Nd were therefore discarded at this stage.

Table 25.3 Weightings for geochemical parameters used to differentiate geological sources. A higher weighting indicates greater analytical reproducibility

Parameter	Weighting
Sr	0.976
χ	0.972
Zn	0.958
Ni	0.956
Zr	0.950
Y	0.947
Cr	0.934
Rb	0.930
Cu	0.927
Ba	0.860
V	0.829
Pb	0.822
La	0.572
Nd	0.219

(iii) Multivariate Discriminant Analysis (MDA) to Select the Composite Fingerprint Parameters

From the parameters that pass Stages 2(i) and 2(ii), MDA is used to select a number of parameters whose combined signatures are capable of successfully differentiating all the source samples. Stepwise selection is achieved by the minimisation of Wilk's lambda, through the parameter with the smallest lambda value being selected at each step. Lambda values close to zero indicate that within-group variability is small compared to total variability, therefore good parameters and composite signatures will be associated with low lambda values.

The results of this process are shown in Table 25.4. Note that due to the very high efficiency of the parameters in differentiating the source groups, very low lambda values are obtained early in the selection process. With only the first four parameters all the source samples are correctly classified (Table 25.4) and this could be used as the composite fingerprint. However, a larger numbers of parameters with contrasting behaviour will improve the reliability of the results (Walling *et al.*, 1993), given the large number of source groups in this study (Table 25.1). With successive parameters added, there is a continued decrease in Wilk's lambda (Table 25.4), which demonstrates the improved discrimination that a larger composite fingerprint affords. Nine parameters (Table 25.4) were therefore selected as the composite fingerprint for this study. XRF and low-frequency magnetic susceptibility data allowed us to effectively differentiate between a range of diverse catchment sources. This could not have been achieved using standard mineralogical analyses such as XRD.

Stage 3: Application of a Multivariate Mixing Model to Determine Quantitative Sediment Source Composition

A similar mixing model approach is taken here to Walling *et al.* (1993). In a linear model it is assumed that the analytical results of the sediment samples are attributable to the relative contributions of the different source groups:

$$B_t = \sum_{s=1}^{S} V_{st} P_s \tag{25.1}$$

Table 25.4   Results of multivariate discriminant analysis (MDA) selection of the composite fingerprint

Parameter	Wilk's lambda	% of samples correctly classified
Cr	0.00789	75.00
χ	0.0000809	75.00
Zr	0.00000449	98.08
Ba	0.000000418	100.00
Sr	0.000000106	100.00
Y	0.0000000679	100.00
Rb	0.0000000312	100.00
Cu	0.0000000141	100.00
V	0.00000000963	100.00

subject to the following set of linear constraints:

$$\sum_{s=1}^{S} P_s = 1 \tag{25.2}$$

$$0 \le P_s \le 1 \tag{25.3}$$

while minimising the error term,

$$E = \sum_{t=1}^{T} \left\{ \left[B_t - \left(\sum_{s=1}^{S} V_{st} P_s \right) \right] / B_t \right\} W_t \tag{25.4}$$

where P_s is the fraction of sediment derived from source type s, V_{st} is the average value of fingerprint property t for source type s, S is the number of source types, B_t is the deposit fine sediment sample value for fingerprint property t, W_t is the weighting for fingerprint property t (Table 25.3), and T is the number of fingerprint properties considered. At the start, the source contributions P_s are set equal and the error term E is calculated. Changes in P_s are carried out that minimise E. This proceeds until no further reduction of E is possible; the change is then halved, and the procedure repeated until the optimum estimate (minimum E) for relative source contributions is achieved (cf. Walling et al., 1993). The values of P_s are therefore used to provide the sediment source results as a percentage value.

Part of the mixing model procedure can be illustrated by comparing the fingerprinting results to the raw XRF and mineral magnetics data. If the mixing model is operating correctly, those samples that are predicted to be dominated by a certain

Table 25.5 Mean source material fingerprint properties for the eight source groups. (Geochemical data are shown in ppm, whilst magnetic susceptibility (X) is in $10^{-8}m^3kg^{-1}$. The percent coefficient of variance is also shown)

Fingerprint property		Limestone	Till	Flysch	Dolomite-enriched flysch	Dolomite	Ophiolite	Terra rossa	Reworked alluvium
Ba	Mean	38	118	302	1034	964	25	429	132
	% CV	59.3	22.1	20.7	32.9		18.4	6.2	71.8
Cr	Mean	7	48	265	97	13	2831	454	47
	% CV	38.6	29.3	18.8	66.7		4.5	37.8	26.7
Cu	Mean	9	24	35	56	16	19	89	24
	% CV	34.9	6.6	48.2	57.0		3.5	27.1	18.8
Rb	Mean	3	26	88	68	11	1	103	19
	% CV	28.7	20.4	48.6	52.7		41.6	7.1	19.2
Sr	Mean	538	482	178	971	203	0	121	514
	% CV	38.3	10.6	31.2	35.3		0.0	57.2	4.9
V	Mean	14	36	120	69	14	72	141	33
	% CV	37.2	27.1	38.2	57.0		12.3	14.4	18.9
Y	Mean	4	11	23	15	11	0	63	9
	% CV	65.3	6.1	22.0	35.9		0.0	17.0	11.6
Zr	Mean	0	32	148	39	12	2	204	13
	% CV	0.0	30.2	16.2	97.0		39.9	16.2	43.7
X	Mean	1.5	3.6	12.8	19.1	1.4	50.1	261.3	5.6
	% CV	133.9	4.3	28.6	52.6		33.5	10.4	72.3

lithology should exhibit similar geochemical and magnetic characteristics to that source type. The main sediment sources have a number of distinctive characteristics. For example, the limestone is high in Sr, but low in most other parameters (Table 25.5). The till is similar to the limestone, with a slightly less pure trace-element geochemistry reflecting the incorporation of a minor flysch component (Table 25.5). The flysch, in contrast, shows relatively high concentrations of Ba, Cr, Rb, V and Zr, and has a higher magnetic susceptibility (Table 25.5). As an example of these contrasts and the mixing model process, Figure 25.7 shows the output for two samples: one alluvial fine matrix sample predicted to have a high limestone and till content (V65a); and one SWD sample from Boila predicted to have a high flysch content (OKB1.9). These are compared to the Cr, Sr and X data, which depict some of the clear differences between these major source groups. A comparison of the data for both samples shows that the fingerprinting predictions are consistent with the geochemical and mineral magnetic signatures, thus providing a simple qualitative check that the mixing model is producing consistent results.

RESULTS AND DISCUSSION

Phases of Aggradation and Incision

Figure 25.8 shows the fingerprinting results for contemporary fluvial fine sediments and the dated alluvial units identified in the main study reach (Figure 25.3). The recent sediments in the catchment (contemporary channel-bed fines and historical floodplain overbank sediments) have a large proportion of flysch-derived sediment. In particular, recent floodplain alluvium is calculated as having an almost 80 percent flysch origin. This undoubtedly reflects the highly erodible nature of the flysch lithology, which shows many signs of intense erosion today. In comparison, the sediments deposited around the cold period of the LGM have very different characteristics, being completely dominated by limestone and till (Figure 25.8). This dramatic contrast in sediment provenance is due to the influx of large amounts of fine sediment produced by glacial erosion in the limestone-dominated mountain headwaters. Frost weathering processes on the gorge walls may also have contributed to the limestone sand fraction of the fine sediment load during cold, glacial phases. These glacial sediment sources appear to have overwhelmed sediment supplied from the flysch slopes at this time.

The alluvial unit dated at 53 000 ± 4000 years BP has a very similar fine sediment provenance signature to that of sediments at the LGM, and can therefore also be interpreted as having been deposited during a cold phase climate. Although the unit dated at 80 000 ± 7000 years BP shows a significant proportion of dolomite, this must be treated with caution considering the low number of dolomite source samples (Table 25.1). It is more significant to note that limestone still dominates the fine fraction, being five times larger than the flysch proportion, and therefore at this time the catchment must also have been experiencing cold phase climate conditions.

The valley fill sediments dated to 113 000 ± 6000 years BP have very different characteristics, with high levels of flysch and reworked soil material. These sediments

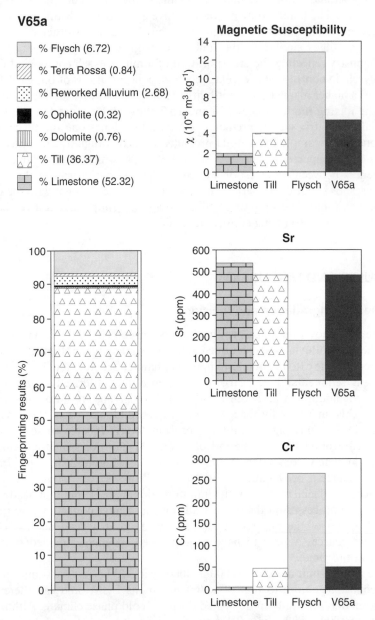

Figure 25.7(a)   Comparison of fingerprinting results with geochemical and magnetic data for (a) sample V65a, which was taken from the fine-grained matrix of a coarse-grained alluvial unit, and (b) sample OKB1.9, which was taken from the Boila rockshelter SWDs

are likely to have been deposited during cool OIS 5d, following the last interglacial (OIS 5e) (Figure 25.4). The extensive thickness of these sediments (Figure 25.3), and their dominantly limestone-derived coarse gravel lithofacies, points to a period of very high sediment supply, which could not have taken place during the warm, densely vegetated environment of the OIS 5e interglacial (Tzedakis, 1994). The high proportion of *terra rossa*, the weathered soil that occurs on the limestone, suggests that it was

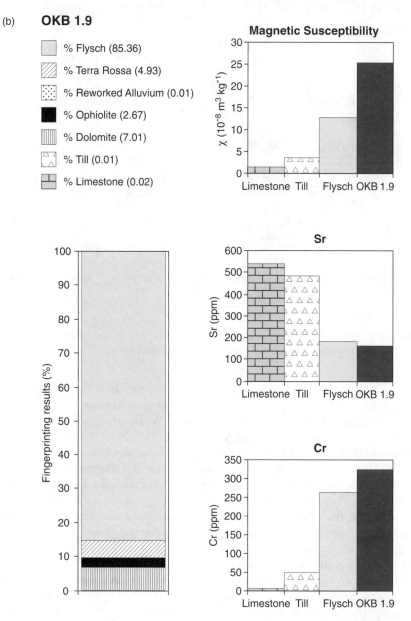

Figure 25.7(b)

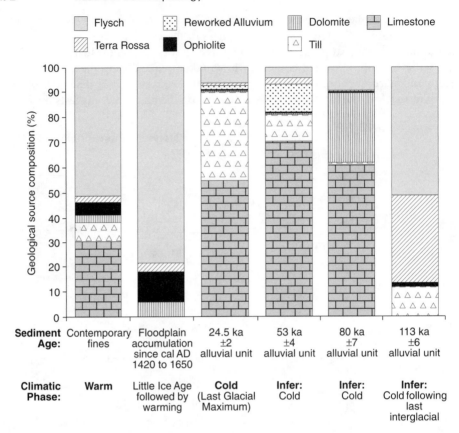

Figure 25.8 Sediment fingerprinting results from the fine matrix of dated valley fill units. (Chronological and proposed climatic information are also shown)

more widespread in the catchment at this time, which is likely considering the potential for extensive soil development during the prolonged warm interglacial conditions (van Andel and Tzedakis, 1996). Weathering and soil development would also have occurred on the flysch outcrops. However, the rapid climatic deterioration of OIS 5d brought a large decrease in vegetation cover over this part of the Balkans (Tzedakis, 1993, 1994). This would have caused slope destabilisation, releasing the pre-existing *terra rossa* and weathered flysch material for entrainment into the fluvial system. These large volumes of flysch and soil could have dominated the fine fraction, whilst the cold OIS 5d conditions would have seen frost weathering of the gorge walls to supply the dominantly limestone-derived gravel that is found in this unit. In addition, flysch erosion at this time might have been encouraged by tectonic uplift, inducing gully incision into this softer lithology. It is also possible that the predicted *terra rossa* component could in part be a secondary signal, having developed from in-situ weathering of these sediments. Establishing the significance of any diagenetic effects would require further mineral magnetic and/or microscopic analysis.

In summary, sediment fingerprinting has proved particularly useful in interpreting the history of fluvial activity. Extensive aggradation during the Late Pleistocene has occurred during periods of cold phase climate. Catchment conditions at OIS 5d, the first cool phase following the last interglacial, were very different in comparison with subsequent phases of deposition, with the fine sediment flux dominated by reworked flysch and soil material. Following that period, geomorphological processes supplying fine sediment from the limestone and till area became dominant during cold Late Pleistocene phases. The volume of flysch-derived sediment runoff probably actually increased during these cold periods, as the low vegetation cover would have made the flysch slopes unstable and susceptible to gully and wash erosion. However, the relative contribution of flysch to the units deposited at *c.* 25 000, 55 000 and 78 000 years BP is suppressed to only *c.* 10 percent (Figure 25.8), which suggests that the volume of flysch-derived material was simply swamped by a vast increase in the volume of sediment supply from glacial erosion and mechanical rock breakdown in the mountainous limestone areas. The evidence for this sediment source change reflecting a considerable increase in sediment delivery is corroborated by the fact that this limestone-dominated sediment fingerprint is found in such thickly aggraded alluvial units. The large increase in sediment supply during cold phases was almost certainly a key stimulus for these periods of aggradation. This demonstrates the large impact of glaciation and cold climate weathering upon this river system during the Late Pleistocene (Woodward *et al.*, 1995). Following cold glacial periods, progressive incision has followed the decline in glacial sediment inputs and the stabilisation of slopes by increasing vegetation density (Turner and Sánchez Goñi, 1997; Willis, 1997). This incision has dominated the post-glacial period, with the only significant phase of aggradation occurring during the Late Holocene following an increase in fine sediment supply from flysch areas and the vertical accretion of the Klithi-type sediments (Lewin *et al.*, 1991).

Periods of high-magnitude flooding

In Figure 25.9 the average source characteristics of several bulk samples taken from sedimentary units within each SWD are shown, with the fingerprints for sediments deposited during warm and cold phase climates included for reference. The SWDs at Boila show very similar characteristics to the sediments that are typical of the modern warm phase climate, with a large proportion of flysch and very little limestone or till. This would suggest that at *c.* 14 000 years BP, when these flood sediments were deposited, glacial sediment inputs were not significant and that environmental conditions may have been similar to those today. However, the characteristics of the sediments in both the Old Klithonia Bridge and Tributary SWDs are very different, with the dominance of limestone and till being comparable to that of the alluvial unit deposited at the LGM. This suggests cold phase deposition of these flood sediments, which is consistent with the date of 21 250 ± 2500 years BP at the Tributary site. Although the timing of deposition at the Old Klithonia Bridge site is not exactly constrained, the sediments show very similar source characteristics to the Tributary site and, given their stratigraphic context discussed above, may therefore have been deposited at a similar time. The results confirm that these floods were not contemporaneous with those which deposited the sediments at Boila. Therefore, even

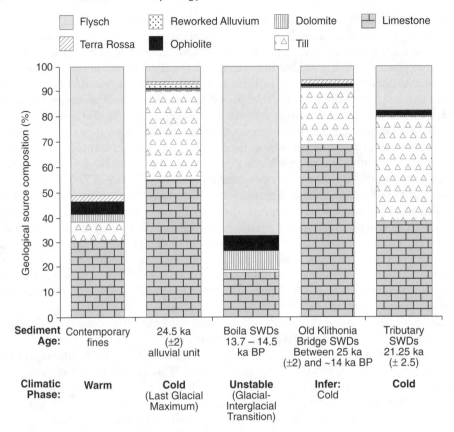

Figure 25.9 Sediment fingerprinting results from the palaeoflood slackwater deposits. (Chronological and proposed climatic information are also shown)

without being able to obtain precise dates for all these deposits, the fingerprinting results enable an interpretation of the timing and the type of environment during which the sediments were deposited. This method shows the considerable potential for provenancing work on SWDs. It is perhaps surprising that the only previous SWD provenance study of note is that of Saynor and Erskine (1993), who successfully employed heavy mineral analysis to differentiate flood sediments from those which contained a proportion of locally derived colluvium. In future work we aim to look in more detail at the provenance of individual flood units within these SWDs. This promises to enable sediments from different flood events to be distinguished with confidence, and to facilitate palaeoenvironmental interpretation of the entire SWD sequences (some of which are over 2.5 m in thickness). In this study, the fingerprinting data provide evidence for two separate periods of high-magnitude flooding during the Late Pleistocene. Extreme flooding around the LGM, represented by SWDs at the Tributary and Old Klithonia Bridge sites, would have coincided with high sediment availability, and would therefore have been important for river aggradation around the LGM (cf. Bull, 1979, 1988). The presence of coarse gravels and small boulders in this LGM alluvial unit also points to high discharges. Such floods probably resulted from intense glacial meltwater in spring. The floods which deposited sediment in Boila, however, occurred during the Late-glacial period where the delivery of limestone-rich glacially comminuted fines had waned, suggesting a reduction in sediment supply and more stable, vegetated slopes (Turner and Sánchez Goñi, 1997). These later floods, occurring at a time of lower sediment availability, may therefore have resulted in considerable incision through the Late Pleistocene deposits (cf. Bull, 1979, 1988).

Reliability of Fingerprinting Results

It is important to assess the reliability of the fingerprinting results. The most extensive lithologies within the catchment (limestone, flysch and till) dominate the fingerprinting predictions, which suggests that the results are realistic. The results are also in accord with previous work based on semi-quantitative XRD analyses (Woodward *et al.*, 1992). However, the difference between measured and predicted parameter values can be used to obtain a mean relative error. In this study these are typically ± 10–15 percent, which is only slightly higher than recent work on contemporary suspended sediments (*c.* ± 10 percent) reported by Collins *et al.* (1997a, 1998). This is encouraging considering some of the potential problems involved in determining provenance over long timescales. However, it is important to appreciate that this is only a crude estimate of error that is partly influenced by between-parameter co-correlation. Further work is required into the improved quantification of fingerprinting uncertainty (e.g. Rowan *et al.*, see Chapter 14).

We have tested the results by comparing the fingerprinting predictions with independent mineralogical data determined by XRD. The distinctive characteristics of the main sediment sources (Table 25.5) reflect their mineralogical differences. The limestone has relatively simple geochemical and mineral magnetic characteristics as it is composed largely of calcite with small amounts of quartz. The till is similar to the limestone, dominated by calcite but including a little more quartz and feldspar. In contrast, high concentrations of many trace elements in the flysch, and its higher magnetic susceptibility, can be explained by a mineralogy dominated by quartz, with a significant amount of feldspar and various clay minerals. As an example, Figure 25.10 shows the mineralogical data for the two samples whose fingerprinting results are displayed in Figure 25.7. For V65a (Figure 25.10) the mineralogy is dominated by calcite, with some quartz but very little clays or feldspar, thus supporting the fingerprinting predictions of approximately 90 percent limestone and till. The agreement of the data is also encouraging for OKB1.9 (Figure 25.10), which has a high quartz content and a significant amount of feldspar and clays, which is consistent with the high flysch content predicted for this sample. Thus in both cases, mineralogical information fully supports the fingerprinting predictions, suggesting that the source ascription is both accurate and reliable.

Nevertheless, there are a number of potential problems associated with fingerprinting Late Pleistocene sediment:

(1) Diagenesis is a potential source of inaccuracy. We have attempted to minimise diagenetic effects through careful sampling to avoid weathered deposits, as well

(a) V65a

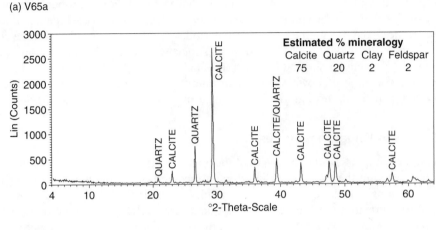

(b) OKB 1.9

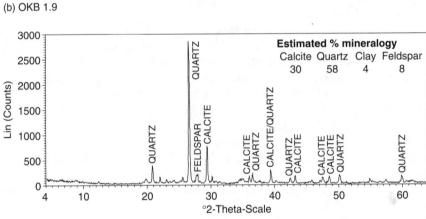

Figure 25.10   Mineralogical characteristics of samples V65a and OKB1.9. The fingerprinting results for these samples are shown in Figure 25.7

as choosing geologically stable fingerprinting parameters. However, this could be explored further through more detailed mineral magnetic or microscopic analysis.

(2)   It is difficult to correct for potential grain-size effects when some of the principal sediment sources are bedrock. Therefore, analysis of standardised fine sediment fractions (powdered <2 mm for XRF, <1 mm for magnetics) was judged to be appropriate.

With these possible sources of error in mind, these results demonstrate the potential of quantitative sediment fingerprinting over Late Pleistocene timescales. Quantitative fine sediment provenancing is a valuable complement to other approaches for the investigation of long-term river behaviour, and offers considerable promise for the interpretation of slackwater sediments.

CONCLUSIONS

(1) A quantitative fingerprinting technique has been successfully applied over long timescales (10^2–10^5 years).

(2) Quantitative information on sediment sources corroborates the radiometric dating results to facilitate a fuller interpretation of Late Pleistocene alluvial history. The approach provides detailed evidence of the changes in sediment delivery processes caused by environmental variations. It has been demonstrated that in the Voidomatis basin, sediment sources have been strongly influenced by climatic controls. Quantitative sediment fingerprinting can therefore be a valuable tool for interpreting long-term river activity.

(3) In the Voidomatis basin, dating of cemented alluvial deposits has been successfully achieved by uranium-series techniques. Extensive river aggradation has occurred over the Late Pleistocene in at least four separate periods, with the cessation of deposition dated to approximately 113 000, 78 000, 55 000 and 25 000 years BP. Fingerprinting results demonstrate that these periods of deposition occurred during a cold phase climate, almost certainly as a result of increased sediment supply from glacial erosion, with a minor contribution from frost weathering processes. This builds considerably on previous work in the area (Lewin *et al.*, 1991), to further demonstrate the dynamic nature of Late Pleistocene river development in Mediterranean mountain environments.

(4) Evidence has been presented for high-magnitude flooding during both glacial (*c.* 21 000 years BP) and Late Glacial (*c.* 14 000 years BP) environments. Such palaeoflood events have played an important role in the Quaternary development of the Voidomatis River. Quantitative sediment fingerprinting is particularly valuable for interpreting palaeoflood slackwater deposits.

ACKNOWLEDGEMENTS

This work was carried out whilst R.H. was in receipt of a Departmental PhD Studentship from the School of Geography, University of Leeds. J.W. and M.G.M. acknowledge the support of NERC Grant GR9/2916. The authors are indebted to Claire Sedgwick for her assistance with the majority of the fieldwork and for many helpful discussions. Adrian McDonald is thanked for his valuable comments. Damien Lanforth and Clair Calvert-Smith provided much assistance with XRF analysis. Simon Robinson is thanked for his helpful advice and for allowing R.H. access to laboratories at the School of Geography, University of Manchester. Beta Analytic Inc. carried out the radiocarbon analysis. IGME (Athens) provided research permits and sample export licences. Eleni Kotjaboupoulou kindly allowed us access to the deposits at Boila. Adrian Harvey and Tim Quine are both thanked for their valuable reviews of this chapter. Finally, we are also grateful for the excellent work of the Graphics Department in the School of Geography, University of Leeds.

REFERENCES

Baker, V. R. 1977. Stream channel response to floods, with examples from central Texas. *Geological Society of America Bulletin*, **88**, 1057–1071.

Baker, V. R. and Pickup, G. 1987. Flood geomorphology of the Katherine Gorge, Northern Territory, Australia. *Geological Society of America Bulletin*, **98**, 653–646.

Baker, V. R., Kochel, R. C., Patton, P. C. and Pickup, G. 1983. Paleohydologic analysis of Holocene flood slack-water sediments. *Special Publications of the International Association of Sedimentology*, **6**, 229–239.

Bailey, G. N., Turner, C., Woodward, J. C. Macklin, M. G., and Lewin, J. 1997. The Voidomatis Basin: an introduction. In Bailey, G. N. (ed.) *Klithi: Palaeolithic Settlement and Quaternary Landscape in Northwest Greece. Volume 2: Klithi in Its Local and Regional Setting*. McDonald Institute, Cambridge, 321–346.

Bard, E., Hamelin, B., Fairbanks, R. G. and Zinder, A. 1990. Calibration of the ^{14}C timescale over the past 30,000 years using mass spectrometric U–Th ages from Barbados corals. *Nature*, **345**, 405–410.

Bard, E., Fairbanks, R. G., Arnold, M. and Hamelin, B. 1992. ^{230}Th/^{234}U and ^{14}C ages obtained by mass spectrometry on corals from Barbados (West Indies), Isabela (Galapagos) and Mururoa (French Polynesia). In Bard, E. and Broecker, W. S. (eds) *The Last Deglaciation: Absolute and Radiocarbon Chronologies*. Springer-Verlag: Berlin, Heidelberg, 103–110.

Benito, G., Machado, M. J., Pérez-González, A. and Sopeña, A. 1998. Palaeoflood hydrology of the Tagus River, Central Spain. In Benito, G., Baker, V. R. and Gregory, K. J. (eds) *Palaeohydrology and Environmental Change*. John Wiley, Chichester, 317–333.

Bischoff, J. L. and Fitzpatrick, J. A. 1991. U-series dating of impure carbonates: an isochron technique using total-sample dissolution. *Geochimica et Cosmochimica Acta*, **55**, 543–554.

Black, S., Macdonald, R. and Kelly, M. 1997. Crustal origin for peralkaline rhyolites from Kenya: evidence from U-series disequilibria and Th-isotopes. *Journal of Petrology*, **38**, 277–297.

Bull, W. B. 1979. Threshold of critical power in streams. *Bulletin of the Geological Society of America*, **90**, 453–464.

Bull, W. B. 1988. Floods: degradation and aggradation. In Baker, V. R. Kochel, R. C. and Patton, P. C. (eds) *Flood Geomorphology*. John Wiley, New York, 157–165.

Collins, A. L., Walling, D. E. and Leeks, G. J. L. 1997a. Source type ascription for fluvial suspended sediment based on a quantitative composite fingerprinting technique. *Catena*, **29**, 1–27.

Collins, A. L., Walling, D. E. and Leeks, G. J. L. 1997b. Use of the geochemical record preserved in floodplain deposits to reconstruct recent changes in river basin sediment sources. *Geomorphology*, **19**, 151–167.

Collins, A. L., Walling, D. E. and Leeks, G. J. L. 1998. Use of composite fingerprints to determine the provenance of the contemporary suspended sediment load transported by rivers. *Earth Surface Processes and Landforms*, **23**, 31–52.

Coxon, P., Owen, L. A. and Mitchell, W. A. 1996. A Late Pleistocene catastrophic flood in the Lahul Himalayas. *Journal of Quaternary Science*, **11**, 495–510.

Dearing, J. A. 1992. Sediment yields and sources in a Welsh upland catchment during the past 800 years. *Earth Surface Processes and Landforms*, **17**, 1–22.

Dearing, J. A. 1994. *Environmental Magnetic Susceptibility: Using the Bartington MS2 System*. Chi Publishing: Kenilworth.

De Boer, D. H. and Crosby, G. 1995. Evaluating the potential of SEM/EDS analysis for fingerprinting suspended sediment derived from two contrasting topsoils. *Catena*, **24**, 243–258.

Ely, L. L. 1997. Response of extreme floods in the southwestern United States to climatic variations in the late Holocene. *Geomorphology*, **19**, 175–201.

Ely, L. L. and Baker, V. R. 1985. Reconstructing paleoflood hydrology with slackwater deposits: Verde River, Arizona. *Physical Geography*, **5**, 103–126.

Farmer, J. G. and Lovell, A. 1984. Massive diagenetic enhancement of manganese in Loch Lomond sediments. *Environmental Technology Letters*, **5**, 257–262.

Foster, I. D. L., Lees, J. A., Owens, P. N. and Walling, D. E. 1998. Mineral magnetic characterization of sediment sources from an analysis of lake and floodplain sediments in the catchments of the Old Mill Reservoir and Slapton Ley, South Devon, UK. *Earth Surface Processes and Landforms*, **23**, 685–703.

Gillieson, D., Ingle Smith, D., Greenaway, M. and Ellaway, M. 1991. Flood history of the limestone ranges in the Kimberly region, Western Australia. *Applied Geography*, **11**, 105–123.

Greenbaum, N., Margalit, A., Schick, A. P., Sharon, D. and Baker, V. R. 1998. A high magnitude storm and flood in a hyperarid catchment, Nahal Zin, Negev Desert, Israel. *Hydrological Processes*, **12**, 1–24.

GRIP (Greenland Ice-core Project) Members 1993. Climate instability during the last inter-glacial period recorded in the GRIP ice core. *Nature*, **364**, 203–207.

Grimshaw, D. L. and Lewin, J. 1980. Source identification for suspended sediments. *Journal of Hydrology*, **47**, 151–161.

Gupta, A. 1983. High-magnitude floods and stream channel response. *Special Publication of the International Association of Sedimentologists* **6**, 219–227.

Haughton, P. D. W., Todd, S. P. and Morton, A. C. 1991. Sedimentary provenance studies. In Morton, A. C., Todd, S. P. and Haughton, P. D. W. (eds) *Developments in Sedimentary Provenance Studies*. Geological Society Special Publication No. 57, The Geological Society, London, 1–13.

Hutchinson, S. M. 1995. Use of magnetic and radiometric measurements to investigate erosion and sedimentation in a British upland catchment. *Earth Surface Processes and Landforms*, **20**, 293–314.

IGME, 1968. *1:50,000 Geological Map of Greece*, Doliana Sheet, Institute of Geological and Mineralogical Research, Athens.

IGME, 1970, *1:50,000 Geological Map of Greece*, Tsepelovon Sheet, Institute of Geological and Mineralogical Research, Athens.

Kochel, R. C. and Baker, V. R. 1982. Paleoflood hydrology. *Science*, **215**, 353–361.

Kochel, R. C. and Baker, V. B. 1988. Paleoflood analysis using slackwater deposits. In Baker, V. R., Kochel, R. C. and Patton P. C. (eds) *Flood Geomorphology*. John Wiley, New York, 357–376.

Kotjaboupoulou, E., Panagopoulou, E. and Adam, E. 1997. The Boila Rockshelter: a pre-liminary report. In Bailey, G. N. (ed.) *Klithi: Palaeolithic Settlement and Quaternary Land-scape in Northwest Greece. Volume 2: Klithi in Its Local and Regional Setting*. McDonald Institute, Cambridge, 427–438.

Kurashige, Y. and Fusejima, Y. 1997. Source identification of suspended sediment from grain size distributions: I. Application of non-parametric statistical tests. *Catena*, **31**, 39–52.

Kuzucuoglu, C., Pastre, J-F., Black, S., Ercan, T., Fontugne, M., Guillou, H., Hatté, C., Karabiyikoglu, M., Orth, P. and Türkecan, A. 1998. Identification and dating of tephras from Quaternary sedimentary sequences of inner Anatolia. *Journal of Volcanology and Geothermal Research*, **85**, 153–172.

Lees, J. 1997. Mineral magnetic properties of mixtures of environmental and synthetic materials: linear additivity and interaction effects. *Geophysics Journal International*, **131**, 335–346.

Lewin, J., Macklin, M. G. and Woodward, J. C. 1991. Late Quaternary fluvial sedimentation in the Voidomatis basin, Epirus, Northwest Greece. *Quaternary Research*, **35**, 103–115.

Macklin, M. G., Lewin, J. and Woodward, J. C. 1997. Quaternary river sedimentary sequences of the Voidomatis basin. In Bailey, G. N. (ed.) *Klithi: Palaeolithic Settlement and Quaternary Landscape in Northwest Greece. Volume 2: Klithi in Its Local and Regional Setting*. Mc-Donald Institute, Cambridge, 347–336.

O'Connor, J. E., Ely, L. L., Wohl, E. E., Stevens, L. E., Melis, T. S., Kale, V. S. and Baker, V. R. 1994. A 4500-year record of large floods on the Colorado River in the Grand Canyon, Arizona. *The Journal of Geology*, **102**, 1–9.

Oldfield, F., Maher, B. A., Donoghue, J. and Pierce, J. 1985. Particle-size related, mineral magnetic source sediment linkages in the Rhode River catchment, Maryland, USA. *Journal of the Geological Society*, **142**, 1035–1046.

Olley, J. M., Murray, A. S., Mackenzie, D. H. and Edwards, K. 1993. Identifying sediment sources in a gullied catchment using natural and anthropogenic radioactivity. *Water Resources Research*, **29**, 1037–1043.

Passmore, D. G. and Macklin, M. G. 1994. Provenance of fine-grained alluvium and late-Holocene land-use change in the Tyne basin, northern England. *Geomorphology*, **9**, 127–142.

Paterne, M., Guichard, F., Labeyrie, J., Gillot, P. Y. and Duplessy, J. C. 1986. Tyrrhenian tephrochronology of the oxygen isotope record for the past 60,000 years. *Marine Geology*, **72**, 259–285

Peart, M. R. and Walling, D. E. 1986. Fingerprinting sediment source: the example of a drainage basin in Devon, UK. *Drainage Basin Sediment Delivery*. IAHS Publication **174**. IAHS Press, Wallingford, 41–55.

Saynor, M. J. and Erskine, W. D. 1993. Characteristics and implications of high-level slack-water deposits in the Fairlight Gorge, Nepean River, Australia. *Australian Journal of Marine and Freshwater Research*, **44**, 735–747.

Smith, G. R., Woodward, J. C., Heywood, D. I. and Gibbard, P. L. 1998. Mapping glaciated karst terrain in a Mediterranean mountain environment using SPOT and TM data. In Burt, P. J. A., Power, C. H. and Zukowskyj, P. M. (eds) *R SS98: Developing International Connections*. Proceedings of the Remote Sensing Society's Annual Meeting, Greenwich, 9–11 September 1998, 457–463.

Thompson, R. and Oldfield, F. 1986. *Environmental Magnetism*. Allen & Unwin, London.

Thouveny, N., de Beaulieu, J.-L., Bonifay, E., Creer, K. M., Guiot, J., Icole, M., Johnsen, S., Jouzel, J., Reille, M., Williams, T. and Williamson, D. 1994. Climate variations in Europe over the past 140 kyr deduced from rock magnetism. *Nature*, **371**, 503–506.

Turner, C. and Sánchez Goñi, M.-F. 1997. Late glacial landscape and vegetation in Epirus. In Bailey, G. N. (ed.) *Klithi: Palaeolithic Settlement and Quaternary Landscape in Northwest Greece. Volume 2: Klithi in Its Local and Regional Setting*. McDonald Institute, Cambridge, 559–585.

Tzedakis, P. C. 1993. Long-term tree populations in northwest Greece through multiple Quaternary climatic cycles. *Nature*, **364**, 437–440.

Tzedakis, P. C. 1994. Vegetation change through glacial–interglacial cycles: a long pollen sequence perspective. *Philosophical Transactions of the Royal Society, London*, Series B, **345**, 403–432.

Tzedakis, P. C., Andrieu, V., de Beaulieu, J.-L., Crowhurst, S., Follieri, M., Hooghiemstra, H., Magri, D., Reille, M., Sadori, L., Shackleton, N. J. and Wijmstra, T. A. 1997. Comparison of terrestrial and marine records of changing climate of the last 500,000 years. *Earth Science and Planetary Science Letters*, **150**, 171–176.

Van Andel, T. H. and Tzedakis, P. C. 1996. Palaeolithic landscapes of Europe and environs, 150,000–25,000 years ago: an overview. *Quaternary Science Reviews*, **15**, 481–500.

Verrucchi, C. and Minisale, A. 1995. Multivariate statistical comparison of Northern Apennines Paleozoic sequences: a case study for the formations of the Monti Romani (Southern Tuscany–Northern Latium, Italy). *Applied Geochemistry*, **10**, 581–598.

Walling, D. E. and Woodward, J. C. 1992. Use of radiometric fingerprints to derive information on suspended sediment sources. In *Erosion and Sediment Transport Monitoring Programmes in River Basins* (Proceedings of the Oslo Symposium, August 1992), IAHS Publication **210**. 153–164.

Walling, D. E. and Woodward, J. C. 1995. Tracing sources of suspended sediment in River Basins: a case study of the River Culm, Devon, UK. *Marine and Freshwater Research*, **46**, 327–336.

Walling, D. E., Peart, M. R., Oldfield, F. and Thompson, R. 1979. Suspended sediment sources identified by magnetic measurements. *Nature*, **281**, 110–113.

Walling, D. E., Woodward, J. C. and Nicholas, A. P. 1993. A multi-parameter approach to fingerprinting suspended sediment sources. In *Tracers in Hydrology* (Proceedings of the

Yokohama Symposium, July 1993), IAHS Publication **215**, IAHS Press, Wallingford. 329–328.

Willis, K. 1997. Vegetational history of the Klithi environment: a palaeoecological viewpoint. In Bailey, G. N. (ed.) *Klithi: Palaeolithic Settlement and Quaternary Landscape in Northwest Greece. Volume 2: Klithi in Its Local and Regional Setting*. McDonald Institute, Cambridge, 395–414.

Wohl, E. E. 1992. Bedrock benches and boulder bars: floods in the Burdekin Gorge of Australia. *Geological Society of America Bulletin*, **104**, 770–778.

Wood, P. A. 1978. Fine-sediment mineralogy of source rocks and suspended sediment, Rother catchment, West Sussex. *Earth Surface Processes*, **3**, 255–263.

Woodward, J. C., Lewin, J. and Macklin, M. G. 1992. Alluvial sediment sources in a glaciated catchment: the Voidomatis basin, Northwest Greece. *Earth Surface Processes and Landforms*, **17**, 205–216.

Woodward, J. C., Macklin, M. G. and Lewin, J. 1994. Pedogenic weathering and relative-age dating of Quaternary alluvial sediments in the Pindus Mountains of Northwest Greece. In Robinson, D. A. and Williams, R. B. G. (eds) *Rock Weathering and Landform Evolution*. John Wiley, Chichester, 259–283.

Woodward, J. C., Lewin, J. and Macklin, M. G. 1995. Glaciation, river behaviour and Palaeolithic settlement in upland northwest Greece. In Lewin, J., Macklin, M. G. and Woodward, J. C. (eds) *Mediterranean Quaternary River Environments*. Balkema, Rotterdam, 115–129.

York, D. 1969. Least squares fitting of a straight line with correlated errors. *Earth and Planetary Science Letters*, **5**, 320–324.

Yu, L. and Oldfield, R. 1989. A multivariate mixing model for identifying sediment source from magnetic measurements. *Quaternary Research*, **32**, 168–181.

Yu, L. and Oldfield, R. 1993. Quantitative sediment source ascription using magnetic measurements in a reservoir-catchment system near Nijar, S.E. Spain. *Earth Surface Processes and Landforms*, **18**, 441–454.

26 Sediment Source Characteristics of the Río Tinto, Huelva, South-west Spain

C. SCHELL,[1] **S. BLACK**[2] **and K. A. HUDSON- EDWARDS**[3]
[1]*School of Geography, University of Leeds, UK*
[2]*PRIS, University of Reading, UK*
[3]*Department of Geology, Birkbeck College, University of London, UK*

INTRODUCTION

In studies of fluvial sediment and associated contaminant dispersal, the identification of sediment transport mechanisms and quantification of suspended sediment sources are fundamental. In particular, knowledge of the source types, their spatial distribution and relative contribution may be important for effective sediment and pollution control strategies (Walling and Woodward, 1995). However, information on sediment transport and sources can be difficult to obtain, especially in the case of diffuse sources or a large drainage basin. Historical data are often lacking, and adequate monitoring of all the variables required for a sediment transport and provenance assessment is operationally difficult, costly and time-consuming (Collins *et al.*, 1997a; Dietrich *et al.*, 1982). These difficulties may be particularly pronounced in anthropogenically disturbed fluvial systems, e.g. those affected by intensive metal mining, because such disturbances often entail significant geomorphological changes and thus more complex fluvial processes (Lewin and Macklin, 1987; Miller, 1997).

Geomorphological tracer techniques provide a viable alternative to detailed long-term monitoring as they generally involve relating physico-chemical characteristics of sediment sources to common benchmarks evident in the sediment properties of, for example, suspended sediment. This means that they are capable of integrating multiple processes, and do not require a disproportionally large number of samples.

In this chapter we present the results of a statistical 'fingerprinting' approach used to ascertain temporal changes in sediment provenance in a highly complex Mediterranean sulphidic river system affected by 4500 years of metal mining. This research is part of a comprehensive project seeking to investigate metal mine waste dispersal in different hydrogeochemical environments, and aims at identifying the impact of historical mining on sediment quality under variable-discharge and acidic river conditions.

Tracers in Geomorphology. Edited by Ian D. L. Foster. © 2000 John Wiley & Sons Ltd.

STUDY AREA

The Río Tinto mining district is part of the Iberian Pyrite Belt which extends for 230 km north-west of Sevilla into Portugal (Figure 26.1; Schermerhorn, 1982) and comprises extensive mineralisation of pyrite, chalcopyrite, galena, sphalerite, gold and silver. The main type of mineralisation is in the form of massive sulphide deposits ranging from 0.1 to 100 million tonnes of which there are between 60 and 70 identified along the entire length of the Iberian Pyrite Belt. These are hosted in a Devonian–Carboniferous age volcanic sedimentary complex (VSC) which comprises steeply dipping slates and igneous porphyries that were formed through submarine exhalation and subsequent burial by sediments on a sea-floor (Schermerhorn, 1982; Velasco *et al.*, 1998). These volcanics and sediments were folded and deformed into an anticline and subsequently eroded, exposing the acidic volcanic central portion with the massive sulphide deposits associated with it (Saez *et al.*, 1996). A heavily weathered gossan cap occurs over most of the deposit which is extremely rich in highly residual metals (e.g. gold and silver, ≥ 1.1 and 83 oz/ton, respectively) and contains in excess of 50 percent Fe. At the base of the gossan immediately above the VSC are some secondary jarosite-bearing deposits which are also extremely rich in gold and silver.

Between commencement of mining for Cu, Au, Ag and pyrite in 2500 BC and the present day, approximately 115 million tonnes of ore has been extracted (Schermerhorn, 1982). At least 90 percent of this production occurred between 1873 and 1954, under British ownership (Strauss *et al.*, 1977). After 1954 the mines were returned to

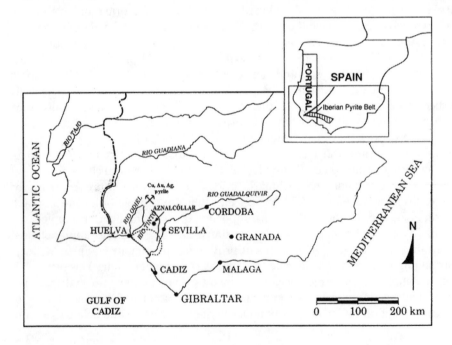

Figure 26.1 Map of southern Spain showing the location of Río Tinto (after Nelson and Lamothe, 1993)

Spain but were past their prime and copper production declined from this point until the present day when only limited extraction is taking place. The copper smelter based at Río Tinto was finally moved to Huelva in 1969, at which time most of the emphasis switched to gold and silver extraction which still carries on today.

The Río Tinto drains the mining area and has a catchment area of 1676 km^2 (Figure 26.2). It derives its name ('red river') from the distinctive deep red colour of its water, which exhibits a consistently low pH (2–3) and very high dissolved Fe, SO_4^{2-} and toxic metal concentrations (Nelson and Lamothe, 1993; Schell, 1998; Hudson-Edwards *et al.*, 1999). River flow is seasonal and concentrated during the high-intensity rainstorms that occur annually between October and March. Alluvial deposits comprise coarse channel bars, and fine-grained channel and channel bank deposits of recent to Pleistocene age, which represent several cycles of aggradation and incision (Schell, 1998). Previous work in the Río Tinto has established that a large proportion of alluvium is highly contaminated with heavy metals and arsenic (Schell, 1998; Hudson-Edwards *et al.*, 1999). The metal dispersal processes are complex, with transport and storage influenced by mineralogical transformations, seasonal river flow, ferricrete formation together with tributary and estuarine inputs (Schell *et al.*, 1996; Schell, 1998; Hudson-Edwards *et al.*, 1999).

The catchment can be divided into three distinct geological subgroups (Figure 26.3). The northern headwaters comprise the Devonian–Carboniferous volcano-sedimentary complex which hosts the ore mineralisation. The middle reaches of the Río Tinto are underlain by a series of Carboniferous slates and greywackes, and southern lithologies are dominated by Tertiary blue marls, biogenic limestone and Quaternary sands and conglomerates. Land use is dominated by the Río Tinto mines at the head of the catchment, with small areas used for goat grazing and honey farms. Further downstream the less steep and more fertile limestone plains allow widespread arable cropping.

METHODOLOGY

The 'fingerprinting' technique has been applied successfully for several years for various purposes and with a wide range of fingerprinting tracers (e.g. Oldfield *et al.*, 1979; Grimshaw and Lewin, 1980; Peart and Walling, 1986, 1988; Walling and Woodward, 1992, 1995; Walling *et al.*, 1993; Collins *et al.*, 1997a,b, 1998; Rowan *et al.*, 1999). It involves the selection of physical or chemical sediment properties that unequivocally differentiate potential sources, and the comparison of measurements of the selected properties in the sediment with the equivalent in source materials. Composite tracer signatures generally offer the most reliable source discrimination (Walling *et al.*, 1993; Collins *et al.*, 1997b).

In this study, the catchment was divided into the three lithological source groups: (i) volcanics; (ii) slates/greywackes; and (iii) the southern lithologies comprising Quaternary sands, limestone and blue marls. Pre-mining alluvial terraces along the Río Tinto, mine-derived alluvium in storage, and the various mine waste and ore rock material in the mining area define three more source groups. Within each of the groups a range of potential sediment source types (83 in total; Table 26.1) were

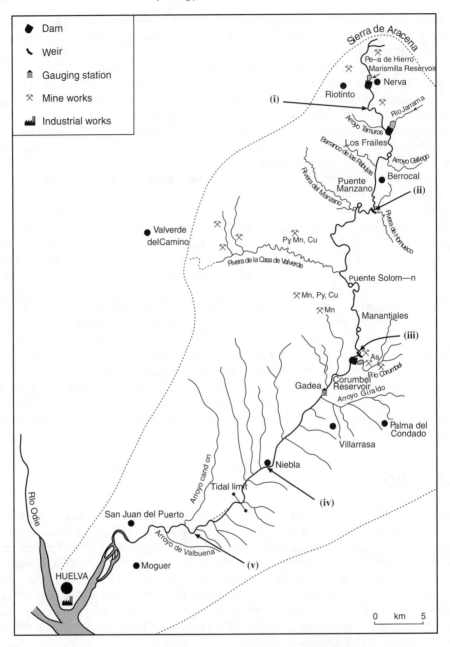

Figure 26.2 Detailed map of the Río Tinto catchment. Sample localities marked are (i) the mining area, where all the waste/ore samples were taken; (ii) Berrocal, where half of the modern sediments were taken together with some of the historical and mine-derived alluvium; (iii) Corumbel, where some of the historical and mine-derived alluvium was collected; (iv) Niebla, where the remainder of the modern sediments and some of the Late Pleistocene samples were taken; (v) Lucena, where the core was taken from the estuary. In addition, samples were taken from all the tributaries marked

sampled, including mine waste tips, river bed and bank material, surface soils and country rock fragments.

After collection, all samples were air-dried. Unconsolidated samples were disaggregated, mixed and sieved to < 2 mm. Concentrations of Ba, Cu, La, Pb, Rb, Sr, V, W, Y, Zn and Zr in the samples were determined by X-ray fluorescence on pressed powder pellets using a Phillips PW 1400 X-ray fluorescence spectrometer after calibration with USGS standard rock powders. Analytical accuracy and precision of at least ±2 percent and ±1.5 percent, respectively, for all elements were routinely obtained. In

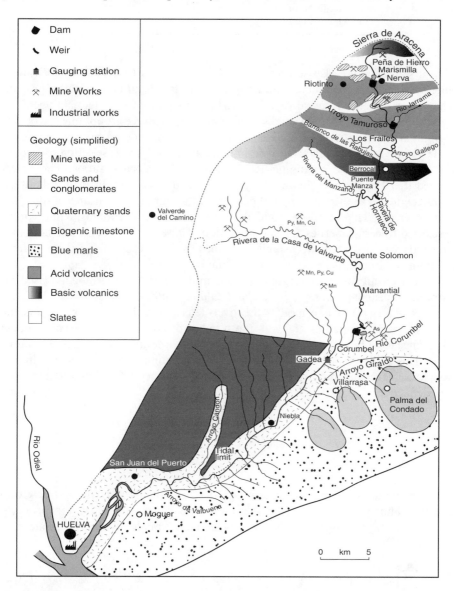

Figure 26.3 Geological sketch map of the Río Tinto catchment

Table 26.1   Number of samples in source groups

Source groups	No. of samples
Volcanics	8
Slates	21
Quaternary/blue marls	16
Historical alluvium	3
Mine-derived alluvium	3
Ore/waste	12
Modern alluvium	20
TOTAL	83

addition, samples were also subjected to X-ray diffraction (XRD) analyses using a Siemens D5000 dual goniometer diffractometer for identification of minerals present.

The non-parametric Kruskal–Wallis *H*-test confirmed that all potential fingerprint properties except W are capable of distinguishing the six source groups unequivocally. The optimal combination of trace elements for the composite fingerprint was ascertained by applying the multivariate discriminant analysis stepwise selection procedure in SPSS (v. 6.1) to the remaining 10 elements and associated ratios. The MDA identified that Pb, Sr, Rb, Nb and Y as sufficient for complete distinction between the source groups as they explain 100 percent of the cumulative variance in the data set; however, Ba, Th and Zn were added to refine the optimal composite fingerprint. The selected tracer properties are a combination of rare earth elements and contaminant metals regarded adequately stable (with the possible exception of Zn) even within the extreme hydrochemical environment of the Río Tinto.

The most recent fingerprinting models use multivariate mixing models based on constrained linear equations (e.g. Yu and Oldfield, 1993; Collins *et al.*, 1997a,b; Rowan *et al.*, 1999). These models are over-determined if $m \geq n$ where m is the number of tracer properties and n is the number of source groups. Hence, model performance must be optimised to compute the relative contributions from the source groups. This study employed a simple spreadsheet optimisation tool (Solver in MS Excel) to achieve maximisation of the explained variance (cf. Nash and Sutcliffe, 1970), defined as:

$$\text{Efficiency} = 1 - \frac{\sum_{i=1}^{m}(\hat{x}_i - x_i)^2}{\sum_{i=1}^{m}(x_i - \bar{x}_i)^2} \tag{26.1}$$

where m is the total number of tracer properties, $\bar{x}_i$ is the mean of the source group properties, and x_i and $\hat{x}_i$ are the measured and predicted values for property $i(i = 1, 2, \ldots, n)$. The algorithm for the linear mixing model used to calculate $\hat{x}_i$ is as follows:

$$\hat{x}_i = \sum_{j=1}^{n}(a_{ij}b_j) \quad \text{subject to the constraints} \quad \sum_{j=1}^{n}b_{j=1} \quad \text{and} \quad 0 \leq b_j \leq 1 \tag{26.2}$$

where n is the number of source groups, a is the mean value of property $i(i = 1, 2, \ldots, m)$ of source group $j(j = 1, 2, \ldots, n)$, and b is the contributory coefficient of source group j.

The above model was applied to the selected set of geochemical properties after the concentrations had been divided by the maximum for each variable (Verruchi and Minissale, 1995). A correction for variations in organic matter content and grain size was deemed unnecessary as there was no significant statistical correlation between trace elements, organic matter and grain size observed in Río Tinto alluvium (Schell, 1998).

The model was run to the optimal efficiency as well as to three 1 percent step efficiency values preceding the optimised solution because recent work has shown that many current models are susceptible to multiple solutions near the end-point (cf. Rowan *et al.*, see Chapter 14). Thereby, the progressive evolution towards the optimised model result could be observed, hence permitting not an absolute but a probable estimate of source contributions, which allows for the likelihood of some source materials being present in the end-member which are not actually detected by the optimised fingerprinting result. Field observations were used to substantiate model results where possible.

RESULTS

Sediment provenance was determined for four overbank sediment samples: modern alluvium from an upstream site at Berrocal and a downstream site at Niebla (Figure 26.2), Quaternary valley fill, and the base layer of an estuarine core which pre-dates the 19th century mining heyday (Schell, 1998). This enabled the comparison of suspended sediment provenance at two locations in the Río Tinto and at three points in time. Figure 26.4 presents sediment source contributions to the four types of overbank alluvium at near-optimum and optimum efficiency.

The optimised mixing model result shows modern flood sediment at the upstream site of Berrocal (Figure 26.2; 14 km downstream of the Río Tinto mines) to be a mixture of source materials from the mining area (mine waste and eroded ore rock), reworked contaminated alluvium and the geological subgroup slates, with ore/waste and mine alluvium contributing approximately equal proportions and dominating the suspended sediment composition (81 percent). This prevalence of mine-derived material in contemporary suspended sediment confirms the substantial impact of historical mine waste on sediment composition and quality in the Río Tinto at some distance downstream of the mine. The large percentage of reworked mine-derived alluvium demonstrates the significance of temporary channel margin storage and subsequent remobilisation of contaminated sediment in this river system. Tributaries draining the slate geological subgroup add 'uncontaminated', or natural background, material to the current flood sediment at Berrocal (19 percent). In addition, the near-optimal model solutions register the possibility of a small contribution from reworked pre-mining alluvium, which is reasonable considering the proximity of medieval terraces to the river at this site. Confirmation of this can be seen more clearly in Tables 26.2–26.4 where XRD results for all the samples are presented. XRD analyses of the

Berrocal flood sediment can also be seen in Figure 26.5. The sample comprises principally pyrite (FeS, 2.71 d Å) and quartz (SiO$_4$; 3.35 d Å) together with small quantities of galena (PbS; 2.97 + 3.43 d Å), arsenopyrite (FeAsS; 2.44 + 2.42 + 2.40 d Å) and feldspar (Na–Ca–KSi$_2$AlO$_4$, 3.18 d Å). The majority of the sample is of a mining-related origin (mine-derived alluvium plus waste/ore). From Table 26.3 both these groups contain abundant quartz, haematite and iron sulphide, with lesser

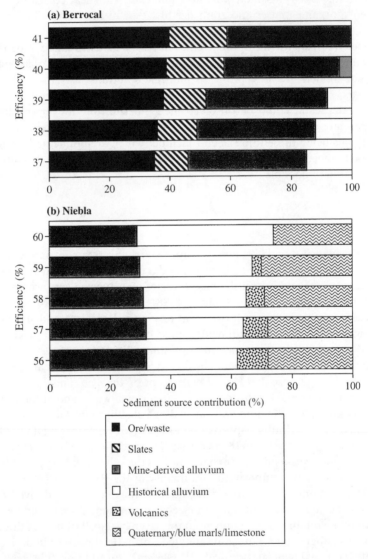

Figure 26.4(a,b) Results of the fingerprinting for the Río Tinto sediments. The optimal efficiency is always at the top of the diagram for each locality, i.e. 41 percent for Berrocal, 60 percent for Niebla, 87 percent for the base of the core and 97 percent for the Late Pleistocene terrace. In addition, the four nearest percentage efficiency steps are also indicated for each locality for comparison

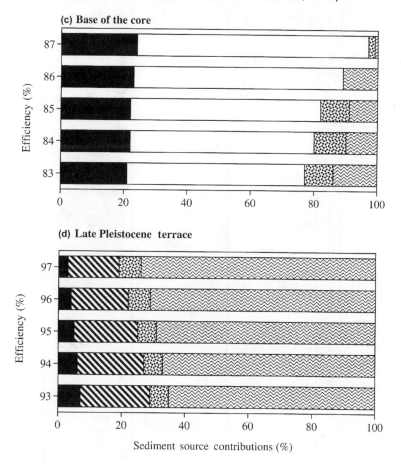

Figure 26.4(c,d)

Table 26.2 Fingerprinting results for the optimal solution

	Berrocal recent flood sediment (41%)	Niebla recent flood sediment (60%)	Base of core (pre-mining) (87%)	Late Pleistocene terrace (97%)
Mine-derived alluvium	40	29	0	0
Ore/waste	41	0	24	3
Slates	19	0	0	16
Historical alluvium	0	44	72	0
Quaternary/blue marls	0	27	2	74
Volcanics	0	0	2	7

Table 26.3 XRD results of the main source groups

	Quartz	Calcite	Clays	Feldspar	Pyroxene	Olivine	Jarosite	Haematite	FeS	PbS	ZnS	FeAsS	CuS
Mine-derived alluvium	***	*	**	**	NP	NP	**	***	***	**	**	**	**
Ore/waste	***	NP	**	**	NP	NP	**	***	****	**	**	**	***
Slates	**	NP	***	**	NP	NP	NP	*	**	*	*	*	*
Historical alluvium	***	**	**	**	NP	NP	*	**	**	*	*	*	*
Quaternary/blue marls	***	****	**	***	NP	NP	*	*	*	*	*	*	*
Volcanics	**	NP	**	***	***	**	NP	**	*	*	NP	NP	**

Table 26.4 XRD results for the four samples fingerprinted

Sample	Quartz	Calcite	Clays	Feldspar	Pyroxene	Olivine	Jarosite	Haematite	FeS	PbS	ZnS	FeAsS	CuS
Berrocal	***	NP	**	**	NP	NP	*	**	***	**	*	**	**
Niebla	***	**	***	**	NP	NP	*	**	**	*	**	**	*
Base of core	***	NP	***	**	NP	NP	NP	*	**	NP	NP	*	*
Late Pleistocene	***	***	**	***	NP	NP	NP	**	*	NP	NP	*	NP

NP = Not Present.

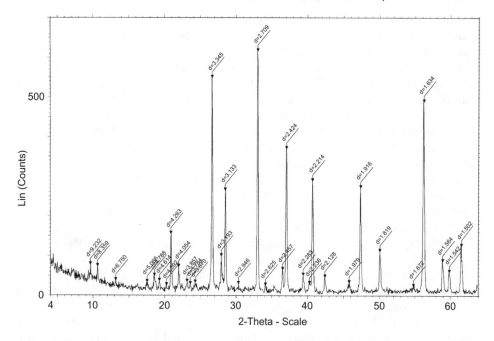

Figure 26.5 X-ray diffraction pattern collected for one of the Berrocal flood sediments. The major peak d-spacings are labelled. Note the major peaks: quartz (3.35 d A°), feldspar (3.18–3.25 d Å), iron sulphide (FeS; 2.71 + 1.63 d A°), galena (PbS; 2.97 + 3.43 d A°), sphalerite (ZnS; 3.12 + 1.91 d Å) and arsenopyrite (FeAsS; 2.44 + 2.42 + 2.40 d A°). See Table 26.2(a)–(c) for more details

amounts of galena, chalcopyrite, arsenopyrite, sphalerite, jarosite and feldspar, very similar to the observed flood sediment. Although only semi-quantitative, the XRD data illustrate that the Berrocal flood sediment comprises principally mine alluvium waste as predicted by the fingerprinting results, thus confirming that our source groups and predicted data are consistent with observations.

By contrast, the optimum model solution for modern flood sediment at Niebla, 53 km downstream of the mines, indicates that it contains a much smaller component (29 percent) of mine-derived sediment, and no material directly sourced from the mining area, only previously stored and reworked alluvium. The remainder is derived from the southern lithologies comprising Quaternary valley fill, blue marls and limestone (27 percent), and from reworked uncontaminated alluvium pre-dating the mining heyday (44 percent). These results indicate that the direct impact of the mine decreases with downstream distance, although reworked mine alluvium remains an important secondary sediment and contaminant source at great distances from the mines. Temporary channel and bank storage therefore promote the preservation of metal contamination within the fluvial system, and the dispersal of mine-derived material over large distances. Reworked alluvium as a whole contributes the majority (73 percent) of modern suspended sediment. The near-optimal model solutions also suggest the likely presence of some volcanic lithologies from the head of the catchment although their contribution is minor compared to reworked alluvium and local Quaternary lithologies. This is again confirmed by the XRD results (Tables 26.2–26.4) where the results for Niebla comprise a mixture of minerals that could be derived from a mixture of historical, mine-derived and Quaternary sequences. Although not unique, the data suggest that the fingerprinting is accurate; for example, no volcanic sources were predicted for this locality, and no pyroxene or olivine was detected (which are minerals typically occurring in basic volcanic lithologies).

The base layer of the core predates the mining heyday from 1873 and 1954 (Schell, 1998), which is reflected in its composition of predominantly reworked uncontaminated alluvium (72 percent), and material from northern as well as southern lithologies (2 percent volcanics, 2 percent Quaternary material). The predominance of reworked alluvium as opposed to direct tributary inputs is again striking and illustrates the importance of this geomorphological process in long-term sediment and metal dispersal. The Río Tinto ore field held a significant influence on alluvial composition even before large-scale mining commenced, as reflected in its contribution of 24 percent. This relatively large contribution may correspond to major anthropogenic activity of the mining area prior to the mining heyday at the turn of the century, including the Roman era and later medieval extraction attempts, which led to the presence of considerable metalliferous waste deposits along the Río Tinto by the 1850s (Avery, 1974). These waste heaps are likely to have been a source of metal contamination to the river at this time. This is supported by the XRD trace for this sample which shows a mixture of principally quartz, feldspar, jarosite, clay minerals and small amounts of pyrite (Tables 26.2–26.4).

Late Pleistocene valley fill exhibits a source composition dominated by the southern lithologies of blue marls and limestone (74 percent) in addition to smaller proportions of volcanics (7 percent) and slates (16 percent) from further upstream in the catchment. The Río Tinto ore field only contributes 3 percent, which reflects the low release rates of metals from the massive ore rock into the river. This is also confirmed by the XRD results which show only quartz, calcite, feldspar, and very small quantities of clay and iron sulphide present.

Optimal efficiency values, which represent the variance of the data explained by the fingerprinting model, range from 41 percent for modern alluvium at Berrocal to 97 percent for Quaternary valley fill, and are generally considered sufficient to permit elementary interpretation of the model result. An efficiency of 41 percent is rather low but in all cases the compositional progression towards the optimum model output as well as corroborating evidence from XRD analysis were taken into account, thereby supporting the interpretation regarding sediment provenance. The relatively low efficiency obtained for the modern alluvium compared to the older alluvial units may be linked to their more complex origin, ensuing variability and more scattered data distribution.

DISCUSSION

The results obtained from the fingerprinting model, though tentative, illustrate the temporal and spatial evolution of sediment source contributions to Río Tinto alluvium. In the absence of mining activity in the Late Pleistocene, inputs from the Río Tinto ore field to suspended sediment in the river were limited by the slow release rates from ore rock by physical and chemical weathering processes. Early alluvium consisted of varying proportions of material from the three major geological sub-areas in the catchment, with local lithologies dominating river sediment at any one site. After more than 4000 years of mineral extraction in the Holocene, including the major Roman exploitation phase, the addition of metalliferous solids to the Río Tinto had increased substantially, as reflected in the base layer of the core which dates from the late 19th century (Schell, 1998). Thus, the disposal of particulate mine waste in the vicinity of the river, comprising large surface area and easily erodible sediments, has resulted in significant contributions of mine-related material to Río Tinto alluvium. This is illustrated clearly in Figure 26.6, with the waste and ore source groups clearly chemically different from the Late Pleistocene and other groups. In addition, the changing source characteristics of the sediments can be seen in their relative positions on the diagram. For example, the recent mud from Niebla is clearly different from the recent flood sediment from Berrocal, as indicated by their different source provenance.

Modern alluvium shows the significant human impact in the Río Tinto catchment in the last century. It derives its source material predominantly from mine-related deposits such as mine waste, ore rock and reworked contaminated alluvium, with 'natural' lithologies only playing a subordinate role in alluvial composition. In comparison with the source contributions to older alluvium, the contributions from mine-derived material exhibit a steady rise over time (Figure 26.6), indicating that the composition of suspended sediment in the Río Tinto is to a large extent determined by mining activity at the head of the catchment, and in particular by the availability and composition of mine waste on the river banks. The historical mine waste heaps on the river banks are at present unconfined and readily erodible, and constitute a persistent contaminant source to the Río Tinto. The high proportion of mine-derived material in modern alluvium is therefore unlikely to decrease and may even increase further in the future.

The influence of the mining area on Río Tinto alluvium declines with downstream distance, with no direct contributions from waste tips and ore rock detectable in alluvium at ~ 53 km downstream, although reworked metals, from mine-derived alluvium, do get into the Gulf of Cadiz (Van Geen et al., 1997). Nevertheless, the immediate effect of the mines and associated waste material extends over distances of tens of kilometres downstream, as reflected in the source composition of modern sediment 14 km from the mines. A significant proportion of modern and historical alluvium in the Río Tinto consists of reworked contaminated and 'clean' pre-mining alluvium, which suggests considerable intermediate storage of alluvium. The propor-tion of reworked alluvium increases downstream as the volume of stored sediment upstream of the end-member sample also becomes greater. A number of studies of fluvial mine waste dispersal in semi-arid and arid environments indicate that metal-rich sediments often accumulate preferentially in the channel and channel margin domain, and may constitute a primary source of suspended sediment (Graf et al., 1991; Graf, 1994). This is in contrast to temperate floodplain systems, where only a small percentage of the annual suspended sediment yield is retained in channel storage (Lambert and Walling, 1986; Miller and Shoemaker, 1986), and mobilisation from

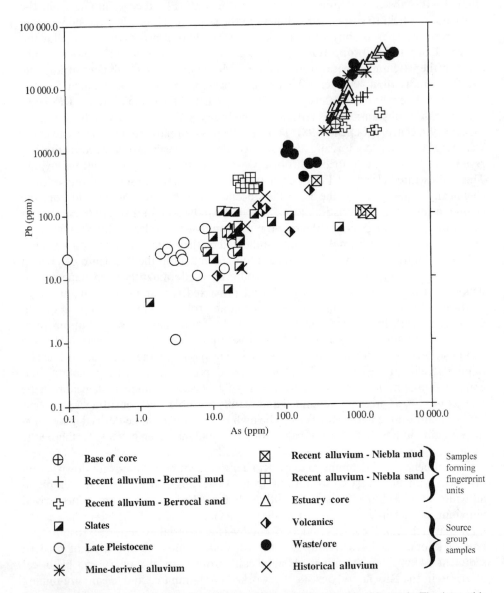

Figure 26.6 Pb–As correlation diagram for all the samples used in this study. The legend is split into two parts: the upper part contains samples that form the fingerprinted units; the lower part contains source group samples. Note the large variation between source groups (on the log–log scale), varying from lowest, the Late Pleistocene group (<20 ppm Pb, <3 ppm As) to the highest, the waste/ore group ($<40\,000$ ppm Pb, <3000 ppm As). The recent alluvium also shows clear, distinctive differences between specific areas, e.g. Berrocal and Niebla, as indicated by the fingerprinting results

channel storage may contribute as little as 5 percent to the annual suspended sediment yield (Miller and Shoemaker, 1986). Alluvial secondary storage has serious implications for the efficiency of contaminant transfer through the Río Tinto fluvial system because it is known to extend the influence of the mine over larger distances as well as over longer timescales (cf. James, 1989, 1991, 1993; Miller, 1997). The substantial amounts of heavy metals stored in alluvium also pose an environmental risk because future climate-induced variations in hydrologic regime and sediment delivery will affect metal mobility and dispersal (Miller, 1997). Recent climate models forecast an increase in temperature and concomitant decrease in precipitation in the Mediterranean basin if global CO_2 levels continue to rise (Verhoog, 1987). This scenario is expected to entail faster rates of reworking due to increased aridity and seasonality and, therefore, higher channel gradient and erosivity of flood events (Schumm, 1971; Verhoog, 1987). The overall effect of global warming on metal contamination in the Río Tinto would thus be temporal and spatial augmentation. Any attempts at impact amelioration must take this important aspect into account.

CONCLUSIONS

(1) The statistical 'fingerprinting' approach has demonstrated a significant deterioration in sediment quality and changes in the composition of modern alluvium as a result of mining activity in the late 19th and 20th centuries. Whilst the Riotinto ore body contributed only 3 percent to the composition of Late Pleistocene alluvium, up to 80 percent of present-day alluvium may consist of mine-derived material. The fingerprinting results also suggest that reworking of contaminated alluvium stored along the course of the river is a considerable secondary source of heavy metal-polluted sediment to the Río Tinto, with between 40 percent and 70 percent of flood sediment derived from alluvial stores. Increased rates of remobilisation of stored sediment, for example as a result of climatic changes, may be expected to entail a concomitant increase in contaminant dispersal rates.

(2) The application of a simple fingerprinting model to the complex Río Tinto fluvial system has yielded robust and realistic estimates of sediment provenance and enabled an alluvial history to be established. The interpretation has been kept tentative in order to avoid over-interpretation of the data, and should be seen as an indicator of approximate sediment source contributions because the present limitations of the technique do not permit error bands to be assigned to the calculated optimal values. However, by using XRD analyses on all samples we were able to support the fingerprinting data with real observations.

(3) The fingerprinting model could be improved by introducing factors that correct for variations in grain size and organic matter, as well as analytical errors (cf. Collins et al., 1997b, 1998), although this is not thought to have a significant effect on the outcome due to the general independence of elements of physicochemical variables in the Río Tinto. The inclusion of additional sediment properties other than trace elements (e.g. mineral magnetics, grain-size distribution and mineralogy) may also help to produce a more accurate fingerprint (e.g. Foster and Walling, 1994; Deboer and Crosby, 1995; Walling and Woodward, 1995).

(4) Following further advances and refinement, particularly the computation of
     errors and confidence limits, fingerprinting may provide a valuable tool to
     environmental planners and mining companies seeking a cost-effective and
     labour-efficient method of assessing the impact of mining activity on alluvial
     composition and contamination.

ACKNOWLEDGEMENTS

C.S. thanks the School of Geography at the University of Leeds for a PhD studentship, and the BGRG for a small grant from the Research and Publications Fund. S.B.
is grateful to the Leverhulme Trust for funding during part of this study. The authors
are grateful to John Rowan and Kirk Nordstrom for helpful reviews of an earlier
version of this manuscript.

REFERENCES

Avery, D. 1974. *Not On Queen Victoria's Birthday: The Story of the Rio Tinto Mines*. Wester-
    ham Press, London.
Collins, A. L., Walling, D. E. and Leeks, G. J. L. 1997a. Use of geochemical records preserved
    in floodplain deposits to reconstruct recent changes in river basin sediment sources. *Geo-
    morphology*, **19**, 151–167.
Collins, A. L., Walling, D. E. and Leeks, G. J. L. 1997b. Source type ascription for fluvial
    suspended sediment based on a quantitative composite fingerprinting technique. *Catena*, **29**,
    1–27.
Collins, A. L., Walling, D. E. and Leeks, G. J. L. 1998. Use of composite fingerprints to
    determine the provenance of the contemporary suspended sediment load transported by
    rivers. *Earth Surface Processes and Landforms*, **23**, 31–52.
Deboer, D. H. and Crosby, G. 1995. Evaluating the potential of SEM/EDS analysis for
    fingerprinting suspended sediment derived from two contrasting topsoils. *Catena*, **24**, 243–
    258.
Dietrich, W. E., Dunne, T., Humphrey, N. F. and Reid, L. M. 1982. Construction of sediment
    budgets for drainage basins. In Swanson, F. J., Janda, R. J., Dunne, T. and Swanston, D. N.
    (eds) *Sediment Budgets and Routing in Forested Drainage Basins*. US Department of Agri-
    culture and Forestry Services General Technical Report PNW-141, 5–23.
Foster, I. D. L. and Walling, D. E. 1994. Using reservoir deposits to reconstruct changing
    sediment yields and sources in the catchment of the Old Mill Reservoir, South Devon, UK
    over the past 50 years. *Hydrological Sciences Journal*, **39**, 347–368.
Graf, W. L. 1994. *Plutonium and the Rio Grande – Environmental Change and Contamination in
    the Nuclear Age*. Oxford University Press, Oxford.
Graf, W. L., Clark, S. L., Kammerer, M. T., Lehman, T., Randall, K. and Schroeder, R. 1991.
    Geomorphology of heavy metals in the sediments of Queen Creek, Arizona, USA. *Catena*,
    18, 567–582.
Grimshaw, D. L. and Lewin, J. 1980. Source identification for suspended sediments. *Journal of
    Hydrology*, **47**, 151–162.
Hudson-Edwards, K. A., Schell, C. and Macklin, M. G. 1999. The mineralogy and geochem-
    istry of alluvium contaminated by metal mining in the Río Tinto area, southwest Spain.
    Applied Geochemistry, **14**, 1015–1030.
James, L. A. 1989. Sustained storage and transport of hydraulic gold mining sediment in the
    Bear River, California. *Annals of the Association of American Geographers*, **79**, 570–592.

James, L. A. 1991. Incision and morphologic evolution of an alluvial channel recovering from hydraulic mining sediment. *Geological Society of America Bulletin*, **103**, 723–736.

James, L. A. 1993. Sustained reworking of hydraulic mining sediment in California: G. K. Gilbert's sediment wave model reconsidered. *Zeitschrift Geomorphology N.F.*, Suppl. Bd. **88**, 49–66.

Lambert, C. P. and Walling, D. E. 1986. Suspended sediment storage in river channels: a case study of the River Exe, Devon, UK. In Hadley, R. F. (ed.) *Drainage Basin Sediment Delivery*. IAHS Publication **159**, IAHS Press, Wallingford, 263–276.

Lewin, J. and Macklin, M. G. 1987. Metal mining and flood plain sedimentation in Britain. In Gardiner, V. (ed.) *International Geomorphology 1986*, Part 1. John Wiley, Chichester, 1009–1028.

Miller, J. R. 1997. The role of fluvial geomorphological processes in the dispersal of heavy metals from mine sites. *Journal of Geochemical Exploration*, **58**, 101–118.

Miller, J. R. and Shoemaker, L. L. 1986. Channel storage of fine-grained sediment in the Potomac river. In Hadley, R. F. (ed.) *Drainage Basin Sediment Delivery*. IAHS Publication **159**, IAHS Press, Wallingford, 287–304.

Nash, J. E. and Sutcliffe, J. V. 1970. River flow forecasting through conceptual models 1: a discussion of principles. *Journal of Hydrology*, **10**, 282–290.

Nelson, C. H. and Lamothe, P. J. 1993. Heavy metal anomalies in the Tinto and Odiel river and estuary system, Spain. *Estuaries*, **16**, 496–511.

Oldfield, F., Rummery, T. A., Thompson, R. and Walling, D. E. 1979. Identification of suspended sediment sources by means of magnetic measurements: some preliminary results. *Water Resources Research*, **15**, 211–218.

Peart, M. R. and Walling, D. E. 1986. Fingerprinting sediment sources: the example of a small drainage basin in Devon, UK. In Hadley, R. F. (ed.) *Drainage Basin Sediment Delivery*. IAHS Publication **159**, IAHS Press, Wallingford, 41–55.

Peart, M. R. and Walling, D. E. 1988. Techniques for establishing suspended sediment sources in two drainage basins in Devon, UK: a comparative assessment. In Bordas, M. P. and Walling, D. E. (eds) *Sediment Budgets*. IAHS Publication **174**, IAHS Press, Wallingford, 269–279.

Rowan, J. S., Black, S. and Schell, C. 1999. Floodplain evolution and sediment provenance reconstructed from channel-fill sequences: the Upper Clyde basin, Scotland. In Brown, A. G. and Quine, T. (eds) *Fluvial Processes and Environmental Change*. John Wiley, Chichester, 223–240.

Saez, R., Almodovar, G. R. and Pascual, E. 1996. Geological constraints on massive sulphide genesis in the Iberian Pyrite Belt. *Ore Geology Reviews*, **11**, 429–451.

Schell, C. 1998. Dispersal of metal mine waste in the Río Tinto, Huelva, south-west Spain. Unpublished PhD Thesis, University of Leeds.

Schell, C., Macklin, M. G. and Hudson-Edwards, K. A. 1996. Flood dispersal and alluvial storage of heavy metals in an acid ephemeral river: Río Tinto, Huelva, SW Spain. *Proceedings of the 4th International Symposium on the Geochemistry of the Earth's Surface*, University of Leeds, Department of Earth Sciences, Leeds, UK, 21–25 July 1996.

Schermerhorn, L. J. G. 1982. Framework and evolution of Hercynian mineralization in the Iberian Meseta. *Comunicacione Geologico Iceo, Portugal*, Allen and Unwin, London, **1**, 91–140.

Schumm, S. A. 1971. Fluvial geomorphology: historical perspective, and channel adjustment and river metamorphosis. In Shen, H. W. (ed.) *River Mechanics*. Australia, Allen and Unwin, 299–310.

Strauss, G. K., Madel, J. and Alonso, F. F. 1977. Exploration practice for strata-bound volcanogenic sulphide deposits in the Spanish–Portugese pyrite belt. In Klemm, D. D. and Scheinder, H.-J. (eds) *Time and Strata-bound Ore Deposits*. Springer Verlag, Berlin, 55–93.

Velasco, F., SanchezEspana, J., Boyce, A. J., Fallick, A. E., Saez, R. and Almodovar, G. R. 1998. A new sulphur isotopic study of some Iberian Pyrite Belt deposits: evidence of a textural control on sulphur isotope composition. *Mineralium Deposita*, **34**, 4–18.

Verhoog, F. H. 1987. Impact of climate change on the morphology of river basins. In Solomon, S. I., Beran, M. and Hogg, W. (eds) *The Influence of Climate Change and Climate Variability on the Hydrologic Regime and Water Resources*. IAHS Publication **168**, IAHS Press, Wallingford, 315–326.

Verruchi, C. and Minissale, A. 1995. Multivariate statistical comparison of Northern Apennines Palaeozoic Sequences: a case study for the formations of Monti Romani (Southern Tuscany–Northern Latium, Italy). *Applied Geochemistry*, **10**, 581–598.

Walling, D. E. and Woodward, J. C. 1995. Tracing sources of suspended sediment in river basins: a case study of the River Culm, Devon, UK. *Marine and Freshwater Ecology*, **46**, 327–336.

Walling, D. E., Woodward, J. C. and Nicholas, A. P. 1993. A multi-parameter approach to fingerprinting suspended sediment sources. In Peters, N. E., Hoehn, E., Leibundgut, C., Tase, N. and Walling, D. E. (eds) *Tracers in Hydrology*. IAHS Publication **215**, IAHS Press, Wallingford, 329–338.

Yu, L. and Oldfield, F. 1993. Quantitative sediment source ascription using magnetic measurements in a reservoir-catchment system near Nijar, S.E. Spain. *Earth Surface Processes and Landforms*, **18**, 441–454.

27 Sediment Sources and Terminal Pleistocene Geomorphological Processes Recorded in Rockshelter Sequences in North-west Greece

J. C. WOODWARD[1] and G. N. BAILEY[2]

[1]*School of Geography, University of Leeds, UK*
[2]*Department of Archaeology, University of Newcastle, UK*

INTRODUCTION

The sedimentary sequences preserved in rockshelters can provide valuable records of past geomorphological processes (e.g. Farrand, 1979; Patton and Dibble, 1982; Gillieson *et al.*, 1986). This chapter is concerned with shallow rockshelters and cave-mouth environments in limestone terrains (rather than deep caves and subterranean karstic caverns) which offer a degree of protection from subaerial processes but which can also trap and retain clastic sediments from a variety of off-site (*allogenic*) sources. Some rockshelters in limestone karst environments have formed long-term depositional sinks for materials derived from a wide range of allogenic and autogenic sources with sedimentary records occasionally extending back to the Middle Pleistocene (see Farrand, 1979; Laville *et al.*, 1980; Butzer, 1981; Huxtable *et al.*, 1992). Thus, as landscape features, many rockshelters demonstrate considerable long-evity and this is commonly due to the presence of a resistant host limestone bedrock. These hard limestones usually contain only very small amounts ($< 1\%$) of insoluble material (Butzer, 1981; Ford and Williams, 1989; Woodward, 1997a) and this purity means that the potential for the on-site production of fine-grained residues from chemical weathering is limited. Furthermore, as such sound rocks tend to liberate few fines – even where frost action is important – allogenic fine sediments often dominate the fine components of the rockshelter sedimentary record (e.g. Butzer, 1981; Woodward, 1990; Bar-Yosef, 1993; Gale *et al.*, 1993). The importance and variety of allogenic sediment inputs within cave and rockshelter sequences have been recognised for many years. For example, Schmid (1969: 156) pointed out the following:

"Observation of the geological conditions in the neighbourhood of a cave enables one to judge from where and how alien substances got into the cave: because a cave impinged upon another rock formation inside the hill or because particles reached the cave floor through crevices and in percolating water from an overlying rock formation, or when gravel, erratic pieces, loess or soil were washed in from the hillside through cracks in the walls of the cave."

Using an appropriate range of sediment properties to trace the provenance of the sediment imports and their variation through time, it is possible to relate them to the operation of particular geomorphological processes in the wider environment and, with good dating control, such rockshelter sediment records can assume considerable palaeoenvironmental significance.

It is well known that rockshelter environments were attractive habitation and activity sites for prehistoric humans, and their sedimentary sequences can contain rich assemblages of faunal remains, stone tools and other artefacts (e.g. Gamble, 1986; Bailey, 1997a). Thus, rockshelter sediments are commonly studied as part of the archaeological excavation of Palaeolithic, Mesolithic and Neolithic sites. The value of sediment studies is widely appreciated and the analysis of rockshelter sediments is now an established part of geoarchaeology. Many geologists and geomorphologists have participated in archaeological projects to provide stratigraphical and environmental context for the archaeological record. Indeed, several analytical and interpretative frameworks have evolved (Farrand, 1975, 1985; Laville, 1976; Collcutt, 1979; Bar-Yosef, 1993; Goldberg *et al.*, 1993) and recent developments in the micromorphological and sedimentological study of rockshelter sequences have allowed detailed reconstructions of past site conditions and human activities (Macphail and Goldberg, 1995; Tsatskin *et al.*, 1995; Woodward, 1997a,b). Nonetheless, it can be argued that many studies of rockshelter sediment records have tended to focus on the *site* sediments without fully considering the wider environmental context and the nature, composition and age of adjacent sedimentary environments and the possibility of material exchanges between them (see Laville *et al.*, 1980). In a discussion of the work on the classic Palaeolithic sites in the Perigord region of south-west France, Vita-Finzi (1978: 44) observed that 'the growing concern with climatic chronology came to divert attention into caves and shelters and away from their environs'.

This chapter aims to highlight the importance of viewing rockshelter sediments as part of wider and dynamic geomorphological systems and to show how careful integration of on-site and off-site sequences – through a consideration of fine sediment provenance – can improve our understanding of rockshelter sediment records and the geomorphological processes involved in their formation. Such an approach can also aid our understanding of the relationship between the cultural record and environmental change, but these archaeological issues will not be pursued here (see Woodward, 1997a). To illustrate these themes and some of the problems involved in the interpretation of rockshelter sediment records, data are presented from two neighbouring yet contrasting rockshelter sites: Klithi and Megalakkos in the Pindus Mountains of north-west Greece. These sites contain important Late Upper Palaeolithic records and sedimentary sequences spanning the last glacial–interglacial transition (Bailey, 1997a,b).

ROCKSHELTER SEDIMENT RECORDS AND ENVIRONMENTAL CHANGE: COARSE AND FINE SEDIMENT COMPONENTS

Rockshelter sediments are typically composed of a range of poorly sorted coarse and fine clastic materials, with the former often produced on-site from mechanical weathering of the host limestone strata (Figure 27.1). However, the palaeoclimatic significance of the sedimentary units containing coarse angular rock fragments (commonly termed *éboulis*) and the significance of any variations in shape and calibre within them has been strongly debated and their interpretation is fraught with uncertainty (see Laville, 1976; Collcutt, 1979; Farrand, 1981, 1985; Bailey and Woodward, 1997).

Figure 27.1 Poorly sorted, coarse-grained angular limestone rock fragments with a fine-grained sediment matrix. These sediments were exposed in the upper 2 m of the Late Pleistocene sedimentary fill at the rockshelter site of Klithi in the Voidomatis Basin. They were exposed in the deep section that was excavated in 1983 (see Bailey and Woodward, 1997 and Figure 27.8)

Some workers have associated their presence with a single geomorphological process such as frost action (e.g. Laville, 1976). However, several authors have observed that a wide range of mechanisms can liberate coarse angular rock fragments from a bedrock wall and it is unwise to read too much into such deposits in terms of palaeoclimatic conditions (Vita-Finzi, 1978; Collcutt, 1979; Farrand, 1981), especially when the frequency of bedding planes and joints is highly variable and rockfall-triggering seismic activity is important (Bailey and Woodward, 1997). This debate will not be elaborated here, but it is relevant because it can be argued that in many contexts the fine sediment fraction within a rockshelter sequence offers a more reliable basis for reconstructing past environments and geomorphological processes for the following reasons:

- Fine sediments within rockshelters may be protected from the subaerial weathering processes that affect surface materials; in certain contexts these deposits can provide a sensitive record of local and regional environmental change.
- As they can be dominated by materials derived from a range of proximal and distal allogenic sources, the fine sediments constitute a potentially important interface between the rockshelter sediment record (and the archaeological data within it) and the off-site Quaternary sedimentary record.
- The fine sediments within the rockshelter sequence may contain evidence of geomorphological events and sediment fluxes that have not been preserved in the wider environment because of erosion and weathering.
- In contrast to the poorly sorted coarse-grained products of host bedrock breakdown, many primary fine sediment properties display limited lateral variation within coeval depositional units.

In addition, silts and clays are commonly abundant throughout rockshelter sequences and recent developments in coring techniques for sampling rockshelter sequences (which yields small samples; see Bailey and Thomas, 1987) and micromorphological analysis (Courty *et al.*, 1989), have focused attention on the significance of the fine sediment fraction (Woodward, 1997a). It is important to bear in mind, however, that both penecontemporaneous and post-depositional (diagenetic) processes (chemical and biological – including human activity) can influence the nature of the fine sediment fraction (Butzer, 1982; Bull, 1983; Bar-Yosef, 1993; Macphail and Goldberg, 1995; Tsatskin *et al.*, 1995). The nature and intensity of these modifications will vary spatially and temporally in accord with the geomorphological context. At sites where prehistoric occupation has been intensive, one of the main challenges facing such investigations is the development of approaches which allow the natural and anthropogenic signatures in the sedimentary record to be decoupled (see Ellwood *et al.*, 1997; Woodward, 1997a). This is rarely straightforward, but the potential for rockshelter fine sediments to provide a bridge between the on-site archaeological record and the off-site sedimentary record (as in alluvial, lacustrine, aeolian, marine or glacial sequences) has yet to be fully exploited; the use of sediment provenance techniques is an important way of exploring this potential.

The value of these environmental archives is enhanced considerably when the fine sediment sources can be identified and related to the operation of particular geomorphological processes in the wider Pleistocene or Holocene environment. As far as the study of rockshelter sediment records is concerned, it is also worth noting that recent developments in AMS radiocarbon, luminescence, ESR and uranium-series dating techniques (e.g. Schwarcz *et al.*, 1988; McDermot *et al.*, 1993; Mercier and Valladas, 1994; Gowlett *et al.*, 1997) have not only allowed issues of temporal resolution, rates of sedimentation and the significance of unconformities to be addressed with much greater precision, but they have also meant that on-site/off-site and inter-site correlations can be established with some confidence (Campy and Chaline, 1993; Bailey 1997a,b). It is important to appreciate that sedimentation rates can be highly variable both within and between individual rockshelter sites (Bailey *et al.*, 1983; Campy and Chaline, 1993) and episodes of erosion can remove portions of the stratigraphic record. Good dating control is therefore essential to identify the significance of any unconformities and the duration of any major gaps in these records (see Farrand, 1993). Furthermore, in order to establish the wider geomorphological and stratigraphical context of a particular rockshelter sequence, and the character of potential sediment sources (and their variation over time), the study of such records has to be accompanied by geological and geomorphological investigations of the surrounding area (see Vita-Finzi, 1978; Farrand, 1985; Woodward *et al.*, 1994, 1995; Macklin *et al.*, 1997).

SOURCES AND PATHWAYS FOR ROCKSHELTER FINE SEDIMENTS

All rockshelters lie within a catchment for both proximal and distal fine sediment source components which is broadly defined by the aspect, geometry and geomorphological setting of the site. Fine sediments can be delivered to the rockshelter via a series of processes and vertical and lateral pathways (Figure 27.2). For a given site this could involve contributions from, for example, aeolian activity, fluvial and colluvial transport, and vertical infiltration through karst conduits. Some examples of these fine sediment delivery mechanisms reported from rockshelter sites in various environments are listed in Table 27.1. The importance of these sources and transfer processes will vary over time and between sites as environmental conditions change. The following sections discuss the nature of rockshelter sediment records and the processes and pathways involved in fine sediment accumulation.

Fine-grained sediments (silts and clays) are the most widespread clastic deposits in caves and rockshelters and their origins are the most diversified (Ford, 1975: Jennings, 1985; Ford and Williams, 1989; Table 27.1). Indeed, it is often helpful to differentiate between proximal and distal sources for allogenic fine sediments (Woodward, 1997a). For example, while aeolian processes can rework local silts and sands from a seasonally dry braided floodplain or littoral zone and deposit them within a neighbouring rockshelter site (Farrand, 1975), wind action can also deliver fine dust and tephra from distant sources hundreds of kilometres away (Vitaliano, *et al.*, 1981; Pye, 1987, 1992; Table 27.1).

At this point it is useful to consider a simple sediment budget for the various potential fine sediment sources and sinks involved in the transfer of material into and through a typical karst environment (Figure 27.2). This approach serves to

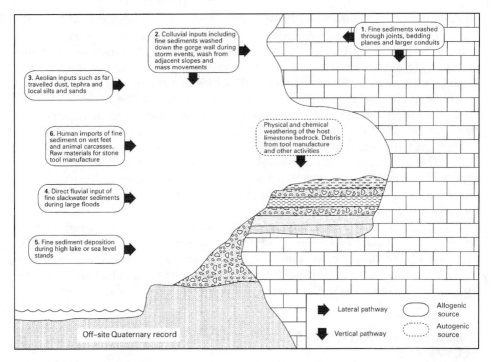

Figure 27.2 Rockshelter sediment sources and pathways and associated geomorphological processes. See text and Table 27.1 for explanation. (This hypothetical example shows a limestone rockshelter in a riparian or littoral setting. Rockshelters and caves can develop in a range of ways and may also form in granites, sandstones and other lithologies)

highlight the dynamic nature of the karst sediment system and this broader framework should be borne in mind when considering the origin of a rockshelter or cave-mouth sequence which includes a considerable fine sediment component. Indeed, much of the fine fraction may represent the end-product of a complex series of sources, pathways, stores and reworking episodes over an unknown timescale. For example, the *subsurface* karst environment forms just one component of the sediment delivery system portrayed in Figure 27.2 for which White (1993) has identified three main sources of fine sediment:

- Sinking streams draining from allogenic recharge areas carry a clastic load which is discharged into swallow holes and carried through the conduit system.
- Weathering debris derived from the insoluble residue of the limestone dissolved at the soil–bedrock contact is flushed into sinkholes and joints to the underlying conduit system.
- The enlargement of caves and conduits by solution in the subsurface also leaves behind the insoluble fraction of the bedrock.

The amount of sediment generated from each of these sources will vary in time and space as environmental conditions and host bedrock properties change. All of these

Table 27.1   Examples of geomorphological (1 to 5) and anthropogenic (6) processes and the fine sediment deposition associated with them in rockshelter sites[a]

---

1. Infiltration   Fine sands, silts and clays can be flushed through joint spacings, enlarged bedding plane partings and conduits such as roof cracks in the limestone bedrock. The sediments are commonly derived from soils and sediments washed into the karst system or from 'stratigraphic leakage' from overlying rocks. Reworked loess and *terra rossa* may be introduced into rockshelters by this process (see Bar-Yosef, 1993). The infiltration mechanism has been reported by several workers including Jancin and Clark (1993), Farrand (1988) and Woodward (1997b).

2. Colluvial processes   These may involve a range of sediment transfer mechanisms including periglacial processes, mass movements and slope wash processes. The latter may include fines washed down gorge walls during storm events which may have exited the karst drainage system via flood-filled conduits. There may be some overlap with infiltration processes. Butzer (1981) identified the erosion and transport of sands and silts by wash processes from external slopes into the rockshelter entrance as an important geomorphological process in Cantabrian Spain.

3. Aeolian processes   Sands, silts and clays can be deposited within rockshelter settings by wind activity. These materials can be derived from a wide range of proximal and distal sources. Tephra and loess can be transported long distances (Pye, 1987) and local deflation zones can be important (Farrand, 1975; Woodward, 1997a). The sedimentary sequence at Franchthi Cave in southern Greece contains the Late Pleistocene Y5 tephra which originated over 800 km away in the Campanian volcanic Province of Italy (Vitaliano *et al.*, 1981).

4. Fluvial processes   Fluvial processes can deposit suspended load and bedload sediments within rockshelter sites and the calibre of the sediments will depend on the magnitude of the flood events and the local geomorphological setting. Rockshelter environments may form important slackwater sedimentation zones and such deposits have been reported at Boila in the Voidomatis basin (Macklin *et al.*, 1988; Hamlin *et al.*, see Chapter 25). At the Arenosa rockshelter in the Pecos River in west Texas, USA, Patton and Dibble (1982) report the presence of slackwater sediments intercalated with aeolian sediments and archaeologically rich layers in a sequence that spans much of the last 10 000 years.

5. Littoral zone processes   Coastal, estuarine, or lake shore environments with short- and long-term fluctuations in water level can inundate rockshelters producing sequences of marine or lacustrine sediments. These may interdigitate with aeolian or colluvial sediments. The basal sediments at Danger Cave in Utah were deposited by Lake Bonneville before it dried out at the end of the last cold stage. The sequence at Kastritsa rockshelter near Lake Ioannina in northwest Greece contains beach sediments dated to the Last Glacial Maximum when the site was located on the lake shore (see Bailey *et al.*, 1983).

6. Human activity   Fine sediments resulting from human activities take a range of forms and are present in many sites. These may include fine alluvial sediments dragged into a site on wet carcasses and the waste products of flint tool manufacture (microdebitage). The latter may also be considered as allogenic materials as the raw materials (flint or chert) would have been imported into the site. Butzer (1981) has identified distinctive sedimentary units in Cantabrian rockshelters where the contribution from flint napping debris is close to 100 percent. Between *c.* 11 500 and 9500 years BP at Franchthi Cave, human imports and debris markedly increased the sedimentation rate (Farrand, 1988). Human occupation will also increase the organic content of the sediments, and ash deposits from hearths can form a significant component of the fine matrix (see Bailey and Woodward, 1997).

---

[a] This list is not exhaustive and coarse sediments (> 2 mm), organic materials and chemical precipitates, for example, are not considered here. The aspect and geometry of the site and the local geomorphological setting are important controls on sediment flux, sediment source and the nature and extent of any post-depositional modifications. It is also important to appreciate that many rockshelters do not contain sedimentary records because of limited sediment supply or removal by erosion.

materials can be deposited within a rockshelter or cave entrance setting by infiltration and washing through the limestone mass or via external pathways such as slopewash down the cliff face above the site or within the suspended load of a resurgent stream. Unweathered fine sediments may also enter the karst drainage system via 'stratigraphic leakage' from unconsolidated clastic rocks which overlie the limestone beds (Jancin and Clark, 1993). For the rockshelter depositional environment – especially in the case of large sites in exposed settings – it is particularly important to consider the non-karstic pathways for sediment transfer (Figure 27.2) and the textural and lithological properties of the rockshelter fine sediment can be used to identify the sources involved (Woodward, 1990; Gale *et al.*, 1993).

A CASE STUDY: THE ROCKSHELTERS OF THE LOWER VIKOS GORGE, NORTH-WEST GREECE

The Klithi and Megalakkos rockshelters are both located at an altitude of *c*. 430 m above sea level in the Lower Vikos Gorge of the Voidomatis River, north-west Greece, in Palaeocene to Eocene limestone bedrock (Figure 27.3). The aspect, geometry and geomorphological setting of the two rockshelters are quite different and their main characteristics are listed in Table 27.2. The headwater catchments of the Voidomatis lie in some of the formerly glaciated limestone terrain of the Northern Pindus Mountains and in lower elevation flysch terrains to the south (Woodward *et al.*, 1995; Smith *et al.*, 2000; Figure 27.3). The geology and geomorphology of this catchment has been described by Bailey *et al.* (1997) and Hamlin *et al.* (see Chapter 25) and will only be presented in outline here. Klithi is located in the main Lower Vikos Gorge at a level approximately 30 m above the right bank of the Voidomatis River (Figures 27.3 and 27.4(a)). The site of Megalakkos is located in a narrow, steep-sided tributary ravine approximately 100 m upstream of the point where it joins the main channel of the Voidomatis River (Figures 27.3 and 27.4(b)). Flysch rocks are present above the limestone bedrock on both sides of the Lower Vikos Gorge and these areas are drained by a series of steep tributary streams (Figure 27.3).

Klithi was excavated between 1983 and 1988 as part of the Klithi Project and less extensive excavations were carried out at Megalakkos in 1986 and 1987 (Bailey, 1997a). The rockshelter site of Boila is located at the downstream end of the Lower Vikos Gorge near the Konitsa Basin (Figure 27.3) and this site was excavated between 1993 and 1997 (Kotjabopoulou *et al.*, 1997). The excavations at Klithi and Megalakkos were accompanied by a wide range of off-site studies into the Quaternary history of the Voidomatis basin and the wider Epirus region (see Bailey, 1997a,b). This included a detailed investigation into the nature and timing of Late Pleistocene river behaviour and alluvial sediment source variations (Lewin *et al.*, 1991; Woodward *et al.*, 1992) and this work has continued as fresh approaches and dating techniques have been developed (Woodward *et al.*, 1994; Macklin *et al.*, 1998; Hamlin *et al.*, see Chapter 25).

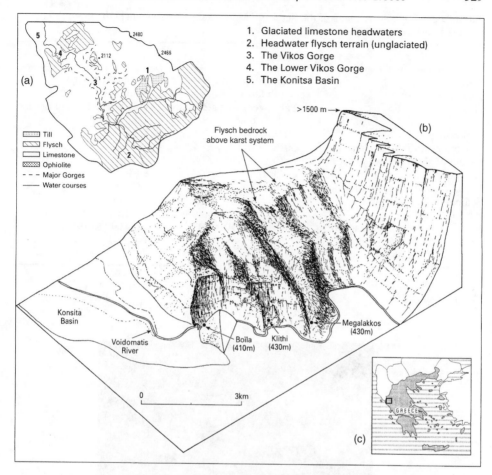

1. Glaciated limestone headwaters
2. Headwater flysch terrain (unglaciated)
3. The Vikos Gorge
4. The Lower Vikos Gorge
5. The Konitsa Basin

Figure 27.3 (a) A simplified map of the Voidomatis River basin showing the major terrain units, the main rock types and the drainage network. (b) The Lower Vikos Gorge in the Voidomatis River basin and the location of the major rockshelters of Klithi, Megalakkos and Boila (modified after Gowlett *et al.*, 1997). Each site contains Late Pleistocene sediments and Late Upper Palaeolithic lithic and faunal materials. The sediments at Boila are being studied by D. Panagiotis Karkanas and this work is still in progress. (c) The location of the study basin in Epirus, north-west Greece

FIELD AND LABORATORY METHODS AND FINE SEDIMENT TRACER PROPERTIES

The Megalakkos sequence was logged, photographed and sampled in the field from both natural exposures and those revealed during the archaeological excavations (Figure 27.4(c)). The sedimentary sequence at Klithi has been described in detail by Bailey and Woodward (1997). The upper 2 m of the Klithi sediments are extremely rich in Late Upper Palaeolithic stone tools, faunal remains and other cultural materials,

Table 27.2   The main features of the Klithi and Megalakkos rockshelters in the Lower Vikos Gorge of the Voidomatis River basin

	Klithi	Megalakkos
Location	Main gorge wall	Tributary ravine
Aspect	South facing	South-west facing
Host bedrock	Palaeocene to Late Eocene limestone	Palaeocene to Late Eocene limestone
Floor area	300 m²	<30m²
Maximum width of shelter	30 m	< 5 m
Height of shelter brow	*c.* 10 m	*c.* 6 m
Altitude (a.s.l.)	*c.* 430 m	*c.*430 m
Elevation above local valley floor	30 m	*c.* 10 m
Sediment thickness	>7 m	>5.5 m
Archaeology	Late Upper Palaeolithic	Late Upper Palaeolithic
Period of occupation (years BP)	*c.* 16 500 to 10 000	*c.* 16 500 to 10 000
Intensive LUP occupation	Yes	No
Number of ¹⁴C dates	24	4

(a)

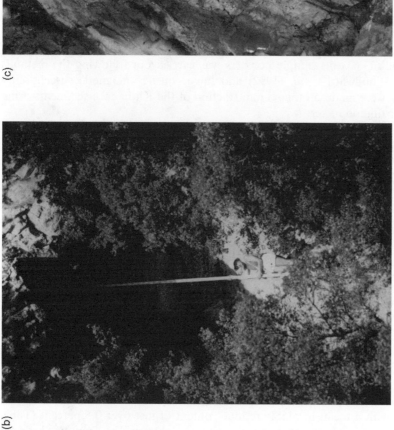

Figure 27.4 The geomorphological setting of Klithi and Megalakkos rockshelters in the Lower Vikos Gorge (see Figure 27.3 and Table 27.2). (a) Klithi is located 30 m above the main channel of the Voidomatis River in the central part of the Lower Vikos Gorge. The river flows to the left beneath the site and the gorge wall is approximately 150 m high at this point. (b) Megalakkos is located in a narrow right bank tributary ravine which joins the main Voidomatis channel upstream of Klithi. (c) Part of the Late Pleistocene sedimentary sequence at Megalakkos. Note that some of the fill has been eroded away

while all the available evidence suggests that the older sediments below (before *c.* 16 500 years BP) are sterile. The Klithi sequence was studied from sections exposed during the archaeological excavations and from a series of sediment cores obtained from drilling programmes conducted in 1986 and 1988. Core Y25 was obtained in 1986 and reached a depth of approximately 7 m (Bailey and Thomas, 1987). This core was subdivided into 56 sediment samples and this sequence is discussed below and compared to the Megalakkos fine sediments.

In order to determine the provenance of the rockshelter fine sediments at each site, a range of analyses were undertaken (Table 27.3). All the sediment samples from Klithi and Megalakkos were screened through a 63 μm sieve to obtain the silt and clay fraction. Particle size analysis of the <63 μm fraction was carried out using the computer-interfaced SediGraph 5000ET system described by Jones *et al.* (1988). This provided detailed information on the silt (63 − 2 μm) fraction at half phi intervals as well as total clay content. All samples were dispersed in calgon and organic materials were removed by hydrogen peroxide using standard methods prior to particle size analysis. The proportion of insoluble residue in bedrock and Quaternary sediment samples was determined by weighing following dissolution in dilute hydrochloric acid (see Gross, 1971; Macleod, 1980; Pye, 1992). The proportion of non-carbonate silt present in the <63 μm fraction was estimated by subtracting the clay content from the insoluble residue fraction for each sample. X-ray diffraction (XRD) analysis of bulk (<63 μm) sample powders was carried out following the methods described by Woodward *et al.* (1992) and low-frequency magnetic susceptibility measurements were made on the <1 mm fraction of the Klithi core sediments using a standard Bartington meter.

Table 27.3 The textural and lithological properties used to establish the source of the fine sediments at Klithi and Megalakkos rockshelters

Sediment property	Size fraction	Method
Particle size modes <63μm		
Coarse silt content	16–63 μm	Jones *et al.* (1988)
Fine silt content	2–8 μm	Jones *et al.* (1988)
Clay content	< 2 μm	Jones *et al.* (1988)
Lithological properties		
Non-carbonate silt content	63–2 μm	Woodward (1997a)
$CaCO_3$ content	< 63 μm	Gross (1971)
Bulk mineralogy (XRD)	< 63 μm	Woodward *et al.* (1992)
Magnetic susceptibility (X)	< 1 mm	Dearing *et al.* (1985)

LOCAL BEDROCK PROPERTIES

The limestone bedrock in the Lower Vikos Gorge can be classified as extremely pure (Bögli, 1980) with a mean insoluble residue content of just over 0.5 percent (Table 27.4). The particle size characteristics of the acid-insoluble residues from two bedrock samples are shown in Figure 27.5(a). This non-carbonate material is dominated by clay and fine silt, and coarse silt is almost absent. The clay content of these bedrock

Table 27.4   The insoluble residue content of host limestone bedrock samples from Klithi and Megalakkos

	Range %	Mean %	n
Klithi	0.32 to 1.04	0.57	7
Megalakkos	0.34 to 0.89	0.54	7

residues from Klithi and Megalakkos is 75.6 percent and 68.5 percent respectively. On either side of the Lower Vikos Gorge the resistant limestone rocks are overlain by Late Tertiary flysch sediments (Figure 27.3) which comprise thinly bedded alternations of hard sandstones and softer, fissile siltstones (Bailey *et al.*, 1997). The flysch rocks are susceptible to gullying and fluvial erosion and are an important source of suspended sediment in the modern river (Woodward *et al.*, 1992). Flysch-derived materials commonly form the matrix in stratified limestone scree deposits. In the Voidomatis River basin, as in many other Mediterranean mountain karst environments, lithological variability is limited and the major rock formations (limestone and flysch in this case) can be differentiated on the basis of simple bulk mineralogy (Figure 27.5(b)) and the particle size characteristics of the non-carbonate fraction. The mineral suite in the flysch rocks includes quartz, plagioclase, calcite and various clay minerals.

THE MEGALAKKOS ROCKSHELTER SEDIMENTS

Megalakkos is a small rockshelter with a narrow opening less than 5 m wide (Figure 27.4(b) and (c)). It contains at least 5.5 m of Late Pleistocene and Holocene sediments.

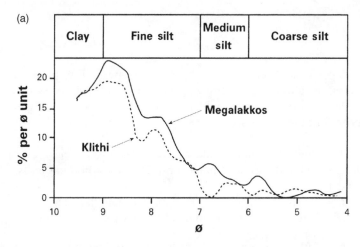

Figure 27.5(a)   The particle size characteristics of the acid-insoluble residue of the host limestone bedrock at the Klithi and Megalakkos rockshelters. Note the virtual absence of coarse silt in these residues. (b) Bulk XRD traces of powdered limestone and flysch bedrock and a sample of fine-grained sediment from Unit 7 in the Megalakkos sequence (see text and Figure 27.6)

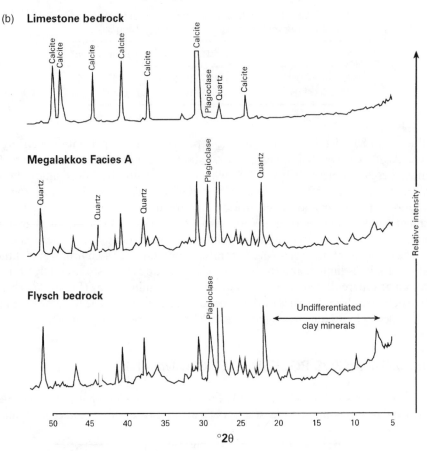

Figure 27.5 (b)

Part of the deposit has been removed by recent slope failure, producing excellent natural sections in almost the entire sequence (Figure 27.4(c)). The sediments at Megalakkos have been subdivided into 12 lithostratigraphic units (Figure 27.6) and these comprise four facies types:

(1) *Facies A*: a finely laminated yellow/brown clayey-silt facies devoid of coarse limestone clasts and forming sharp contracts with adjoining units. Units 3, 5, 7, 10 and 12.
(2) *Facies B*: a poorly stratified, matrix-supported, angular limestone debris facies characterised by poorly sorted, angular clasts with a significant proportion of clayey-silt matrix. Units 1, 4, 6 and 9.
(3) *Facies C*: a well-sorted, stratified deposit of angular limestone clasts, partly clast-supported with a yellow/brown silty matrix. Units 2 and 11.
(4) *Facies D*: a fine fluvial gravel facies of well-rounded flysch clasts up to 25 mm (b axis). Some evidence of imbrication and largely matrix-supported. Unit 8.

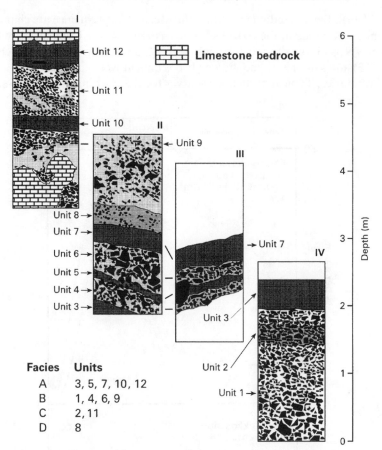

Facies	Units
A	3, 5, 7, 10, 12
B	1, 4, 6, 9
C	2, 11
D	8

Figure 27.6 Schematic log of the Late Pleistocene sedimentary sequence at Megalakkos (after Woodward, 1997b). The fine-grained units (Facies A) are archaeologically sterile. Units 4, 6 and 9 (Facies B) contain most of the Late Upper Palaeolithic lithic and faunal remains which are very similar to the cultural record at Klithi. Four radiocarbon dates have been obtained from the Megalakkos sequence (see Bailey and Woodward, 1997, for discussion) and, in conjunction with the archaeological data, these indicate that the use of the site was much less intensive, but broadly contemporaneous with occupation at Klithi

The Megalakkos sequence is noteworthy because it contains sediments characterised by coarse angular limestone clasts that are rich in Late Upper Palaeolithic materials (Units 4, 6 and 9) interbedded with distinctive, fine-grained sedimentary units that are archaeologically sterile (Figure 27.6). The fine-grained Facies A sediments are of particular interest and a number of approaches were employed to establish their depositional context and source.

The lithological and textural characteristics of the fine sediments at Megalakkos indicate that they are derived from the local flysch rocks and flysch-derived soils that lie above the limestone karst mass (Figure 27.3). XRD traces illustrate the close similarity between the mineralogy of the flysch bedrock and the fine-grained sediments at Megalakkos (Figure 27.5(b)). In contrast to the limestone bedrock residues

(Figure 27.5(a)), the fine sediments within the Megalakkos sequence are characterised by a pronounced mode in the coarse silt range (Figure 27.7). The main tributaries in the Lower Vikos Gorge drain catchments with headwaters in the local flysch terrain (Figure 27.3) and samples of fine-grained bed sediment were collected from a number of these tributaries. Typical particle size curves for these sediments are also shown in

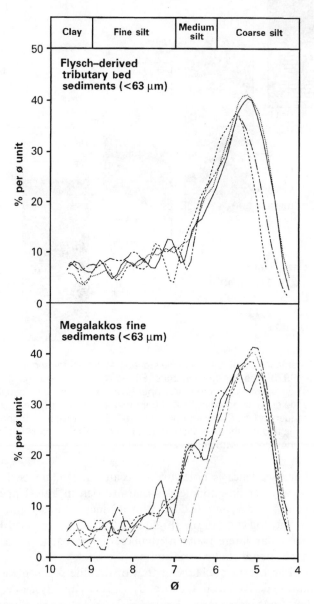

Figure 27.7   Typical particle size curves for the fine-grained sediments at Megalakkos and for contemporary fine bed sediments from several tributaries in the Lower Vikos Gorge (compare with Figure 27.5)

Figure 27.7. These samples have been sorted by fluvial action and the average clay content is *c.* 9 percent. For each of these samples, the proportion of fine, medium and coarse silt is very similar to the fine sediments at Megalakkos (Figure 27.7).

It could be argued that the Facies A sediments have much in common with typical loess sediments as described by Pye (1987), but Megalakkos is in a deep, sheltered ravine which is not conducive to the receipt of aeolian materials. Moreover, field observations and micromorphological analysis of the fine-grained units at Megalakkos have shown that these sediments are finely laminated and were deposited from suspension in a shallow, low-energy ponded environment (see Bretz, 1942; Bull, 1981). These flysch-derived sediments are infiltrates and were washed through the karst system by sediment-laden waters into shallow pools of standing water (Woodward, 1997b). Infiltrated fine sediments are present in limestone conduits in and around the site of Megalakkos and throughout the Lower Vikos Gorge. The absence of coarse sands and fine gravels within all the Facies A units and the low-energy conditions indicated by the laminated microstructures argue against deposition by the tributary stream. Moreover, Unit 12 at the top of the Megalakkos sequence is Holocene in age (and is probably still accumulating today) and is approximately 15 m above the present bed of the tributary channel. The only evidence of direct fluvial deposition within the site by the Megalakkos tributary stream is provided by Unit 8 (Facies D) which contains some imbricated rounded flysch gravel. This unit represents either an extreme flood event or a temporary damming of the tributary ravine during a high flow event. The sedimentary sequence at Megalakkos forms a valuable source of comparison with the main rockshelter site of Klithi and the nature of the fine sediments at Megalakkos will be discussed further below.

In summary, the Megalakkos clayey-silts in the Facies A and B sediments are the product of a low-energy sedimentation mechanism associated with local hydrological processes within the karst drainage system. This flysch-derived material is transferred through the surrounding limestone strata via sediment-laden groundwater flows. Megalakkos has served as an effective location for the deposition and store of externally derived fine sediments for a considerable part of Late-glacial and Holocene time. The sharp contacts between the sedimentary units at Megalakkos and the repeated alternation of markedly different facies provides evidence for fluctuating climatic conditions in the region after the Last Glacial Maximum between *c.* 16 500 and 10 000 years BP (Woodward, 1997b). The Facies A sediments were deposited during humid and warm conditions not unlike the present day, while the Facies B deposits (which include coarse angular limestone clasts) probably indicate cooler conditions (see Woodward (1997b) for a discussion).

THE KLITHI ROCKSHELTER SEDIMENTS

Klithi faces almost due south and this aspect, together with the overhang of the limestone roof, provides excellent shelter from the elements (Bailey, 1997c). The shelter was probably formed by fluvial erosion of the bedrock wall when the Voidomatis flowed at a considerably higher level than at present (Figure 27.4(a)). The sediment fill takes the form of a cone with its apex towards the western end of the shelter. This reflects intra-site variation in host bedrock properties as the western portion of the shelter wall contains numerous joints and beds and is more prone to breakdown – whereas the shelter overhang is more massive and resistant (Bailey and Woodward, 1997). The upper part of the fill has been truncated and levelled, forming a flat, near-horizontal surface, with a 10 cm layer of modern goat dung resting unconformably on the Palaeolithic deposits.

Figure 27.8 shows a section exposed in 1983 during the first season of excavation at Klithi. The sediments evidence a wide range of particle sizes, and boulder- and cobble-sized clasts are present throughout the sequence. These materials are roughly stratified screes incorporating alternations of coarse and finer grade angular limestone debris. There is considerable vertical and lateral variation in particle calibre, sorting, sediment fabric and void ratio. A closed, clast-supported fabric is common (Figure 27.1) as these rockshelter sediments contain a significant proportion of fine-grained (sand, silt and clay) matrix (Bailey and Woodward, 1997; see Figure 27.1). Archaeological materials are extremely abundant throughout the upper part the Klithi fill. This sequence contrasts markedly with the sedimentary record at Megalakkos where the units containing archaeological material are separated by distinctive fine-grained archaeologically sterile units (Facies A). The sequence shown in Figure 27.8 represents the upper 2 m or so of the Klithi deposit, which is associated with the Late Upper Palaeolithic use of this site. The main period of occupation took place between *c.* 16 500 and 13 500 uncalibrated radiocarbon years BP, perhaps extending at latest to about 12 400 BP, with traces of re-occupation at about 10 400 years BP (Bailey and Woodward, 1997). Below *c.* 2 m, the rockshelter sediments are archaeologically sterile and the coring programme provided the only available windows into the nature of the earlier sedimentary record.

The source of the fine (<63 μm) sediments in the full Klithi sequence and the processes involved in their deposition have been studied in detail from the Y25 core samples using the suite of textural and lithological parameters described above (Table 27.3). The sequence from Y25 is shown in Figure 27.9 and this can be subdivided into three main sections. The upper part of the core (0–2.5 m) corresponds to the Late Upper Palaeolithic occupation of the site and is believed to be broadly time equivalent to Units 2 to 11 at Megalakkos (Figure 27.6). It can be shown that most of the fine sediment within the core sequence is derived from allogenic sources. The textural and lithological parameters shown in Figure 27.9 all refer to the sediment fraction <63 μm while the magnetic susceptibility data refer to the fraction <1 mm. Thus, the lower core (4.2–7 m) is dominated by $CaCO_3$-rich silts and clay-grade material with mean values of 73 percent and 22 percent respectively. Indeed, clay-grade (<2 μm) material accounts for much of the insoluble residue fraction in this part of the sequence. An important feature of the Klithi core sequence is the divergence of the $CaCO_3$ and clay curves above 4.2 m and this persists throughout the rest of the record. This pattern marks a major change in sediment source as *non-carbonate silt* becomes an important component of the fine sediments in the site. The non-carbonate silt curve is shown in Figure 27.9.

In order to explain the marked lithological and textural changes in the Late Pleistocene fine sediment record at Klithi, the relationship between key sedimentological parameters was investigated and several of these are shown in Figure 27.10.

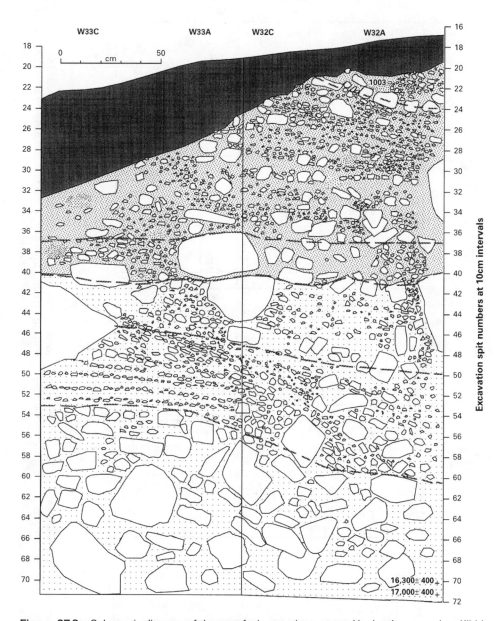

Figure 27.8 Schematic diagram of the east-facing section exposed in the deep trench at Klithi based on an original field drawing by Janusz Kozlowski and Colette Roubet. This figure highlights the spatial and temporal variations in coarse sediment calibre, fabric and sorting evident in the Klithi sequence (after Bailey and Woodward, 1997). The upper unit (shaded) represents contaminated layers affected by goat trampling in modern times. Compare with Figure 27.6. W33C, W33A, W32C and W32A refer to the site excavation grid (see Bailey, 1997c)

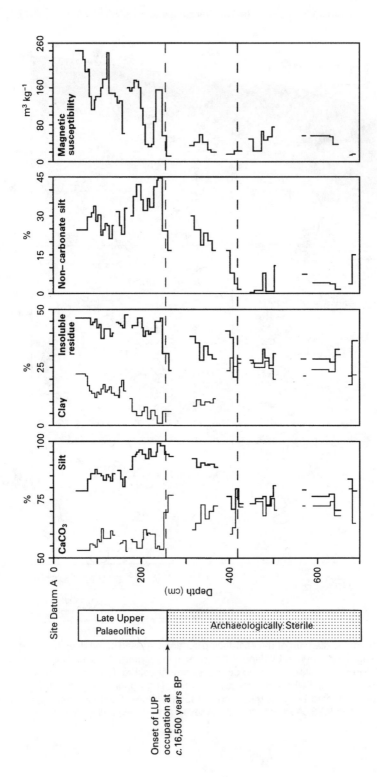

Figure 27.9 Textural and lithological data from the Y25 core at Klithi. The non-carbonate silt parameter has been estimated by subtracting the clay content from the insoluble residue of the fraction < 63 μm. The full suite of radiocarbon dates from Klithi is discussed in detail by Bailey and Woodward (1997). Note that the depths given in this figure are relative to the site datum and not the top of the core

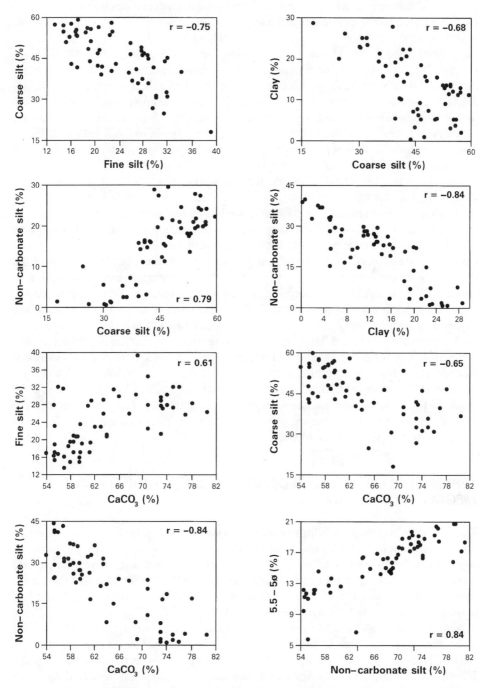

Figure 27.10 The relationship between various lithological and textural parameters from the Y25 (Klithi) core sediments. All values are percentages of the fraction $< 63 \, \mu m$ ($n = 56$)

In terms of sediment sources, a number of important points can be gleaned from these plots. First, the strong positive relationship between non-carbonate silt and coarse silt ($r = 0.79$) and the 5.5–5 ϕ size range ($r = 0.84$) indicate that this is flysch-derived material (Figure 27.7). The negative relationship between fine silt and coarse silt ($r = -0.75$) suggests that these materials derive from different sources. In contrast, the fine silt is rich in $CaCO_3$ ($r = 0.61$) and is likely to contain a significant limestone-derived component. This is discussed further in the next section. The strong negative relationship between non-carbonate silt and clay is also of interest and could indicate that flysch-derived silts and clays are sorted during transport and follow different transfer pathways. This relationship may also suggest that other (proximal and/or distal) sources of clay material are important.

The magnetic susceptibility profile for the upper core shown in Figure 27.9 is independent of the other sedimentological parameters discussed above as it reflects the impact of human activity in the site through sediment firing. This enhancement in magnetic susceptibility due to burning in the archaeologically rich sediments *within* the rockshelter contrasts markedly with the magnetic susceptibility profiles observed from Late Pleistocene alluvial soils in the vicinity of the site. In the pedogenic weathering environment off-site (where the parent materials are very similar in composition and age), high susceptibility values are associated with strongly weath- ered, *decalcified* horizons (Woodward *et al.*, 1994), whereas the rockshelter sediments evidence high susceptibilities in association with $CaCO_3$-rich sediments (Figure 27.9). Such comparisons between the on-site and off-site environments are important as they can help to decouple the natural and anthropogenic signals in rockshelter sediment records. It is also worth nothing that several samples in the middle and lower core sections evidence susceptibility values that are higher ($>20\,m^3kg^{-1}$) than would be expected for unweathered, archaeologically sterile materials. These features and the rest of the sedimentary sequence from Core Y25 are discussed further below and compared to the Megalakkos record.

TERMINAL PLEISTOCENE SEDIMENT SOURCES AT KLITHI AND MEGALAKKOS

The Y25 core sequence at Klithi reveals a series of major changes in fine sediment texture and lithology during the course of the Late Pleistocene. The major sediment sources can be characterised on the basis of lithology and distinctive particle size assemblages. These changes can be related to shifts in the dominant off-site sediment sources which reflect the dramatic environmental changes associated with the last glacial–interglacial transition in the region (Macklin *et al.*, 1997).

Aeolian sediments have been reported in many rockshelter sites (Table 27.1) and Pleistocene loess has been identified in the Doliana basin between 5 and 10 km to the west of Klithi. A number of sedimentological comparisons were made between the Klithi and Megalakkos sediments and samples of yellow loess collected from an exposure in the Doliana basin (Woodward, 1990). Figure 27.11(a) illustrates the relationship between $CaCO_3$ content, clay content and non-carbonate silt for these materials. These parameters show a clear separation between the three sample groups.

The Megalakkos deposits are generally finer grained (richer in clay) and contain more $CaCO_3$ than the Doliana loess sediments. The provenance of the Doliana basin loess is not known and it contrasts markedly with the Klithi fine sediments.

During the last cold stage the suspended sediment load of the Voidomatis River was dominated by limestone-derived fine sediments from the glaciated limestone head-waters (Woodward *et al.*, 1992; Hamlin *et al.*, see Chapter 25). During periods of low flow, this fine material (local loess) was blown into the rockshelter site from the wide, braided channel system beneath the site (Woodward, 1997a). The cold stage river was at least 10 m higher than at present. Experimental work has shown that hard lime-stone lithologies like the bedrock at Klithi do not produce silt and clay-grade material during mechanical weathering (see Lautridou, 1988). Furthermore, the fine matrix of cold stage alluvial gravels preserved in the Lower Vikos Gorge is rich in calcareous silt (Lewin *et al.*, 1991). It is interesting to note that micromorphological observations of the Facies A sediments at Megalakkos have provided evidence of some secondary deposition of calcium carbonate (Woodward, 1997b). Since the average $CaCO_3$

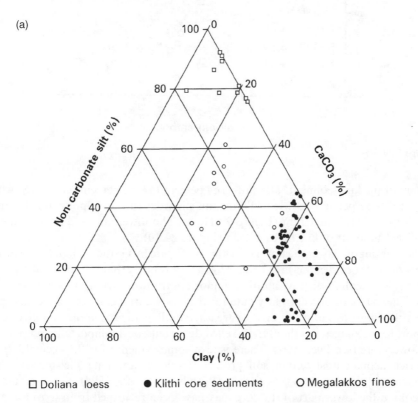

□ Doliana loess   ● Klithi core sediments   O Megalakkos fines

Figure 27.11 (a) Ternary plot showing the lithological and textural composition of the fine sediments from Klithi and Megalakkos and samples of loess from the Doliana Basin. (b) Plot highlighting the lithological and textural groupings evident within the Klithi fine sediments of Core Y25 and the changes in fine sediment composition that took place during the Late Pleisto-cene (see Figure 27.9). Fine sediment samples from the four facies at Megalakkos are also shown. Note the high clay content of the Facies A units

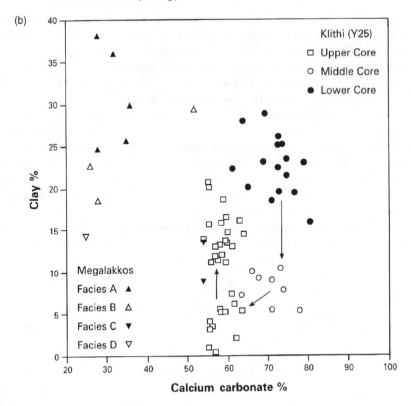

Figure 27.11 (b)

content of the fine sediments at Klithi is 64 percent (54–80 percent) compared with 36 percent (25–54 percent) at Megalakkos, this provides further support for the import of calcareous loess at Klithi (Figure 27.11(a)). The lower core sediments contain a significant amount of fine (2–8 μm) silt (Figure 27.10).

The amount of clay in the lower core (mean 22 percent) is too high to be the product of *in-situ* host rock weathering (Table 27.4) and it is likely that this material represents far-travelled aeolian dust with only a minor contribution from local limestone bedrock. The magnetic susceptibility data for the lower core discussed above may partly reflect the input of naturally fired or weathered allogenic materials and give further support to the suggestion that far-travelled dust could be an important component of the lower core (see Pye, 1992). There is also some evidence for the accumulation of clay-rich aeolian dust within soil profiles developed on Late Pleistocene alluvial sediments in the Lower Vikos Gorge (Woodward *et al.*, 1994). Some of the clay material in the lower part of the core may have been reworked from *terra rossa* soils developed on the limestone strata and its deposition in the site may result from a combination of aeolian and infiltration processes. Ford and Williams (1989) report winnowing of *terra rossa* in southern France during the infiltration process to form clay-rich cave earths. Rockshelters may form important natural long-term dust traps and the precise mineralogical and elemental composition of this clay-grade material is currently under investigation. Apart from the clay material at Klithi discussed above, much of the fine sediment at both sites appears to be dominated by local off-site sources.

The middle section of Y25 (2.5–4.2 m) at Klithi marks a major change in fine sediment sources as the clay component declines significantly (to almost zero) and non-carbonate silts begin to dominate the insoluble residue. These non-carbonate silts are similar in composition to the silts at Megalakkos and they appear to be derived mainly from flysch sediments (Figure 27.11(a)). The hydrogeological settings of the two rockshelters and the nature of the stratigraphic records indicate that infiltration processes are dominant at Megalakkos. The flysch-derived silts were introduced into Klithi after the Last Glacial Maximum in two main ways. The suspended sediment load of the Voidomatis River contained a higher flysch component at this time as the input of glacially comminuted limestone silts declined (Woodward *et al.*, 1992). It is likely that this material could also have been redistributed by aeolian activity in the Lower Vikos Gorge and the significant $CaCO_3$ content (>50 percent) of the Klithi fine sediments provides additional support for this mechanism during this period. At present, flysch-derived fines are commonly washed down the walls of the Lower Vikos Gorge during heavy rainstorms, and high-velocity sediment-laden flows have been observed exiting conduits in cliff faces. It is not possible to disentangle the roles of infiltration and external bedrock slope wash processes in the middle and upper core with any certainty, but each is likely to have been an increasingly important source of non-carbonate silt material as the main channel of the Voidomatis incised after *c.* 16 000 years BP (Macklin *et al.*, 1997) and the valley floor was colonised by vegetation. These changes on the valley floor would have limited the production of fine sediment by aeolian activity from the floodplain system. Figure 27.11(b) shows that the Y25 sediments fall into three or four main groupings, highlighting both the contrast in the nature of the major sediment inputs to Klithi and Megalakkos and how the geomorphological changes of the Late-glacial in the Voidomatis are recorded at Klithi.

The upper part of the Y25 sequence is broadly contemporaneous with the bulk of the sedimentary sequence (Units 2 to 11) at Megalakkos (Woodward, 1997a,b) and can be further divided into two sections based on the increase in clay content at the top of the core (Figures 27.9 and 27.11(b)). This could indicate drier conditions and an increase in aeolian dust transport at the end of the Late-glacial period (Woodward, 1997a). This may be equivalent to the Younger Dryas cooling, but poor dating resolution at the top of the sequence does not allow this to be confirmed. At present, the dating control at Megalakkos is not good enough to allow a high-resolution comparison between the two sites, but it is clear from the marked facies changes in the Megalakkos sequence that the Late Upper Palaeolithic occupation of the Lower Vikos Gorge was accompanied by significant environmental changes (Figures 27.6 and 27.11(b)). While the sedimentary record at Klithi is rather longer, the sequences at Klithi and Megalakkos do demonstrate that such changes will be recorded in different ways at different rockshelter sites only hundreds of metres apart (Table 27.5). The major changes in fine sediment sources and geomorphological processes during the Late Pleistocene discussed above are summarised in Figure 27.12 and Table 27.5.

Table 27.5 The processes responsible for fine sediment delivery to the Klithi and Megalakkos rockshelters during the Late Pleistocene[a]

Sediment body	Fine sediment delivery mechanisms
Klithi rockshelter	
Upper Core	Slope wash/infiltration, aeolian processes (D) and human activity (CALC)
Middle Core	Slope wash/infiltration, aeolian processes (D and P) (CALC)
Lower Core	Aeolian processes (D and P), infiltration (CALC)
Megalakkos rockshelter	
Facies A	Infiltration
Facies B	Infiltration and human activity (CALC)
Facies C	Infiltration and human activity (CALC)[b]
Facies D	Fluvial processes (large flood)

[a] Each process is listed in Table 27.1. For aeolian inputs: D = distal source and P = proximal source. CALC denotes the presence of coarse angular limestone clasts and these may form the bulk of the sediment body as at Klithi (Figure 27.1).
[b] At Megalakkos, debris from human activity is only present in one of the Facies C units – the upper part of Unit 2.

CONCLUSIONS

Rockshelter sequences are composed of sediments derived from a range of allogenic (off-site) and autogenic (on-site) sources and their relative significance will vary over time and between sites as environmental conditions change. The fine-grained components of the sedimentary records preserved in limestone rockshelters are commonly dominated by allogenic materials. It is therefore important to place such sequences in a wider context by establishing the nature and age of off-site sedimentary sequences and to consider all potential sediment sources and transfer mechanisms. Furthermore, sedimentological data from the off-site Quaternary record are often crucial for the palaeoenvironmental interpretation of on-site sediments (Woodward, 1997a). If the provenance of the fine sediment fraction can be established, well-dated rockshelter sequences can provide an important source of information on past geomorphological processes.

In the Lower Vikos Gorge of north-west Greece, marked contrasts in Late Pleistocene sediment sources and depositional processes are apparent between the nearby rockshelter sites of Klithi and Megalakkos. While the fine sediment fraction at each site is dominated by off-site sources, there are marked, stratigraphical, lithological and textural contrasts between the two sites which indicate that the contributions from aeolian, infiltration and other processes have varied between the two sites and over time (Table 27.5). The rockshelter site of Boila is located at the end of the Lower Vikos Gorge only 10 m above the present river channel (Figure 27.3). It provides an additional contrast to Klithi as the Boila record contains slackwater sediments following repeated inundation of this site prior to the onset of Late Upper Palaeolithic occupation (Macklin *et al.*, 1988; Hamlin *et al.*, see Chapter 25). The geomorphological and hydrogeological settings of the rockshelters are particularly important controls on fine sediment provenance and depositional processes. These conclusions are based on a relatively simple approach at Klithi and Megalakkos using textural and

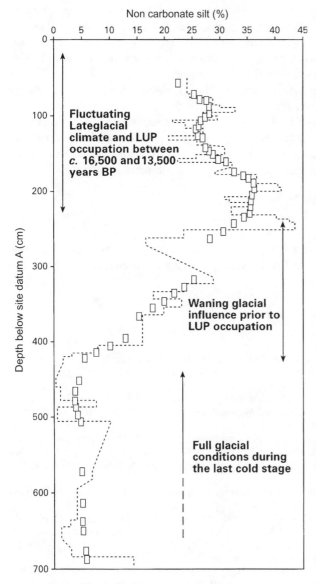

Non carbonate silt (%)

Fluctuating
Lateglacial
climate and LUP
occupation between
c. 16,500 and 13,500
years BP

Waning glacial
influence prior to
LUP occupation

Full glacial
conditions during
the last cold stage

Depth below site datum A (cm)

c. 16,500 to 10,000 years BP
The upper core is marked by a
gradual decline in non-carbonate
silt material which could signal
a reduction in runoff. This is
compensated by the increase in
clay content at the top of the core
which may indicate an increase
in aeolian dust transport towards
the end of the Late-glacial period.
Marked facies changes are
recorded at Megalakkos throughout
this period with warm and humid
conditions (Facies A) alternating
with cold and humid conditions
(Facies B).

c. 20,000 to 16,500 years BP
The central section marks a major
change in fine sediment sources
as the clay component declines
to almost zero and non-carbonate
silts begin to dominate the insoluble
residue. These sediments are
similar in composition to the fines at
Megalakkos as they appear to be
derived mainly from flysch sediments.
The glacial influence appears to be
waning as the transfer of local flysch
fines becomes important. These
sediments were introduced into Klithi
after the Last Glacial Maximum by local
aeolian activity and infiltration and
runoff down the gorge wall.

Before c. 20,000 years BP
The calcareous silts were blown into
the rockshelter site during the last cold
stage from the broad, braided channel
beneath the site. The headwaters of
the Voidomatis were glaciated at this
time and large volumes of limestone-
derived fine sediment were fed into
the river. The cold glacial climate may
have increased host rock breakdown
but the production of fines would have
been limited. The clay in the lower core
probably represents far-travelled aeolian
dust with a contribution from locally
sourced materials such as *terra rossa*.

Figure 27.12 Summary diagram showing the non-carbonate silt curve to highlight the
changes in terminal Pleistocene sediment sources recorded at Klithi and the associated geomorphological processes. This curve has also been replicated in Core CC27 that was obtained in
1988 (see Woodward, 1997a). The open boxes represent a five-point moving average and provide a useful summary of the major changes that took place during the last glacial–interglacial
transition

lithological parameters, but there is also considerable scope for using multi-parameter tracing methods to produce quantitative estimates of sediment source contributions (see Walling and Woodward, 1995) in such rockshelter sediment records (see Hamlin *et al.*, see Chapter 25). Nonetheless, the data from the Lower Vikos Gorge sites presented here demonstrate that a simple classification based on fine sediment provenance provides a valuable framework for the interpretation of rockshelter sediment records as it serves to highlight their position within a wider geomorphological system and underscores the fact that such sites can provide sinks for a wide variety of sediment imports. When viewed in this way, rockshelter sediment records constitute important geomorphological archives and they offer a direct and valuable link between the archaeological data recovered from excavation and the off-site proxy climate records derived from other Pleistocene and Holocene sequences.

ACKNOWLEDGEMENTS

We are grateful to John Lewin, Chris Hunt and David Keen for providing valuable comments on this chapter and to David Appleyard for his patience in producing the figures.

REFERENCES

Bailey, G. N. 1997a. *Klithi: Palaeolithic Settlement and Quaternary Landscapes in Northwest Greece. Volume 1: Excavation and Intra-site Analysis at Klithi*. McDonald Institute for Archaeological Research, Cambridge.

Bailey, G. N. 1997b. *Klithi: Palaeolithic Settlement and Quaternary Landscapes in Northwest Greece. Volume 2: Klithi in Its Local and Regional Setting*. McDonald Institute for Archaeological Research, Cambridge.

Bailey, G. N. 1997c. Klithi excavations: aims and methods. In Bailey, G. N. (ed.) *Klithi: Palaeolithic Settlement and Quaternary Landscapes in Northwest Greece. Volume 1: Excavation and Intra-site Analysis at Klithi*. McDonald Institute for Archaeological Research, Cambridge, 43–60.

Bailey, G. N. and Thomas, G. 1987. The use of percussion drilling to obtain core samples from rockshelter deposits. *Antiquity*, **61**, 433–439.

Bailey, G. N. and Woodward, J. C. 1997. The Klithi deposits: sedimentology, stratigraphy and chronology. In Bailey, G. N. (ed.) *Klithi: Palaeolithic Settlement and Quaternary Landscapes in Northwest Greece. Volume 1: Excavation and Intra-site Analysis at Klithi*. McDonald Institute for Archaeological Research, Cambridge, 61–94.

Bailey, G. N., Carter, P. L., Gamble, C. and Higgs, H. P. 1983. Asprochaliko and Kastritsa: further investigations of Palaeolithic settlement and economy in Epirus (North-west Greece). *Proceedings of the Prehistoric Society*, **49**, 15–42.

Bailey, G. N., Turner, C., Woodward, J. C., Macklin, M. G. and Lewin, J. 1997. The Voidomatis Basin: an introduction. In Bailey, G. N. (ed.) *Klithi: Palaeolithic Settlement and Quaternary Landscapes in Northwest Greece. Volume 2: Klithi in Its Local and Regional Setting*. McDonald Institute for Archaeological Research, Cambridge, 321–345.

Bar-Yosef, O. 1993. Site formation processes from a Levantine viewpoint. In Goldberg, P., Nash, D. T. and Petraglia, M. D. (eds) *Formation Processes in Archaeological Context*. Monographs in World Archaeology No. 17, Prehistory Press, Madison, Wisconsin, 13–32.

Bögli, A. 1980. *Karst Hydrology and Physical Speleology*. Springer-Verlag, Berlin and Heidelberg.

Bretz, J. H. 1942. Vadose and phreatic features of limestone caverns. *Journal of Geology*, **50**, 675–811.

Bull, P. A. 1981. Some fine-grained sedimentation phenomena in caves. *Earth Surface Processes and Landforms*, **6**, 11–22.

Bull, P. A. 1983. Chemical sedimentation in caves. In Goudie, A. S. and Pye, K. (eds) *Chemical Sediments and Geomorphology*. Academic Press, London, 301–319.

Butzer, K. W. 1981. Cave sediments, Upper Pleistocene stratigraphy and Mousterian facies in Cantabrian Spain. *Journal of Archaeological Science*, **8**, 133–183.

Butzer, K. W. 1982. *Archaeology as Human Ecology*. Cambridge University Press, Cambridge.

Campy, M. and Chaline, J. 1993. Missing records and depositional breaks in French Late Pleistocene cave sediments. *Quaternary Research*, **40**, 318–331.

Collcutt, S. N. 1979. The analysis of Quaternary cave sediments. *World Archaeology*, **10**, 290–301.

Courty, M. A., Goldberg, P. and Macphail, R. 1989. *Soils and Micromorphology in Archaeology*. Cambridge University Press, Cambridge.

Dearing, J. A., Maher, B. A. and Oldfield, F. 1985. Geomorphological linkages between soils and sediments: the role of magnetic measurements. In Richards, K. S., Arnett, R. R. and Ellis, S. (eds) *Geomorphology and Soils*. George Allen and Unwin, London, 245–266.

Ellwood, B. B., Petruso, K. M., Harrold, F. B. and Schuldenrein, J. 1997. High-resolution palaeoclimatic records for the Holocene identified using magnetic susceptibility data from archaeological excavations in caves. *Journal of Archaeological Science*, **24**, 569–573.

Farrand, W. R. 1975. Sediment analysis of a prehistoric rockshelter: the Abri Pataud. *Quaternary Research*, **5**, 1–26.

Farrand, W. R. 1979. Chronology and palaeoenvironment of Levantine prehistoric sites as seen from sediment studies. *Journal of Archaeological Science*, **6**, 369–392.

Farrand, W. R. 1981. Pluvial climates and frost action during the Last Glacial Cycle in the Eastern Mediterranean: evidence from archaeological sites. In Mahaney, W. C. (ed.) *Quaternary Palaeoclimate*. Geoabstracts, Norwich, 393–410.

Farrand, W. R. 1985. Rockshelter and cave sediments. In Stein, J. K. and Farrand, W. R. (eds) *Archaeological Sediments in Context* (Peopling Americas Edited Volume Series, 1). Centre for the Study of Early Man, Institute of Quaternary Studies, University of Maine Orono, 21–40.

Farrand, W. R. 1988. Integration of Late Quaternary climatic records from France and Greece: cave sediments, pollen and marine events. In Dibble, H. L. and Montet-White, A. (eds) *Upper Pleistocene Prehistory of Western Eurasia*. University of Pennsylvania, Philadelphia, 305–319.

Farrand, W. R. 1993. Discontinuity in the stratigraphic record: snapshots from Franchthi Cave. In Goldberg, P., Nash, D. T. and Petraglia, M. D. (eds) *Formation Processes in Archaeological Context*. Monographs in World Archaeology No. 17, Prehistory Press, Madison, Wisconsin, 85–96.

Ford, T. D. 1975. Sediments in caves. *Transactions of the British Cave Research Association*, **2**(1), 41–46.

Ford, D. C. and Williams, P. W. 1989. *Karst Geomorphology and Hydrology*. Unwin Hyman, London.

Gale, S. J., Gilbertson, D. D., Hoare, P. G., Hunt, C. O., Jenkinson, R. D. S., Lamble, A. P., O'Toole, C., van der Veen, M. and Yates, G. 1993. Late Holocene environmental change in the Libyan pre-desert. *Journal of Arid Environments*, **24**, 1–19.

Gamble, C. S. 1986. *The Palaeolithic Settlement of Europe*. Cambridge University Press, Cambridge.

Gillieson, D., Oldfield, F. and Krawieki, A. 1986. Records of prehistoric soil erosion from rockshelter sites in Papua New Guinea. *Mountain Research and Development*, **6**, 315–324.

Goldberg, P., Nash, D. T. and Petraglia, M. D. (eds) 1993. *Formation Processes in Archaeological Context*. Monographs in World Archaeology No. 17, Prehistory Press, Madison, Wisconsin, 13–32.

Gowlett, J. A. J., Hedges, R. E. M. and Housley, R. A. 1997. Klithi: the AMS radiocarbon dating programme for the site and its environs. In Bailey, G. N. (ed.) *Klithi: Palaeolithic*

Settlement and Quaternary Landscapes in Northwest Greece. Volume 1: Excavation and Intra-site Analysis at Klithi. McDonald Institute for Archaeological Research, Cambridge, 27–40.

Gross, M. G. 1971. Carbon determination. In Carver, R. E. (ed.) *Procedures in Sedimentary Petrology*. John Wiley, New York, 541–569.

Huxtable, J., Gowlett, J. A. J., Bailey, G. N., Carter, P. L. and Papaconstantinou, V. 1992. Thermoluminescence dates and a new analysis of the Early Mousterian from Asprochaliko. *Current Anthropology*, **33**, 109–114.

Jancin, M. and Clark, D. D. 1993. Subsidence sink hole development in light of mud infiltrate structures within interstratal karst of the Coastal Plain, southeast of United States. *Environmental Geology*, **22**, 330–336.

Jennings, J. N. 1985. *Karst Geomorphology* (2nd edition) Basil Blackwell, Oxford.

Jones, K. P. N., McCave, I. N. and Patel, P. D. 1988. A computer-interfaced sedigraph for modal size analysis of fine-grained sediment. *Sedimentology*, **35**, 163–172.

Kotjabopoulou, E., Panagopoulou, E. and Adam, E. 1997. The Boila rockshelter: a preliminary report. In Bailey, G. N. (ed.) *Klithi: Palaeolithic Settlement and Quaternary Landscapes in Northwest Greece. Volume 2: Klithi in Its Local and Regional Setting*. McDonald Institute for Archaeological Research, Cambridge, 427–437.

Lautridou, J. P. 1988. Recent advances in cryogenic weathering. In Clark, M. G. (ed.) *Advances in Periglacial Geomorphology*. John Wiley, Chichester, 33–47.

Laville, H. 1976. Deposits in calcareous rock shelters: analytical methods and climatic interpretation. In Davidson, D. A. and Shackley, M. L. (eds) *Geoarchaeology*. Duckworth, London, 137–155.

Laville, H., Rigaud, J. P. and Sackett, J. 1980. *Rockshelters of the Perigord*. Academic Press, New York.

Lewin, J., Macklin, M. G. and Woodward, J. C. 1991. Late Quaternary fluvial sedimentation in the Voidomatis Basin, Epirus, northwest Greece. *Quaternary Research*, **35**, 103–115.

Macklin, M. G., Woodward, J. C. and Lewin, J. 1988. *Catastrophic Flooding and Environmental Change in Mediterranean Mountain Environments*. Final Report to the UK Natural Environment Research Council (NERC), December 1988.

Macklin, M. G., Lewin, J. and Woodward, J. C. 1997. Quaternary river sedimentary sequences of the Voidomatis basin. In Bailey, G. N. (ed.) *Klithi: Palaeolithic Settlement and Quaternary Landscapes in Northwest Greece. Volume 2: Klithi in Its Local and Regional Setting*. McDonald Institute for Archaeological Research, Cambridge, 347–359.

Macleod, D. A. 1980. The origin of the red Mediterranean Soils in Epirus, Greece. *Journal of Soil Science*, **31**, 125–136.

Macphail, R. L. and Goldberg, P. 1995. Recent advances in micromorphological interpretations of soils and sediments from archaeological sites. In Barham, A. J. and Macphail, R. I. (eds) *Archaeological Sediments and Soils: Analysis, Interpretation and Management*. Institute of Archaeology, University College, London.

McDermot, F., Grün, R., Stringer, C. B. and Hawkesworth, C. J. 1993. Mass-spectrometric U-series dates for Israeli Neanderthal/early modern hominid sites. *Nature*, **363**, 252–254.

Mercier, N. and Valladas, H. 1994. Thermoluminescence dates for the Palaeolithic Levant. In Bar Yosef, O. and Kra, R. S. (eds) *Late Quaternary Chronology and Palaeoclimates of the Eastern Mediterranean*. Radiocarbon, Department of Geosciences, The University of Arizona, Tuscon, Arizona, 13–20.

Patton, P. C. and Dibble, D. S. 1982. Archaeologic and geomorphological evidence for the paleohydrologic record of the Pecos River in West Texas. *American Journal of Science*, **282**, 97–121.

Pye, K. 1987. *Aeolian Dust and Dust Deposits*. Academic Press, London.

Pye, K. 1992. Aeolian dust transport and deposition over Crete and adjacent parts of the Mediterranean Sea. *Earth Surface Processes and Landforms*, **17**, 271–288.

Schmid, E. 1969. Cave sediments and prehistory. In Brothwell, D. R. and Higgs, E. S. (eds) *Science in Archaeology* (2nd edition). Thames and Hudson, London, 151–166.

Schwarcz, H. P., Grün, R., Vandermeersch, B., Bar-Yosef, O., Valladas, H. and Tchernov, E. 1988. ESR dates for the hominid burial site of Qafzeh in Israel. *Journal of Human Evolution*, **17**, 733–737.

Smith, G. R., Woodward, J. C., Heywood, D. I. and Gibbard, P. L. 2000. Interpreting Pleistocene glacial features from SPOT HRV data using fuzzy techniques. *Computers and Geosciences*, **26**, 479–490.

Tsatskin, A., Weinstein-Evron, M. and Ronen, A. 1995. Weathering and pedogenesis of wind-blown sediments in the Mount Carmel Caves, Israel. In Derbyshire, E. (ed.) *Wind Blown Sediments in the Quaternary Record*. Quaternary Proceedings No. 4, John Wiley, Chichester, 83–93.

Vita-Finzi, C. 1978. *Archaeological Sites in their Setting*. Thames and Hudson, London.

Vitaliano, C. J., Taylor, S. R., Farrand, W. R. and Jacobsen, T. W. 1981. Tephra layer in Franchthi Cave, Peloponnesos, Greece. In Self, S. and Sparks, R. S. J. (eds) *Tephra Studies*. Riedel, Amsterdam, 373–379.

Walling, D. E. and Woodward, J. C. 1995. Tracing suspended sediment sources in river basins: a case study of the River Culm, Devon, UK. *Marine and Freshwater Research*, **46**, 327–336.

White, W. B. 1993. Analysis of Karst aquifers. In Alley, W. M. (ed.) *Regional Ground Water Quality*. Van Nostrand Reinhold, New York, 471–489.

Woodward, J. C. 1990. Late Quaternary sedimentary environments in the Voidomatis Basin, northwest Greece. PhD Thesis, University of Cambridge.

Woodward, J. C. 1997a. Late Pleistocene rockshelter sedimentation at Klithi. In Bailey, G. N. (ed.) *Klithi: Palaeolithic Settlement and Quaternary Landscapes in Northwest Greece. Volume 2: Klithi in Its Local and Regional Setting*. McDonald Institute for Archaeological Research, Cambridge, 361–376.

Woodward, J. C. 1997b. Late Pleistocene rockshelter sedimentation at Megalakkos. In Bailey, G. N. (ed.) *Klithi: Palaeolithic Settlement and Quaternary Landscapes in Northwest Greece. Volume 2: Klithi in Its Local and Regional Setting*. McDonald Institute for Archaeological Research, Cambridge, 377–393.

Woodward, J. C., Lewin, J. and Macklin, M. G. 1992. Alluvial sediment sources in a glaciated catchment: the Voidomatis Basin, northwest Greece. *Earth Surface Processes and Landforms*, **17**, 205–216.

Woodward, J. C., Macklin, M. G. and Lewin, J. 1994. Pedogenic weathering and relative-age dating of Quaternary alluvial sediments in the Pindus Mountains of northwest Greece. In Robinson, D. A. and Williams, R. B. G. (eds) *Rock Weathering and Landform Evolution*. John Wiley, Chichester, 259–283.

Woodward, J. C., Lewin, J. and Macklin, M. G. 1995. Glaciation, river behaviour and Palaeolithic settlement in upland northwest Greece. In Lewin, J., Macklin, M. G. and Woodward, J. C. (eds) *Mediterranean Quaternary River Environments*. A. A. Balkema, Rotterdam, 115–129.

Index

(all rivers are given as R. name)